LTE – The UMTS Long Term Evolution

LTE – The UMTS Long Term Evolution

From Theory to Practice

Stefania Sesia

ST-NXP Wireless/ETSI, France

Issam Toufik

ST-NXP Wireless, France

Matthew Baker

Philips Research, UK

A John Wiley and Sons, Ltd, Publication

This edition first published 2009
© 2009 John Wiley & Sons Ltd.

Registered office
John Wiley & Sons Ltd, The Atrium, Southern Gate, Chichester, West Sussex, PO19 8SQ,
United Kingdom

For details of our global editorial offices, for customer services and for information about how to apply
for permission to reuse the copyright material in this book please see our website at www.wiley.com.

Library of Congress Cataloging-in-Publication Data

Sesia, Stefania.
    LTE–the UMTS long term evolution : from theory to practice / Stefania Sesia, Matthew Baker, and
Issam Toufik.
        p. cm.
    Includes bibliographical references and index.
    ISBN 978-0-470-69716-0 (cloth)
1. Universal Mobile Telecommunications System. I. Baker, Matthew (Matthew P. J.) II. Toufik, Issam.
III. Title.
    TK5103.4883.S47 2009
    621.3845'6–dc22

2008041823

A catalogue record for this book is available from the British Library.

ISBN 978-0-470-69716-0 (H/B)

Set in 10/12pt Times by Sunrise Setting Ltd, Torquay, UK.
Printed in Great Britain by CPI Antony Rowe, Chippenham, Wiltshire

Dedication

To my family.
Stefania Sesia

To my parents for their sacrifices and unconditional love. To my brother and sisters for their love and continual support. To my friends for being what they are.
Issam Toufik

To the glory of God, who 'so loved the world that He gave His only Son, that whoever believes in Him shall not perish but have eternal life'. — The Bible.
Matthew Baker

Contents

14 Broadcast Operation   323

Olivier Hus and Matthew Baker

Part III   Physical Layer for Uplink   343

Part IV   Practical Deployment Aspects   473

21 The Radio Propagation Environment   475
Juha Ylitalo and Tommi Jämsä

22 Radio Frequency Aspects   501
*Tony Sayers, Adrian Payne, Stefania Sesia, Robert Love, Vijay Nangia and
Gunnar Nitsche*

Editors' Biographies

Matthew Baker holds degrees in Engineering and Electrical and Information Sciences from the University of Cambridge. He has over 10 years' experience of conducting leading-edge research into a variety of wireless communication systems and techniques with Philips Research, including propagation modelling, DECT,[1] Hiperlan and UMTS. He has been actively participating in the standardization of both UMTS WCDMA and LTE in 3GPP since 1999, where he has been active in RAN working groups 1, 2, 4 and 5, contributing several hundred proposals and leading the Philips RAN standardization team. He is the author of several international conference papers and inventor of numerous patents. He is a Chartered Engineer and Member of the Institution of Engineering and Technology.

Stefania Sesia received her Ph.D. degree in Communication Systems and Coding Theory from the Eurecom, Sophia Antipolis/ENST-Paris, France in 2005. From 2002 to 2005 she worked at Motorola Research Labs, Paris, towards her Ph.D. thesis. In June 2005 she joined Philips/NXP Semiconductors (now ST-NXP Wireless) Research and Development Centre in Sophia Antipolis, France where she was technical leader and responsible for the High Speed Downlink Packet Access algorithm development. She has been participating in 3GPP RAN working groups 1 and 4 standardization meetings, and since 2007 she has been on secondment from NXP Semiconductors to the European Telecommunications Standard Institute (ETSI) acting as Working Group Technical Officer. She is the author of several international IEEE conference and journal papers and contributions to 3GPP, and inventor of numerous US and European patents.

Issam Toufik graduated in Telecommunications Engineering (majored in Mobile Communication Systems) in 2002 from both ENST-Bretagne (Brest, France) and Eurecom (Sophia Antipolis, France). In 2006, he received his Ph.D. degree in Communication Systems from Eurecom/ENST-Paris, France. From June to August 2005 he worked for Samsung Advanced Institute of Technology (SAIT), South Korea, as a Research Engineer on LTE. In January 2007, he joined NXP semiconductors (now ST-NXP Wireless), Sophia Antipolis, France, as a Research and Development Engineer for UMTS and LTE algorithm development. He is the author of several international IEEE conference and journal papers and contributions to 3GPP, and inventor of numerous patents.

---

[1]Digital Enhanced Cordless Telecommunications.

List of Contributors

Ancora, Andrea, ST-NXP Wireless
e-mail: andrea.ancora@stnwireless.com

Anderson, Nickolas, NextWave Wireless
e-mail: nanderson@ipwireless.com

Baker, Matthew, Philips
e-mail: m.p.j.baker.92@cantab.net, matthew.baker@philips.com

Bertrand, Pierre, Texas Instruments
e-mail: p-bertrand@ti.com

Boixadera, Francesc, MStar Semiconductor
e-mail: francesc.boixadera@mstarsemi.com

Bury, Andreas, ST-NXP Wireless
e-mail: andreas.bury@stnwireless.com

Chun, SungDuck, LG Electronics
e-mail: duckychun@lge.com

Classon, Brian, Huawei
e-mail: brian.classon@huawei.com

Courau, François, Alcatel-Lucent
e-mail: francois.courau@alcatel-lucent.fr

Fischer, Patrick, Bouygues Telecom
e-mail: pfischer@bouyguestelecom.fr

Gesbert, David, Eurecom
e-mail: david.gesbert@eurecom.fr

Godin, Philippe, Alcatel-Lucent
e-mail: philippe.godin@alcatel-lucent.fr

Hus, Olivier, Philips
e-mail: olivier.hus@philips.com

Jämsä, Tommi, Elektrobit
e-mail: tommi.jamsa@elektrobit.com

Jiang, Jing, Texas Instruments
e-mail: jing.jiang@ti.com

Kaltenberger, Florian, Eurecom
e-mail: florian.kaltenberger@eurecom.fr

Knopp, Raymond, Eurecom
e-mail: raymond.knopp@eurecom.fr

Lee, YoungDae, LG Electronics
e-mail: leego@lge.com

Love, Robert, Motorola
e-mail: robert.love@motorola.com

Moulsley, Tim, Philips
e-mail: t.moulsley@btopenworld.com

Nangia, Vijay, Motorola
e-mail: vijay.nangia@motorola.com

Nimbalker, Ajit, Motorola
e-mail: aijt.nimbalker@motorola.com

Nitsche, Gunnar, ST-NXP Wireless
e-mail: gunnar.nitsche@stnwireless.com

Palat, K. Sudeep, Alcatel-Lucent
e-mail: spalat@alcatel-lucent.com

Payne, Adrian, NXP Semiconductors
e-mail: adrian.payne@email.com

Sälzer, Thomas, Orange-France Telecom
e-mail: thomas.salzer@orange-ftgroup.com, thomas.salzer@gmx.de

Sayers, Tony, NXP Semiconductors
e-mail: tony.sayers@talktalk.net

Sesia, Stefania, ST-NXP Wireless
e-mail: stefania.sesia@stnwireless.com, stefania.sesia@etsi.org

Slock, Dirk, Eurecom
e-mail: dirk.slock@eurecom.fr

Tomatis, Fabrizio, ST-NXP Wireless
e-mail: fabrizio.tomatis@stnwireless.com

Tosato, Filippo, Toshiba
e-mail: filippo.tosato@toshiba-trel.com

Toufik, Issam, ST-NXP Wireless
e-mail: issam.toufik@stnwireless.com, issam.toufik@eurecom.fr

van der Velde, Himke, Samsung
e-mail: himke.vandervelde@samsung.com

van Rensburg, Cornelius, Huawei
e-mail: cdvanren@ieee.org

Yi, SeungJune, LG Electronics
e-mail: seungjune@lge.com

Ylitalo, Juha, Elektrobit
e-mail: juha.ylitalo@elektrobit.com

Foreword

A GSM, and its evolution through GPRS, EDGE, WCDMA and HSPA, is the technology stream of choice for the vast majority of the world's mobile operators. Today's commercial offerings which are based on this technology evolution typically offer downlink speeds in the order of 7 Mbps, with the expectation that 14 Mbps will become widely available in the near future. With such an improvement in 3rd Generation (3G) capabilities, there are obvious questions to be asked about what should happen next. From a standardization perspective 3G work is now well-advanced and, while improvements continue to be made to leverage the maximum performance from currently deployed systems, there is a limit to the extent to which further enhancements will be effective. If the only aim were to deliver higher performance, then this in itself would be relatively easy to achieve. The added complexity is that such improved performance must be delivered through systems which are cheaper to install and maintain. Users have experienced a dramatic reduction in telecommunications charges and they now expect to pay less but to receive more. Therefore, in deciding the next standardization step, there must be a dual approach: seeking considerable performance improvement but at reduced cost. LTE is that next step and will be the basis on which future mobile telecommunications systems will be built.

Many articles have already been published on the subject of LTE, varying from doctoral theses to network operator analyses and manufacturers' product literature. By their very nature, those publications have viewed the subject from one particular perspective, be it academic, operational or promotional. A very different approach has been taken with this book. The authors come from a number of different spheres within the mobile telecommunications ecosystem and collectively bring a refreshing variety of views. What binds the authors together is a thorough knowledge of the subject material which they have derived from their long experience within the standards-setting environment, 3rd Generation Partnership Project (3GPP). LTE discussions started within 3GPP in 2004 and so it is not really a particularly new subject. In order to fully appreciate the thinking that conceived this technology, however, it is necessary to have followed the subject from the very beginning and to have witnessed the discussions that took place from the outset. Moreover, it is important to understand the thread that links academia, through research to standardization since it is widely acknowledged that by this route impossible dreams become market realities. Considerable research work has taken place to prove the viability of the technical basis on which LTE is founded and it is essential to draw on that research if any attempt is made to explain LTE to a wider audience. The authors of this book have not only followed the LTE story from the beginning but many have also been active players in WCDMA and its predecessors, in which LTE has its roots.

This book provides a thorough, authoritative and complete tutorial of the LTE system. It gives a detailed explanation of the advances made in our theoretical understanding and the practical techniques that will ensure the success of this ground-breaking new radio access technology. Where this book is exceptional is that the reader will not just learn how LTE works but why it works.

I am confident that this book will earn its rightful place on the desk of anyone who needs a thorough understanding of the LTE technology, the basis of the world's mobile telecommunications systems for the next decade.

Adrian Scrase ETSI Vice-President,
International Partnership Projects

Preface

Research workers and engineers toil unceasingly on the development of wireless telegraphy. Where this development can lead, we know not. However, with the results already achieved, telegraphy over wires has been extended by this invention in the most fortunate way. Independent of fixed conductor routes and independent of space, we can produce connections between far-distant places, over far-reaching waters and deserts. This is the magnificent practical invention which has flowered upon one of the most brilliant scientific discoveries of our time!

These words accompanied the presentation of the Nobel Prize for Physics to Guglielmo Marconi in December 1909.

Marconi's success was the practical and commercial realization of wireless telegraphy – the art of sending messages without wires – thus exploiting for the first time the amazing capability for wireless communication built into our Universe. While others worked on wireless telephony – the transmission of audio signals for voice communication – Marconi interestingly saw no need for this. He believed that the transmission of short text messages was entirely sufficient for keeping in touch.

One could be forgiven for thinking that the explosion of wireless voice communication in the intervening years has proved Marconi wrong; but the resurgence of wireless data transmission at the close of the twentieth century, beginning with the mobile text messaging phenomenon, or 'SMS', reveals in part the depth of insight Marconi possessed.

Nearly 100 years after Marconi received his Nobel prize, the involvement of thousands of engineers around the world in major standardization initiatives such as the 3rd Generation Partnership Project (3GPP) is evidence that the same unceasing toil of research workers and engineers continues apace.

While the first mobile communications standards focused primarily on voice communication, the emphasis now has returned to the provision of systems optimized for data. This trend began with the 3rd Generation Wideband Code Division Multiple Access (WCDMA) system designed in the 3GPP, and is now reaching fulfilment in its successor, known as the 'Long-Term Evolution' (LTE). LTE is the first cellular communication system optimized from the outset to support packet-switched data services, within which packetized voice communications are just one part. Thus LTE can truly be said to be the heir to Marconi's heritage – the system, unknown indeed to the luminaries of his day, to which his developments have led.

LTE is an enabler. It is not technology for technology's sake, but technology with a purpose, connecting people and information to enable greater things to be achieved. It will provide higher data rates than ever previously achieved in mobile communications, combined with wide-area coverage and seamless support for mobility without regard for the type of data being transmitted. To provide this level of functionality and flexibility, it is inevitable that the complexities of the LTE system have far surpassed anything Marconi could have imagined.

One aim of this book, therefore, is to chart an explanatory course through the LTE specifications, to support those who will design the equipment to bring LTE to fruition.

The LTE specification documents themselves do not tell the whole story. Essentially they are a record of decisions taken – decisions which are often compromises between performance and cost, theoretical possibility and practical constraints.

We also aim therefore to give the reader a detailed insight into the evaluations and trade-offs which lie behind the technology choices inherent in LTE.

Above all, it is vital to remember that if recent years had not given rise to major advances in the fundamental science and theoretical understanding underlying mobile communications, there would have been no LTE.

The thousands of engineers active in the standardization committees of 3GPP are just the tip of the iceberg of the ongoing 'unceasing toil'. Behind these thousands work many thousands, even tens of thousands, more, in the research divisions of companies, in universities the world over, and other public and private research institutes, inventing, understanding, testing and explaining new theories and techniques which have eventually been exploited in LTE.

It is particularly these advances in the underlying theory and academic understanding, without which LTE would never have been possible, which this books seeks to highlight.

As an example, for decades the famous Shannon capacity formula for a single radio link was considered the upper-bound on data rates which could be transmitted in a given bandwidth. While previous standards such as WCDMA have come close to achieving this thanks to advances in coding, much recent effort has been expended on extending this theory to communication involving a multiplicity of antennas in order to push the bounds of feasible data rates still further. LTE is the first mobile communication system to have so-called Multiple-Input Multiple-Output (MIMO) antenna transmission designed from the start as an integral part of the original system.

In selecting the technologies to include in LTE, an important consideration has been the trade-off between practical benefit and cost of implementation. Fundamental to this assessment, therefore, has been a much-enhanced understanding of the radio propagation environment and scenarios in which LTE will be deployed and used. This has been built on significant advances in radio-channel modelling and simulation capabilities.

Moreover, while theoretical understanding has advanced, the practicalities of what is feasible in a cost-effective implementation have also moved on. Developments in integrated circuit technology and signal processing power have rendered feasible techniques which would have been unthinkable only a few years ago.

Other influences on the design of LTE have included changes in the commercial and regulatory fields. Global roaming requires global spectrum allocation, while higher data rates require ever wider bandwidths to be available. This results in the need for LTE to be adaptable, capable of being scaled and deployed in a wide range of different spectrum bands and bandwidths.

With this breadth and depth in mind, the authors of the chapters in this book are drawn from a variety of fields of the ecosystem of research and development which has underpinned the design of LTE. They work in the 3GPP standardization itself, in the R & D departments of companies active in LTE, for network operators as well as equipment manufacturers, in universities and in other collaborative research projects. They are uniquely placed to share their insights from the full range of perspectives.

To borrow Marconi's words, where LTE will lead, we know not; but we can be sure that it will not be the last development in wireless telegraphy.

Matthew Baker, *Stefania Sesia* and *Issam Toufik*

Acknowledgements

This book is first and foremost the fruit of a significant team effort, which would not have been successful without the expertise and professionalism displayed by all the contributors, as well as the support of their companies. The dedication of all the co-authors to their task, their patience and flexibility in allowing us to modify and move certain parts of their material for harmonization purposes, are hereby gratefully acknowledged. Particular thanks are due to ST-NXP Wireless, Philips and ETSI for giving us the encouragements and working environment to facilitate such a time-consuming project. The help provided by ETSI, 3GPP and others in authorizing us to reproduce certain copyrighted material is also gratefully acknowledged. We would like to express our gratitude to the many experts who kindly provided advice, feedback, reviews and other valuable assistance. We believe their input in all its forms has made this book a more accurate, valuable and even enjoyable resource. These experts include Kevin Baum, Keith Blankenship, Yufei Blankenship, Kevin Boyle, Sarah Boumendil, Paul Bucknell, Richard Burbidge, Aaron Byman, Emilio Calvanese Strinati, Choo Chiap Chiau, Anand Dabak, Peter Darwood, Merouane Debbah, Vip Desai, Marko Falck, Jeremy Gosteau, Lajos Hanzo, Lassi Hentilä, Shin Horng Wong, Paul Howard, Howard Huang, Alan Jones, Achilles Kogiantis, Pekka Kyösti, Thierry Lestable, Gert-Jan van Lieshout, Andrew Lillie, Matti Limingoja, Huiheng Mai, Darren McNamara, Juha Meinilä, Tarik Muharemovic, Jukka-Pekka Nuutinen, SungJun Park, Roope Parviainen, Safouane Sfar, Zukang Shen, Ken Stewart, Ludo Tolhuizen, Li Wang and Tim Wilkinson.

We would also like to acknowledge the efforts of all participants in 3GPP who, through innumerable contributions and intense discussions often late into the night, have facilitated the completion of the LTE specifications in such a short space of time.

Finally, we would like to thank the publishing team at John Wiley & Sons, especially Tiina Ruonamaa, Sarah Hinton, Anna Smart and Sarah Tilley for their professionalism and extensive support and encouragement throughout the project to bring this work to fruition.

The Editors

List of Acronyms

3GPP 3rd Generation Partnership Project

3GPP2 3rd Generation Partnership Project 2

AC Access Class

ACI Adjacent Channel Interference

ACIR Adjacent Channel Interference Ratio

ACK Acknowledgement

ACLR Adjacent Channel Leakage Ratio

ACS Adjacent Channel Selectivity

ADC Analogue to Digital Converter

ADSL Asymmetric Digital Subscriber Line

AGI Antenna Gain Imbalance

AM Acknowledged Mode

AMC Adaptive Modulation and Coding

AMPS Analogue Mobile Phone System

AMR Adaptive MultiRate

ANR Automatic Neighbour Relation

ANRF Automatic Neighbour Relation Function

AoA Angle-of-Arrival

AoD Angle-of-Departure

APN Access Point Name

APP A-Posteriori Probability

ARFCN Absolute Radio Frequency Channel Number

ARIB Association of Radio Industries and Businesses

ARP Almost Regular Permutation*

ARP Allocation and Retention Priority*

ARQ Automatic Repeat reQuest

AS Access Stratum*

AS Angular Spread*

A-SEM Additional SEM

ATDMA Advanced TDMA

ATIS Alliance for Telecommunications Industry Solutions

AuC Authentication Centre

AWGN Additive White Gaussian Noise

BCC Base station Colour Code

BCH Broadcast CHannel

BCCH Broadcast Control CHannel

BCJR Algorithm named after its inventors, Bahl, Cocke, Jelinek and Raviv

BER Bit Error Rate

BLER BLock Error Rate

BM-SC Broadcast-Multicast Service Centre

BP Belief Propagation

BPRE Bits Per Resource Element

bps bits per second

BPSK Binary Phase Shift Keying

BSIC Base Station Identification Code

BSR Buffer Status Reports

CAZAC Constant Amplitude Zero AutoCorrelation

CB Circular Buffer

CCCH Common Control CHannel

CCE Control Channel Element

CCI Co-Channel Interference

CCO Cell Change Order

CCSA China Communications Standards Association

CDD Cyclic Delay Diversity

CDF Cumulative Distribution Function

CDL Clustered Delay Line

CDM Code Division Multiplex(ed/ing)

CDMA Code Division Multiple Access

C/I Carrier-to-Interference ratio

CF Contention-Free

CFI Control Format Indicator

CFO Carrier Frequency Offset

CINR Carrier-to-Interference-and-Noise Ratio

CIR Channel Impulse Response

CM Cubic Metric

CMHH Constant Modulus HouseHolder

CN Core Network

CODIT UMTS Code DIvision Testbed

COFDM Coded OFDM

CP Cyclic Prefix

CPICH Common PIlot CHannel

CPR Common Phase Rotation

CPT Control PDU Type

CQI Channel Quality Indicator

CRC Cyclic Redundancy Check

C-RNTI Cell Radio Network Temporary Identifier

CS Circuit-Switched

CSG Closed Subscriber Group

CSI Channel State Information

CSIT Channel State Information at the Transmitter

CTF Channel Transfer Function

CVA Circular Viterbi Algorithm

CVQ Channel Vector Quantization

CW Continuous-Wave

DAB Digital Audio Broadcasting

DAC Digital to Analogue Converter

dB deci-Bel

d.c. direct current

DCCH Dedicated Control CHannel

DCFB Direct Channel FeedBack

DCI Downlink Control Information

DFT Discrete Fourier Transform

DFT-S-OFDM DFT-Spread OFDM

Diffserv Differentiated Services

DL DownLink

DL-SCH DownLink Shared CHannel

DMB Digital Mobile Broadcasting

DM RS DeModulation RS

DOA Direction Of Arrival

DPC Dirty-Paper Coding

DRB Data Radio Bearer

DRX Discontinuous Reception

DS-CDMA Direct-Sequence Code Division Multiple Access

DSP Digital Signal Processor

DTCH Dedicated Traffic CHannel

DTX Discontinuous Transmission

DVB-H Digital Video Broadcasting – Handheld

DVB-T Digital Video Broadcasting – Terrestrial

DwPTS Downlink Pilot TimeSlot

ECM EPS Connection Management

EDGE Enhanced Data rates for GSM Evolution

EESM Exponential Effective SINR Mapping

EMM EPS Mobility Management

eNodeB evolved NodeB

EPA Extended Pedestrian A

EPC Evolved Packet Core

EPS Evolved Packet System

ESP Encapsulating Security Payload

ETSI European Telecommunications Standards Institute

ETU Extended Typical Urban

E-UTRA Evolved-UTRA

E-UTRAN Evolved-UTRAN

EVA Extended Vehicular A

EVM Error Vector Magnitude

FACH Forward Access CHannel

FB Frequency Burst

FCCH Frequency Control CHannel

FDD Frequency Division Duplex

FDE Frequency Domain Equalizer

FDM Frequency Division Multiplexing

FDMA Frequency Division Multiple Access

FDSS Frequency Domain Spectral Shaping

FFT Fast Fourier Transform

FI Framing Info

FIR Finite Impulse Response

FMS First Missing SDU

FSTD Frequency Switched Transmit Diversity

FTP File Transfer Protocol

FTTH Fibre-To-The-Home

GBR Guaranteed Bit Rate

GCL Generalized Chirp-Like

GERAN GSM EDGE Radio Access Network

GGSN Gateway GPRS Support Node

GMSK Gaussian Minimum-Shift Keying

GPRS General Packet Radio Service

GPS Global Positioning System

GSM Global System for Mobile communications

GT Guard Time

GTP GPRS Tunnelling Protocol

GTP-U GTP-User plane

HARQ Hybrid Automatic Repeat reQuest

HD-FDD Half-Duplex FDD

HFN Hyper Frame Number

HII High Interference Indicator

HLR Home Location Register

HRPD High Rate Packet Data

HSDPA High Speed Downlink Packet Access

HSPA High Speed Packet Access

HSPA+ High Speed Packet Access Evolution

HSS Home Subscriber Server

HSUPA High Speed Uplink Packet Access

HTTP HyperText Transfer Protocol

ICI InterCarrier Interference

ICIC InterCell Interference Coordination

IDFT Inverse Discrete Fourier Transform

IETF Internet Engineering Task Force

IFDMA Interleaved Frequency Division Multiple Access

IFFT Inverse Fast Fourier Transform

i.i.d. Independently identically distributed

IM Implementation Margin

IMD InterModulation Distortion

IMS IP Multimedia Subsystem

IMSI International Mobile Subscriber Identity

IMT International Mobile Telecommunications

IP Internet Protocol

IR Incremental Redundancy

IRC Interference Rejection Combining

ISD InterSite Distance

ISI InterSymbol Interference

IST-WINNER Information Society Technologies-Wireless world INitiative NEw Radio

ITU International Telecommunication Union

ITU-R ITU Radiocommunication sector

J-TACS Japanese Total Access Communication System

LA Local Area

LB Long Block

LBP Layered Belief Propagation

LBRM Limited Buffer Rate Matching

LCID Logical Channel ID

LDPC Low-Density Parity Check

LI Length Indicator

LLR Log-Likelihood Ratio

LMMSE Linear MMSE

LNA Low Noise Amplifier

LO Local Oscillator

LOS Line-Of-Sight

LS Least Squares

LSF Last Segment Flag

LTE Long-Term Evolution

MA Metropolitan Area

MAC Medium Access Control

MAC-I Message Authentication Code for Integrity

MAN Metropolitan Area Network

MAP Maximum A posteriori Probability

MBMS Multimedia Broadcast/Multicast Service

MBMS GW MBMS GateWay

MBR Maximum Bit Rate

MBSFN Multimedia Broadcast Single Frequency Network

MCCH Multicast Control CHannel

MCE Multicell/Multicast Coordination Entity

MCH Multicast CHannel

MCL Minimum Coupling Loss

MCS Modulation and Coding Scheme

Mcps Megachips per second

MDS Minimum Discernible Signal

MediaFLO Media Forward Link Only

MIB Master Information Block

MIMO Multiple-Input Multiple-Output

MIP Mobile Internet Protocol

MISO Multiple-Input Single-Output

ML Maximum Likelihood

MLD Maximum Likelihood Detector

MME Mobility Management Entity

MMSE Minimum MSE

MO Mobile Originated

M-PSK M-ary Phase-Shift Keying

MQE Minimum Quantization Error

MRC Maximum Ratio Combining

MSAP MCH Subframe Allocation Pattern

MSB Most Significant Bit

MSE Minimum Squared Error

MSR Maximum Sensitivity Reduction

MTCH Multicast Traffic CHannel

MU-MIMO Multi-User MIMO

NACC Network Assisted Cell Change

NACK Negative ACKnowledgement

NACS NonAdjacent Channel Selectivity

NAS Non Access Stratum

NCC Network Colour Code

NCL Neighbour Cell List

NDI New Data Indicator

NF Noise Figure

NGMN Next Generation Mobile Networks

NLMS Normalized Least-Mean-Square

NLOS Non-Line-Of-Sight

NMT Nordic Mobile Telephone

NNSF NAS Node Selection Function

NodeB The base station in WCDMA systems

O&M Operation and Maintenance

OBPD Occupied Bandwidth Power De-rating

OBW Occupied BandWidth

OFDM Orthogonal Frequency Division Multiplexing

OFDMA Orthogonal Frequency Division Multiple Access

OI Overload Indicator

OOB Out-Of-Band

P/S Parallel-to-Serial

PA Power Amplifier

PAN Personal Area Network

PAPR Peak-to-Average Power Ratio

PBCH Physical Broadcast CHannel

PBR Prioritized Bit Rate

PCC Policy Control and Charging

PCCH Paging Control CHannel

P-CCPCH Primary Common Control Physical CHannel

PCEF Policy Control Enforcement Function

PCFICH Physical Control Format Indicator CHannel

PCG Project Coordination Group

PCH Paging CHannel

PCI Physical Cell Identity

P-CPICH Primary Common PIlot CHannel

PCRF Policy Control and charging Rules Function

PDCCH Physical Downlink Control CHannel

PDCP Packet Data Convergence Protocol

PDN Packet Data Network

PDP Power Delay Profile

PDSCH Physical Downlink Shared CHannel

PDU Protocol Data Unit

PF Paging Frame

PFS Proportional Fair Scheduling

P-GW PDN GateWay

PHICH Physical Hybrid ARQ Indicator CHannel

PLL Phase-Locked Loop

PLMN Public Land Mobile Network

P-MCCH Primary MCCH

PMCH Physical Multicast CHannel

PMI Precoding Matrix Indicators

PMIP Proxy MIP

PN Pseudo-Noise

PO Paging Occasion

PRACH Physical Random Access CHannel

PRB Physical Resource Block

P-RNTI Paging RNTI

PS Packet-Switched

P-SCH Primary Synchronization CHannel

PSD Power Spectral Density

PSS Primary Synchronization Signal

PUCCH Physical Uplink Control CHannel

PUSCH Physical Uplink Shared CHannel

PVI Precoding Vector Indicator

QAM Quadrature Amplitude Modulation

QCI QoS Class Identifier

QoS Quality-of-Service

QPP Quadratic Permutation Polynomial

QPSK Quadrature Phase Shift Keying

RA Random Access

RACH Random Access CHannel

RAN Radio Access Network

RAR Random Access Response

RA-RNTI Random Access Radio Network Temporary Identifier

RAT Radio Access Technology

RB Resource Block

RE Resource Element

REG Resource Element Group

RF Radio Frequency

RFC Request For Comments

RI Rank Indicator

RLC Radio Link Control

RLS Recursive Least Squares

RM Rate Matching

RNC Radio Network Controller

RNTI Radio Network Temporary Identifier

RNTP Relative Narrowband Transmit Power

ROHC RObust Header Compression

RoT Rise over Thermal

RPF RePetition Factor

R-PLMN Registered PLMN

RRC Radio Resource Control*

RRC Root-Raised-Cosine*

RRM Radio Resource Management

RS Reference Signal

RSCP Received Signal Code Power

RSRP Reference Signal Received Power

RSRQ Reference Signal Received Quality

RSSI Received Signal Strength Indicator

RTCP Real-time Transport Control Protocol

RTD Round-Trip Delay

RTP Real-time Transport Protocol

RTT Round-Trip Time

RV Redundancy Version

S/P Serial-to-Parallel

S1AP S1 Application Protocol

SAE System Architecture Evolution

SAP Service Access Point

SAW Stop-And-Wait

SB Short Block*

SB Synchronization Burst*

SBP Systematic Bit Puncturing

SC-FDMA Single-Carrier Frequency Division Multiple Access

SCH Synchronization CHannel

SCM Spatial Channel Model

SCME Spatial Channel Model Extension

SCTP Stream Control Transmission Protocol

SDMA Spatial Division Multiple Access

SDO Standards Development Organization

SDU Service Data Unit

SEM Spectrum Emission Mask

SFBC Space-Frequency Block Code

SFDR Spurious-Free Dynamic Range

SFN System Frame Number

SGSN Serving GPRS Support Node

S-GW Serving GateWay

SI System Information

SIB System Information Block

SIC Successive Interference Cancellation

SIMO Single-Input Multiple-Output

SINR Signal-to-Interference plus Noise Ratio

SIP Session Initiation Protocol

SIR Signal-to-Interference Ratio

SI-RNTI System Information Radio Network Temporary Identifier

SISO Single-Input Single-Output*

SISO Soft-Input Soft-Output*

S-MCCH Secondary MCCH

SMS Short Message Service

SN Sequence Number

SNR Signal-to-Noise Ratio

SO Segmentation Offset

SON Self-Optimizing Networks

SPA Sum-Product Algorithm

SPS Semi-Persistent Scheduling

SPS-C-RNTI Semi-Persistent Scheduling C-RNTI

SR Scheduling Request

SRB Signalling Radio Bearer

SRNS Serving Radio Network Subsystem

SRS Sounding Reference Signal

S-SCH Secondary Syncronization CHannel

SSS Secondary Synchronization Signal

STBC Space-Time Block Code

S-TMSI SAE-Temporary Mobile Subscriber Identity

STTD Space-Time Transmit Diversity

SU-MIMO Single-User MIMO

SVD Singular-Value Decomposition

TA Tracking Area

TACS Total Access Communication System

TB Transport Block

TCP Transmission Control Protocol

TDD Time Division Duplex

TDL Tapped Delay Line

TDMA Time Division Multiple Access

TD-SCDMA Time Division Synchronous Code Division Multiple Access

TEID Tunnelling End ID

TF Transport Format

TFT Traffic Flow Template

TM Transparent Mode

TMD Transparent Mode Data

TNL Transport Network Layer

TNMSE Truncated Normalized Mean-Squared Error

TPC Transmitter Power Control

TPD Total Power De-rating

TR Tone Reservation

TSC Training Sequence Code

TSG Technical Specification Group

TTA Telecommunications Technology Association

TTC Telecommunications Technology Committee

TTI Transmission Time Interval

TU Typical Urban

UDP User Datagram Protocol

UE User Equipment

UL UpLink

UL-SCH UpLink Shared CHannel

UM Unacknowledged Mode

UMB Ultra-Mobile Broadband

UMTS Universal Mobile Telecommunications System

UP Unitary Precoding

UpPTS Uplink Pilot TimeSlot

US Uncorrelated-Scattered

USIM Universal Subscriber Identity Module

UTRA Universal Terrestrial Radio Access

UTRAN Universal Terrestrial Radio Access Network

VA Viterbi Algorithm

VCB Virtual Circular Buffer

VCO Voltage-Controlled Oscillator

VoIP Voice-over-IP

VRB Virtual Resource Block

WA Wide Area

WAN Wide Area Network

WCDMA Wideband Code Division Multiple Access

WFT Winograd Fourier Transform

WG Working Group

WiMAX Worldwide interoperability for Microwave Access

WINNER Wireless world INitiative NEw Radio

WLAN Wireless Local Area Network

WPD Waveform Power De-rating

WRC World Radiocommunication Conference

WSS Wide-Sense Stationary

WSSUS Wide-Sense Stationary Uncorrelated Scattering

ZC Zadoff–Chu

ZCZ Zero Correlation Zone

ZF Zero-Forcing

ZFEP Zero-Forcing Equal Power

*This acronym can have different meanings depending on the context. The meaning is clearly indicated in the chapter when used.

1

Introduction and Background

Thomas Sälzer and Matthew Baker

1.1 The Context for the Long Term Evolution of UMTS

1.1.1 Historical Context

The Long Term Evolution of UMTS is just one of the latest steps in an advancing series of mobile telecommunications systems.

Arguably, at least for land-based systems, the series began in 1947 with the development of the concept of *cells* by the famous Bell Labs of the USA. The use of cells enabled the capacity of a mobile communications network to be increased substantially, by dividing the coverage area up into small cells each with its own base station operating on a different frequency.

The early systems were confined within national boundaries. They attracted only a small number of users, as the equipment on which they relied was expensive, cumbersome and power-hungry, and therefore was only really practical in a car.

The first mobile communication systems to see large-scale commercial growth arrived in the 1980s and became known as the 'First Generation' systems. The First Generation comprised a number of independently-developed systems worldwide (e.g. AMPS (Analogue Mobile Phone System, used in America), TACS (Total Access Communication System, used in parts of Europe), NMT (Nordic Mobile Telephone, used in parts of Europe) and J-TACS (Japanese Total Access Communication System, used in Japan and Hong Kong)), using analogue technology.

Global roaming first became a possibility with the development of the digital 'Second Generation' system known as GSM (Global System for Mobile Communications). The success of GSM was due in part to the collaborative spirit in which it was developed. By harnessing the creative expertise of a number of companies working together under the

LTE – The UMTS Long Term Evolution: from Theory to Practice   Stefania Sesia, Issam Toufik and Matthew Baker
© 2009 John Wiley & Sons, Ltd

auspices of the European Telecommunications Standards Institute (ETSI), GSM became a robust, interoperable and widely-accepted standard.

Fuelled by advances in mobile handset technology, which resulted in small, fashionable terminals with a long battery life, the widespread acceptance of the GSM standard exceeded initial expectations and helped to create a vast new market. The resulting near-universal penetration of GSM phones in the developed world provided an ease of communication never previously possible, first by voice and text message, and later also by more advanced data services. Meanwhile in the developing world, GSM technology had begun to connect communities and individuals in remote regions where fixed-line connectivity was non-existent and would be prohibitively expensive to deploy.

This ubiquitous availability of user-friendly mobile communications, together with increasing consumer familiarity with such technology and practical reliance on it, thus provides the context for new systems with more advanced capabilities. In the following section, the series of progressions which have succeeded GSM is outlined, culminating in the development of the system currently known as LTE – the Long Term Evolution of UMTS (Universal Mobile Telecommunications System).

1.1.2 LTE in the Mobile Radio Landscape

In contrast to transmission technologies using media such as copper lines and optical fibres, the radio spectrum is a medium shared between different, and potentially interfering, technologies.

As a consequence, regulatory bodies – in particular, ITU-R (International Telecommunication Union, Radio Communication Sector) [1], but also regional and national regulators – play a key role in the evolution of radio technologies since they decide which parts of the spectrum and how much bandwidth may be used by particular types of service and technology. This role is facilitated by the *standardization* of families of radio technologies – a process which not only provides specified interfaces to ensure interoperability between equipment from a multiplicity of vendors, but also aims to ensure that the allocated spectrum is used as efficiently as possible, so as to provide an attractive user experience and innovative services.

The complementary functions of the regulatory authorities and the standardization organizations can be summarized broadly by the following relationship:

$$\text{Aggregated data rate} = \underbrace{\text{bandwidth}}_{\substack{\text{regulation and licences} \\ \text{(ITU-R, regional regulators)}}} \times \underbrace{\text{spectral efficiency}}_{\substack{\text{technology} \\ \text{and standards}}}$$

On a worldwide basis, ITU-R defines technology families and associates specific parts of the spectrum with these families. Facilitated by ITU-R, spectrum for mobile radio technologies is identified for the radio technologies which meet the ITU-R's requirements to be designated as members of the *International Mobile Telecommunications* (IMT) family. Effectively, the IMT family comprises systems known as 'Third Generation' (for the first time providing data rates up to 2 Mbps) and beyond.

From the technology and standards angle, there are currently three main organizations responsible for developing the standards meeting IMT requirements, and which are continuing to shape the landscape of mobile radio systems, as shown in Figure 1.1.

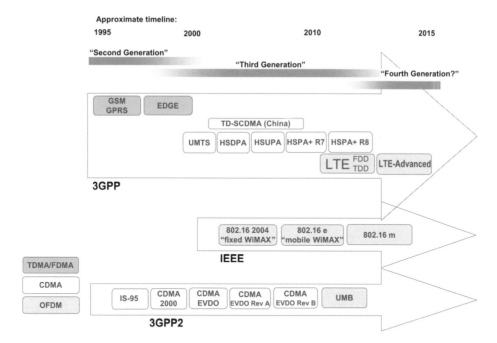

Figure 1.1 Approximate timeline of the mobile communications standards landscape.

The uppermost evolution track shown in Figure 1.1 is that developed in the 3rd Generation Partnership Project (3GPP), which is currently the dominant standards development group for mobile radio systems and is described in more detail below.

Within the 3GPP evolution track, three multiple access technologies are evident: the 'Second Generation' GSM/GPRS/EDGE family[1] was based on Time- and Frequency-Division Multiple Access (TDMA/FDMA); the 'Third Generation' UMTS family marked the entry of Code Division Multiple Access (CDMA) into the 3GPP evolution track, becoming known as *Wideband* CDMA (owing to its 5 MHz carrier bandwidth) or simply WCDMA; finally LTE has adopted Orthogonal Frequency-Division Multiplexing (OFDM), which is the access technology dominating the latest evolutions of all mobile radio standards.

In continuing the technology progression from the GSM and UMTS technology families within 3GPP, the LTE system can be seen as completing the trend of expansion of service provision beyond voice calls towards a multiservice air interface. This was already a key aim of UMTS and GPRS/EDGE, but LTE was designed from the start with the goal of evolving the radio access technology under the assumption that all services would be packet-switched, rather than following the circuit-switched model of earlier systems. Furthermore, LTE is accompanied by an evolution of the non-radio aspects of the complete system, under the term 'System Architecture Evolution' (SAE) which includes the Evolved Packet Core (EPC) network. Together, LTE and SAE comprise the Evolved Packet System (EPS), where both the core network and the radio access are fully packet-switched.

---

[1]The maintenance and development of specifications for the GSM family was passed to 3GPP from ETSI.

The standardization of LTE and EPS does not mean that further development of the other radio access technologies in 3GPP has ceased. In particular, the enhancement of UMTS with new releases of the specifications continues in 3GPP, to the greatest extent possible while ensuring backward compatibility with earlier releases: the original 'Release 99' specifications of UMTS have been extended with high-speed downlink and uplink enhancements (HSDPA[2] and HSUPA[3] in Releases 5 and 6 respectively), known collectively as 'HSPA' (High-Speed Packet Access). HSPA has been further enhanced in Release 7 (becoming known as HSPA+) with higher-order modulation and, for the first time in a cellular communication system, multistream 'MIMO' operation (Multiple-Input Multiple-Output antenna system). Further enhancements of HSPA+ are being introduced in Release 8 in parallel to the first release of LTE (which for consistency is also termed Release 8). These backward-compatible enhancements will enable network operators who have invested heavily in the WCDMA technology of UMTS to generate new revenues from new features while still providing service to their existing subscribers using legacy terminals.

LTE is able to benefit from the latest understanding and technology developments from HSPA and HSPA+, especially in relation to optimizations of the protocol stack, while also being free to adopt radical new technology without the constraints of backward compatibility or a 5 MHz carrier bandwidth. However, LTE also has to satisfy new demands, for example in relation to spectrum flexibility for deployment. LTE can operate in Frequency-Division Duplex (FDD) and Time-Division Duplex (TDD) modes in a harmonized framework designed also to support the evolution of TD-SCDMA (Time-Division Synchronous Code Division Multiple Access), which has been developed in 3GPP as an additional branch of the UMTS technology path, essentially for the Chinese market.

The second path of evolution has emerged from the IEEE 802 LAN/MAN[4] standards committee, which created the '802.16' family as a broadband wireless access standard. This family is also fully packet-oriented. It is often referred to as *WiMAX*, on the basis of a so-called 'System Profile' assembled from the 802.16 standard and promoted by the WiMAX Forum. The WiMAX Forum also ensures the corresponding product certification. While the first version known as 802.16-2004 was restricted to fixed access, the following version 802.16e includes basic support of mobility and is therefore often referred to as 'mobile WiMAX'. However, it can be noted that in general the WiMAX family has not been designed with the same emphasis on mobility and compatibility with operators' core networks as the 3GPP technology family, which includes core network evolutions in addition to the radio access network evolution. Nevertheless, the latest generation currently under development by the IEEE, known as 802.16m, has similar targets to the likely future enhancements to LTE which are outlined in the concluding chapter of this book, Chapter 24.

A third evolution track shown in Figure 1.1 is led by a partnership organization similar to 3GPP and known as 3GPP2. Based on the American 'IS95' standard, which was the first mobile cellular communication system to use CDMA technology, CDMA2000 was developed and deployed mainly in the USA, Korea and Japan. Standardization in 3GPP2 has continued with parallel evolution tracks towards data-oriented systems (EV-DO), to a certain extent taking a similar path to the evolutions in 3GPP. Mirroring LTE, 3GPP2's latest

---

[2]High-Speed Downlink Packet Access.
[3]High-Speed Uplink Packet Access.
[4]Local Area Network/Metropolitan Area Network.

evolution is a new OFDM-based system called Ultra-Mobile Broadband (UMB), derived in part from a proprietary system known as 'Flash OFDM'.

The overall pattern is of an evolution of mobile radio towards flexible, packet-oriented, multiservice systems. The aim of all these systems is towards offering a mobile broadband user experience that can approach that of current fixed access networks such as Asymmetric Digital Subscriber Line (ADSL) and Fibre-To-The-Home (FTTH).

1.1.3 The Standardization Process in 3GPP

The collaborative standardization model which so successfully produced the GSM system became the basis for the development of the UMTS system. In the interests of producing truly global standards, the collaboration for both GSM and UMTS was expanded beyond ETSI to encompass regional Standards Development Organizations (SDOs) from Japan (ARIB and TTC), Korea (TTA), North America (ATIS) and China (CCSA), as shown in Figure 1.2.

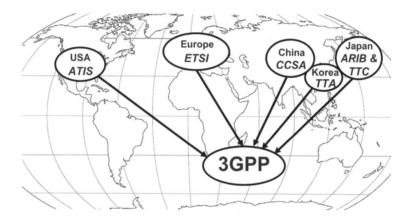

Figure 1.2 3GPP is a global partnership of six regional SDOs.

So the 3GPP was born, which by 2008 boasted over 300 individual member companies.

The successful creation of such a large and complex system specification as that for UMTS or LTE requires a well-structured organization with pragmatic working procedures. 3GPP is divided into four Technical Specification Groups (TSGs), each of which is comprised of a number of Working Groups (WGs) with responsibility for a specific aspect of the specifications as shown in Figure 1.3.

A distinctive feature of the working methods of these groups is the consensus-driven approach to decision-making. This facilitates open discussion and iterative improvement of technical proposals, frequently leading to merging of proposals from multiple companies in the quest for the optimal solution.

All documents submitted to 3GPP are publicly available on the 3GPP website,[5] including contributions from individual companies, technical reports and technical specifications.

---

[5] http://www.3gpp.org.

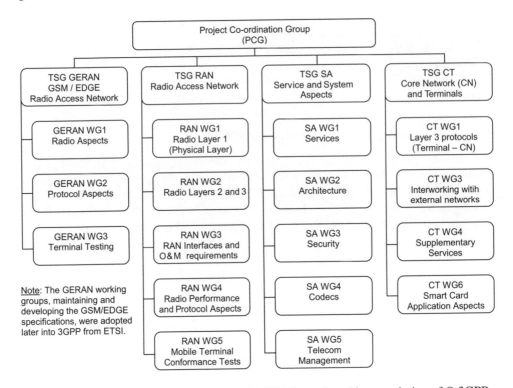

Figure 1.3 The Working Group structure of 3GPP. Reproduced by permission of © 3GPP.

In reaching consensus around a technology, the WGs take into account a variety of considerations, including but not limited to performance, implementation cost, complexity and compatibility with earlier versions or deployments. Simulations are frequently used to compare performance of different techniques, especially in the WGs focusing on the physical layer of the air interface and on performance requirements. This requires consensus first to be reached around the simulation assumptions to be used for the comparison, including, in particular, understanding and defining the scenarios of interest to network operators.

Formal voting is therefore rare in 3GPP, thus for the most part avoiding polarization of the contributing companies into factions, or bureaucratic stalemate situations which sometimes occur in standardization efforts.

The LTE standardization process was inaugurated at a workshop in Toronto in November 2004, when a broad range of companies involved in the mobile communications business presented their visions for the future evolution of the specifications to be developed in 3GPP. These visions included both initial perceptions of the *requirements* which needed to be satisfied, and proposals for *suitable technologies* to meet those requirements.

The requirements are reviewed in detail in Section 1.2, while the key technologies are introduced in Section 1.3.

1.2 Requirements and Targets for the Long Term Evolution

Discussion of the key requirements for the new LTE system led to the creation of a formal 'Study Item' in 3GPP with the specific aim of 'evolving' the 3GPP radio access technology to ensure competitiveness over a 10-year time-frame. Under the auspices of this Study Item, the requirements for LTE were refined and crystallized, being finalized in June 2005.

They can be summarized as follows:

- reduced delays, in terms of both connection establishment and transmission latency;

- increased user data rates;

- increased cell-edge bit-rate, for uniformity of service provision;

- reduced cost per bit, implying improved spectral efficiency;

- greater flexibility of spectrum usage, in both new and pre-existing bands;

- simplified network architecture;

- seamless mobility, including between different radio-access technologies;

- reasonable power consumption for the mobile terminal.

It can also be noted that network operator requirements for next generation mobile systems were formulated by the Next Generation Mobile Networks (NGMN) alliance of network operators [2], which served as an additional reference for the development and assessment of the LTE design. Such operator-driven requirements will also guide the development of the next phase of LTE, namely LTE-Advanced (see Chapter 24).

To address these objectives, the LTE system design covers both the radio interface and the radio network architecture. The main chapters of this book describe the technologies by which these targets are achieved, and even exceeded in some aspects, by the first version of the LTE system.

1.2.1 System Performance Requirements

Improved system performance compared to existing systems is one of the main requirements from network operators, to ensure the competitiveness of LTE and hence to arouse market interest. In this section, we highlight the main performance metrics used in the definition of the LTE requirements and its performance assessment.

Table 1.1 summarizes the main performance requirements to which the first release of LTE was designed. Many of the figures are given relative to the performance of the most advanced available version of UMTS, which at the time of the definition of the LTE requirements was HSDPA/HSUPA Release 6 – referred to here as the *reference baseline*. It can be seen that the target requirements for LTE represent a significant step from the capacity and user experience offered by the 'Third Generation' mobile communications systems which were being deployed at the time when LTE was being developed.

Table 1.1 Summary of key performance requirement targets for LTE.

		Absolute requirement	Comparison to Release 6	Comment
Downlink	Peak transmission rate	> 100 Mbps	7 × 14.4 Mbps	LTE in 20 MHz FDD, 2 × 2 spatial multiplexing. Reference: HSDPA in 5 MHz FDD, single antenna transmission
	Peak spectral efficiency	> 5 bps/Hz	3 bps/Hz	
	Average cell spectral efficiency	> 1.6 – 2.1 bps/Hz/cell	3 – 4 × 0.53 bps/Hz/cell	LTE: 2 × 2 spatial multiplexing, Interference Rejection Combining (IRC) receiver [3]. Reference: HSDPA, Rake receiver [4], 2 receive antennas
	Cell edge spectral efficiency	> 0.04 – 0.06 bps/Hz/user	2-3 × 0.02 bps/Hz	As above, 10 users assumed per cell
	Broadcast spectral efficiency	> 1 bps/Hz	N/A	Dedicated carrier for broadcast mode
Uplink	Peak transmission rate	> 50 Mbps	5 × 11 Mbps	LTE in 20 MHz FDD, single antenna transmission. Reference: HSUPA in 5 MHz FDD, single antenna transmission
	Peak spectral efficiency	> 2.5 bps/Hz	2 bps/Hz	
	Average cell spectral efficiency	> 0.66 – 1.0 bps/Hz/cell	2 – 3 × 0.33 bps/Hz	LTE: single antenna transmission, IRC receiver [3]. Reference: HSUPA, Rake receiver [4], 2 receive antennas
	Cell edge spectral efficiency	> 0.02 – 0.03 bps/Hz/user	2 – 3 × 0.01 bps/Hz	As above, 10 users assumed per cell
System	User plane latency (two way radio delay)	< 10 ms	One fifth	
	Connection set-up latency	< 100 ms		Idle state → active state
	Operating bandwidth	1.4 – 20 MHz	5 MHz	(initial requirement started at 1.25 MHz)
	VoIP capacity	NGMN preferred target expressed in [2] is > 60 sessions/MHz/cell		

As mentioned above, HSPA technologies are also continuing to be developed to offer higher spectral efficiencies than were assumed for the reference baseline case. However, LTE has been able to benefit from avoiding the constraints of backward compatibility, enabling the inclusion of advanced MIMO schemes in the system design from the beginning, and highly flexible spectrum usage built around new multiple access schemes.

The requirements shown in Table 1.1 are discussed and explained in more detail below.

1.2.1.1 Peak Rates and Peak Spectral Efficiency

For marketing purposes, the first parameter by which different radio access technologies are usually compared is the peak per-user data rate which can be achieved. This peak data rate generally scales according to amount of spectrum used, and, for MIMO systems, according to the minimum of the number of transmit and receive antennas (see Section 11.1).

...he maximum throughput per user assuming the whole ...ser with the highest modulation and coding scheme ...ported. Typical radio interface overhead (control ...s estimated and taken into account for a given ...e is generally calculated for the downlink and ...to obtain a single value independent of the ...n that is agnostic of the duplex mode. The ...ly by dividing the peak rate by the used

...in the LTE system were set at 100 Mbps ...dth,[6] corresponding to respective peak ...ng assumption here is that the terminal ...e number of antennas used at the base ...ator, and the first version of the LTE ...wnlink MIMO operation with up to ...ues enabling high peak data rates are described...

When ...io communication technologies, great emphasi... ... on the peak data rate capabilities. While this is one indicator of how technologically advanced a system is and can be obtained by simple calculations, it may not be a key differentiator in the usage scenarios for a mobile communication system in practical deployment. Moreover, it is relatively easy to design a system that can provide very high peak data rates for users close to the base station, where interference from other cells is low and techniques such as MIMO can be used to their greatest extent. It is much more challenging to provide high data rates with good coverage and mobility, but it is exactly these latter aspects which contribute most strongly to user satisfaction.

In typical deployments, individual users are located at varying distances from the base stations, the propagation conditions for radio signals to individual users are rarely ideal, and moreover the available resources must be shared between many users. Consequently, although the claimed peak data rates of a system are genuinely achievable in the right conditions, it is rare for a single user to be able to experience the peak data rates for a sustained period, and the envisaged applications do not usually require this level of performance.

A differentiator of the LTE system design compared to some other systems has been the recognition of these 'typical deployment constraints' from the beginning. During the design process, emphasis was therefore placed not only on providing a competitive peak data rate for use when conditions allow, but also importantly on *system level performance*, which was evaluated during several performance verification steps.

System-level evaluations are based on simulations of multicell configurations where data transmission from/to a population of mobiles is considered in a typical deployment scenario. The paragraphs below describe the main metrics used as requirements for system level performance. In order to make these metrics meaningful, parameters such as the deployment scenario, traffic models, channel models and system configuration need to be thoroughly defined.

---

[6]Four times the bandwidth of a WCDMA carrier.

The key definitions used for the system evaluations of LTE can be found in an operator input document addressing the performance verification milestone in the LTE development process [5]. This document takes into account deployment scenarios and channel models agreed during the LTE Study Item [6], and is based on an evaluation methodology elaborated by NGMN NGMN operators in [7]. The reference deployment scenarios which were given special consideration for the LTE performance evaluation covered macrocells with base station separations of between 500 m and 1.7 km, as well as microcells using MIMO with base station separations of 130 m. A range of mobile terminal speeds were studied, focusing particularly on the range 3–30 km/h, although higher mobile speeds were also considered important.

1.2.1.2   Cell Throughput and Spectral Efficiency

Performance at the cell level is an important criterion, as it relates directly to the number of cell sites that a network operator requires, and hence to the capital cost of deploying the system. For LTE, it was chosen to assess the cell level performance with full-queue traffic models (i.e. assuming that there is never a shortage of data to transmit if a user is given the opportunity) and a relatively high system load, typically 10 users per cell.

The requirements at the cell level were defined in terms of the following metrics:

- Average cell throughput [bps/cell] and spectral efficiency [bps/Hz/cell].

- Average user throughput [bps/user] and spectral efficiency [bps/Hz/user].

- Cell-edge user throughput [bps/user] and spectral efficiency [bps/Hz/user]. The metric used for this assessment is the 5-percentile user throughput, obtained from the cumulative distribution function of the user throughput.

For the UMTS Release 6 reference baseline, it was assumed that both the terminal and the base station use a single transmit antenna and two receive antennas; for the terminal receiver the assumed performance corresponds to a two-branch Rake receiver [4] with linear combining of the signals from the two antennas.

For the LTE system, the use of two transmit and receive antennas was assumed at the base station. At the terminal two receive antennas were assumed, but still only a single transmit antenna. The receiver for both downlink and uplink is assumed to be a linear receiver with optimum combining of the signals from the antenna branches [3]. In the uplink, higher per-user throughput should be achievable by also using multiple transmit antennas at the terminal, which will be considered for future releases of LTE.

The original requirements for the cell level metrics were only expressed as relative gains compared to the Release 6 reference baseline. The absolute values provided in Table 1.1 are based on evaluations of the reference system performance that can be found in [8] and [9] for downlink and uplink respectively.

1.2.1.3   Voice Capacity

Unlike full queue traffic (such as file download) which is typically delay-tolerant and does not require a guaranteed bit-rate, real-time traffic such as Voice over IP (VoIP) has tight delay constraints. It is important to set system capacity requirements for such services – a particular challenge in fully packet-based systems like LTE which rely on adaptive scheduling.

The system capacity requirement is defined as the number of satisfied VoIP users, given a particular traffic model and delay constraints. The details of the traffic model used for evaluating LTE can be found in [5]. Here, a VoIP user is considered to be in outage (i.e. not satisfied) if more than 2% of the VoIP packets do not arrive successfully at the radio receiver within 50 ms and are therefore discarded. This assumes an overall end-to-end delay (from mobile terminal to mobile terminal) below 200 ms. The system capacity for VoIP can then be defined as the number of users present per cell when more than 95% of the users are satisfied.

The NGMN group of network operators expressed a preference for the ability to support 60 satisfied VoIP sessions per MHz – an increase of two to four times what can typically be achieved in the Release 6 reference case. This is an area where there is scope for further enhancement of LTE in later releases.

1.2.1.4 Mobility and Cell Ranges

In terms of mobility, the LTE system is required to support communication with terminals moving at speeds of up to 350 km/h, or even up to 500 km/h depending on the frequency band. The primary scenario for operation at such high speeds is usage on high-speed trains – a scenario which is increasing in importance across the world as the number of high-speed rail lines increases and train operators aim to offer an attractive working environment to their passengers. These requirements mean that handover between cells has to be possible without interruption – in other words, with imperceptible delay and packet loss for voice calls, and with reliable transmission for data services.

These targets are to be achieved by the LTE system in typical cells of radius up to 5 km, while operation should continue to be possible for cell ranges of up to 100 km to enable wide-area deployments.

1.2.1.5 Broadcast Mode Performance

Although not available in the first release due to higher prioritization of other service modes, LTE is required to integrate an efficient broadcast mode for high rate Multimedia Broadcast/Multicast Services (MBMS) such as Mobile TV, based on a Single Frequency Network mode of operation as explained in detail in Chapter 14. This mode is able to operate either on a shared carrier frequency together with unicast transmissions, or on a dedicated broadcast carrier. To ensure efficient broadcast performance a requirement was defined for the dedicated carrier case.

In broadcast systems, the system throughput is limited to what is achievable for the users in the worst conditions. Consequently, the broadcast performance requirement was defined in terms of an achievable system throughput (bps) and spectral efficiency (bps/Hz) assuming a coverage of 98% of the nominal coverage area of the system. This means that only 2% of the locations in the nominal coverage area are in outage – where outage for broadcast services is defined as experiencing a packet error rate higher than 1%.

This broadcast spectral efficiency requirement was set to 1 bps/Hz [10].

1.2.1.6 User Plane Latency

User plane latency is an important performance metric for real-time and interactive services. On the radio interface, the minimum user plane latency can be calculated based on signalling analysis for the case of an unloaded system. It is defined as the average time between the first transmission of a data packet and the reception of a physical layer Acknowledgement (ACK). The calculation should include typical HARQ[7] retransmission rates (e.g. 0–30%). This definition therefore considers the capability of the system design, without being distorted by the scheduling delays that would appear in the case of a loaded system. The round-trip latency is obtained simply by multiplying the one-way user plane latency by a factor of two.

The LTE system is also required to be able to operate with an IP-layer one-way data-packet latency across the radio access network as low as 5 ms in optimal conditions. However, it is recognized that the actual delay experienced in a practical system will be dependent on system loading and radio propagation conditions. For example, HARQ plays a key role in maximizing spectral efficiency at the expense of increased delay while retransmissions take place, whereas maximal spectral efficiency may not be essential in situations when minimum latency is required.

1.2.1.7   Control Plane Latency and Capacity

In addition to the user plane latency requirement, call setup delay is required to be significantly reduced compared to existing cellular systems. This not only enables a good user experience but also affects the battery life of terminals, since a system design which allows a fast transition from an idle state to an active state enables terminals to spend more time in the low-power idle state.

Control plane latency is measured as the time required for performing the transitions between different LTE states. LTE is based on only two main states, 'RRC_IDLE' and 'RRC_CONNECTED' (i.e. 'active').

The LTE system is required to support transition from idle to active in less than 100 ms (excluding paging delay and Non-Access Stratum (NAS) signalling delay).

The LTE system capacity is dependent not only on the supportable throughput but also on the number of users simultaneously located within a cell which can be supported by the control signalling. For the latter aspect, the LTE system is required to support at least 200 active-state users per cell for spectrum allocations up to 5 MHz, and at least 400 users per cell for wider spectrum allocations; only a small subset of these users would be actively receiving or transmitting data at any given time instant, depending, for example, on the availability of data to transmit and the prevailing radio channel conditions. An even larger number of non-active users may also be present in each cell, and therefore able to be paged or to start transmitting data with low latency.

1.2.2   Deployment Cost and Interoperability

Besides the system performance aspects, a number of other considerations are important for network operators. These include reduced deployment cost, spectrum flexibility and enhanced interoperability with legacy systems – essential requirements to enable deployment of LTE networks in a variety of scenarios and to facilitate migration to LTE.

---

[7]Hybrid Automatic Repeat reQuest – see Section 10.3.2.5.

1.2.2.1 Spectrum Allocations and Duplex Modes

As demand for suitable radio spectrum for mobile communications increases, LTE is required to be able to operate in a wide range of frequency bands and sizes of spectrum allocations in both uplink and downlink. LTE can use spectrum allocations ranging from 1.4 to 20 MHz with a single carrier and addresses all frequency bands currently identified for IMT systems by ITU-R [1] including those below 1 GHz.

This will in due course include deploying LTE in spectrum currently occupied by older radio access technologies – a practice often known as 'spectrum refarming'.

The ability to operate in both paired and unpaired spectrum is required, depending on spectrum availability. LTE provides support for FDD, TDD and half-duplex FDD operation in a unified design, ensuring a high degree of commonality which facilitates implementation of multimode terminals and allows worldwide roaming.

1.2.2.2 Inter-Working with Other Radio Access Technologies

Flexible interoperation with other radio access technologies is essential for service continuity, especially during the migration phase in early deployments of LTE with partial coverage, where handover to legacy systems will often occur.

LTE relies on an evolved packet core network which allows interoperation with various access technologies, in particular earlier 3GPP technologies (GSM/EDGE and UTRAN) as well as non-3GPP technologies (e.g. WiFi, CDMA2000 and WiMAX).

However, service continuity and short interruption times can only be guaranteed if measurements of the signals from other systems and fast handover mechanisms are integrated in the LTE radio access design. In its first releases LTE will thus support tight inter-working with all legacy 3GPP technologies and some non-3GPP technologies such as CDMA2000.

1.2.2.3 Terminal Complexity and Cost

A key consideration for competitive deployment of LTE is the availability of low-cost terminals with long battery life, both in stand-by and during activity. Therefore, low terminal complexity has been taken into account where relevant throughout the LTE system, as well as designing the system wherever possible to support low terminal power consumption.

1.2.2.4 Network Architecture Requirements

LTE is required to allow a cost-effective deployment by an improved radio access network architecture design including:

- flat architecture consisting of just one type of node, the base station, known in LTE as the *eNodeB*;

- effective protocols for the support of packet-switched services;

- open interfaces and support of multivendor equipment interoperability;

- efficient mechanisms for operation and maintenance, including self-optimization functionalities;

- support of easy deployment and configuration, for example for so-called home base stations (otherwise known as femto-cells).

1.3   Technologies for the Long Term Evolution

The fulfilment of the extensive range of requirements outlined above is only possible thanks to advances in the underlying mobile radio technology. As an overview, we outline here three fundamental technologies that have shaped the LTE radio interface design: *multicarrier* technology, *multiple-antenna* technology, and the application of *packet-switching* to the radio interface. Finally, we summarize the combinations of capabilities that are supported by different categories of LTE mobile terminal.

1.3.1   Multicarrier Technology

Adopting a multicarrier approach for multiple access in LTE was the first major design choice. After initial consolidation of proposals, the candidate schemes for the downlink were Orthogonal Frequency-Division Multiple Access (OFDMA)[8] and Multiple WCDMA, while the candidate schemes for the uplink were Single-Carrier Frequency-Division Multiple Access (SC-FDMA), OFDMA and Multiple WCDMA. The choice of multiple-access schemes was made in December 2005, with OFDMA being selected for the downlink, and SC-FDMA for the uplink. Both of these schemes open up the frequency domain as a new dimension of flexibility in the system, as illustrated schematically in Figure 1.4.

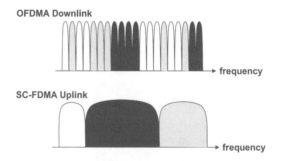

Figure 1.4 Frequency-domain view of the LTE multiple-access technologies.

OFDMA extends the multicarrier technology OFDM to provide a very flexible multiple-access scheme. OFDM subdivides the bandwidth available for signal transmission into a multitude of narrowband subcarriers, arranged to be mutually orthogonal, which either individually or in groups can carry independent information streams; in OFDMA, this

---

[8]OFDM technology was already well understood in 3GPP as a result of an earlier study of the technology in 2003–4.

subdivision of the available bandwidth is exploited in sharing the subcarriers among multiple users.[9]

This resulting flexibility can be used in various ways:

- Different spectrum bandwidths can be utilized without changing the fundamental system parameters or equipment design.

- Transmission resources of variable bandwidth can be allocated to different users and scheduled freely in the frequency domain.

- Fractional frequency re-use and interference coordination between cells are facilitated.

Extensive experience with OFDM has been gained in recent years from deployment of digital audio and video broadcasting systems such as DAB, DVB and DMB.[10] This experience has highlighted some of the key advantages of OFDM, which include:

- robustness to time-dispersive radio channels, thanks to the subdivision of the wide-band transmitted signal into multiple narrowband subcarriers, enabling inter-symbol interference to be largely constrained within a guard interval at the beginning of each symbol;

- low-complexity receivers, by exploiting frequency-domain equalization;

- simple combining of signals from multiple transmitters in broadcast networks.

These advantages, and how they arise from the OFDM signal design, are explained in detail in Chapter 5.

By contrast, the transmitter design for OFDM is more costly, as the Peak-to-Average Power Ratio (PAPR) of an OFDM signal is relatively high, resulting in a need for a highly-linear RF power amplifier. However, this limitation is not inconsistent with the use of OFDM for *downlink* transmissions, as low-cost implementation has a lower priority for the base station than for the mobile terminal.

In the uplink, however, the high PAPR of OFDM is difficult to tolerate for the transmitter of the mobile terminal, since it is necessary to compromise between the output power required for good outdoor coverage, the power consumption, and the cost of the power amplifier. SC-FDMA, which is explained in detail in Chapter 15, provides a multiple-access technology which has much in common with OFDMA – in particular the flexibility in the frequency domain, and the incorporation of a guard interval at the start of each transmitted symbol to facilitate low-complexity frequency-domain equalization at the receiver. At the same time, SC-FDMA has a significantly lower PAPR. It therefore resolves to some extent the dilemma of how the uplink can benefit from the advantages of multicarrier technology while avoiding excessive cost for the mobile terminal transmitter and retaining a reasonable degree of commonality between uplink and downlink technologies.

As mentioned above, during the early stages of the development of LTE another multicarrier based solution to the multiple access scheme was also actively considered – namely multiple WCDMA carriers. This would have had the advantage of reusing existing

---

[9]The use of the frequency domain comes in addition to the well-known time-division multiplexing which continues to play an important role in LTE.

[10]Digital Audio Broadcasting, Digital Video Broadcasting and Digital Mobile Broadcasting.

technology from the established UMTS systems. However, as the LTE system is intended to remain competitive for many years into the future, the initial benefits of technology reuse from UMTS become less advantageous in the long-term; continuation with the same technology would have missed the opportunity to embrace new possibilities and to benefit from OFDM with its flexibility, low receiver complexity and high performance in time-dispersive channels.

1.3.2 Multiple Antenna Technology

The use of multiple antenna technology allows the exploitation of the spatial-domain as another new dimension. This becomes essential in the quest for higher spectral efficiencies. As will be detailed in Chapter 11, with the use of multiple antennas the theoretically-achievable spectral efficiency scales linearly with the minimum of the number of transmit and receive antennas employed, at least in suitable radio propagation environments.

Multiple antenna technology opens the door to a large variety of features, but not all of them easily deliver their theoretical promises when it comes to implementation in practical systems. Multiple antennas can be used in a variety of ways, mainly based on three fundamental principles, schematically illustrated in Figure 1.5:

- **Diversity gain.** Use of the space-diversity provided by the multiple antennas to improve the robustness of the transmission against multipath fading.

- **Array gain.** Concentration of energy in one or more given directions via precoding or beamforming. This also allows multiple users located in different directions to be served simultaneously (so-called multi-user MIMO).

- **Spatial multiplexing gain.** Transmission of multiple signal streams to a single user on multiple spatial layers created by combinations of the available antennas.

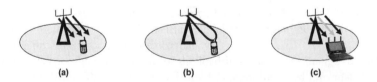

(a)   (b)   (c)

Figure 1.5 Three fundamental benefits of multiple antennas: (a) diversity gain; (b) array gain; (c) spatial multiplexing gain.

A large part of the LTE 'Study Item' phase was therefore dedicated to the selection and design of the various multiple antenna features to be included in LTE. The final system includes several complementary options which allow for adaptability according to the deployment and the propagation conditions of the different users.

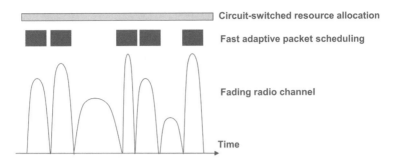

Figure 1.6 Fast scheduling and link adaptation.

1.3.3   Packet-Switched Radio Interface

As has already been noted, LTE has been designed as a completely packet-oriented multi-service system, without the reliance on circuit-switched connection-oriented protocols prevalent in its predecessors. In LTE, this philosophy is applied across all the layers of the protocol stack.

The route towards fast packet scheduling over the radio interface was already opened by HSDPA, which allowed the transmission of short packets having a duration of the same order of magnitude as the coherence time of the fast fading channel, as shown in Figure 1.6. This calls for a joint optimization of the physical layer configuration and the resource management carried out by the link layer protocols according to the prevailing propagation conditions. This aspect of HSDPA involves tight coupling between the lower two layers of the protocol stack – the MAC (Medium Access Control layer; see Chapter 4) and the physical layer. In HSDPA, this coupling already includes features such as fast channel state feedback, dynamic link adaptation, scheduling exploiting multi-user diversity, and fast retransmission protocols.

In LTE, in order to improve the system latency the packet duration was further reduced from the 2 ms used in HSDPA down to just 1 ms. This short transmission interval, together with the new dimensions of frequency and space, has further extended the field of cross-layer techniques between the MAC and physical layers to include the following techniques in LTE:

- adaptive scheduling in both the frequency and spatial dimensions;

- adaptation of the MIMO configuration including the selection of the number of spatial layers transmitted simultaneously;

- link adaptation of modulation and code-rate, including the number of transmitted codewords;

- several modes of fast channel state reporting.

These different levels of optimization are combined with very sophisticated control signalling, which proved to be one of the significant challenges in turning the LTE concept into a working system.

1.3.4   User Equipment Capabilities

The whole LTE system is built around the three fundamental technologies outlined above, combined with a new flat network architecture. Together, these technologies enable the targets set out in Section 1.2 to be met. By exploiting these technologies to the full, it would be possible for all LTE terminals, known as User Equipment (UE), to reach performance exceeding the peak transmission rates and spectral efficiencies.

However, in practice it is important to recognize that the market for UEs is large and diverse, and there is therefore a need for LTE to support a range of categories of UE with different capabilities to satisfy different market segments. In general, each market segment attaches different priorities to aspects such as peak data rate, UE size, cost and battery life. Some typical trade-offs include the following:

- Support for the highest data rates is key to the success of some applications, but generally requires large amounts of memory for data processing, which increases the cost of the UE.

- UEs which may be embedded in large devices such as laptop computers are often not significantly constrained in terms of acceptable power consumption or the number of antennas which may be used; on the other hand, other market segments require ultra-slim hand-held terminals which have little space for multiple antennas or large batteries.

The wider the range of UE categories supported, the closer the match which may be made between a UE's capabilities and the requirements of a particular market segment. However, support for a large number of UE categories also has drawbacks in terms of the signalling overhead required for each UE to inform the network about its capabilities, as well as increased costs due to loss of economies of scale and increased complexity for testing the interoperability of many different configurations.

The LTE system has therefore been designed to support a compact set of five categories of UE, ranging from relatively low-cost terminals with similar capabilities to UMTS HSPA, up to very high-capability terminals which exploit the LTE technology to the maximum extent possible and exceed the peak data rate targets.

The capabilities of the five categories are summarized in Table 1.2.

It can be seen that the highest category of LTE UE possesses peak data rate capabilities far exceeding the LTE targets.

1.4   From Theory to Practice

As a result of intense activity by a larger number of contributing companies than ever before in 3GPP, the study phase finally closed in September 2006, just two years after the LTE inauguration workshop. It had been shown that fulfilment of the agreed requirements for LTE was feasible, and the process of finalizing the technical choices and drafting a complete version of the specifications for the LTE was able to begin. By December 2007, although significant areas of detail remained to be defined, the specifications had reached a sufficient level of completeness to enable LTE to be submitted to the ITU-R as a member of the IMT family of radio access technologies, and therefore able to be deployed in IMT-designated spectrum.

Table 1.2 Categories of LTE user equipment.

	UE category				
	1	2	3	4	5
Maximum downlink data rate (Mbps)	10	50	100	150	300
Maximum uplink data rate (Mbps)	5	25	50	50	75
Number of receive antennas required	2	2	2	2	4
Number of downlink MIMO streams supported	1	2	2	2	4
Support for 64QAM modulation in downlink	✔	✔	✔	✔	✔
Support for 64QAM modulation in uplink	✘	✘	✘	✘	✔
Relative memory requirement for physical layer processing (normalized to category 1 level)	1	4.9	4.9	7.3	14.6

Thus the advances in theoretical understanding and technology which underpin the LTE specifications are destined for practical exploitation. This book is written with the primary aim of illuminating the transition from this underlying academic progress to the realization of useful advances in the provision of mobile communication services. Particular focus is given to the physical layer of the Radio Access Network (RAN), as it is here that many of the most dramatic technical advances are manifested. This should enable the reader to develop an understanding of the background to the technology choices in the LTE system, and hence to understand better the LTE specifications and how to implement them.[11]

Part 1 of the book sets the radio interface in the context of the network architecture and protocols, as well as explaining the new developments in these areas which distinguish LTE from previous systems.

In Part 2, the physical layer of the RAN downlink is covered in detail, beginning with an explanation of the theory of the new downlink multiple access technology, OFDMA, in Chapter 5. This sets the context for the details of the LTE downlink design in Chapters 6 to 9. As coding, link adaptation and multiple antenna operation are of fundamental importance in fulfilling the LTE requirements, two chapters are then devoted to these topics, covering both the theory and the practical implementation in LTE.

Chapters 12 and 13 show how these techniques can be applied to the system-level operation of the LTE system, focusing on applying the new degrees of freedom to multi-user scheduling, interference coordination and radio resource management.

Finally for the downlink, Chapter 14 covers broadcast operation – a mode which has its own unique challenges in a cellular system but which is nonetheless important in enabling a range of services to be provided to the end user.

Part 3 addresses the physical layer of the RAN uplink, beginning in Chapter 15 with an introduction to the theory behind the new uplink multiple access technology, SC-FDMA. This is followed in Chapters 16 to 20 with an analysis of the detailed uplink structure

---

[11]The explanations in this book are based on the first version of the LTE specifications, known as Release 8, as at the time of writing. Although most aspects of the specifications were stable at this time, it should be noted that the specifications are regularly updated, and the reader should always refer to the specification documents themselves for the definitive details.

and operation, including the design of the associated procedures for random access, timing control and power control which are essential to the efficient operation of the uplink.

This leads on to Part 4, which examines a number of aspects of the LTE system which arise specifically as a result of it being a mobile cellular system. Chapter 21 provides a thorough analysis of the characteristics of the radio propagation environments in which LTE systems will be deployed, since an understanding of the propagation environment underpins much of the technology adopted for the LTE specifications. The new technologies and bandwidths adopted in LTE also have implications for the radio-frequency implementation of the mobile terminals in particular, and some of these are analysed in Chapter 22. The LTE system is designed to operate not just in wide bandwidths but also in a diverse range of spectrum allocation scenarios, and Chapter 23 therefore addresses the different duplex modes applicable to LTE and the effects that these may have on system design and operation.

Finally, Part 5 recognizes that the initial version of the LTE system will not terminate the long process of advancement of mobile communications. Chapter 24 takes us beyond the initial version of LTE to consider some of the ways in which the evolution is already continuing towards LTE-Advanced, in response to the latest challenges posed by the ITU-R and by the ever-higher expectations of end-users.

References[13]

[1] ITU, 'International Telecommunications Union', www.itu.int/itu-r.

[2] NGMN, 'Next Generation Mobile Networks Beyond HSPA & EVDO – A white paper', www.ngmn.org, December 2006.

[3] J. H. Winters, Optimum Combining in Digital Mobile Radio with Cochannel Interference. *IEEE Journal on Selected Areas in Communications*, Vol. 2, July 1984.

[4] R. Price and P. E. Green, 'A Communication Technique for Multipath Channels' in *Proceedings of the IRE*, Vol. 46, March 1958.

[5] Orange, China Mobile, KPN, NTT DoCoMo, Sprint, T-Mobile, Vodafone, and Telecom Italia, 'R1-070674: LTE Physical Layer Framework for Performance Verification', www.3gpp.org, 3GPP TSG RAN WG1, meeting 48, St Louis, USA, February 2007.

[6] 3GPP Technical Report 25.814, 'Physical Layer Aspects for Evolved UTRA (Release 7)', www.3gpp.org.

[7] NGMN, 'Next Generation Mobile Networks Radio Access Performance Evaluation Methodology', www.ngmn.org, June 2007.

[8] Ericsson, 'R1-072578: Summary of Downlink Performance Evaluation', www.3gpp.org, 3GPP TSG RAN WG1, meeting 49, Kobe, Japan, May 2007.

[9] Nokia, 'R1-072261: LTE Performance Evaluation – Uplink Summary', www.3gpp.org, 3GPP TSG RAN WG1, meeting 49, Kobe, Japan, May 2007.

[10] 3GPP Technical Report 25.913, 'Requirements for Evolved UTRA (E-UTRA) and Evolved UTRAN (E-UTRAN) (Release 7)', www.3gpp.org.

---

[13] All web sites confirmed 18th December 2008.

Part I

Network Architecture and Protocols

2

Network Architecture

Sudeep Palat and Philippe Godin

2.1 Introduction

As mentioned in the preceding chapter, LTE has been designed to support only packet-switched services, in contrast to the circuit-switched model of previous cellular systems. It aims to provide seamless Internet Protocol (IP) connectivity between User Equipment (UE) and the Packet Data Network (PDN), without any disruption to the end users' applications during mobility. While the term 'LTE' encompasses the evolution of the radio access through the Evolved-UTRAN (E-UTRAN), it is accompanied by an evolution of the non-radio aspects under the term 'System Architecture Evolution' (SAE) which includes the Evolved Packet Core (EPC) network. Together LTE and SAE comprise the Evolved Packet System (EPS).

EPS uses the concept of *EPS bearers* to route IP traffic from a gateway in the PDN to the UE. A bearer is an IP packet flow with a defined Quality of Service (QoS) between the gateway and the UE. The E-UTRAN and EPC together set up and release bearers as required by applications.

In this chapter, we present the overall EPS network architecture, giving an overview of the functions provided by the Core Network (CN) and E-UTRAN. The protocol stack across the different interfaces is then explained, along with an overview of the functions provided by the different protocol layers. Section 2.4 outlines the end-to-end bearer path including QoS aspects and provides details of a typical procedure for establishing a bearer. The remainder of the chapter presents the network interfaces in detail, with particular focus on the E-UTRAN interfaces and the procedures used across these interfaces, including those for the support of user mobility. The network elements and interfaces used solely to support broadcast services are covered in Chapter 14.

LTE – The UMTS Long Term Evolution: from Theory to Practice   Stefania Sesia, Issam Toufik and Matthew Baker
© 2009 John Wiley & Sons, Ltd

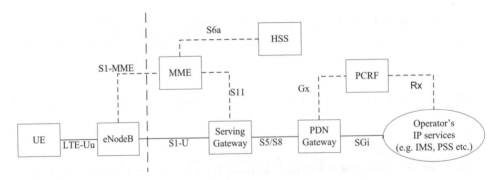

Figure 2.1 The EPS network elements.

2.2 Overall Architectural Overview

EPS provides the user with IP connectivity to a PDN for accessing the Internet, as well as for running services such as Voice over IP (VoIP). An EPS bearer is typically associated with a QoS. Multiple bearers can be established for a user in order to provide different QoS streams or connectivity to different PDNs. For example, a user might be engaged in a voice (VoIP) call while at the same time performing web browsing or File Transfer Protocol (FTP) download. A VoIP bearer would provide the necessary QoS for the voice call, while a best-effort bearer would be suitable for the web browsing or FTP session.

The network must also provide sufficient security and privacy for the user and protection for the network against fraudulent use.

This is achieved by means of several EPS network elements which have different roles. Figure 2.1 shows the overall network architecture including the network elements and the standardized interfaces. At a high level, the network is comprised of the CN (EPC) and the access network (E-UTRAN). While the CN consists of many logical nodes, the access network is made up of essentially just one node, the evolved NodeB (eNodeB), which connects to the UEs. Each of these network elements is inter-connected by means of interfaces which are standardized in order to allow multivendor interoperability. This gives network operators the possibility to source different network elements from different vendors. In fact, network operators may choose in their physical implementations to split or merge these logical network elements depending on commercial considerations. The functional split between the EPC and E-UTRAN is shown in Figure 2.2. The EPC and E-UTRAN network elements are described in more detail below.

2.2.1 The Core Network

The CN (called EPC in SAE) is responsible for the overall control of the UE and establishment of the bearers. The main logical nodes of the EPC are:

- PDN Gateway (P-GW);
- Serving Gateway (S-GW);
- Mobility Management Entity (MME).

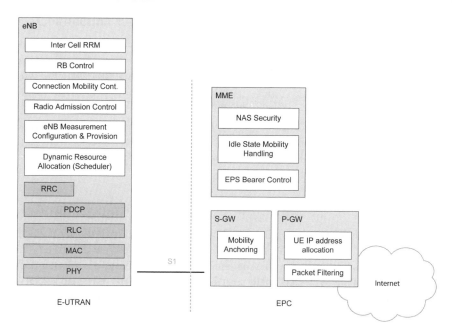

Figure 2.2   Functional split between E-UTRAN and EPC. Reproduced by permission of © 3GPP.

In addition to these nodes, EPC also includes other logical nodes and functions such as the Home Subscriber Server (HSS) and the Policy Control and Charging Rules Function (PCRF). Since the EPS only provides a bearer path of a certain QoS, control of multimedia applications such as VoIP is provided by the IP Multimedia Subsystem (IMS) which is considered to be outside the EPS itself.

The logical CN nodes (specified in [1]) are shown in Figure 2.1 and discussed in more detail in the following.

- **PCRF.** It is responsible for policy control decision-making, as well as for controlling the flow-based charging functionalities in the Policy Control Enforcement Function (PCEF) which resides in the P-GW. The PCRF provides the QoS authorization (QoS class identifier and bitrates) that decides how a certain data flow will be treated in the PCEF and ensures that this is in accordance with the user's subscription profile.

- **Home Location Register (HLR).** The HLR contains users' SAE subscription data such as the EPS-subscribed QoS profile and any access restrictions for roaming (see Section 2.2.3). It also holds information about the PDNs to which the user can connect. This could be in the form of an Access Point Name (APN) (which is a label according to DNS[1] naming conventions describing the access point to the PDN), or a PDN Address (indicating subscribed IP address(es)). In addition the HLR holds dynamic information such as the identity of the MME to which the user is currently attached

---

[1]Domain Name System.

or registered. The HLR may also integrate the Authentication Centre (AuC) which generates the vectors for authentication and security keys.

- **P-GW.** The P-GW is responsible for IP address allocation for the UE, as well as QoS enforcement and flow-based charging according to rules from the PCRF. The P-GW is responsible for the filtering of downlink user IP packets into the different QoS based bearers. This is performed based on Traffic Flow Templates (TFTs) (see Section 2.4). The P-GW performs QoS enforcement for Guaranteed Bit Rate (GBR) bearers. It also serves as the mobility anchor for inter-working with non-3GPP technologies such as CDMA2000 and WiMAX networks (see Section 2.2.4 and Chapter 13 for more information about mobility).

- **S-GW.** All user IP packets are transferred through the S-GW, which serves as the local mobility anchor for the data bearers when the UE moves between eNodeBs. It also retains the information about the bearers when the UE is in idle state (known as ECM-IDLE, see Section 2.2.1.1) and temporarily buffers downlink data while the MME initiates paging of the UE to re-establish the bearers. In addition, the S-GW performs some administrative functions in the visited network such as collecting information for charging (e.g. the volume of data sent to or received from the user), and legal interception. It also serves as the mobility anchor for inter-working with other 3GPP technologies such as GPRS and UMTS (see Section 2.2.4 and Chapter 13 for more information about mobility).

- **MME.** The MME is the control node which processes the signalling between the UE and the CN. The protocols running between the UE and the CN are known as the *Non-Access Stratum* (NAS) protocols.

  The main functions supported by the MME are classified as:

  Functions related to bearer management. This includes the establishment, maintenance and release of the bearers, and is handled by the session management layer in the NAS protocol.

  Functions related to connection management. This includes the establishment of the connection and security between the network and UE, and is handled by the connection or mobility management layer in the NAS protocol layer.

  NAS control procedures are specified in [1] and are discussed in more detail in the following section.

2.2.1.1   Non-Access Stratum (NAS) Procedures

The NAS procedures, especially the connection management procedures, are fundamentally similar to UMTS. The main change from UMTS is that EPS allows concatenation of some procedures to allow faster establishment of the connection and the bearers.

The MME creates a *UE context* when a UE is turned on and attaches to the network. It assigns a unique short temporary identity termed the SAE-Temporary Mobile Subscriber Identity (S-TMSI) to the UE which identifies the UE context in the MME. This UE context holds user subscription information downloaded from the HSS. The local storage of subscription data in the MME allows faster execution of procedures such as bearer establishment since it removes the need to consult the HSS every time. In addition, the UE context also holds dynamic information such as the list of bearers that are established and the terminal capabilities.

To reduce the overhead in the E-UTRAN and processing in the UE, all UE-related information in the access network can be released during long periods of data inactivity. This state is called EPS Connection Management IDLE (ECM-IDLE). The MME retains the UE context and the information about the established bearers during these idle periods.

To allow the network to contact an ECM-IDLE UE, the UE updates the network as to its new location whenever it moves out of its current Tracking Area (TA); this procedure is called a 'Tracking Area Update'. The MME is responsible for keeping track of the user location while the UE is in ECM-IDLE.

When there is a need to deliver downlink data to an ECM-IDLE UE, the MME sends a paging message to all the eNodeBs in its current TA, and the eNodeBs page the UE over the radio interface. On receipt of a paging message, the UE performs a service request procedure which results in moving the UE to ECM-CONNECTED state. UE-related information is thereby created in the E-UTRAN, and the bearers are re-established. The MME is responsible for the re-establishment of the radio bearers and updating the UE context in the eNodeB. This transition between the UE states is called an idle-to-active transition. To speed up the idle-to-active transition and bearer establishment, EPS supports concatenation of the NAS and AS procedures for bearer activation (see also Section 2.4.1). Some inter-relationship between the NAS and AS protocols is intentionally used to allow procedures to run simultaneously rather than sequentially, as in UMTS. For example, the bearer establishment procedure can be executed by the network without waiting for the completion of the security procedure.

Security functions are the responsibility of the MME for both signalling and user data. When a UE attaches with the network, a mutual authentication of the UE and the network is performed between the UE and the MME/HSS. This authentication function also establishes the security keys which are used for encryption of the bearers, as explained in Section 3.2.3.1. The security architecture for SAE is specified in [2].

2.2.2 The Access Network

The Access Network of LTE, E-UTRAN, simply consists of a network of eNodeBs, as illustrated in Figure 2.3. For normal user traffic (as opposed to broadcast), there is no centralized controller in E-UTRAN; hence the E-UTRAN architecture is said to be flat.

The eNodeBs are normally inter-connected with each other by means of an interface known as *X2*, and to the EPC by means of the *S1* interface – more specifically, to the MME by means of the S1-MME interface and to the S-GW by means of the S1-U interface.

The protocols which run between the eNodeBs and the UE are known as the *Access Stratum* (AS) protocols.

The E-UTRAN is responsible for all radio-related functions, which can be summarized briefly as:

- **Radio Resource Management.** This covers all functions related to the radio bearers, such as radio bearer control, radio admission control, radio mobility control, scheduling and dynamic allocation of resources to UEs in both uplink and downlink (see Chapter 13).

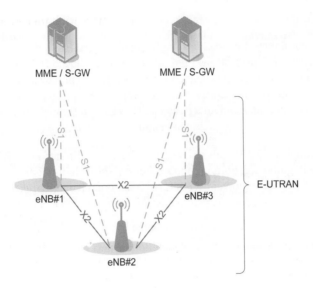

Figure 2.3 Overall E-UTRAN architecture. Reproduced by permission of © 3GPP.

- **Header Compression.** This helps to ensure efficient use of the radio interface by compressing the IP packet headers which could otherwise represent a significant overhead, especially for small packets such as VoIP (see Section 4.2.2).

- **Security.** All data sent over the radio interface is encrypted (see Sections 3.2.3.1 and 4.2.3).

- **Connectivity to the EPC.** This consists of the signalling towards the MME and the bearer path towards the S-GW.

On the network side, all of these functions reside in the eNodeBs, each of which can be responsible for managing multiple cells. Unlike some of the previous second- and third-generation technologies, LTE integrates the radio controller function into the eNodeB. This allows tight interaction between the different protocol layers of the radio access network, thus reducing latency and improving efficiency. Such distributed control eliminates the need for a high-availability, processing-intensive controller, which in turn has the potential to reduce costs and avoid 'single points of failure'. Furthermore, as LTE does not support soft handover there is no need for a centralized data-combining function in the network.

One consequence of the lack of a centralized controller node is that, as the UE moves, the network must transfer all information related to a UE, i.e. the UE context, together with any buffered data, from one eNodeB to another. As discussed in Section 2.3.1.1, mechanisms are therefore needed to avoid data loss during handover. The operation of the X2 interface for this purpose is explained in more detail in Section 2.6.

An important feature of the S1 interface linking the Access Network to the CN is known as *S1-flex*. This is a concept whereby multiple CN nodes (MME/S-GWs) can serve a common geographical area, being connected by a mesh network to the set of eNodeBs in that area (see Section 2.5). An eNodeB may thus be served by multiple MME/S-GWs, as is the case for

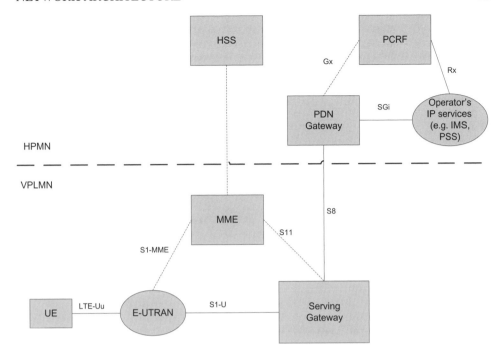

Figure 2.4 Roaming architecture for 3GPP accesses with P-GW in home network.

eNodeB#2 in Figure 2.3. The set of MME/S-GW nodes which serves a common area is called an *MME/S-GW pool*, and the area covered by such a pool of MME/S-GWs is called a *pool area*. This concept allows UEs in the cell(s) controlled by one eNodeB to be shared between multiple CN nodes, thereby providing a possibility for load sharing and also eliminating single points of failure for the CN nodes. The UE context normally remains with the same MME as long as the UE is located within the pool area.

2.2.3 Roaming Architecture

A network run by one operator in one country is known as a Public Land Mobile Network (PLMN). Roaming, where users are allowed to connect to PLMNs other than those to which they are directly subscribed, is a powerful feature for mobile networks, and LTE/SAE is no exception. A roaming user is connected to the E-UTRAN, MME and S-GW of the visited LTE network. However, LTE/SAE allows the P-GW of either the visited or the home network to be used, as shown in Figure 2.4. Using the home network's P-GW allows the user to access the home operator's services even while in a visited network. A P-GW in the visited network allows a 'local breakout' to the Internet in the visited network.

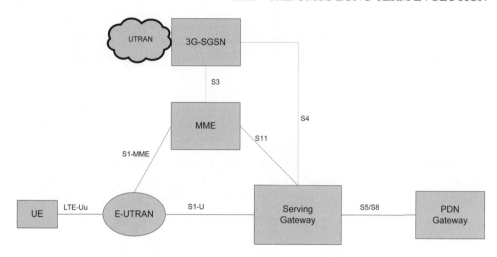

Figure 2.5 Architecture for 3G UMTS interworking.

2.2.4   Inter-Working with other Networks

EPS also supports inter-working and mobility (handover) with networks using other Radio Access Technologies (RATs), notably GSM, UMTS, CDMA2000 and WiMAX. The architecture for inter-working with 2G and 3G GPRS/UMTS networks is shown in Figure 2.5. The S-GW acts as the mobility anchor for inter-working with other 3GPP technologies such as GSM and UMTS, while the P-GW serves as an anchor allowing seamless mobility to non-3GPP networks such as CDMA2000 or WiMAX. The P-GW may also support a Proxy Mobile Internet Protocol (PMIP) based interface. More details of the radio interface procedures for inter-working are specified in [3] and are also covered in Sections 2.5.6.2 and 3.2.4.

2.3   Protocol Architecture

We outline here the radio protocol architecture of E-UTRAN.

2.3.1   User Plane

An IP packet for a UE is encapsulated in an EPC-specific protocol and tunnelled between the P-GW and the eNodeB for transmission to the UE. Different tunnelling protocols are used across different interfaces. A 3GPP-specific tunnelling protocol called the GPRS Tunnelling Protocol (GTP) [4] is used over the core network interfaces, S1 and S5/S8.[2]

The E-UTRAN user plane protocol stack is shown greyed in Figure 2.6, consisting of the PDCP (Packet Data Convergence Protocol), RLC (Radio Link Control) and MAC

---

[2]SAE also provides an option to use PMIP on S5/S8. More details on the MIP-based S5/S8 interface can be found in [3].

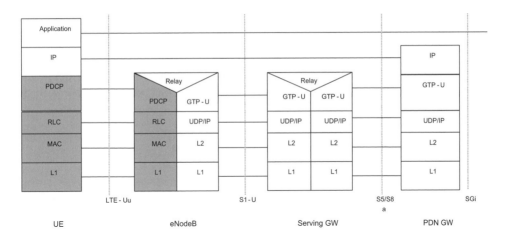

Figure 2.6 The E-UTRAN user plane protocol stack. Reproduced by permission of © 3GPP.

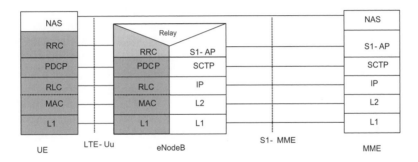

Figure 2.7 Control-plane protocol stack. Reproduced by permission of © 3GPP.

(Medium Access Control) sublayers which are terminated in the eNodeB on the network side. The respective roles of each of these layers are explained in detail in Chapter 4.

2.3.1.1   Data Handling During Handover

In the absence of any centralized controller node, data buffering during handover due to user mobility in the E-UTRAN must be performed in the eNodeB itself. Data protection during handover is a responsibility of the PDCP layer and is explained in detail in Section 4.2.4.

The RLC and MAC layers both start afresh in a new cell after handover.

2.3.2   Control Plane

The protocol stack for the control plane between the UE and MME is shown in Figure 2.7.

The greyed region of the stack indicates the access stratum protocols. The lower layers perform the same functions as for the user plane with the exception that there is no header compression function for control plane.

The RRC protocol is known as 'Layer 3' in the access stratum protocol stack. It is the main controlling function in the access stratum, being responsible for establishing the radio bearers and configuring all the lower layers using RRC signalling between the eNodeB and the UE. These functions are detailed in Section 3.2.

2.4   Quality of Service and EPS Bearers

In a typical case, multiple applications may be running in a UE at any time, each one having different QoS requirements. For example, a UE can be engaged in a VoIP call while at the same time browsing a web page or downloading an FTP file. VoIP has more stringent requirements for QoS in terms of delay and delay jitter than web browsing and FTP, while the latter requires a much lower packet loss rate. In order to support multiple QoS requirements, different bearers are set up within EPS, each being associated with a QoS.

Broadly, bearers can be classified into two categories based on the nature of the QoS they provide:

- **Minimum Guaranteed Bit Rate (GBR) bearers** which can be used for applications such as VoIP. These have an associated GBR value for which dedicated transmission resources are permanently allocated (e.g. by an admission control function in the eNodeB) at bearer establishment/modification. Bit rates higher than the GBR may be allowed for a GBR bearer if resources are available. In such cases, a Maximum Bit Rate (MBR) parameter, which can also be associated with a GBR bearer, sets an upper limit on the bit rate which can be expected from a GBR bearer.

- **Non-GBR bearers** which do not guarantee any particular bit rate. These can be used for applications such as web browsing or FTP transfer. For these bearers, no bandwidth resources are allocated permanently to the bearer.

In the access network, it is the responsibility of the eNodeB to ensure the necessary QoS for a bearer over the radio interface. Each bearer has an associated QoS Class Identifier (QCI), and an Allocation and Retention Priority (ARP).

Each QCI is characterized by priority, packet delay budget and acceptable packet loss rate. The QCI label for a bearer determines how it is handled in the eNodeB. Only a dozen such QCIs have been standardized so that vendors can all have the same understanding of the underlying service characteristics and thus provide the corresponding treatment, including queue management, conditioning and policing strategy. This ensures that an LTE operator can expect uniform traffic handling behaviour throughout the network regardless of the manufacturers of the eNodeB equipment. The set of standardized QCIs and their characteristics (from which the PCRF in an EPS can select) is provided in Table 2.1 (from Section 6.1.7, in [5]). The QCI table specifies values for the priority handling, acceptable delay budget and packet error loss rate for each QCI label.

The priority and packet delay budget (and to some extent the acceptable packet loss rate) from the QCI label determine the RLC mode configuration (see Section 4.3.1), and how the scheduler in the MAC (Section 4.4.2.1) handles packets sent over the bearer (e.g. in terms of

Table 2.1 Standardized QoS Class Identifiers (QCIs) for LTE.

QCI	Resource type	Priority	Packet delay budget (ms)	Packet error loss rate	Example services
1	GBR	2	100	10^{-2}	Conversational voice
2	GBR	4	150	10^{-3}	Conversational video (live streaming)
3	GBR	5	300	10^{-6}	Non-conversational video (buffered streaming)
4	GBR	3	50	10^{-3}	Real time gaming
5	Non-GBR	1	100	10^{-6}	IMS signalling
6	Non-GBR	7	100	10^{-3}	Voice, video (live streaming), interactive gaming
7	Non-GBR	6	300	10^{-6}	Video (buffered streaming)
8	Non-GBR	8	300	10^{-6}	TCP-based (e.g. WWW, e-mail) chat, FTP, p2p file sharing, progressive video, etc.
9	Non-GBR	9	300	10^{-6}	

scheduling policy, queue management policy and rate shaping policy). For example, a packet with a higher priority can be expected to be scheduled before a packet with lower priority. For bearers with a low acceptable loss rate, an Acknowledged Mode (AM) can be used within the RLC protocol layer to ensure that packets are delivered successfully across the radio interface (see Section 4.3.1.3).

The ARP of a bearer is used for call admission control – i.e. to decide whether or not the requested bearer should be established in case of radio congestion. It also governs the prioritization of the bearer for pre-emption with respect to a new bearer establishment request. Once successfully established, a bearer's ARP does not have any impact on the bearer-level packet forwarding treatment (e.g. for scheduling and rate control). Such packet forwarding treatment should be solely determined by the other bearer level QoS parameters such as QCI, GBR and MBR.

An EPS bearer has to cross multiple interfaces as shown in Figure 2.8 – the S5/S8 interface from the P-GW to the S-GW, the S1 interface from the S-GW to the eNodeB, and the radio interface (also known as the LTE-Uu interface) from the eNodeB to the UE. Across each interface, the EPS bearer is mapped onto a lower layer bearer, each with its own bearer identity. Each node must keep track of the binding between the bearer IDs across its different interfaces.

An S5/S8 bearer transports the packets of an EPS bearer between a P-GW and a S-GW. The S-GW stores a one-to-one mapping between an S1 bearer and an S5/S8 bearer. The bearer is identified by the GTP tunnel ID across both interfaces.

An S1 bearer transports the packets of an EPS bearer between a S-GW and an eNodeB. A radio bearer [6] transports the packets of an EPS bearer between a UE and an eNodeB.

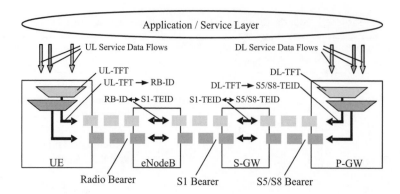

Figure 2.8 LTE/SAE bearers across the different interfaces. Reproduced by permission of © 3GPP.

An eNodeB stores a one-to-one mapping between a radio bearer ID and an S1 bearer to create the mapping between the two.

IP packets mapped to the same EPS bearer receive the same bearer-level packet forwarding treatment (e.g. scheduling policy, queue management policy, rate shaping policy, RLC configuration). Providing different bearer-level QoS thus requires that a separate EPS bearer is established for each QoS flow, and user IP packets must be filtered into the different EPS bearers.

Packet filtering into different bearers is based on Traffic Flow Templates (TFTs). The TFTs use IP header information such as source and destination IP addresses and Transmission Control Protocol (TCP) port numbers to filter packets such as VoIP from web browsing traffic so that each can be sent down the respective bearers with appropriate QoS. An UpLink TFT (UL TFT) associated with each bearer in the UE filters IP packets to EPS bearers in the uplink direction. A DownLink TFT (DL TFT) in the P-GW is a similar set of downlink packet filters.

As part of the procedure by which a UE attaches to the network, the UE is assigned an IP address by the P-GW and at least one bearer is established. This is called the default bearer, and it remains established throughout the lifetime of the PDN connection in order to provide the UE with always-on IP connectivity to that PDN. The initial bearer-level QoS parameter values of the default bearer are assigned by the MME, based on subscription data retrieved from the HSS. The PCEF may change these values in interaction with the PCRF or according to local configuration. Additional bearers called dedicated bearers can also be established at any time during or after completion of the attach procedure. A dedicated bearer can be either a GBR or a non-GBR bearer, (the default bearer always has to be a non-GBR bearer since it is permanently established). The distinction between default and dedicated bearers should be transparent to the access network (e.g. E-UTRAN). Each bearer has an associated QoS, and if more than one bearer is established for a given UE, then each bearer must also be associated with appropriate TFTs. These dedicated bearers could be established by the network, based for example on a trigger from the IMS domain, or they could be requested by the UE. The dedicated bearers for a UE may be provided by one or more P-GWs.

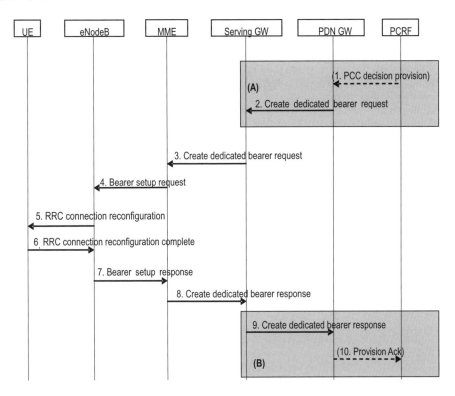

Figure 2.9 An example message flow for a LTE/SAE bearer establishment. Reproduced by permission of © 3GPP.

The bearer-level QoS parameter values for dedicated bearers are received by the P-GW from the PCRF and forwarded to the S-GW. The MME only transparently forwards those values received from the S-GW over the S11 reference point to the E-UTRAN.

2.4.1 Bearer Establishment Procedure

This section describes an example of the end-to-end bearer establishment procedure across the network nodes using the functionality described in the above sections.

A typical bearer establishment flow is shown in Figure 2.9. Each of the messages is described below.

When a bearer is established, the bearers across each of the interfaces discussed above are established.

The PCRF sends a 'PCC[3] Decision Provision' message indicating the required QoS for the bearer to the P-GW. The P-GW uses this QoS policy to assign the bearer-level QoS parameters. The P-GW then sends a 'Create Dedicated Bearer Request' message including the QoS and UL TFT to be used in the UE to the S-GW.

---

[3]Policy Control and Charging.

The S-GW forwards the Create Dedicated Bearer Request message (including bearer QoS, UL TFT and S1-bearer ID) to the MME (message 3 in Figure 2.9).

The MME then builds a set of session management configuration information including the UL TFT and the EPS bearer identity, and includes it in the 'Bearer Setup Request' message which it sends to the eNodeB (message 4 in Figure 2.9). The session management configuration is NAS information and is therefore sent transparently by the eNodeB to the UE.

The Bearer Setup Request also provides the QoS of the bearer to the eNodeB; this information is used by the eNodeB for call admission control and also to ensure the necessary QoS by appropriate scheduling of the user's IP packets. The eNodeB maps the EPS bearer QoS to the radio bearer QoS. It then signals a 'RRC Connection Reconfiguration' message (including the radio bearer QoS, session management configuration and EPS radio bearer identity) to the UE to set up the radio bearer (message 5 in Figure 2.9). The RRC Connection Reconfiguration message contains all the configuration parameters for the radio interface. This is mainly for the configuration of the Layer 2 (the PDCP, RLC and MAC parameters), but also the Layer 1 parameters required for the UE to initialize the protocol stack.

Messages 6 to 10 are the corresponding response messages to confirm that the bearers have been set up correctly.

2.5   The E-UTRAN Network Interfaces: S1 Interface

The provision of Self-Optimizing Networks (SONs) is one of the key objectives of LTE. Indeed, self-optimization of the network is a high priority for network operators, as a tool to derive the best performance from the network in a cost-effective manner, especially in changing radio propagation environments. Therefore SON has been placed as a cornerstone from the beginning around which all X2 and S1 procedures have been designed.

The S1 interface connects the eNodeB to the EPC. It is split into two interfaces, one for the control plane and the other for the user plane. The protocol structure for the S1 and the functionality provided over S1 are discussed in more detail below.

2.5.1   Protocol Structure Over S1

The protocol structure over S1 is based on a full IP transport stack with no dependency on legacy SS7[4] network configuration as used in GSM or UMTS networks. This simplification provides one expected area of savings on operational expenditure when LTE networks are deployed.

2.5.1.1   Control Plane

Figure 2.10 shows the protocol structure of the S1 control plane which is based on the well-known Stream Control Transmission Protocol / IP (SCTP/IP) stack.

---

[4]Signalling System #7 (SS7) is a communications protocol defined by the International Telecommunication Union (ITU) Telecommunication Standardization Sector (ITU-T) with a main purpose of setting up and tearing down telephone calls. Other uses include Short Message Service (SMS), number translation, prepaid billing mechanisms, and many other services.

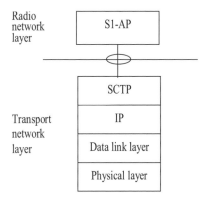

Figure 2.10 S1-MME control plane protocol stack. Reproduced by permission of © 3GPP.

The SCTP protocol is well known for its advanced features inherited from TCP which ensure the required reliable delivery of the signalling messages. In addition it makes it possible to benefit from improved features such as the handling of multistreams to implement transport network redundancy easily and avoid head-of-line blocking or multihoming (see 'IETF RFC4960' [7]).

A further simplification in LTE (compared to the UMTS Iu interface, for example) is the direct mapping of S1-AP (S1 Application Protocol) on top of SCTP. This results in a simplified protocol stack compared to UMTS with no intermediate connection management protocol. The individual connections are directly handled at the application layer. Multiplexing takes place between S1-AP and SCTP whereby each stream of an SCTP association is multiplexed with the signalling traffic of multiple individual connections.

One further area of flexibility brought with LTE lies in the lower layer protocols for which full optionality has been left regarding the choice of the IP version and the choice of the data link layer. For example, this enables the operator to start deployment using IP version 4 with the data link tailored to the network deployment scenario.

2.5.1.2 User Plane

Figure 2.11 gives the protocol structure of the S1 user plane, which is based on the GTP/UDP[5]/IP stack which is already well known from UMTS networks.

One of the advantages of using GTP-User plane (GTP-U) is its inherent facility to identify tunnels and also to facilitate intra-3GPP mobility.

The IP version number and the data link layer have been left fully optional, as for the control plane stack.

A transport bearer is identified by the GTP tunnel endpoints and the IP address (source Tunnelling End ID (TEID), destination TEID, source IP address, destination IP address).

The S-GW sends downlink packets of a given bearer to the eNodeB IP address (received in S1-AP) associated to that particular bearer. Similarly, the eNodeB sends upstream packets of a given bearer to the EPC IP address (received in S1-AP) associated to that particular bearer.

---

[5]User Datagram Protocol.

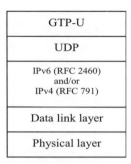

GTP-U
UDP
IPv6 (RFC 2460) and/or IPv4 (RFC 791)
Data link layer
Physical layer

Figure 2.11 S1-U user plane protocol stack. Reproduced by permission of © 3GPP.

Vendor-specific traffic categories (e.g. real-time traffic) can be mapped onto Differentiated Services (Diffserv) code points (e.g. expedited forwarding) by network O&M (Operation and Maintenance) configuration to manage QoS differentiation between the bearers.

2.5.2   Initiation Over S1

The initialization of the S1-MME control plane interface starts with the identification of the MMEs to which the eNodeB must connect, followed by the setting up of the Transport Network Layer (TNL).

With the support of the S1-flex function in LTE, an eNodeB must initiate an S1 interface towards each MME node of the pool area to which it belongs. This list of MME nodes of the pool together with an initial corresponding remote IP address can be directly configured in the eNodeB at deployment (although other means may also be used). The eNodeB then initiates the TNL establishment with that IP address. Only one SCTP association is established between one eNodeB and one MME.

During the establishment of the SCTP association, the two nodes negotiate the maximum number of streams which will be used over that association. However, multiple pairs of streams (note that a stream is unidirectional and therefore pairs must be used) are typically used in order to avoid the head-of-line blocking issue mentioned above. Among these pairs of streams, one particular pair must be reserved by the two nodes for the signalling of the common procedures (i.e. those which are not specific to one UE). The other streams are used for the sole purpose of the dedicated procedures (i.e. those which are specific to one UE).

Once the TNL has been established, some basic application-level configuration data for the system operation is automatically exchanged between the eNodeB and the MME through an 'S1 SETUP' procedure initiated by the eNodeB. This procedure constitutes one example of a network self-configuration process provided in LTE to reduce the configuration effort for network operators compared to the more usual manual configuration procedures of earlier systems.

An example of such basic application data which can be configured automatically via the S1 SETUP procedure is the tracking area identities. These identities are very important for the system operation because the tracking areas correspond to the zones in which UEs are paged, and their mapping to eNodeBs must remain consistent between the E-UTRAN and the EPC. Thus, once all the tracking area identities which are to be broadcast over the radio interface have been configured within each and every eNodeB, they are sent automatically to all the relevant MME nodes of the pool area within the S1 SETUP REQUEST message of this procedure. The same applies for the broadcast list of PLMNs which is used in the case of a network being shared by several operators (each having its own PLMN ID which needs to be broadcast for the UEs to recognize it). This saves a significant amount of configuration effort in the core network, avoids the risk of human error, and ensures that the E-UTRAN and EPC configurations regarding tracking areas and PLMNs are aligned.

Once the S1 SETUP procedure has been completed, the S1 interface is operational.

2.5.3 Context Management Over S1

Within each pool area, a UE is associated to one particular MME for all its communications during its stay in this pool area. This creates a context in this MME for the UE. This particular MME is selected by the NAS Node Selection Function (NNSF) in the first eNodeB from which the UE entered the pool.

Whenever the UE becomes active (i.e. makes a transition from idle to active mode) under the coverage of a particular eNodeB in the pool area, the MME provides the UE context information to this eNodeB using the 'INITIAL CONTEXT SETUP REQUEST' message (see Figure 2.12). This enables the eNodeB in turn to create a context and manage the UE for the duration of its activity in active mode.

Even though the setup of bearers is otherwise relevant to a dedicated 'Bearer Management' procedure described below, the creation of the eNodeB context by the INITIAL CONTEXT SETUP procedure also includes the creation of one or several bearers including the default bearer.

At the next transition back to idle mode following a 'UE CONTEXT RELEASE' message sent from the MME, the eNodeB context is erased and only the MME context remains.

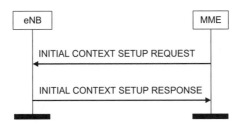

Figure 2.12 Initial context setup procedure. Reproduced by permission of © 3GPP.

2.5.4 Bearer Management Over S1

LTE uses independent dedicated procedures respectively covering the setup, modification and release of bearers. For each bearer requested to be set up, the transport layer address and the tunnel endpoint are provided to the eNodeB in the 'BEARER SETUP REQUEST' message to indicate the termination of the bearer in the S-GW where uplink user plane data must be sent. Conversely, the eNodeB indicates in the 'BEARER SETUP RESPONSE' message the termination of the bearer in the eNodeB where the downlink user plane data must be sent.

For each bearer, the QoS parameters (see Section 2.4 above) requested for the bearer are also indicated. Independently of the standardized QCI values, it is also still possible to use extra proprietary labels for the fast introduction of new services if vendors and operators agree upon them.

2.5.5   Paging Over S1

As mentioned in Section 2.5.3, in order to re-establish a connection towards a UE in idle mode, the MME distributes a paging request to the relevant eNodeBs based on the tracking areas where the UE is expected to be located. When receiving the 'PAGING REQUEST' message, the eNodeB sends a page over the radio interface in the cells which are contained within one of the tracking areas provided in that message.

The UE is normally paged using its SAE-Temporary Mobile Subscriber Identity (S-TMSI). The 'PAGING REQUEST' message also contains a UE identity index value in order for the eNodeB to calculate the paging occasions at which the UE will switch on its receiver to listen for paging messages (see Section 3.4).

2.5.6   Mobility Over S1

LTE/SAE supports mobility within LTE/SAE, and also mobility to other systems using both 3GPP specified and non-3GPP technologies. The mobility procedures over the radio interface are defined in Section 3.2. These mobility procedures also involve the network interfaces. The sections below discuss the procedures over S1 to support mobility. Mobility procedures from the point of view of the UE are outlined in Chapter 13.

2.5.6.1   Intra-LTE Mobility

There are two types of handover procedure in LTE for UEs in active mode: the S1-handover procedure and the X2-handover procedure.

For intra-LTE mobility, the X2-handover procedure is normally used for the inter-eNodeB handover (described in Section 2.6.3). However, when there is no X2 interface between the two eNodeBs, or if the source eNodeB has been configured to initiate handover towards a particular target eNodeB via the S1 interface, then an S1-handover will be triggered.

The S1-handover procedure has been designed in a very similar way to the UMTS Serving Radio Network Subsystem (SRNS) relocation procedure and is shown in Figure 2.13: it consists of a preparation phase involving the core network, where the resources are first prepared at the target side (steps 2 to 8), followed by an execution phase (steps 8 to 12) and a completion phase (after step 13).

Compared to UMTS, the main difference is the introduction of the 'STATUS TRANSFER' message sent by the source eNodeB (steps 10 and 11). This message has been added in order to carry some PDCP status information that is needed at the target eNodeB in cases when PDCP status preservation applies for the S1-handover (see Section 4.2.4); this is in alignment with the information which is sent within the X2 'STATUS TRANSFER' message used for the X2-handover (see below). As a result of this alignment, the handling of the handover by

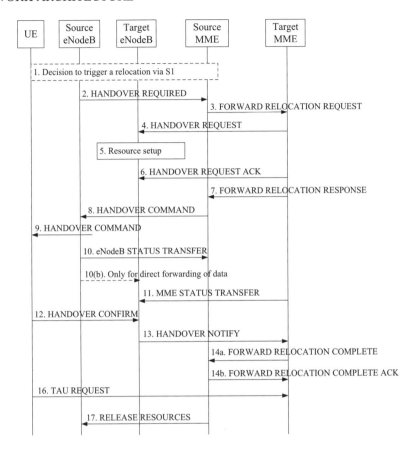

Figure 2.13 S1-based handover procedure. Reproduced by permission of © 3GPP.

the target eNodeB as seen from the UE is exactly the same, regardless of the type of handover (S1 or X2) the network had decided to use.

The Status Transfer procedure is assumed to be triggered in parallel with the start of data forwarding after the source eNodeB has received the 'HANDOVER COMMAND' message from the source MME. This data forwarding can be either direct or indirect, depending on the availability of a direct path for the user plane data between the source eNodeB and the target eNodeB.

The 'HANDOVER NOTIFY' message (step 13), which is sent later by the target eNodeB when the arrival of the UE at the target side is confirmed, is forwarded by the MME to trigger the update of the path switch in the S-GW towards the target eNodeB. In contrast to the X2-handover, the message is not acknowledged and the resources at the source side are released later upon reception of a 'RELEASE RESOURCE' message directly triggered from the source MME (step 17 in Figure 2.13).

2.5.6.2   Inter-Radio Access Technologies (RAT) Mobility

One key element of the design of the first release of LTE is the need to co-exist with other technologies.

For mobility from LTE towards UMTS, the handover process can reuse the S1-handover procedures described above, with the exception of the STATUS TRANSFER message which is not needed at steps 10 and 11 since no PDCP context is continued.

For mobility towards CDMA2000, dedicated uplink and downlink procedures have been introduced in LTE. They essentially aim at tunnelling the CDMA2000 signalling between the UE and the CDMA2000 system over the S1 interface, without being interpreted by the eNodeB on the way. The UPLINK S1 CDMA2000 TUNNELLING message presented in Figure 2.14 also includes the RAT type in order to identify which CDMA2000 RAT the tunnelled CDMA2000 message is associated with in order for the message to be routed to the correct node within the CDMA2000 system.

Figure 2.14   Uplink S1 CDMA2000 tunnelling procedure. Reproduced by permission of © 3GPP.

2.5.7   Load Management Over S1

Three types of load management procedures apply over S1: a normal 'load balancing' procedure to distribute the traffic, an 'overload' procedure to overcome a sudden peak in the loading and a 'load rebalancing' procedure to partially/fully offload an MME.

The MME load balancing procedure aims to distribute the traffic to the MMEs in the pool evenly according to their respective capacities. To achieve that goal, the procedure relies on the normal NNSF present in each eNodeB as part of the S1-flex function. Provided that suitable weight factors corresponding to the capacity of each MME node are available in the eNodeBs beforehand, a weighted NNSF done by each and every eNodeB in the network normally achieves a statistically balanced distribution of load among the MME nodes without further action. However, specific actions are still required for some particular scenarios:

- If a new MME node is introduced (or removed), it may be necessary temporarily to increase (or decrease) the weight factor normally corresponding to the capacity of this node in order to make it catch more (or less) traffic at the beginning until it reaches an adequate level of load.

• In case of an unexpected peak in the loading, an OVERLOAD message can be sent over the S1 interface by the overloaded MME. When received by an eNodeB, this message calls for a temporary restriction of a certain type of traffic. An MME can adjust the reduction of traffic it desires by defining the number of eNodeBs to which it sends the OVERLOAD message and by defining the types of traffic subject to restriction.

• Finally, if the MME wants to force rapidly the offload of part or all of its UEs, it will use the rebalancing function. This function forces the UEs to reattach to another MME by using a specific 'cause value' in the UE Release Command S1 message. In a first step it applies to idle mode UEs and in a second step it may also apply to UEs in connected mode (if the full MME offload is desired, e.g. for maintenance reasons).

2.6 The E-UTRAN Network Interfaces: X2 Interface

The X2 interface is used to inter-connect eNodeBs. The protocol structure for the X2 interface and the functionality provided over X2 are discussed below.

2.6.1 Protocol Structure Over X2

The control plane and user plane protocol stacks over the X2 interface are the same as over the S1 interface, as shown in Figures 2.15 and 2.16 respectively (with the exception that in Figure 2.15 the X2-AP is substituted for the S1-AP). This also means again that the choice of the IP version and the data link layer are fully optional. The use of the same protocol structure over both interfaces provides advantages such as simplifying the data forwarding operation.

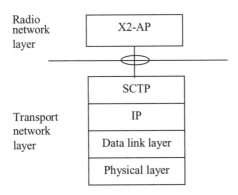

Figure 2.15 X2 signalling bearer protocol stack. Reproduced by permission of © 3GPP.

2.6.2 Initiation Over X2

The X2 interface may be established between one eNodeB and some of its neighbour eNodeBs in order to exchange signalling information when needed. However, a full mesh

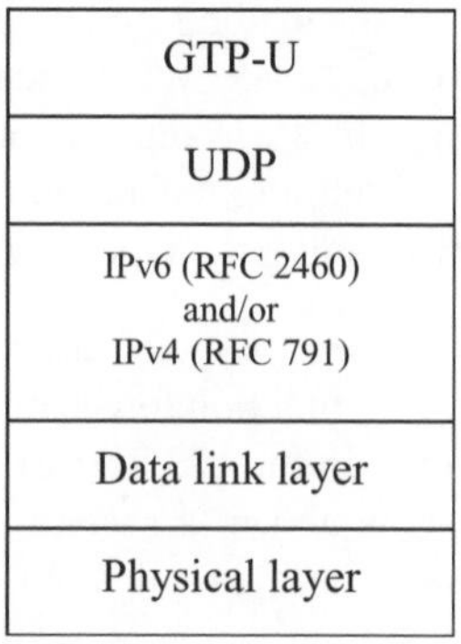

GTP-U
UDP
IPv6 (RFC 2460)   and/or   IPv4 (RFC 791)
Data link layer
Physical layer

Figure 2.16 Transport network layer for data streams over X2. Reproduced by permission of © 3GPP.

is not mandated in an E-UTRAN network. Two types of information may typically need to be exchanged over X2 to drive the establishment of an X2 interface between two eNodeBs: load or interference related information (see Section 2.6.4) and handover related information (see mobility in Section 2.6.3).

Because these two types of information are fully independent of one another, it is possible that an X2 interface may be present between two eNodeBs for the purpose of exchanging load or interference information, even though the X2-handover procedure is not used to handover UEs between those eNodeBs. (In such a case, the S1-handover procedure is used instead.)

The initialization of the X2 interface starts with the identification of a suitable neighbour followed by the setting up of the TNL.

The identification of a suitable neighbour may be done by configuration, or alternatively a function known as the Automatic Neighbour Relation Function (ANRF) may be used. This function makes use of the UEs to identify the useful neighbour eNodeBs: an eNodeB may ask a UE to read the global cell identity from the broadcast information of another eNodeB for which the UE has identified the physical cell identity during the new cell identification procedure (see Section 7.2).

The ANRF is another example of a SON process introduced successfully in LTE. Through this self-optimizing process, UEs and eNodeB measurements are used to auto-tune the network.

Once a suitable neighbour has been identified, the initiating eNodeB can further set up the TNL using the transport layer address of this neighbour – either as retrieved from the network or locally configured.

Once the TNL has been set up, the initiating eNodeB must trigger the X2 setup procedure. This procedure enables an automatic exchange of application level configuration data relevant to the X2 interface, similar to the S1 setup procedure already described in Section 2.5.2. For example, each eNodeB reports within the X2 SETUP REQUEST message to a neighbour eNodeB information about each cell it manages, such as the cell's physical identity, the frequency band, the tracking area identity and/or the associated PLMNs.

This automatic data exchange in the X2 setup procedure is also the core of another SON feature: the automatic self-configuration of the Physical Cell Identities (PCIs). Under this new SON feature, the O&M system can provide the eNodeBs with either a list of possible

PCI values to use or a specific PCI value. In the first case, in order to avoid collisions, the eNodeB should use a PCI which is not already used in its neighbourhood. Because the PCI information is included in the LTE X2 setup procedure, while

detecting a neighbour cell by the ANR function an eNodeB can also discover all the PCI values used in the neighbourhood of that cell and consequently eliminate those values from the list of suitable PCIs to start with.

Once the X2 setup procedure has been completed, the X2 interface is operational.

2.6.3 Mobility Over X2

Handover via the X2 interface is triggered by default unless there is no X2 interface established or the source eNodeB is configured to use the S1-handover instead.

The X2-handover procedure is illustrated in Figure 2.17. Like the S1-handover, it is also composed of a preparation phase (steps 4 to 6), an execution phase (steps 7 to 9) and a completion phase (after step 9).

The key features of the X2-handover for intra-LTE handover are:

- The handover is directly performed between two eNodeBs. This makes the preparation phase quick.

- Data forwarding may be operated per bearer in order to minimize data loss.

- The MME is only informed at the end of the handover procedure once the handover is successful, in order to trigger the path switch.

- The release of resources at the source side is directly triggered from the target eNodeB.

For those bearers for which in-sequence delivery of packets is required, the STATUS TRANSFER message (step 8) provides the Sequence Number (SN) and the Hyper Frame Number (HFN) which the target eNodeB should assign to the first packet with no sequence number yet assigned that it must deliver. This first packet can either be one received over the target S1 path or one received over X2 if data forwarding over X2 is used (see below). When it sends the STATUS TRANSFER message, the source eNodeB freezes its transmitter/receiver status – i.e. it stops assigning PDCP SNs to downlink packets and stops delivering uplink packets to the EPC.

Mobility over X2 can be categorized according to its resilience to packet loss: the handover can be said 'seamless' if it minimizes the interruption time during the move of the UE, or 'lossless' if it tolerates no loss of packets at all. These two modes use data forwarding of user plane downlink packets. The source eNodeB may decide to operate one of these two modes on a per-EPS-bearer basis, based on the QoS received over S1 for this bearer (see Section 2.5.4) and the service at stake.

2.6.3.1 Seamless Handover

If the source eNodeB selects the seamless mode for one bearer, it proposes to the target eNodeB in the HANDOVER REQUEST message to establish a GTP tunnel to operate the downlink data forwarding. If the target eNodeB accepts, it indicates in the HANDOVER REQUEST ACK message the tunnel endpoint where the forwarded data is expected to

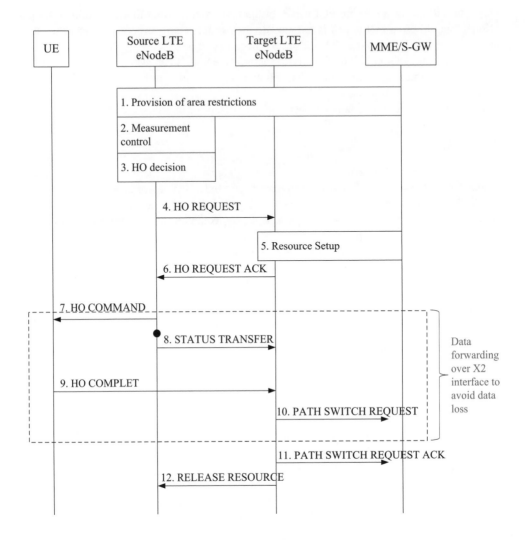

Figure 2.17 X2-based handover procedure.

be received. This tunnel endpoint may be different from the one set up as the termination point of the new bearer established over the target S1.

Upon reception of the HANDOVER REQUEST ACK message, the source eNodeB can start to forward the data freshly arriving over the source S1 path towards the indicated tunnel endpoint in parallel to sending the handover trigger to the UE over the radio interface. This forwarded data is thus available at the target eNodeB to be delivered to the UE as early as possible.

When forwarding is in operation and in-sequence delivery of packets is required, the target eNodeB is assumed to deliver first the packets forwarded over X2 before delivering the first ones received over the target S1 path once the S1 path switch has been done. The end of the forwarding is signalled over X2 to the target eNodeB by the reception of some 'special GTP packets' which the S-GW has inserted over the source S1 path just before switching this S1 path; these are then forwarded by the source eNodeB over X2 like any other regular packets.

2.6.3.2 Lossless Handover

If the source eNodeB selects the lossless mode for one bearer, it will additionally forward over X2 those user plane downlink packets which it has PDCP processed but are still buffered locally because they have not yet been delivered and acknowledged by the UE. These packets are forwarded together with their assigned PDCP SN included in a GTP extension header field. They are sent over X2 prior to the fresh arriving packets from the source S1 path. The same mechanisms described above for the seamless handover are used for the GTP tunnel establishment. The end of forwarding is also handled in the same way, since in-sequence packet delivery applies to lossless handovers. In addition, the target eNodeB must ensure that all the packets – including the ones received with sequence number over X2 – are delivered in-sequence at the target side. Further details of seamless and lossless handover are described in Section 4.2.

Selective retransmission. A new feature in LTE is the optimization of the radio by selective retransmission. When lossless handover is operated, the target eNodeB may, however, not deliver over the radio interface some of the forwarded downlink packets received over X2 if it is informed by the UE that those packets have already been received at the source side (see Section 4.2.6). This is called downlink selective retransmission.

Similarly in the uplink, the target eNodeB may desire that the UE does not retransmit packets already received earlier at the source side by the source eNodeB, for example to avoid wasting radio resources. To operate this uplink selective retransmission scheme for one bearer, it is necessary that the source eNodeB forwards to the target eNodeB, over another new GTP tunnel, those user plane uplink packets which it has received out of sequence. The target eNodeB must first request the source eNodeB to establish this new forwarding tunnel by including in the HANDOVER REQUEST ACK message a GTP tunnel endpoint where it expects the forwarded uplink packets to be received. The source eNodeB must, if possible, then indicate in the STATUS TRANSFER message for this bearer the list of SNs corresponding to the forwarded packets which are to be expected. This list helps the target eNodeB to inform the UE earlier of the packets not to be retransmitted, making the uplink selective retransmission overall scheme faster (see also Section 4.2.6).

2.6.3.3 Multiple Preparation

Another new feature of the LTE handover procedure is 'multiple preparation'. This feature enables the source eNodeB to trigger the handover preparation procedure towards multiple candidate target eNodeBs. Even though only one of the candidates is indicated as target to the UE, this makes recovery faster in case the UE fails on this target and connects to one of the other prepared candidate eNodeBs. The source eNodeB receives only one RELEASE RESOURCE message from the final selected eNodeB.

Regardless of whether multiple or single preparation is used, the handover can be cancelled during or after the preparation phase. If the multiple preparation feature is operated, it is recommended that upon reception of the RELEASE RESOURCE message the source eNodeB triggers a 'cancel' procedure towards each of the non-selected prepared eNodeBs.

2.6.4   Load and Interference Management Over X2

The exchange of load information between eNodeBs is of key importance in the flat architecture used in LTE, as there is no central RRM node as was the case, for example, in UMTS with the RNC.

The exchange of load information falls into two categories depending on the purpose it serves.

- The exchange of load information can serve for the (X2) load balancing process in which case the relevant frequency of exchange is rather low (in the order of seconds);

- The exchange of load information can serve to optimize some RRM processes such as interference coordination (as discussed in Section 12.5), in which case the frequency of exchange is rather high (in the order of tens of milliseconds).

2.6.4.1   Load Balancing

Like the ANRF SON function described in Section 2.6.2, load balancing is another aspect of SON built into the design of LTE. The objective of load balancing is to counteract local traffic load imbalance between neighbouring cells with the aim of improving the overall system capacity. One solution is to optimize the cell reselection/handover parameters (such as thresholds and hysteresis) between neighbouring cells autonomously upon detection of an imbalance (see also Section 13.6).

In order to detect an imbalance, it is necessary to compare the load of the cells and therefore to exchange information about them between neighbouring eNodeBs.

The cell load information exchanged can be of different types: radio measurements corresponding to the usage of physical resource blocks, possibly partitioned into real-time and non-real-time traffic; or generic measurements representing non-radio-related resource usage such as processing or hardware occupancy.

A client-server mechanism is used for the load information exchange: the RESOURCE STATUS RESPONSE/UPDATE message is used to report the load information over the X2 interface between one requesting eNodeB (client) and the eNodeBs which have subscribed to this request (servers). The reporting of the load is periodic and according to the periodicity expressed in the RESOURCE STATUS REQUEST message.

2.6.4.2   Interference Management

A separate Load Indication procedure is used over the X2 interface for the exchange of load information related to interference management as shown in Figure 2.18. As these measurements have direct influence on some RRM real-time processes, the frequency of reporting via this procedure may be high.

Figure 2.18   The LOAD INDICATION over X2 interface. Reproduced by permission of © 3GPP.

For the uplink interference, two indicators can be provided within the LOAD INDICA-TION message: a 'High Interference Indicator' and an 'Overload Indicator'. The usage of these indicators is explained in Section 12.5.

2.6.5   UE Historical Information Over X2

Historical UE information constitutes another example of a feature designed to support SON which is embedded in the design of LTE. It is part of the X2-handover procedure.

Historical UE information consists, for example, of the last few cells visited by the UE, together with the time spent in each one. This information is propagated from one eNodeB to another within the HANDOVER REQUEST messages and can be used to determine the occurrence of ping-pong between two or three cells for instance. The length of the history information can be configured for more flexibility.

More generally, the Historical UE information consists of some RRM information which is passed from the source eNodeB to the target eNodeB within the HANDOVER REQUEST message to assist the RRM management of a UE. The information can be partitioned into two types:

- UE RRM-related information, passed over X2 within the RRC transparent container;

- Cell RRM-related information, passed over X2 directly as an information element of the X2 AP HANDOVER REQUEST message itself.

2.7   Summary

The EPS provides UEs with IP connectivity to the packet data network. The EPS supports multiple data flows with different quality of service per UE for applications that need guaranteed delay and bit rate such as VoIP as well as best effort web browsing.

In this chapter we have seen an overview of the EPS network architecture, including the functionalities provided by the E-UTRAN access network and the evolved packet CN.

It can be seen that the concept of EPS bearers, together with their associated quality of service attributes, provide a powerful tool for the provision of a variety of simultaneous services to the end user.

From the perspective of the network operator, the LTE system is also breaking new ground in terms of its degree of support for self-optimization and self-configuration of the network via the X2, S1 and Uu interfaces, to facilitate deployment.

References[6]

[1] 3GPP Technical Specification 24.301, 'Non-Access-Stratum (NAS) protocol for Evolved Packet System (EPS); Stage 3 (Release 8)', www.3gpp.org.

[2] 3GPP Technical Specification 33.401, 'System Architecture Evolution (SAE): Security Architecture (Release 8)', www.3gpp.org.

[3] 3GPP Technical Specification 23.402, 'Architecture enhancements for non-3GPP accesses (Release 8)', www.3gpp.org.

[4] 3GPP Technical Specification 29.060, 'General Packet Radio Service (GPRS); GPRS Tunnelling Protocol (GTP) across the Gn and Gp interface (Release 8)', www.3gpp.org.

[5] 3GPP Technical Specification 23.203, 'Policy and charging control architecture (Release 8)', www.3gpp.org.

[6] 3GPP Technical Specification 36.300, 'Evolved Universal Terrestrial Radio Access (E-UTRA) and Evolved Universal Terrestrial Radio Access Network (E-UTRAN); Overall description; Stage 2 (Release 8)', www.3gpp.org.

[7] Request for Comments 4960 The Internet Engineering Task Force (IETF), Network Working Group, 'Stream Control Transmission Protocol', http://www.ietf.org.

---

[6]All web sites confirmed 18th December 2008.

3

Control Plane Protocols

Himke van der Velde

3.1 Introduction

As introduced in Section 2.2.2, the Control Plane of the Access Stratum (AS) handles
radio-specific functionality. The AS interacts with the Non-Access Stratum (NAS), which
is also referred to as the 'upper layers'. Among other functions, the NAS control protocols
handle Public Land Mobile Network[1] (PLMN) selection, tracking area update, paging,
authentication and Evolved Packet System (EPS) bearer establishment, modification and
release.

The applicable AS-related procedures largely depend on the Radio Resource Control (RRC) state of the User Equipment (UE), which is either RRC_IDLE or RRC_
CONNECTED.

A UE in RRC_IDLE performs cell selection and reselection – in other words, it decides
on which cell to camp. The cell (re)selection process takes into account the priority of each
applicable frequency of each applicable Radio Access Technology (RAT), the radio link
quality and the cell status (i.e. whether a cell is barred or reserved). An RRC_IDLE UE
monitors a paging channel to detect incoming calls, and also acquires system information.
The system information mainly consists of parameters by which the network (E-UTRAN)
can control the cell (re)selection process.

In RRC_CONNECTED, the E-UTRAN allocates radio resources to the UE to facilitate
the transfer of (unicast) data via shared data channels.[2] To support this operation, the UE
monitors an associated control channel[3] which is used to indicate the dynamic allocation

---

[1]The network of one operator in one country.

[2]The Physical Downlink Shared Channel (PDSCH) and Physical Uplink Shared Channel – see Sections 9.2.2
and 17.2 respectively.

[3]The Physical Downlink Control Channel (PDCCH) – see Section 9.3.2.2.

LTE – The UMTS Long Term Evolution: from Theory to Practice Stefania Sesia, Issam Toufik and Matthew Baker
© 2009 John Wiley & Sons, Ltd

of the shared transmission resources in time and frequency. The UE provides the network with reports of its buffer status and of the downlink channel quality, as well as neighbouring cell measurement information to enable E-UTRAN to select the most appropriate cell for the UE. These measurement reports include cells using other frequencies or RATs. The UE also receives system information, consisting mainly of information required to use the transmission channels. To extend its battery lifetime, a UE in RRC_CONNECTED may be configured with a Discontinuous Reception (DRX) cycle.

RRC, as specified in [1], is the protocol by which the E-UTRAN controls the UE behaviour in RRC_CONNECTED. RRC also specifies the control signalling applicable for a UE in RRC_IDLE, namely paging and system information. The UE behaviour in RRC_IDLE is specified in [2].

Chapter 13 gives some further details of the UE measurements which support the mobility procedures.

Functionality related to Multimedia Broadcast/Multicast Services (MBMSs) is covered separately in Chapter 14.

3.2   Radio Resource Control (RRC)

3.2.1   Introduction

The RRC protocol supports the transfer of *common* NAS information (i.e. NAS information which is applicable to all UEs) as well as *dedicated* NAS information (which is applicable only to a specific UE). In addition, for UEs in RRC_IDLE, RRC supports notification of incoming calls.

The RRC protocol covers a number of functional areas.

- **System information** handles the broadcasting of system information, which includes NAS common information. Some of the system information is applicable only for UEs in RRC_IDLE while other system information is also applicable for UEs in RRC_CONNECTED.

- **RRC connection control** covers all procedures related to the establishment, modification and release of an RRC connection, including paging, initial security activation, establishment of Signalling Radio Bearers (SRBs) and of radio bearers carrying user data (Data Radio Bearers, DRBs), handover within LTE (including transfer of UE RRC context information[4]), configuration of the lower protocol layers,[5] access class barring and radio link failure.

- **Network controlled inter-RAT mobility** includes (besides the mobility procedures) security activation and transfer of UE RRC context information.

- **Measurement configuration and reporting** for intra-frequency, inter-frequency and inter-RAT mobility, includes configuration and activation of measurement gaps.

---

[4]This UE context information includes the radio resource configuration including local settings not configured across the radio interface, UE capabilities and radio resource management information.

[5]Packet Data Convergence Protocol (PDCP), Radio Link Control (RLC), Medium Access Control (MAC), all of which are explained in detail in Chapter 4, and the physical layer which is explained in Chapters 5–11 and 15–20.

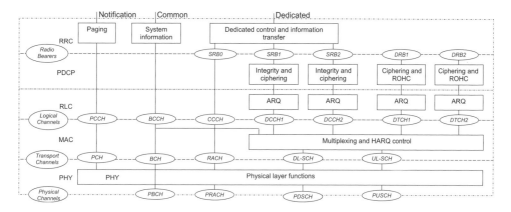

Figure 3.1 Radio architecture.

- **Miscellaneous functions** including, for example, transfer of dedicated NAS information and transfer of UE radio access capability information.

Dedicated RRC messages are transferred across SRBs, which are mapped via the PDCP
and RLC layers onto logical channels – either the Common Control Channel (CCCH) during
connection establishment or a Dedicated Control Channel (DCCH) in RRC_CONNECTED.
System Information and Paging messages are mapped directly to logical channels – the
Broadcast Control Channel (BCCH) and Paging Control Channel (PCCH) respectively. The
various logical channels are described in more detail in Section 4.4.1.2.

SRB0 is used for RRC messages which use the CCCH, SRB1 is for RRC messages using
DCCH, and SRB2 is for the (lower-priority) RRC messages using DCCH which only include
NAS dedicated information.[6] All RRC messages using DCCH are integrity-protected and
ciphered by the PDCP layer (after security activation) and use Automatic Repeat reQuest
(ARQ) protocols for reliable delivery through the RLC layer. The RRC messages using
CCCH are not integrity-protected and do not use ARQ in the RLC layer.

It should also be noted that the NAS independently applies integrity protection and
ciphering.

Figure 3.1 illustrates the overall radio protocol architecture as well as the use of radio
bearers, logical channels, transport channels and physical channels.

For parameters for which low transfer delay is more important than reliable transfer (i.e.
for which the use of ARQ is inappropriate due to the additional delay it incurs), MAC
signalling is used provided that there are no security concerns (integrity protection and
ciphering are not applicable for MAC signalling).

---

[6]Prior to SRB2 establishment, SRB1 is also used for RRC messages which only include NAS dedicated
information. In addition, SRB1 is used for higher priority RRC messages which only include NAS dedicated
information.

3.2.2 System Information

System information is structured by means of System Information Blocks (SIBs), each of which contains a set of functionally-related parameters. The SIB types that have been defined include:

- **The Master Information Block** (MIB), which includes a limited number of the most frequently transmitted parameters which are essential for a UE's initial access to the network.

- **System Information Block Type 1** (SIB1), which contains parameters needed to determine if a cell is suitable for cell selection, as well as information about the time-domain scheduling of the other SIBs.

- **System Information Block Type 2** (SIB2), which includes common and shared channel information.

- **SIB3–SIB8**, which include parameters used to control intra-frequency, inter-frequency and inter-RAT cell reselection.

Three types of RRC message are used to transfer system information: the MIB message, the SIB1 message and System Information (SI) messages. An SI message, of which there may be several, includes one or more SIBs which have the same scheduling requirements (i.e. the same transmission periodicity). Table 3.1 provides an example of a possible system information scheduling configuration, also showing which SIBs the UE has to acquire in the idle and connected states. The physical channels used for carrying the system information are explained in Section 9.2.1.

Table 3.1 Example of system information scheduling configuration.

Message	Content	Period (ms)	Applicability
MIB	Most essential parameters	40	Idle and connected
SIB1	Cell access related parameters, scheduling information	80	Idle and connected
1st SI	SIB2: Common and shared channel configuration	160	Idle and connected
2nd SI	SIB3: Common cell reselection information and intra-frequency cell reselection parameters other than the neighbouring cell information SIB4: Intra-frequency neighbouring cell information	320	Idle only
3rd SI	SIB5: Inter-frequency cell reselection information	640	Idle only
4th SI	SIB6: UTRA cell reselection information SIB7: GERAN cell reselection information	640	Idle only, depending on UE support of UMTS or GERAN

3.2.2.1 Time-Domain Scheduling of System Information

The time-domain scheduling of the MIB and SIB1 messages is fixed: they have periodicities of 40 ms and 80 ms respectively, as explained in Sections 9.2.1 and 9.2.2.2.

The time-domain scheduling of the SI messages is dynamically flexible: each SI message is transmitted in a defined periodically-occurring time-domain window, while physical layer control signalling[7] indicates in which subframes[8] within this window the SI is actually scheduled. The scheduling windows of the different SIs (referred to as SI-windows) are consecutive (i.e. there are neither overlaps nor gaps between them) and have a common length that is configurable. SI-windows can include subframes in which it is not possible to transmit SI messages, such as subframes used for SIB1, and subframes used for the uplink in TDD.

Figure 3.2 illustrates an example of the time-domain scheduling of system information, showing the subframes used to transfer the MIB, SIB1 and four SI messages. The example uses an SI-window of length 10 subframes, and shows a higher number of transmissions being used for the larger SI messages.

SI messages may have different periodicities. Consequently, in some clusters of SI windows all the SI messages are scheduled, while in other clusters of SI windows only the SIs with shorter repetition periods are transmitted. For the example of Table 3.1, the cluster of SI windows beginning at System Frame Number (SFN) 0 contains all the SI messages, the cluster beginning at SFN160 contains only the first SI message, that beginning at SFN320 contains the first and second SI messages, and that beginning at SFN480 contains only the first SI message.

Note that Figure 3.2 shows a cluster of SI windows where all the SI messages are transmitted. At occasions where a given SI is not transmitted (due to a longer repetition period), its corresponding SI-window is not used.

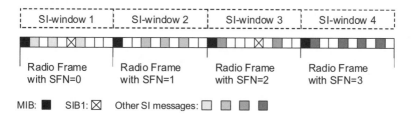

Figure 3.2 System information scheduling example.

3.2.2.2 Validity and Change Notification

System information normally changes only at specific radio frames whose System Frame Number is given by SFN mod $N = 0$, where N is configurable and defines the period between two radio frames at which a change may occur, known as the *modification period*. Prior to

---

[7]The Physical Downlink Control Channel – PDCCH; see Section 9.3.2.2.

[8]A subframe in LTE has a duration of 1 ms; see Section 6.2.

Figure 3.3 System information modification periods. Reproduced by permission of © 3GPP.

performing a change of the system information, the E-UTRAN notifies the UEs by means of a *Paging* message including a *SystemInfoModification* flag. Figure 3.3 illustrates the change of system information, with different shading indicating different content.

LTE provides two mechanisms for indicating that system information has changed:

1. A paging message including a flag indicating whether or not system information has changed.

2. A value tag in SIB1 which is incremented every time one or more SI messages change.

UEs in RRC_IDLE use the first mechanism, while UEs in RRC_CONNECTED can use either mechanism. The second mechanism can, for example, be useful in cases when a UE was unable to receive the paging messages.

UEs in RRC_IDLE are only required to receive the paging message at their normal paging occasions – i.e. no additional wake-ups are expected to detect changes of system information. In order to ensure reliability of reception, the change notification paging message is normally repeated a number of times during the BCCH modification period preceding that in which the new system information is first transmitted. Correspondingly, the modification period is expressed as a multiple of the cell-specific default paging cycle.

UEs in RRC_CONNECTED are expected to try receiving a paging message the same number of times per modification period as UEs in RRC_IDLE using the default paging cycle. The exact times at which UEs in RRC_CONNECTED which are using this method have to try to receive a paging message are not specified; the UE may perform these tries at convenient times, such as upon wake-up from DRX, using any of the subframes which are configured for paging during the modification period. Since the eNodeB anyway has to notify all the UEs in RRC_IDLE, it has to send a paging message in all subframes which are configured for paging (up to a maximum of four subframes per radio frame) during an entire modification period. Connected mode UEs can utilize any of these subframes. The overhead of transmitting paging messages to notify UEs of a change of SI is considered marginal, since changes to system information are expected to be infrequent – at worst once every few hours.

If the UE receives a notification of a change of system information, it considers all system information to be invalid from the start of the next modification period. This means that UE operations may be restricted until the UE has re-acquired the most essential system information, especially in RRC_CONNECTED. It is, however, assumed that system information change does not occur frequently.

If the UE returns to a cell, it is allowed to consider the SI previously acquired from the cell to remain valid if it was received less than 3 hours previously and the value tag matches.

3.2.3 Connection Control within LTE

Connection control involves:

- Security activation;

- Connection establishment, modification and release;

- DRB establishment, modification and release;

- Mobility within LTE.

3.2.3.1 Security Key Management

Security is a very important feature of all 3GPP radio access technologies. LTE provides security in a similar way to its predecessors UMTS and GSM.

Two functions are provided for the maintenance of security: *ciphering* of both control plane (RRC) data (i.e. SRBs 1 and 2) and user plane data (i.e. all DRBs), and *integrity protection* which is used for control plane (RRC) data only. Ciphering is used in order to protect the data streams from being received by a third party, while integrity protection allows the receiver to detect packet insertion or replacement. RRC always activates both functions together, either following connection establishment or as part of the handover to LTE.

The hierarchy of keys by which the AS security keys are generated is illustrated in Figure 3.4. The process is based on a common secret key K_{ASME} (Access Security Management Entity) which is available only in the Authentication Centre in the Home Subscriber Server (HSS) (see Section 2.2.1) and in a secure part of the Universal Subscriber Identity Module (USIM) in the UE. A set of keys and checksums are generated at the Authentication Centre using this secret key and a random number. The generated keys, checksums and random number are transferred to the Mobility Management Entity (MME) (see Section 2.2.1), which passes one of the generated checksums and the random number to the UE. The USIM in the UE then computes the same set of keys using the random number and the secret key. Mutual authentication is performed by verifying the computed checksums in the UE and network using NAS protocols.

Upon connection establishment, the AS derives an *AS base-key* K_{eNB}, which is eNodeB-specific, from K_{ASME}. K_{eNB} is used to generate three further security keys known as the *AS derived-keys*: one for integrity protection of the RRC signalling (SRBs), one for ciphering of the RRC signalling and one for ciphering of user data (DRBs).

In case of handover within E-UTRAN, a new AS base-key and new AS derived-keys are computed from the AS base-key used in the source cell. For handover to E-UTRAN from UTRAN or GERAN, the AS base-key is derived from integrity and ciphering keys used in the UTRAN or GERAN. Handover within LTE may be used to take a new K_{ASME} into account, i.e. following a re-authentication by NAS.

The use of the security keys for the integrity protection and ciphering functions is handled by the PDCP layer, as described in Section 4.2.3.

The security functions are never deactivated, although it is possible to apply a 'NULL' ciphering algorithm. The 'NULL' algorithm may also be used in certain special cases, such as for making an emergency call without a USIM.

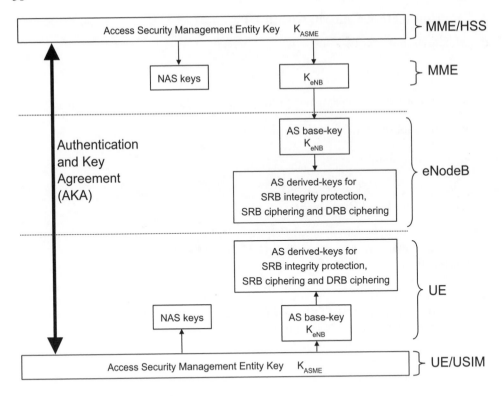

Figure 3.4 Security key derivation.

3.2.3.2 Connection Establishment and Release

Two levels of NAS states reflect the state of a UE in respect of connection establishment: the EPS Mobility Management (EMM) state (EMM-DEREGISTERED or EMM-REGISTERED) reflects whether the UE is registered in the MME, and the EPS Connection Management (ECM) state (ECM-IDLE or ECM-CONNECTED) reflects the connectivity of the UE with the Evolved Packet Core (EPC – see Chapter 2).

The NAS states, and their relationship to the AS RRC states, are illustrated in Figure 3.5.

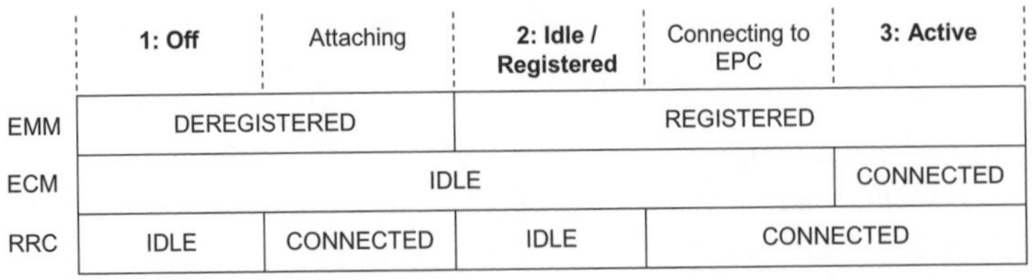

	1: Off	Attaching	2: Idle / Registered	Connecting to EPC	3: Active
EMM	DEREGISTERED		REGISTERED		
ECM	IDLE				CONNECTED
RRC	IDLE	CONNECTED	IDLE	CONNECTED	

Figure 3.5 Possible combinations of NAS and AS states.

The transition from ECM-IDLE to ECM-CONNECTED not only involves establishment of the RRC connection but also includes establishment of the S1-connection (see Section 2.5). RRC connection establishment is initiated by the NAS and is completed prior to S1-connection establishment, which means that connectivity in RRC_CONNECTED is initially limited to the exchange of control information between UE and E-UTRAN.

UEs are typically moved to ECM-CONNECTED when becoming active. It should be noted, however, that in LTE the transition from ECM-IDLE to ECM-CONNECTED is performed within 100 ms. Hence, UEs engaged in intermittent data transfer need not be kept in ECM-CONNECTED if the ongoing services can tolerate such transfer delays. In any case, an aim in the design of LTE was to support similar battery power consumption levels for UEs in RRC_CONNECTED as for UEs in RRC_IDLE.

RRC connection release is initiated by the eNodeB following release of the S1 connection between the eNodeB and the Core Network (CN).

Connection establishment message sequence. RRC connection establishment involves the establishment of SRB1 and the transfer of the initial uplink NAS message. This NAS message triggers the establishment of the S1 connection, which normally initiates a subsequent step during which E-UTRAN activates AS-security and establishes SRB2 and one or more DRBs (corresponding to the default and optionally dedicated EPS bearers).

Figure 3.6 illustrates the RRC connection establishment procedure, including the subsequent step of initial security activation and radio bearer establishment.

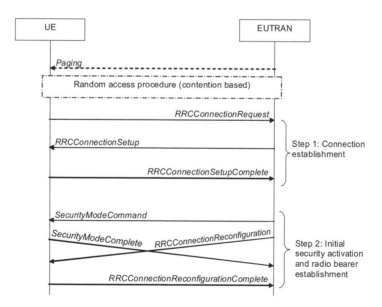

Figure 3.6 Connection establishment.

Step 1: Connection establishment

- Upper layers in the UE trigger connection establishment, which may be in response to paging. The UE checks if access is barred (see Section 3.3.4.6). If this is not the case, the lower layers in the UE perform a contention-based random access procedure as described in Section 19.3, and the UE starts a timer (known as *T300*) and sends the RRCConnectionRequest message. This message includes an initial identity (S-TMSI[9] or a random number) and an establishment cause.

- If E-UTRAN accepts the connection, it returns the RRCConnectionSetup message that includes the initial radio resource configuration including SRB1. Instead of signalling each individual parameter, E-UTRAN may order the UE to apply a default configuration – i.e. a configuration for which the parameter values are specified in the RRC specification [1].

- The UE returns the RRCConnectionSetupComplete message and includes the NAS message, an identifier of the selected PLMN (used to support network sharing) and, if provided by upper layers, an identifier of the registered MME. Based on the last two parameters, the eNodeB decides on the CN node to which it should establish the S1-connection.

Step 2: Initial security activation and radio bearer establishment

- E-UTRAN sends the SecurityModeCommand message to activate integrity protection and ciphering. This message, which is integrity-protected but not ciphered, indicates which algorithms shall be used.

- The UE verifies the integrity protection of the SecurityModeControl message, and, if this succeeds, it configures lower layers to apply integrity protection and ciphering to all subsequent messages (with the exception that ciphering is not applied to the response message, i.e. the SecurityModeComplete (or SecurityModeFailure) message).

- E-UTRAN sends the RRCConnectionReconfiguration message including a radio resource configuration used to establish SRB2 and one or more DRBs. This message may also include other information such as a piggybacked NAS message or a measurement configuration. E-UTRAN may send the RRCConnectionReconfiguration message prior to receiving the SecurityModeComplete message. In this case, E-UTRAN should release the connection when one (or both) procedures fail (because the two procedures result from a single S1-procedure, which does not support partial success).

- The UE finally returns the RRCConnectionReconfigurationComplete message.

A connection establishment may fail for a number of reasons, such as the following:

- Access may be barred (see Section 3.3.4.6).

---

[9]S-Temporary Mobile Subscriber Identity.

- In case cell re-selection occurs during connection establishment, the UE aborts the procedure and informs upper layers of the failure to establish the connection.

- E-UTRAN may temporarily reject the connection establishment by including a wait timer, in which case the UE rejects any connection establishment request until the wait time has elapsed.

- The NAS may abort an ongoing RRC connection establishment, for example upon NAS timer expiry.

3.2.3.3 DRB Establishment

To establish, modify or release DRBs, E-UTRAN applies the RRC connection reconfiguration procedure as described in Section 3.2.3.2.

When establishing a DRB, E-UTRAN decides how to transfer the packets of an EPS bearer across the radio interface. An EPS bearer is mapped (1-to-1) to a DRB, a DRB is mapped (1-to-1) to a DTCH (Dedicated Traffic Channel – see Section 4.4.1.2) logical channel, all logical channels are mapped (n-to-1) to the Downlink or Uplink Shared Transport Channel (DL-SCH or UL-SCH), which are mapped (1-to-1) to the corresponding Physical Downlink or Uplink Shared Channel (PDSCH or PUSCH). This radio bearer mapping is illustrated in Figure 3.1.

The radio resource configuration covers the configuration of the PDCP, RLC, MAC and physical layers. The main configuration parameters / options include the following:

- For services using small packet sizes (e.g. VoIP), PDCP may be configured to apply header compression to significantly reduce the signalling overhead.

- The RLC Mode is selected from those listed in Section 4.3.1. RLC Acknowledged Mode (AM) is applicable, except for services which require a very low transfer delay and for which reliable transfer is less important.

- E-UTRAN assigns priorities and Prioritized Bit-Rates (PBRs) to control how the UE divides the granted uplink resources between the different radio bearers (see Section 4.4.2.6).

- Unless the transfer delay requirements for any of the ongoing services are very strict, the UE may be configured with a DRX cycle (see Section 4.4.2.5).

- For services involving a semi-static packet rate (e.g. VoIP), semi-persistent scheduling may be configured to reduce the control signalling overhead (see Section 4.4.2.1). Specific resources may also be configured for reporting buffer status and radio link quality.

- Services tolerating higher transfer delays may be configured with a Hybrid ARQ (HARQ) profile involving a higher average number of HARQ transmissions.

3.2.3.4 Mobility Control in RRC_IDLE and RRC_CONNECTED

Mobility control in RRC_IDLE is UE-controlled (cell-reselection), while in RRC_ CONNECTED it is controlled by the E-UTRAN (handover). However, the mechanisms used in the two states need to be consistent so as to avoid ping-pong between cells upon state transitions. The mobility mechanisms are designed to support a wide variety of scenarios including network sharing, country borders, home deployment and varying cell ranges and subscriber densities; an operator may, for example, deploy its own radio access network in populated areas and make use of another operator's network in rural areas.

If a UE were to access a cell which does not have the best radio link quality of the available cells on a given frequency, it may create significant interference to the other cells. Hence, as for most technologies, radio link quality is the primary criterion for selecting a cell on an LTE frequency. When choosing between cells on different frequencies or RATs the interference concern does not apply. Hence, for inter-frequency and inter-RAT cell reselection other criteria may be considered such as UE capability, subscriber type and call type. As an example, UEs with no (or limited) capability for data transmission may be preferably handled on GSM, while home customers or 'premium subscribers' might be given preferential access to the frequency or RAT supporting the highest data rates. Furthermore, in some LTE deployment scenarios, voice services may initially be provided by a legacy RAT only (as a Circuit-switched (CS) application), in which case the UE needs to be moved to the legacy RAT upon establishing a voice call (also referred to as *CS fallback*).

E-UTRAN provides a list of neighbouring frequencies and cells which the UE should consider for cell reselection and for reporting of measurements. In general, such a list is referred to as a *white-list* if the UE is to consider only the listed frequencies or cells – i.e. other frequencies or cells are not available; conversely, in the case of a *black-list* being provided, a UE may consider any *un*listed frequencies or cells. In LTE, white-listing is used to indicate all the neighbouring frequencies of each RAT that the UE is to consider. On the other hand, E-UTRAN is not required to indicate all the neighbouring cells that the UE shall consider. Which cells the UE is required to detect by itself depends on the UE state as well as on the RAT, as explained below.

Note that for GERAN, typically no information is provided about individual cells. Only in specific cases, such as at country borders, is signalling[10] provided to indicate the group of cells that the UE is to consider – i.e. a white cell list.

Mobility in idle mode. In RRC_IDLE, cell re-selection between frequencies is based on absolute priorities, where each frequency has an associated priority. Cell-specific default values of the priorities are provided via system information. In addition, E-UTRAN may assign UE-specific values upon connection release, taking into account factors such as UE capability or subscriber type. In case equal priorities are assigned to multiple cells, the cells are ranked based on radio link quality. Equal priorities are not applicable between frequencies of different RATs. The UE does not consider frequencies for which it does not have an associated priority; this is useful in situations such as when a neighbouring frequency is applicable only for UEs of one of the sharing networks.

---

[10]The 'NCC-permitted' parameter – see GERAN specifications.

Table 3.2   List of system information parameters which may be used to control cell reselection.

	Intra-Freq.	Inter-Freq.	UMTS	GERAN	CDMA2000
Frequency list					
White frequency list	n/a	+	+	+	+
Frequency specific reselection info[a]	Priority	Priority Qoffset, ThresX-High, ThresX-Low	Priority ThresX-High, ThresX-Low	Priority ThresX-High, ThresX-Low	Priority ThresX-High, ThresX-Low
Frequency specific suitability info[b]			Q-RxLevMin, MaxTxPower, Q-QualMin	Q-RxLevMin	
Cell list					
White cell list	–	–	–	NCC permitted[c]	–
Black cell list	+	+	–	–	–
List of cells with specific info[d]	Qoffset	Qoffset	–	–	–

[a]See Section 3.3.4.2; [b]see Section 3.3.3; [c]see GERAN specifications; [d]see Section 3.3.4.3.

Table 3.2 provides an overview of the system information parameters which E-UTRAN may use to control cell reselection (excluding serving-cell-specific parameters and RAT-specific parameters). Other than the priority of a frequency, no idle mode mobility-related parameters may be assigned via dedicated signalling. Further details of the parameters listed are provided in Section 3.3.

Mobility in connected mode.   In RRC_CONNECTED, the E-UTRAN decides to which cell a UE should hand over in order to maintain the radio link. As with RRC_IDLE, E-UTRAN may take into account not only the radio link quality but also factors such as UE capability, subscriber type and access restrictions. Although E-UTRAN may trigger handover without measurement information (blind handover), normally it configures the UE to report measurements of the candidate target cells. Table 3.3 provides an overview of the frequency- and cell-specific information which E-UTRAN can configure.

In LTE the UE always connects to a single cell only – in other words, the switching of a UE's connection from a source cell to a target cell is a *hard* handover. The hard handover process is normally a 'backward' one, whereby the eNodeB which controls the source cell requests the target eNodeB to prepare for the handover. The target eNodeB subsequently generates the RRC message to order the UE to perform the handover, and the message is transparently forwarded by the source eNodeB to the UE. LTE also supports a kind of 'forward' handover, in which the UE by itself decides to connect to the target cell,

Table 3.3 Frequency- and cell-specific information which can be configured in connected mode.

	Intra-Freq.	Inter-Freq.	UTRA	GERAN	CDMA2000
Frequency list					
White frequency list Frequency specific info[a]	n/a Qoffset	+ Qoffset	+ Qoffset	+ Qoffset	+ Qoffset
Cell list					
White cell list Black cell list List of cells with specific info.	− + Qoffset	− + Qoffset	+ − −	NCC permitted[b] − −	+ − −

[a]See Section 3.3.4.3; [b]see GERAN specifications.

where it then requests that the connection be continued. The UE applies this connection re-establishment procedure only after loss of the connection to the source cell; the procedure only succeeds if the target cell has been prepared in advance for the handover.

Besides the handover procedure, LTE also provides for a UE to be redirected to another frequency or RAT upon connection release. This redirection may also be performed if AS-security has not been activated. Redirection during connection establishment is not supported, since at that time the E-UTRAN may not yet be in possession of all the relevant information such as the capabilities of the UE and the type of subscriber.

Message sequence for handover within LTE. In RRC_CONNECTED, the E-UTRAN controls mobility by ordering the UE to perform handover to another cell, which may be on the same frequency ('intra-frequency') or a different frequency ('inter-frequency'). Inter-frequency measurements may require the configuration of measurement gaps, depending on the capabilities of the UE (e.g. whether it has a dual receiver).

The E-UTRAN may also use the handover procedures for completely different purposes, such as to change the security keys to a new set (see Section 3.2.3.1), or to perform a 'synchronized reconfiguration' in which the E-UTRAN and the UE apply the new configuration simultaneously.

The message sequence for the procedure for handover within LTE is shown in Figure 3.7. The sequence is as follows:

1. The UE may send a MeasurementReport message (see Section 3.2.5).

2. Before sending the handover command to the UE, the source eNodeB requests one or more target cells to prepare for the handover. As part of this 'handover preparation request', the source eNodeB provides UE RRC context information[11] about the UE capabilities, the current AS-configuration and UE-specific Radio Resource Management (RRM) information. In response, the eNodeB controlling the target cell generates the 'handover command'. The source eNodeB will forward this to the UE in

---

[11]This UE context information includes the radio resource configuration including local settings not configured across the radio interface, UE capabilities and radio resource management information.

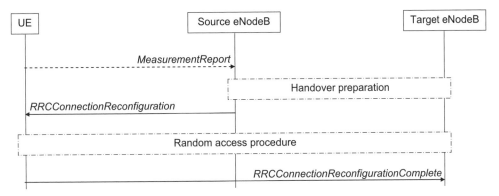

Figure 3.7 Handover within LTE.

the RRCConnectionReconfiguration message. This is done transparently (apart from performing integrity protection and ciphering) – i.e. the source eNodeB does not add or modify the protocol information contained in the message.

3. The source eNodeB sends the RRCConnectionReconfiguration message to the UE. This is the message which orders the UE to perform handover, and it includes mobility control information (namely the identity, and optionally the frequency, of the target cell) and the radio resource configuration information which is common to all UEs in the target cell (e.g. information required to perform random access). The message also includes the dedicated radio resource configuration, the security configuration and the C-RNTI[12] to be used in the target cell. Although the message may optionally include the measurement configuration, the E-UTRAN is likely to use another reconfiguration procedure for re-activating measurements, in order to avoid the RRCConnectionReconfiguration message becoming excessively large. If no measurement configuration information is included in the message used to perform inter-frequency handover, the UE stops any inter-frequency and inter-RAT measurements and deactivates the measurement gap configuration.

4. If the UE is able to comply with the configuration included in the received RRC-ConnectionReconfiguration message, the UE starts a timer, known as *T304*, and initiates a random access procedure (see Section 19.3), using the received RACH configuration, to the target cell at the first available occasion.[13] It is important to note that the UE does not need to acquire system information directly from the target cell prior to initiating random access and resuming data communication. However, the UE may be unable to use some parts of the physical layer configuration from the very start (e.g. semi-persistent scheduling (see Section 4.4.2.1), the PUCCH (see Section 17.3) and the Sounding Reference Signal (SRS) (see Section 16.6)). The UE derives new security keys and applies the received configuration in the target cell.

---

[12]The Cell Radio Network Temporary Identifier is the RNTI to be used by a given UE while it is in a particular cell.

[13]The target cell does not specify when the UE is to initiate random access in that cell. Hence, the handover process is sometimes described as *asynchronous*.

5. Upon successful completion of the random access procedure, the UE stops the timer
   T304. The AS informs the upper layers in the UE about any uplink NAS messages for
   which transmission may not have completed successfully, so that the NAS can take
   appropriate action.

3.2.3.5 Connection Re-Establishment Procedure

In a number of failure cases (e.g. radio link failure, handover failure, RLC unrecover-
able error, reconfiguration compliance failure), the UE initiates the RRC connection re-
establishment procedure, provided that security is active. If security is not active when one
of the indicated failures occurs, the UE moves to RRC_IDLE instead.

To attempt RRC connection re-establishment, the UE starts a timer known as *T311*
and performs cell selection. The UE should prioritize searching on LTE frequencies.
However, no requirements are specified as to for how long the UE shall refrain from
searching for other RATs. Upon finding a suitable cell on an LTE frequency, the UE
stops the timer T311, starts the timer T301 and initiates a contention based random access
procedure to enable the RRCConnectionReestablishmentRequest message to be sent. In the
RRCConnectionReestablishmentRequest message, the UE includes the identity used in the
cell in which the failure occurred, the identity of that cell, a short Message Authentication
Code and a cause.

The E-UTRAN uses the re-establishment procedure to continue SRB1 and to re-
activate security without changing algorithms. A subsequent RRC connection reconfiguration
procedure is used to resume operation on radio bearers other than SRB1 and to re-activate
measurements. If the cell in which the UE initiates the re-establishment is not prepared (i.e.
does not have a context for that UE), the E-UTRAN will reject the procedure, causing the
UE to move to RRC_IDLE.

3.2.4   Connected Mode Inter-RAT Mobility

The overall procedure for the control of mobility is explained in this section; some further
details can be found in Chapter 13.

3.2.4.1 Handover to LTE

The procedure for handover to LTE is largely the same as the procedure for handover within
LTE, so it is not necessary to repeat the details here. The main difference is that upon
handover to LTE the entire AS-configuration needs to be signalled, whereas within LTE it is
possible to use 'delta signalling', whereby only the changes to the configuration are signalled.

If ciphering had not yet been activated in the previous RAT, the E-UTRAN activates
ciphering, possibly using the NULL algorithm, as part of the handover procedure. The
E-UTRAN also establishes SRB1, SRB2 and one or more DRBs (i.e. at least the DRB
associated with the default EPS bearer).

3.2.4.2 Mobility from LTE

The procedure for mobility from LTE to another RAT supports both handover and Cell
Change Order (CCO), possibly with Network Assistance (NACC – Network Assisted Cell

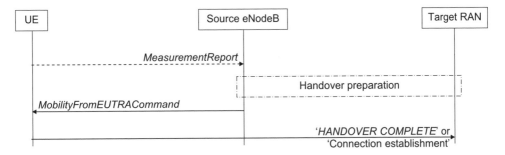

Figure 3.8 Mobility from LTE.

Change). The CCO/NACC procedure is applicable only for mobility to GERAN. Mobility from LTE is performed only after security has been activated.

The procedure is illustrated in Figure 3.8.

1. The UE may send a MeasurementReport message (see Section 3.2.5 for further details).

2. In case of handover (as opposed to CCO), the source eNodeB requests the target RAN node to prepare for the handover. As part of the 'handover preparation request' the source eNodeB provides information about the applicable inter-RAT UE capabilities as well as information about the currently-established bearers. In response, the target RAN generates the 'handover command' and returns this to the source eNodeB.

3. The source eNodeB sends a MobilityFromEUTRACommand message to the UE, which includes either the inter-RAT message received from the target (in case of handover), or the target cell/frequency and a few inter-RAT parameters (in case of CCO).

4. Upon receiving the MobilityFromEUTRACommand message, the UE starts the timer T304 and connects to the target node, either by using the received radio configuration (handover) or by initiating connection establishment (CCO) in accordance with the applicable specifications of the target RAT.

3.2.4.3 CDMA2000

For CDMA2000, additional procedures have been defined to support the transfer of dedicated information from the CDMA2000 upper layers, which are used to register the UE's presence in the target core network prior to performing the handover (referred to as preregistration). These procedures use SRB1.

3.2.5   Measurements

3.2.5.1   Measurement Configuration

The E-UTRAN can configure the UE to report measurement information to support the control of UE mobility. The following measurement configuration elements can be signalled via the RRCConnectionReconfiguration message.

1. **Measurement objects.** A measurement object defines on what the UE should perform the measurements – such as a carrier frequency. The measurement object may include a list of cells to be considered (white-list or black-list) as well as associated parameters, e.g. frequency- or cell-specific offsets.

2. **Reporting configurations.** A reporting configuration consists of the (periodic or event-triggered) criteria which cause the UE to send a measurement report, as well as the details of what information the UE is expected to report (e.g. the quantities, such as Received Signal Code Power (RSCP) for UMTS or Reference Signal Received Power (RSRP) (see Sections 13.4.2.1 and 13.4.1.1) for LTE, and the number of cells).

3. **Measurement identities.** These identify a measurement and define the applicable measurement object and reporting configuration.

4. **Quantity configurations**. The quantity configuration defines the filtering to be used on each measurement.

5. **Measurement gaps.** Measurement gaps define time periods when no uplink or down-link transmissions will be scheduled, so that the UE may perform the measurements. The measurement gaps are common for all gap-assisted measurements. Further details of the measurement gaps are discussed in Section 13.6.1.

The details of the above parameters depend on whether the measurement relates to an LTE, UMTS, GERAN or CDMA2000 frequency. Further details of the measurements performed by the UE are explained in Section 13.4. The E-UTRAN configures only a single measurement object for a given frequency, but more than one measurement identity may use the same measurement object. The identifiers used for the measurement object and reporting configuration are unique across all measurement types. An example of a set of measurement objects and their corresponding reporting configurations is shown in Figure 3.9.

In LTE it is possible to configure the quantity which triggers the report (RSCP or RSRP) for each reporting configuration. The UE may be configured to report either the trigger quantity or both quantities.

3.2.5.2   Measurement Report Triggering

Depending on the measurement type, the UE may measure and report any of the following:

- The serving cell;

- Listed cells (i.e. cells indicated as part of the measurement object);

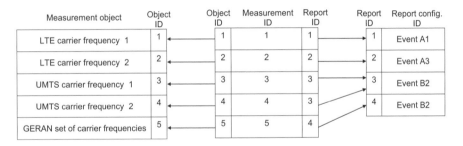

Figure 3.9 Example measurement configuration.

- Detected cells on a listed frequency (i.e. cells which are not listed cells but are detected by the UE).

For some RATs, the UE measures and reports listed cells only (i.e. the list is a white-list), while for other RATs the UE also reports detected cells. For further details, see Table 3.3.

For LTE, the following event-triggered reporting criteria are specified:

- **Event A1.** Serving cell becomes better than absolute threshold.

- **Event A2.** Serving cell becomes worse than absolute threshold.

- **Event A3.** Neighbour cell becomes better than an offset relative to the serving cell.

- **Event A4.** Neighbour cell becomes better than absolute threshold.

- **Event A5.** Serving cell becomes worse than one absolute threshold and neighbour cell becomes better than another absolute threshold.

For inter-RAT mobility, the following event-triggered reporting criteria are specified:

- **Event B1.** Neighbour cell becomes better than absolute threshold.

- **Event B2.** Serving cell becomes worse than one absolute threshold and neighbour cell becomes better than another absolute threshold.

The UE triggers an event when one or more cells meets a specified 'entry condition'. The E-UTRAN can influence the entry condition by setting the value of some configurable parameters used in these conditions – for example, one or more thresholds, an offset, and/or a hysteresis. The entry condition must be met for at least a duration corresponding to a 'timeToTrigger' parameter configured by the E-UTRAN in order for the event to be triggered. The UE scales the timeToTrigger parameter depending on its speed (see Section 3.3 for further detail).

Figure 3.10 illustrates the triggering of event A3 when a timeToTrigger and an offset are configured.

The UE may be configured to provide a number of periodic reports after having triggered an event. This 'event-triggered periodic reporting' is configured by means of parameters 'reportAmount' and 'reportInterval', which specify respectively the number of periodic

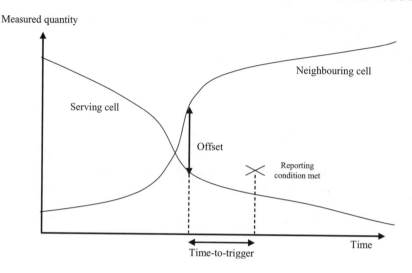

Figure 3.10 Event triggered report condition (Event A3).

reports and the time period between them. If event-triggered periodic reporting is configured, the UE's count of the number of reports sent is reset to zero whenever a new cell meets the entry condition. The same cell cannot then trigger a new set of periodic reports unless it first meets a specified 'leaving condition'.

In addition to event-triggered reporting, the UE may be configured to perform periodic measurement reporting. In this case, the same parameters may be configured as for event-triggered reporting, except that the UE starts reporting immediately rather than only after the occurrence of an event.

3.2.5.3 Measurement Reporting

In a MeasurementReport message, the UE only includes measurement results related to a single measurement – in other words, measurements are not combined for reporting purposes. If multiple cells triggered the report, the UE includes the cells in order of decreasing value of the reporting quantity – i.e. the best cell is reported first. The number of cells the UE includes in a MeasurementReport may be limited by a parameter 'maxReportCells'.

3.2.6 Other RRC Signalling Aspects

3.2.6.1 UE Capability Transfer

In order to avoid signalling of the UE radio access capabilities across the radio interface upon each transition from RRC_IDLE to RRC_CONNECTED, the core network stores the AS capabilities while the UE is in RRC_IDLE/EMM-REGISTERED. Upon S1 connection establishment, the core network provides the capabilities to the E-UTRAN. If the E-UTRAN does not receive the (required) capabilities from the core network (e.g. due to the UE being in EMM-DEREGISTERED), it requests the UE to provide its capabilities using the UE capability transfer procedure. The E-UTRAN can indicate for each RAT (LTE, UMTS, GERAN) whether it wants to receive the associated capabilities. The UE provides the requested capabilities using a separate container for each RAT. Dynamic change of UE capabilities is not supported.

3.2.6.2   Uplink/Downlink Information Transfer

The uplink/downlink information transfer procedures are used to transfer only upper layer information (i.e. no RRC control information is included). The procedure supports the transfer of 3GPP NAS dedicated information as well as CDMA2000 dedicated information.

In order to reduce latency, NAS information may also be included in the RRCConnection-SetupComplete and RRCConnectionReconfiguration messages. For the latter message, NAS information is only included if the AS and NAS procedures are dependent (i.e. they jointly succeed or fail). This applies for EPS bearer establishment, modification and release.

As noted earlier, some additional NAS information transfer procedures have also been defined for CDMA2000 for preregistration.

3.3   PLMN and Cell Selection

3.3.1   Introduction

After a UE has selected a PLMN, it performs *cell selection* – in other words, it searches for a suitable cell on which to camp (see Chapter 7). While camping on the chosen cell, the UE acquires the system information that is broadcast (see Section 9.2.1). Subsequently, the UE registers its presence in the tracking area, after which it can receive paging information which is used to notify UEs of incoming calls. The UE may establish an RRC connection, for example to establish a call or to perform a tracking area update.

When camped on a cell, the UE regularly verifies if there is a better cell; this is known as performing *cell reselection*.

LTE cells are classified according to the service level the UE obtains on them: a *suitable cell* is a cell on which the UE obtains normal service. If the UE is unable to find a suitable cell, but manages to camp on a cell belonging to another PLMN, the cell is said to be an *acceptable cell*, and the UE enters a 'limited service' state in which it can only perform emergency calls – as is also the case when no USIM is present in the UE. Finally, some cells may indicate via their system information that they are barred or reserved; a UE can obtain no service on such a cell.

A category called 'operator service' is also supported in LTE, which provides normal service but is applicable only for UEs with special access rights.

Figure 3.11 provides a high-level overview of the states and the cell (re)selection procedures.

3.3.2   PLMN Selection

The NAS handles PLMN selection based on a list of available PLMNs provided by the AS. The NAS indicates the selected PLMN together with a list of equivalent PLMNs, if available. After successful registration, the selected PLMN becomes the *Registered* PLMN (R-PLMN).

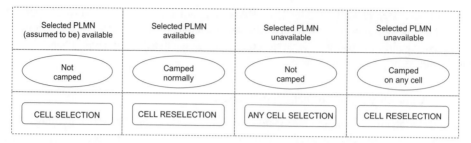

Figure 3.11 Idle mode states and procedures.

The AS may autonomously indicate available PLMNs. In addition, NAS may request the AS to perform a full search for available PLMNs. In the latter case, the UE searches for the strongest cell on each carrier frequency. For these cells, the UE retrieves the PLMN identities from system information. If the quality of a cell satisfies a defined radio criterion, the corresponding PLMNs are marked as *high quality*; otherwise, the PLMNs are reported together with their quality.

3.3.3   Cell Selection

Cell selection consists of the UE searching for the strongest cell on all supported carrier frequencies of each supported RAT until it finds a suitable cell. The main requirement for cell selection is that it should not take too long, which becomes more challenging with the ever increasing number of frequencies and RATs to be searched. The NAS can speed up the search process by indicating the RATs associated with the selected PLMN. In addition, the UE may use information stored from a previous access.

The cell selection criterion is known as the *S-criterion* and is fulfilled when the cell-selection receive level value $S_{\text{rxlev}} > 0$ dB, where

$$S_{\text{rxlev}} = Q_{\text{rxlevmeas}} - (Q_{\text{rxlevmin}} - Q_{\text{rxlevminoffset}})$$

in which $Q_{\text{rxlevmeas}}$ is the measured cell receive level value, also known as the RSRP (see Section 13.4.1.1), and Q_{rxlevmin} is the minimum required receive level in the cell. $Q_{\text{rxlevminoffset}}$ is an offset which may be configured to prevent ping-pong between PLMNs, which may otherwise occur due to fluctuating radio conditions. The offset is taken into account only when performing a periodic search for a higher priority PLMN while camped on a suitable cell in a visited PLMN.

The cell selection related parameters are broadcast within the SystemInformationBlock-Type1 message.

For some specific cases, additional requirements are defined:

- Upon leaving connected mode, the UE should normally attempt to select the cell to which it was connected. However, the connection release message may include information directing the UE to search for a cell on a particular frequency.

- When performing 'any cell selection', the UE tries to find an acceptable cell of any PLMN by searching all supported frequencies on all supported RATs. The UE may

stop searching upon finding a cell that meets the 'high quality' criterion applicable for
that RAT.

Note that the UE only verifies the suitability of the strongest cell on a given frequency. In
order to avoid the UE needing to acquire system information from a candidate cell that does
not meet the S-criterion, suitability information is provided for inter-RAT neighbouring cells.

3.3.4 Cell Reselection

Once the UE camps on a suitable cell, it starts cell reselection. This process aims to move
the UE to the 'best' cell of the selected PLMN and of its equivalent PLMNs, if any. As
described in Section 3.2.3.4, cell reselection between frequencies and RATs is primarily
based on absolute priorities. Hence, the UE first evaluates the frequencies of all RATs based
on their priorities. Secondly, the UE compares the cells on the relevant frequencies based
on radio link quality, using a ranking criterion. Finally, upon reselecting to the target cell
the UE verifies the cell's accessibility. Further rules have also been defined to allow the UE
to limit the frequencies to be measured, to speed up the process and save battery power, as
discussed in Section 3.3.4.1. Figure 3.12 provides a high-level overview of the cell reselection
procedure.

It should be noted that the UE performs cell reselection only after having camped for at
least one second on the current serving cell.

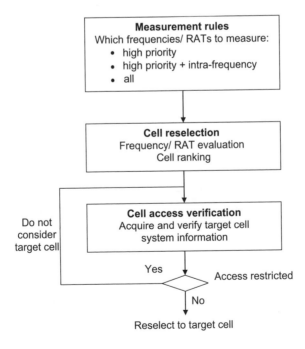

Figure 3.12 Cell reselection.

3.3.4.1 Measurement Rules

To enable the UE to save battery power, rules have been defined which limit the measurements the UE is required to perform. Firstly, the UE is required to perform intra-frequency measurements only when the quality of the serving cell is below or equal to a threshold ('SintraSearch'). Furthermore, the UE is required to measure other frequencies/RATs of lower or equal priority only when the quality of the serving cell is below or equal to another threshold ('SnonintraSearch'). The UE is always required to measure frequencies and RATs of higher priority. The required performance (i.e. how often the UE is expected to make the measurements, and to what extent this depends on, for example, the serving cell quality) is specified in [3].

3.3.4.2 Frequency/RAT Evaluation

E-UTRAN configures an absolute priority for all applicable frequencies of each RAT. In addition to the cell-specific priorities which are optionally provided via system information, E-UTRAN can assign UE-specific priorities via dedicated signalling. Of the frequencies that are indicated in the system information, the UE is expected to consider for cell reselection only those for which it has priorities. Equal priorities are not applicable for inter-RAT cell reselection.

The UE reselects to a cell on a higher priority frequency if the S-criterion (see Section 3.3.3) of the concerned target cell exceeds a high threshold ($Thresh_{X\text{-}High}$) for longer than a certain duration $T_{reselection}$. The UE reselects to a cell on a lower-priority frequency if the S-criterion of the serving cell is below a low threshold ($Thresh_{Serving\text{-}Low}$) while the S-criterion of the target cell on a lower-priority frequency (possibly on another RAT) exceeds a low threshold ($Thresh_{X\text{-}Low}$) during the time interval $T_{reselection}$, while no cell on a higher-priority frequency is available. Figure 3.13 illustrates the condition(s) to be met for reselecting to a cell on a higher-priority frequency (light grey bar) and to a cell on a lower priority frequency (dark grey bars).

When reselecting to a frequency, possibly on another RAT, which has a different priority, the UE reselects to the highest-ranked cell on the frequency concerned (see Section 3.3.4.3).

Note that, as indicated in Section 3.2.3.4, thresholds and priorities are configured per frequency, while $T_{reselection}$ is configured per RAT.

From Release-8 onwards, UMTS and GERAN support the same priority-based cell reselection as provided in LTE, with a priority per frequency. Release-8 radio access networks will continue to handle legacy UEs by means of offset-based ranking. Likewise, Release-8 UEs should apply the ranking based on radio link quality (with offsets) unless UMTS or GERAN indicate support for priority-based reselection.

3.3.4.3 Cell Ranking

The UE ranks the intra-frequency cells and the cells on other frequencies having equal priority which fulfil the S-criterion using a criterion known as the *R-criterion*. The R-criterion generates rankings R_s and R_n for the serving cell and neighbour cells respectively:

$$\text{For the serving cell:} \quad R_s = Q_{meas,s} + Q_{hyst,s}$$

$$\text{For neighbour cells:} \quad R_n = Q_{meas,n} + Q_{off\ s,n}$$

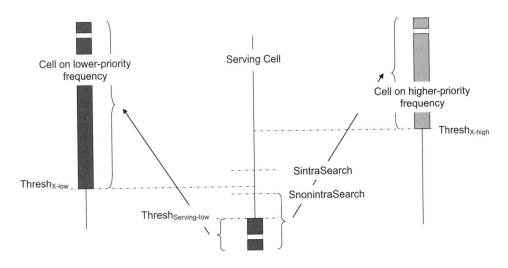

Figure 3.13 Frequency/RAT evaluation.

where Q_{meas} is the measured cell received quality (RSRP) (see Section 13.4.1.1), $Q_{hyst,s}$ is a parameter controlling the degree of hysteresis for the ranking, and $Q_{off\,s,n}$ is an offset applicable between serving and neighbouring cells on frequencies of equal priority (the sum of the cell-specific and frequency-specific offsets).

The UE reselects to the highest-ranked candidate cell provided that it is better ranked than the serving cell for at least the duration of $T_{reselection}$. The UE scales the parameters $T_{reselection}$ and Q_{hyst}, depending on the UE speed (see Section 3.3.4.5 below).

3.3.4.4 Accessibility Verification

If the best cell on an LTE frequency is barred or reserved, the UE is required to exclude this cell from the list of cell reselection candidates. In this case, the UE may consider other cells on the same frequency unless the barred cell indicates (by means of a broadcast bit in the SI) that intra-frequency reselection is not allowed for a certain duration (except for Closed Subscriber Group (CSG) cells). If, however, the best cell is unsuitable for some other specific reason (e.g. because it belongs to a forbidden tracking area or to another non-equivalent PLMN), the UE is not permitted to consider any cell on the concerned frequency as a cell reselection candidate for a maximum of 300 s.

3.3.4.5 Speed Dependent Scaling

The UE scales the cell reselection parameters depending on its speed. This applies both in idle mode ($T_{reselection}$ and Q_{hyst}) and in connected mode (timeToTrigger). The UE speed is categorized by a mobility state (high, normal or low), which the UE determines based on the number of cell reselections/handovers which occur within a defined period, excluding consecutive reselections/handovers between the same two cells. The state is determined by comparing the count with thresholds for medium and high state, while applying some hysteresis. For idle and connected modes, separate sets of control parameters are used, signalled in SIB3 and within the measurement configuration respectively.

3.3.4.6   Cell Access Restrictions

Access barring is performed during connection establishment (see Section 3.2.3.2) and provides a means to control the load introduced by UE-originated traffic. There are separate means for controlling Mobile Originated (MO) calls and MO signalling.

Each UE belongs to an Access Class (AC) in the range 0–9. In addition, some UEs may belong to one or more high-priority ACs in the range 11–15, which are reserved for specific uses (e.g. security services, public utilities, emergency services, PLMN staff). AC10 is used for emergency access. Further details, for example regarding in which PLMN the high priority ACs apply, are provided in [4]. The UE considers access to be barred if access is barred for all its applicable ACs.

SIB2 may include a set of AC barring parameters for MO calls and/or MO signalling. This set of parameters comprises a probability factor and a barring timer for AC0–9 and a list of barring bits for AC11–15. For AC0–9, if the UE initiates a MO call and the relevant AC barring parameters are included, the UE draws a random number. If this number exceeds the probability factor, access is not barred. Otherwise access is barred for a duration which is randomly selected centred on the broadcast barring timer value. For AC11–15, if the UE initiates a MO call and the relevant AC barring parameters are included, access is barred whenever the bit corresponding to all of the UE's ACs is set. The behaviour is similar in the case of UE-initiated MO signalling.

For cell (re)selection, the UE is expected to consider cells which are neither barred nor reserved for operator or future use. In addition, a UE with an access class in the range 11–15 shall consider a cell that is (only) reserved for operator use and part of its home PLMN (or an equivalent) as a candidate for cell reselection. The UE is not even allowed to perform emergency access on a cell which is not considered to be a candidate for cell reselection.

3.3.4.7   Any Cell Selection

When the UE is unable to find a suitable cell of the selected PLMN, it performs 'any cell selection'. In this case, the UE performs normal idle mode operation: monitoring paging, acquiring system information, performing cell reselection. In addition, the UE regularly attempts to find a suitable cell on other frequencies or RATs (i.e. not listed in system information). The UE is not allowed to receive MBMS in this state.

3.3.4.8   Closed Subscriber Group

LTE supports the existence of cells which are accessible only for a limited set of UEs – a Closed Subscriber Group (CSG). In order to prevent UEs from attempting to register on a CSG cell on which they do not have access, the UE maintains a CSG white list, i.e. a list of CSG identities for which access has been granted to the UE. The CSG white list can be transferred to the UE by upper layers, or updated upon successful access of a CSG cell. To facilitate the latter, UEs support 'manual selection' of CSG cells which are not in the CSG white list. The manual selection may be requested by the upper layers, based on a text string broadcast by the cell.

3.4 Paging

To receive paging messages from E-UTRAN, UEs in idle mode monitor the PDCCH channel for an RNTI value used to indicate paging: the P-RNTI (see Section 9.2.2.2). The UE only needs to monitor the PDCCH channel at certain UE-specific occasions (i.e. at specific subframes within specific radio frames – see Section 6.2 for an introduction to the LTE radio frame structure.). At other times, the UE may apply DRX, meaning that it can switch off its receiver to preserve battery power.

The E-UTRAN configures which of the radio frames and subframes are used for paging. Each cell broadcasts a default paging cycle. In addition, upper layers may use dedicated signalling to configure a UE-specific paging cycle. If both are configured, the UE applies the lowest value. The UE calculates the radio frame (the Paging Frame (PF)) and the subframe within that PF (the Paging Occasion (PO)), which E-UTRAN applies to page the UE as follows:

$$\text{SFN mod } T = (T/N) \times (\text{UE\_ID mod } N)$$

$$i\_s = (\text{UE\_ID}/N) \text{ mod Ns}$$

$$T = \min(T_\text{c}, T_\text{ue})$$

$$N = \min(T, \text{ number of paging subframes per frame} \times T)$$

$$\text{Ns} = \max(1, \text{ number of paging subframes per frame}) \tag{3.1}$$

where:

T_c is the cell-specific default paging cycle $\{32, 64, 128, 256\}$ radio frames,
T_ue is the UE-specific paging cycle $\{32, 64, 128, 256\}$ radio frames,
N is the number of paging frames with the paging cycle of the UE,
UE_ID is the IMSI mod 4096, with IMSI being the decimal rather than the binary number,
i_s is an index pointing to a pre-defined table defining the corresponding subframe,
Ns is the number of 'paging subframes' in a radio frame that is used for paging.

Table 3.4 includes a number of examples to illustrate the calculation of the paging radio frames and subframes.

Table 3.4 Examples for calculation of paging frames and subframes.

Case	UE_ID	T_c	T_ue	T	nPS	N	Ns	PF	i_s
A	147	256	256	256	0.25	64	1	76	0
B	147	256	128	128	0.25	32	1	76	0
C	147	256	128	128	2	128	2	0	1

In cases A and B in Table 3.4, one out of every four radio frames is used for paging, using one subframe in each of those radio frames. For case B, there are 32 paging frames within the UE's paging cycle, across which the UEs are distributed based on the UE-identity. In case C, two subframes in each radio frame are used for paging, i.e. Ns = 2. In this case, there are 128 paging frames within the UE's paging cycle and the UEs are also distributed across the two subframes within the paging frame. The LTE specifications include a table that indicates the subframe applicable for each combination of Ns and i_s, which is the index that follows from Equation (3.1). Figure 3.14 illustrates cases B and C. The non-blank subframes are used for paging, while the grey ones are applicable for the UE with the indicated identity.

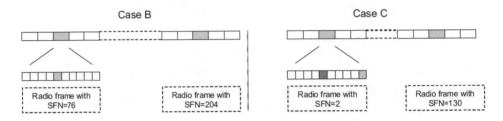

Figure 3.14 Paging frame and paging occasion examples.

3.5   Summary

The main aspects of the Control Plane protocols in LTE can be broken down into the Cell Selection and Reselection Procedures when the UE is in Idle Mode, and the RRC protocol when the UE is in Connected Mode.

The roles of these protocols include supporting security, mobility both between different LTE cells and between LTE and other radio systems, and establishment and reconfiguration of the radio bearers which carry control information and user data.

References[14]

[1] 3GPP Technical Specification 36.331, 'Evolved Universal Terrestrial Radio Access (E-UTRA); Radio Resource Control (RRC); Protocol specification (Release 8)', www.3gpp.org.

[2] 3GPP Technical Specification 36.304, 'Evolved Universal Terrestrial Radio Access (E-UTRA); User Equipment (UE) procedures in idle mode (Release 8)', www.3gpp.org.

[3] 3GPP Technical Specification 36.133, 'Evolved Universal Terrestrial Radio Access (E-UTRA); Requirements for support of radio resource management (Release 8)', www.3gpp.org.

[4] 3GPP Technical Specification 22.011, 'Service accessibility (Release 8)', www.3gpp.org.

---

[14]All web sites confirmed 18th December 2008.

4

User Plane Protocols

Patrick Fischer, SeungJune Yi, SungDuck Chun and YoungDae Lee

4.1 Introduction to the User Plane Protocol Stack

The LTE Layer 2 user-plane protocol stack is composed of three sublayers, as shown in Figure 4.1:

- **The Packet Data Convergence Protocol (PDCP) layer [1]:** This layer processes Radio Resource Control (RRC) messages in the control plane and Internet Protocol (IP) packets in the user plane. Depending on the radio bearer, the main functions of the PDCP layer are header compression, security (integrity protection and ciphering), and support for reordering and retransmission during handover. There is one PDCP entity per radio bearer.

- **The Radio Link Control (RLC) layer [2]:** The main functions of the RLC layer are segmentation and reassembly of upper layer packets in order to adapt them to the size which can actually be transmitted over the radio interface. For radio bearers which need error-free transmission, the RLC layer also performs retransmission to recover from packet losses. Additionally, the RLC layer performs reordering to compensate for out-of-order reception due to Hybrid Automatic Repeat reQuest (HARQ) operation in the layer below. There is one RLC entity per radio bearer.

- **The Medium Access Control (MAC) layer [3]:** This layer performs multiplexing of data from different radio bearers. Therefore there is only one MAC entity per UE. By deciding the amount of data that can be transmitted from each radio bearer and instructing the RLC layer as to the size of packets to provide, the MAC layer aims to achieve the negotiated Quality of Service (QoS) for each radio bearer. For the

LTE – The UMTS Long Term Evolution: from Theory to Practice   Stefania Sesia, Issam Toufik and Matthew Baker
© 2009 John Wiley & Sons, Ltd

uplink, this process includes reporting to the eNodeB the amount of buffered data for transmission.

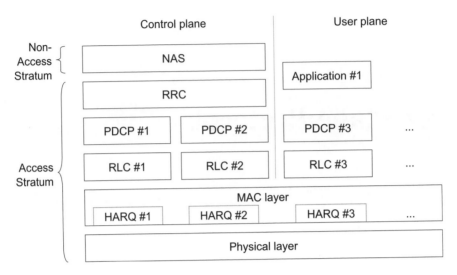

Figure 4.1 Overview of user-plane architecture.

At the transmitting side, each layer receives a Service Data Unit (SDU) from a higher layer, for which the layer provides a service, and outputs a Protocol Data Unit (PDU) to the layer below. As an example, the RLC layer receives packets from the PDCP layer. These packets are called PDCP PDUs from a PDCP point of view and represent RLC SDUs from an RLC point of view. The RLC layer creates packets which are provided to the layer below, i.e. the MAC layer. The packets which RLC provides to the MAC layer are RLC PDUs from an RLC point of view, and MAC SDUs from a MAC point of view. At the receiving side, the process is reversed, with each layer passing SDUs up to the layer above, where they are received as PDUs.

An important design feature of the LTE protocol stack is that all the PDUs and SDUs are *byte aligned*.[1] This is to facilitate handling by microprocessors, which are normally defined to handle packets in units of bytes. In order to further reduce the processing requirements of the user plane protocol stack in LTE, the headers created by each of the PDCP, RLC and MAC layers are also byte-aligned. This implies that sometimes unused padding bits are needed in the headers, and thus the cost of designing for efficient processing is that a small amount of potentially-available capacity is wasted.

4.2 Packet Data Convergence Protocol

4.2.1 Functions and Architecture

The PDCP layer performs the following functions:

---

[1] Byte alignment means that the lengths of the PDUs and SDUs are multiples of 8 bits.

- Header compression and decompression for user plane data.

- Security functions:

    – ciphering and deciphering for user plane and control plane data;

    – integrity protection and verification for control plane data.

- Handover support functions:

    – in-sequence delivery and reordering of PDUs for the layer above at handover;

    – lossless handover for user plane data mapped on RLC Acknowledged Mode (see Section 4.3.1).

- Discard for user plane data due to timeout.

The PDCP layer manages data streams for the user plane, as well as for the control plane (i.e. the RRC protocol – see Section 3.2). The architecture of the PDCP layer differs for user plane data and control plane data, as shown in Figures 4.2 and 4.3 respectively.

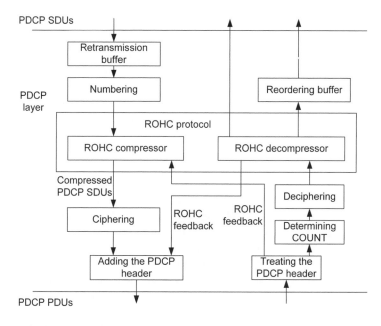

Figure 4.2 Overview of user-plane PDCP. Reproduced by permission of © 3GPP.

Each of the main functions is explained in the following subsections. Two different types of PDCP PDU are defined in LTE: PDCP Data PDUs and PDCP Control PDUs. PDCP Data PDUs are used for both control and user plane data. PDCP Control PDUs are only used to transport the feedback information for header compression, and for PDCP status reports which are used in case of handover (see Section 4.2.6) and hence are only used within the user plane.

PDCP SDUs

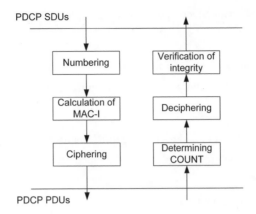

PDCP PDUs

Figure 4.3 Overview of control-plane PDCP. Reproduced by permission of © 3GPP.

4.2.2 Header Compression

One of the main functions of PDCP is header compression using the RObust Header Compression (ROHC) protocol defined by the IETF (Internet Engineering Task Force). In LTE, header compression is very important because there is no support for the transport of voice services via the Circuit-Switched (CS) domain. Thus, in order to provide voice services on the Packet-Switched (PS) domain in a way that comes close to the efficiency normally associated with CS services it is necessary to compress the IP/UDP/RTP[2] header which is typically used for Voice over IP (VoIP) services.

The IETF specifies in 'RFC 4995'[3] a framework which supports a number of different header compression 'profiles' (i.e. sets of rules and parameters for performing the compression). The header compression profiles supported for LTE are shown in Table 4.1. This means that a UE may implement one or more of these ROHC profiles. It is important to notice that the profiles already defined in the IETF's earlier 'RFC 3095' have been redefined in RFC 4995 in order to increase robustness in some cases. The efficiency of RFC 3095 and RFC 4995 is similar, and UMTS supports only RFC 3095.

The support of ROHC is not mandatory for the UE, except for those UEs which support VoIP. UEs which support VoIP have to support at least one profile for compression of RTP, UDP and IP.[4] The eNodeB controls by RRC signalling which of the ROHC profiles supported by the UE are allowed to be used. The ROHC compressors in the UE and the eNodeB then dynamically detect IP flows that use a certain IP header configuration and choose a suitable compression profile from the allowed and supported profiles.

ROHC header compression operates by allowing both the sender and the receiver to store the static parts of the header (e.g. the IP addresses of the sender/receiver), and to update these only when they change. Furthermore, dynamic parts (as, for example, the timestamp in

---

[2]Internet Protocol / User Datagram Protocol / Real-Time Transport Protocol.

[3]Requests for Comments (RFCs) capture much of the output of the IETF.

[4]ROHC is required for VoIP supported via the IP Multimedia Subsystem (IMS); in theory it could be possible to support raw IP VoIP without implementing ROHC.

Table 4.1 Supported header compression protocols.

Reference	Usage
RFC 4995	No compression
RFC 3095, RFC 4815	RTP/UDP/IP
RFC 3095, RFC 4815	UDP/IP
RFC 3095, RFC 4815	ESP/IP
RFC 3843, RFC 4815	IP
RFC 4996	TCP/IP
RFC 5225	RTP/UDP/IP
RFC 5225	UDP/IP
RFC 5225	ESP/IP
RFC 5225	IP

the RTP header) are compressed by transmitting only the difference from a reference clock maintained in both the transmitter and the receiver.

As the non-changing parts of the headers are thus transmitted only once, successful decompression depends on their correct reception. Feedback is therefore used in order to confirm the correct reception of initialization information for the header decompression. Furthermore, the correct decompression of the received PDCP PDUs is confirmed periodically, depending on the experienced packet losses.

As noted above, the most important use case for ROHC is VoIP. Typically, for the transport of a VoIP packet which contains a payload of 32 bytes, the header added will be 60 bytes for the case of IPv6 and 40 bytes for the case of IPv4[5] – i.e. an overhead of 188% and 125% respectively. By means of ROHC, after the initialization of the header compression entities this overhead can be compressed to four to six bytes, and thus to a relative overhead of 12.5% to 18.8%. This calculation is valid during the active periods, but during silence periods the payload size is smaller so the relative overhead is higher.

4.2.3 Security

The security architecture of LTE was introduced in Section 3.2.3.1. The implementation of security, by ciphering (of both control plane (RRC) data and user plane data) and integrity protection (for control plane (RRC) data only), is the responsibility of the PDCP layer.

A PDCP Data PDU counter (known as 'COUNT' in the LTE specifications) is used as an input to the security algorithms. The COUNT value increments for each PDCP Data PDU during an RRC connection; it has a length of 32 bits in order to allow an acceptable duration for an RRC connection.

During an RRC connection, the COUNT value is maintained by both the UE and the eNodeB by counting each transmitted/received PDCP Data PDU. In order to provide robustness against lost packets, each protected PDCP Data PDU includes a PDCP Sequence

---

[5]IPv6 is the successor to the original IPv4, for many years the dominant version of IP used on the Internet, and introduces a significantly expanded address space.

Number (SN) which corresponds to the least significant bits of the COUNT value.[6] Thus if one or more packets are lost, the correct COUNT value of a newly received packet can be determined using the PDCP SN. This means that the associated COUNT value is the next highest COUNT value for which the least significant bits correspond to the PDCP SN. A loss of synchronization of the COUNT value between the UE and eNodeB can then only occur if a number of packets corresponding to the maximum SN are lost consecutively. In principle, the probability of this kind of loss of synchronization occurring could be minimized by increasing the length of the SN, even to the extent of transmitting the whole COUNT value in every PDCP Data PDU. However, this would cause a high overhead, and therefore only the least significant bits are used as the SN; the actual SN length depends on the configuration and type of PDU, as explained in the description of the PDCP PDU formats in Section 4.2.6.

This use of a counter is designed to protect against a type of attack known as a *replay attack*, where the attacker tries to resend a packet that has been intercepted previously; the use of the COUNT value also provides protection against attacks which aim at deriving the used key or ciphering pattern by comparing successive patterns. Due to the use of the COUNT value, even if the same packet is transmitted twice, the ciphering pattern will be completely uncorrelated between the two transmissions, thus preventing possible security breaches.

Integrity protection is realized by adding a field known as 'Message Authentication Code for Integrity' (MAC-I)[7] to each RRC message. This code is calculated based on the AS keys (see Section 3.2.3.1), the message itself, the radio bearer ID, the direction (i.e. uplink or downlink) and the COUNT value.

If the integrity check fails, the message will be discarded and treated as if it had not been received.

Ciphering is realized by performing an XOR operation with the message and a ciphering stream that is generated by the ciphering algorithm based on the Access Stratum (AS) derived keys (see Section 3.2.3.1), the radio bearer ID, the direction (i.e. uplink or downlink), and the COUNT value.

Ciphering can only be applied to PDCP Data PDUs. Control PDUs (such as ROHC feedback or PDCP status reports) are neither ciphered nor integrity protected.

Except for identical retransmissions, the same COUNT value is not allowed to be used more than once for a given security key. The eNodeB is responsible for avoiding reuse of the COUNT with the same combination of radio bearer ID, AS base-key and algorithm. In order to avoid such reuse, the eNodeB may for example use different radio bearer IDs for successive radio bearer establishments, trigger an intracell handover or trigger a UE state transition from connected to idle and back to connected again (see Section 3.2).

4.2.4 Handover

Handover is performed when the UE moves from the coverage of one cell to the coverage of another cell. Depending on the required QoS, a seamless or a lossless handover is performed as appropriate for each radio bearer.

---

[6]In order to avoid excessive overhead, the most significant bits of the COUNT value, also referred to as the Hyper Frame Number (HFN), are not signalled but derived from counting overflows of the PDCP SN.

[7]Note that the MAC-I has no relation to the MAC layer.

4.2.4.1   Seamless Handover

Seamless handover is applied to all radio bearers carrying control plane data, and for user plane radio bearers mapped on RLC Unacknowledged Mode (UM, see Section 4.3.1). These types of data are typically reasonably tolerant of losses but less tolerant of delay (e.g. voice services). Seamless handover is therefore designed to minimize complexity and delay, but may result in loss of some SDUs.

At handover, for radio bearers to which seamless handover applies, the PDCP entities including the header compression contexts are reset, and the COUNT values are set to zero. As a new key is anyway generated at handover, there is no security reason to maintain the COUNT values. PDCP SDUs in the UE for which the transmission has not yet started will be transmitted after handover to the target cell. In the eNodeB PDCP SDUs that have not yet been transmitted can be forwarded via the X2 interface[8] to the target eNodeB. PDCP SDUs for which the transmission has already started but that have not been successfully received will be lost. This minimizes the complexity because no context (i.e. configuration information) has to be transferred between the source and the target eNodeB at handover.

4.2.4.2   Lossless Handover

Based on the sequence number that is added to PDCP Data PDUs it is possible to ensure in-sequence delivery during handover, and even provide a fully lossless handover functionality, performing retransmission of PDCP PDUs for which reception has not yet been acknowledged prior to the handover. This lossless handover function is used mainly for delay-tolerant services such as file downloads where the loss of one PDCP SDU can result in a drastic reduction in the data rate due to the reaction of the Transmission Control Protocol (TCP).

Lossless handover is applied for radio bearers that are mapped on RLC Acknowledged Mode (AM, see Section 4.3.1).

For lossless handover, the header compression protocol is reset in the UE because the header compression context is not forwarded from the source eNodeB to the target eNodeB. However, the PDCP sequence numbers and the COUNT values associated with PDCP SDUs are maintained. For simplicity reasons, inter-eNodeB handover and intra-eNodeB handover are handled in the same way in LTE.

In normal transmission while the UE is not handing over from one cell to another, the RLC layer in the UE and the eNodeB ensures in-sequence delivery. PDCP PDUs that are retransmitted by the RLC protocol, or that arrive out of sequence due to the variable delay in the HARQ transmission, are reordered based on the RLC Sequence Number. At handover, the RLC layer in the UE and in the eNodeB will deliver all PDCP PDUs that have already been received to the PDCP layer in order to have them decompressed before the header compression protocol is reset. Because some PDCP SDUs may not be available at this point, the PDCP SDUs that are not available in-sequence are not delivered immediately to higher layers in the UE or to the gateway in the network. In the PDCP layer, the PDCP SDUs received out of order are stored in the reordering buffer (see Figure 4.2). PDCP SDUs that have been transmitted but not yet been acknowledged by the RLC layer are stored in a retransmission buffer in the PDCP layer.

---

[8]For details of the X2 interface, see Section 2.6.

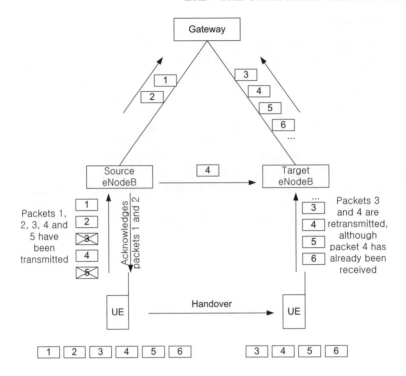

Figure 4.4 Lossless handover in the uplink.

In order to ensure lossless handover in the *uplink*, the UE retransmits the PDCP SDUs stored in the PDCP retransmission buffer. This is illustrated in Figure 4.4. In this example the PDCP entity has initiated transmission for the PDCP PDUs with the sequence numbers 1 to 5; the packets with the sequence numbers 3 and 5 have not been received by the source eNodeB, for example due to the handover interrupting the HARQ retransmissions. After the handover, the UE restarts the transmission of the PDCP SDUs for which successful transmission has not yet been acknowledged, to the target eNodeB. In the example in Figure 4.4 only the PDCP PDUs 1 and 2 have been acknowledged prior to the handover. Therefore, after the handover the UE will retransmit the packets 3, 4 and 5, although the network had already received packet 4.

In order to ensure in-sequence delivery in the uplink, the source eNodeB, after decompression, delivers the PDCP SDUs that are received in-sequence to the gateway, and forwards the PDCP SDUs that are received out-of-sequence to the target eNodeB. Thus, the target eNodeB can reorder the decompressed PDCP SDUs received from the source eNodeB and the retransmitted PDCP SDUs received from the UE based on the PDCP SNs which are maintained during the handover, and deliver them to the gateway in the correct sequence.

In order to ensure lossless handover in the *downlink*, the source eNodeB forwards the uncompressed PDCP SDUs for which reception has not yet been acknowledged by the UE to the target eNodeB for retransmission in the downlink. The source eNodeB receives an indication from the gateway that indicates the last packet sent to the source eNodeB.

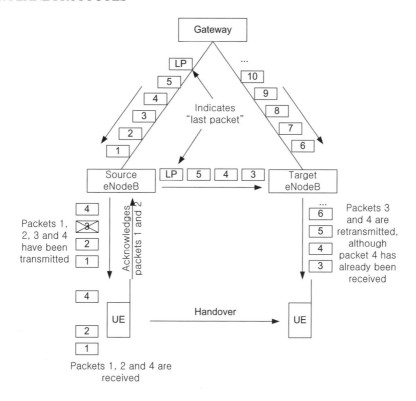

Figure 4.5 Lossless handover in the downlink.

The source eNodeB also forwards this indication to the target eNodeB such that the target eNodeB can know when it can start transmission of packets received from the gateway. In the example in Figure 4.5, the source eNodeB has started the transmission of the PDCP PDUs 1 to 4; due to, for example, the handover occurring prior to the HARQ retransmissions of packet 3, packet 3 will not be received by the UE from the source eNodeB. Furthermore the UE has only sent an acknowledgment for packets 1 and 2, although packet 4 has been received by the UE. The target eNodeB then ensures that the PDCP SDUs that have not yet been acknowledged in the source eNodeB are sent to the UE. Thus, the UE can reorder the received PDCP SDUs and the PDCP SDUs that are stored in the reordering buffer, and deliver them to higher layers in sequential order.

The UE will expect the packets from the target eNodeB in ascending order of sequence numbers. In the case of a packet not being forwarded from the source eNodeB to the target eNodeB, i.e. when one of the packets that the UE expects is missing during the handover operation, the UE can immediately conclude that the packet is lost and can forward the packets which have already been received in sequence to higher layers. This avoids the UE having to retain already-received packets in order to wait for a potential retransmission. Thus the forwarding of the packets in the network can be decided without informing the UE.

In some cases it may happen that a PDCP SDU has been successfully received, but a corresponding RLC acknowledgement has not. In this case, after the handover there may be unnecessary retransmissions initiated by the UE or the target eNodeB based on the incorrect status received by the RLC layer. In order to avoid these unnecessary retransmissions a PDCP status report can be sent from the eNodeB to the UE and from the UE to the eNodeB as described in Section 4.2.6. Additionally, a PDCP status report can request retransmission of PDCP SDUs which were correctly received but failed in header decompression. Whether to send a PDCP status report after handover is configured independently for each radio bearer.

4.2.5 Discard of Data Packets

Typically, the data rate that is available on the radio interface is smaller than the data rate available on the network interfaces (e.g. S1[9]). Thus, when the data rate of a given service is higher than the data rate provided by the LTE radio interface, this leads to buffering in the UE and in the eNodeB. This buffering allows some freedom to the scheduler in the MAC layer – in other words, it allows the scheduler to vary the instantaneous data rate at the physical layer in order to adapt to the current radio channel conditions. Due to the buffering, the variations in the instantaneous data rate are then seen by the application only as some jitter in the transfer delay.

However, when the data rate provided by the application exceeds the data rate provided by the radio interface for a long period, large amounts of buffered data can result. This may lead to a large loss of data at handover if lossless handover is not applied to the bearer, or to an excessive delay for real time applications.

In the fixed internet, one of the roles typically performed by the routers is to drop packets when the data rate of an application exceeds the available data rate in a part of the internet. An application may then detect this loss of packets and adapt its data rate to the available data rate. A typical example is the TCP transmit window handling, where the transmit window of TCP is reduced when a lost packet is detected, thus adapting to the available data rate. Other applications such as video or voice calls via IP can also detect lost packets, for example via RTCP (Real-time Transport Control Protocol) feedback, and can adapt the data rate accordingly.

In order to allow these mechanisms to work, and in order to prevent excessive delay, a discard function has been included in the PDCP layer for LTE. This discard function is based on a timer, where for each PDCP SDU received from the higher layers in the transmitter a timer is started, and when the transmission of the PDCP SDU has not yet been initiated in the UE at the expiry of this timer the PDCP SDU is discarded. If the timer is set to an appropriate value for the required QoS of the radio bearer, this discard mechanism can prevent excessive delay and queuing in the transmitter.

4.2.6 PDCP PDU Formats

PDCP PDUs for user plane data comprise a 'D/C' field in order to distinguish Data and Control PDUs, the formats of which are shown in Figures 4.6 and 4.7 respectively. PDCP Data PDUs comprise a 7- or 12-bit SN as shown in Table 4.2. PDCP Data PDUs for user plane data contain either an uncompressed (if header compression is not used) or a compressed IP packet.

---

[9]For details of the S1 interface, see Section 2.5.

D/C	PDCP SN	Data	MAC-I

Figure 4.6 Key features of PDCP Data PDU format. See Table 4.2 for presence of D/C and MAC-I fields.

Table 4.2 PDCP Data PDU formats.

PDU type	D/C field	Sequence number length	MAC-I	Applicable RLC Modes (see Section 4.3.1)
User plane long SN	Present	12 bits	Absent	AM / UM
User plane short SN	Present	7 bits	Absent	UM
Control plane	Absent	5 bits	32 bits	AM / UM

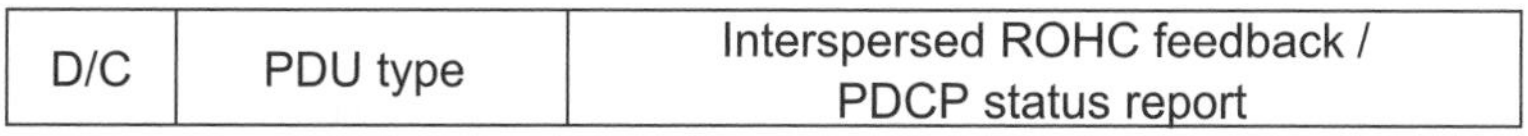

D/C	PDU type	Interspersed ROHC feedback / PDCP status report

Figure 4.7 Key features of PDCP Control PDU format.

PDCP Data PDUs for control plane data (i.e. RRC signalling) comprise a MAC-I field of 32-bit length for integrity protection. PDCP Data PDUs for control plane data contain one complete RRC message.

As can be seen in Table 4.2 there are three types of PDCP Data PDU, distinguished mainly by the length of the PDCP SN and the presence of the MAC-I field. As mentioned in Section 4.2.3, the length of the PDCP SN in relation to the data rate, the packet size and the packet inter-arrival rate determines the maximum possible interruption time without desynchronizing the COUNT value which is used for ciphering and integrity protection.

The PDCP Data PDU for user plane data using the long SN allows longer interruption times, and makes it possible, when it is mapped on RLC Acknowledged Mode (see Section 4.3.1.3), to perform lossless handover as described in Section 4.2.4, but implies a higher overhead. Therefore it is mainly used for data applications with a large IP packet size where the overhead compared to the packet size is not too significant, for example for file transfer, web browsing, or e-mail traffic.

The PDCP Data PDU for user plane data using the short SN is mapped on RLC Unacknowledged Mode (see Section 4.3.1.2) and is typically used for VoIP services, where only seamless handover is used and retransmission is not necessary.

PDCP Control PDUs are used by PDCP entities handling user plane data (see Figure 4.1). There are two types of PDCP Control PDU, distinguished by the PDU type field in the PDCP header. PDCP Control PDUs carry either PDCP Status Reports for the case of lossless handover, or ROHC feedback created by the ROHC header compression protocol.

In order to reduce complexity, a PDCP Control PDU carrying ROHC feedback carries exactly one ROHC feedback packet – there is no possibility to transmit several ROHC feedback packets in one PDCP PDU.

A PDCP Control PDU carrying a PDCP Status Report for the case of lossless handover is used in order to prevent the retransmission of already-correctly-received PDCP SDUs, and also to request retransmission of PDCP SDUs which were correctly received but for which header decompression failed. This PDCP Control PDU contains a bitmap indicating which PDCP SDUs need to be retransmitted and a reference sequence number, the First Missing SDU (FMS). In the case that all PDCP SDUs have been received in sequence this field indicates the next expected sequence number, and no bitmap is included.

4.3 Radio Link Control (RLC)

The RLC layer is located between the PDCP layer (the 'upper' layer) and the MAC layer (the 'lower' layer). It communicates with the PDCP layer through a Service Access Point (SAP), and with the MAC layer via logical channels. The RLC layer reformats PDCP PDUs in order to fit them into the size indicated by the MAC layer; that is, the RLC transmitter segments and/or concatenates the PDCP PDUs, and the RLC receiver reassembles the RLC PDUs to reconstruct the PDCP PDUs.

In addition, the RLC reorders the RLC PDUs if they are received out of sequence due to the HARQ operation performed in the MAC layer. This is the key difference from UMTS, where the HARQ reordering is performed in the MAC layer. The advantage of HARQ reordering in RLC is that no additional SN and reception buffer are needed for HARQ reordering. In LTE, the RLC SN and RLC reception buffer are used for both HARQ reordering and RLC-level ARQ related operations.

The functions of the RLC layer are performed by 'RLC entities'. An RLC entity is configured in one of three data transmission modes: Transparent Mode (TM), Unacknowledged Mode (UM), and Acknowledged Mode (AM). In AM, special functions are defined to support retransmission. When UM or AM is used, the choice between the two modes is made by the eNodeB during the RRC radio bearer setup procedure (see Section 3.2.3.3), based on the QoS requirements of the EPS bearer.[10] The three RLC modes are described in detail in the following sections.

4.3.1 RLC Entities

4.3.1.1 Transparent Mode (TM) RLC Entity

As the name indicates, the TM RLC entity is transparent to the PDUs that pass through it – no functions are performed and no RLC overhead is added. Since no overhead is added, an RLC SDU is directly mapped to an RLC PDU and vice versa.

Therefore, the use of TM RLC is very restricted. Only RRC messages which do not need RLC configuration can utilize the TM RLC, such as broadcast system information messages, paging messages, and RRC messages which are sent when no Signalling Radio Bearers (SRBs) other than SRB0 (see Section 3.2.1) are available. TM RLC is not used for user plane data transmission in LTE.

---

[10]Evolved Packet System – see Section 2.

TM RLC provides a unidirectional data transfer service – in other words, a single TM RLC entity is configured either as a transmitting TM RLC entity or as a receiving TM RLC entity.

4.3.1.2 Unacknowledged Mode (UM) RLC Entity

UM RLC provides a unidirectional data transfer service like TM RLC. UM RLC is mainly utilized by delay-sensitive and error-tolerant real-time applications, especially VoIP, and other delay-sensitive streaming services. Point-to-multipoint services such as MBMS (Multimedia Broadcast/Multicast Service) also use UM RLC – since no feedback path is available in the case of point-to-multipoint services, AM RLC cannot be utilized by these services.

A block diagram of the UM RLC entity is shown in Figure 4.8.

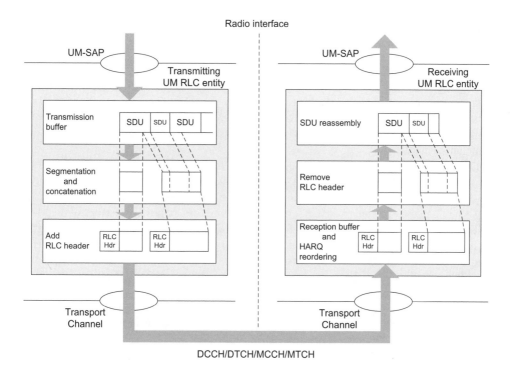

Figure 4.8 Model of UM RLC entities. Reproduced by permission of © 3GPP.

The main functions of UM RLC can be summarized as follows:

- Segmentation and concatenation of RLC SDUs;

- Reordering of RLC PDUs;

- Duplicate detection of RLC PDUs;

- Reassembly of RLC SDUs.

Segmentation and concatenation. The transmitting UM RLC entity performs segmentation and/or concatenation on RLC SDUs received from upper layers, to form RLC PDUs. The size of the RLC PDU at each transmission opportunity is decided and notified by the MAC layer depending on the radio channel conditions and the available transmission resources; therefore, the size of each transmitted RLC PDU can be different.

The transmitting UM RLC entity includes RLC SDUs into an RLC PDU in the order in which they arrive at the UM RLC entity. Therefore, a single RLC PDU can contain RLC SDUs or segments of RLC SDUs according to the following pattern:

(zero or one) SDU segment + (zero or more) SDUs + (zero or one) SDU segment.

The constructed RLC PDU is always byte-aligned and has no padding.

After segmentation and/or concatenation of RLC SDUs, the transmitting UM RLC entity includes relevant UM RLC headers in the RLC PDU to indicate the sequence number[11] of the RLC PDU, and additionally the size and boundary of each included RLC SDU or RLC SDU segment.

Reordering, duplicate detection, and reassembly. When the receiving UM RLC entity receives RLC PDUs, it first reorders them if they are received out of sequence. Out-of-sequence reception is unavoidable due to the fact that the HARQ operation in the MAC layer uses multiple HARQ processes (see Section 4.4). Any RLC PDUs received out of sequence are stored in the reception buffer until all the previous RLC PDUs are received and delivered to the upper layer.

During the reordering process, any duplicate RLC PDUs received are detected by checking the sequence numbers and discarded. This ensures that the upper layer receives upper layer PDUs only once. The most common cause of receiving duplicates is HARQ ACKs for MAC PDUs being misinterpreted as NACKs, resulting in unnecessary retransmissions of the MAC PDUs, which causes duplication in the RLC layer.

To detect reception failures and avoid excessive reordering delays, a reordering timer is used in the receiving UM RLC entity to set the maximum time to wait for the reception of RLC PDUs that have not been received in sequence. The receiving UM RLC entity starts the reordering timer when a missing RLC PDU is detected, and it waits for the missing RLC PDUs until the timer expires. When the timer expires, the receiving UM RLC entity declares the missing RLC PDUs as lost and starts to reassemble the next available RLC SDUs from the RLC PDUs stored in the reception buffer.

The reassembly function is performed on an RLC SDU basis; only RLC SDUs for which all segments are available are reassembled from the stored RLC PDUs and delivered to the upper layers. RLC SDUs that have at least one missing segment are simply discarded and not reassembled. If RLC SDUs were concatenated in an RLC PDU, the reassembly function in the RLC receiver separates them into their original RLC SDUs. The RLC receiver delivers reassembled RLC SDUs to the upper layers in increasing order of sequence numbers.

An example scenario of a lost RLC PDU with HARQ reordering is shown in Figure 4.9. A reordering timer is started when the RLC receiver receives PDU#8. If PDU#7 has not been received before the timer expires, the RLC receiver decides that the PDU#7 is lost, and starts to reassemble RLC SDUs from the next received RLC PDU. In this example, SDU#22 and

---

[11]Note that the RLC sequence number is independent from the sequence number added by PDCP.

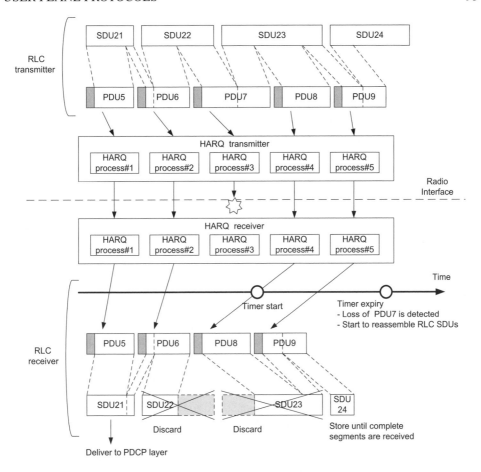

Figure 4.9 Example of PDU loss detection with HARQ reordering.

SDU#23 are discarded because they are not completely received, and SDU#24 is kept in the reception buffer until all segments are received. Only SDU#21 is completely received, so it is delivered up to the PDCP layer.

4.3.1.3   Acknowledged Mode (AM) RLC Entity

Contrary to the other RLC transmission modes, AM RLC provides a bidirectional data transfer service. Therefore, a single AM RLC entity is configured with the ability both to transmit and to receive – we refer to the corresponding parts of the AM RLC entity as the *transmitting side* and the *receiving side* respectively.

The most important feature of AM RLC is 'retransmission'. An Automatic Repeat reQuest (ARQ) operation is performed to support error-free transmission. Since transmission errors are corrected by retransmissions, AM RLC is mainly utilized by error-sensitive and delay-tolerant non-real-time applications. Examples of such applications include most

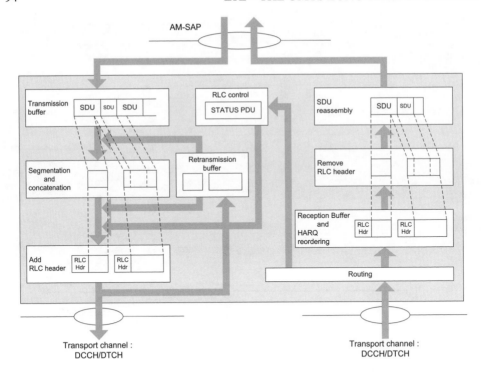

Figure 4.10 Model of AM RLC entity. Reproduced by permission of © 3GPP.

of the interactive/background type services, such as web browsing and file downloading. Streaming-type services also frequently use AM RLC if the delay requirement is not too stringent. In the control plane, RRC messages typically utilize the AM RLC in order to take advantage of RLC acknowledgements and retransmissions to ensure reliability.

A block diagram of the AM RLC entity is shown in Figure 4.10.

Although the AM RLC block diagram looks complicated at first glance, the transmitting and receiving sides are similar to the UM RLC transmitting and receiving entities respectively, except for the retransmission-related blocks. Therefore, most of the UM RLC behaviour described in the previous section applies to AM RLC in the same manner. The transmitting side of the AM RLC entity performs segmentation and/or concatenation of RLC SDUs received from upper layers to form RLC PDUs together with relevant AM RLC headers, and the receiving side of the AM RLC entity reassembles RLC SDUs from the received RLC PDUs after HARQ reordering.

In addition to performing the functions of UM RLC, the main functions of AM RLC can be summarized as follows:

- Retransmission of RLC Data PDUs;

- Re-segmentation of retransmitted RLC Data PDUs;

- Polling;

- Status reporting;

- Status prohibit.

Retransmission and resegmentation. As mentioned before, the most important function of AM RLC is *retransmission*. In order that the transmitting side retransmits only the missing RLC PDUs, the receiving side provides a 'status report' to the transmitting side indicating ACK and/or NACK information for the RLC PDUs. Status reports are sent by the transmitting side of the AM RLC entity whose receiving side received the corresponding RLC PDUs. Hence, the AM RLC transmitting side is able to transmit two types of RLC PDU, namely RLC Data PDUs containing data received from upper layers and RLC Control PDUs generated in the AM RLC entity itself. To differentiate between Data and Control PDUs, a 1-bit flag is included in the AM RLC header (see Section 4.3.2.3).

When the transmitting side transmits RLC Data PDUs, it stores the PDUs in the retransmission buffer for possible retransmission if requested by the receiver through a status report. In case of retransmission, the transmitter can resegment the original RLC Data PDUs into smaller PDU segments if the MAC layer indicates a size that is smaller than the original RLC Data PDU size.

An example of RLC re-segmentation is shown in Figure 4.11. In this example, an original PDU of 600 bytes is resegmented into two PDU segments of 200 and 400 bytes at retransmission.

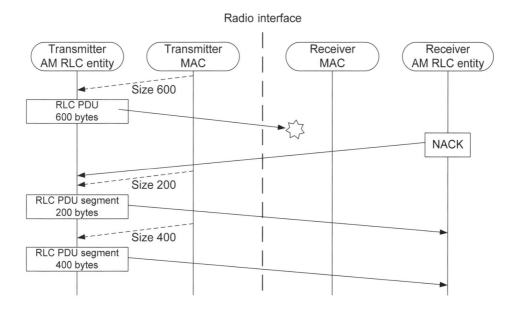

Figure 4.11 Example of RLC resegmentation.

The identification between the original RLC PDU and the retransmitted segments is achieved by another 1-bit flag in the AM RLC header: in the case of a retransmitted segment, some more fields are included in the AM RLC header to indicate resegmentation related information. The receiver can use status reports to indicate the status of individual retransmitted segments, not just full PDUs.

Polling, status report and status prohibit.   The transmitting side of the AM RLC entity can request a status report from the peer receiving side, by means of a 1-bit polling indicator included in the AM RLC header. This function is called 'polling' and it allows the transmitting side actively to get the receiver status report. The transmitting side can then use the status reports to select the RLC Data PDUs to be retransmitted, and manage transmission and retransmission buffers efficiently. Typical circumstances in which the transmitting side may initiate a poll include, for example, the last PDU in the transmitting side having been transmitted, or a predefined number of PDUs or data bytes having been transmitted.

When the receiving side of the AM RLC entity receives a poll from the peer transmitting side, it checks the reception buffer status and transmits a status report at the earliest transmission opportunity.

The receiving side can also generate a status report of its own accord if it detects a reception failure of an AM RLC PDU. For the detection of a reception failure, a similar mechanism is used as in the case of UM RLC in relation to the HARQ reordering delay. In AM RLC, however, the detection of a reception failure triggers a status report instead of considering the relevant RLC PDUs as permanently lost.

Note that the transmission of status reports needs to be carefully controlled according to the trade-off between transmission delay and radio efficiency. To reduce the transmission delay, status reports need to be transmitted frequently, but on the other hand frequent transmission of status reports wastes radio resources. Moreover, if further status reports are sent whilst the retransmissions triggered by a previous status report have not yet been received, unnecessary retransmissions may result, thus consuming further radio resources; in AM RLC this is in fact a second cause of duplicate PDUs occurring which have to be discarded by the duplicate-detection functionality. Therefore, to control the frequency of status reporting in an effective way, a 'status prohibit' function is available in AM RLC, whereby the transmission of new status reports is prohibited while a timer is running.

4.3.2   RLC PDU Formats

As mentioned before, the RLC layer provides two types of PDU, namely the RLC Data PDU and the RLC Control PDU. The RLC Data PDU is used to transmit PDCP PDUs and is defined in all RLC transmission modes. The RLC Control PDU delivers control information between peer RLC entities and is defined only in AM RLC. The RLC PDUs used in each RLC transmission mode are summarized in Table 4.3.

In the following subsections, each of the RLC PDU formats is explained.

4.3.2.1   Transparent Mode Data PDU Format

The Transparent Mode Data (TMD) PDU consists only of a data field and does not have any RLC headers. Since no segmentation or concatenation is performed, an RLC SDU is directly mapped to a TMD PDU.

Table 4.3 PDU types used in RLC.

RLC Mode	Data PDU	Control PDU
TM	TMD (TM Data)	N/A
UM	UMD (UM Data)	N/A
AM	AMD (AM Data)/AMD segment	STATUS

4.3.2.2 Unacknowledged Mode Data PDU Format

The Unacknowledged Mode Data (UMD) PDU (Figure 4.12) consists of a data field and UMD PDU header. PDCP PDUs (i.e. RLC SDUs) can be segmented and/or concatenated into the data field. The UMD PDU header is further categorized into a fixed part (included in each UMD PDU) and an extension part (included only when the data field contains more than one SDU or SDU segment – i.e. only when the data field contains any SDU borders).

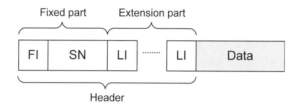

Figure 4.12 Key features of UMD PDU format.

- **Framing Info (FI).** This 2-bit field indicates whether the first and the last data field elements are complete SDUs or partial SDUs (i.e. whether the receiving RLC entity needs to receive multiple RLC PDUs in order to reassemble the corresponding RLC SDU).

- **Sequence Number (SN).** For UMD PDUs, either a short (5 bits) or a long (10 bits) SN field can be used. This field allows the receiving RLC entity unambiguously to identify a UMD PDU, which allows reordering and duplicate-detection to take place.

- **Length Indicator (LI).** This 11-bit field indicates the length of the corresponding data field element present in the UMD PDU. There is a one-to-one correspondence between each LI and a data field element, except for the last data field element for which the LI field is omitted because the length of the last data field element can be deduced from the UMD PDU size.

4.3.2.3 Acknowledged Mode Data PDU Format

In addition to the UMD PDU header fields, the Acknowledged Mode Data (AMD) PDU header (Figure 4.13) contains additional fields to support the RLC ARQ mechanism. The only difference in the PDU fields is that only the long SN field (10 bits) is used for AMD PDUs. The additional fields are as follows:

- **Data/Control (D/C).** This 1-bit field indicates whether the RLC PDU is an RLC Data PDU or an RLC Control PDU. It is present in all types of PDU used in AM RLC.

- **Resegmentation Flag (RF).** This 1-bit field indicates whether the RLC PDU is an AMD PDU or an AMD PDU segment.

- **Polling (P).** This 1-bit field is used to request a status report from the peer receiving side.

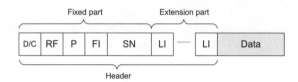

Figure 4.13 Key features of AMD PDU format.

4.3.2.4   AMD PDU Segment Format

The AMD PDU segment format (Figure 4.14) is used in case of resegmented retransmissions (when the available resource for retransmission is smaller than the original PDU size), as described in Section 4.3.1.3.

If the RF field indicates that the RLC PDU is an AMD PDU segment, the following additional resegmentation related fields are included in the fixed part of the AMD PDU header to enable correct reassembly:

- **Last Segment Flag (LSF).** This 1-bit field indicates whether or not this AMD PDU segment is the last segment of an AMD PDU.

- **Segmentation Offset (SO).** This 15-bit field indicates the starting position of the AMD PDU segment within the original AMD PDU.

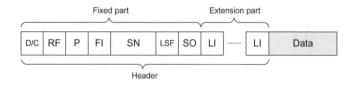

Figure 4.14 Key features of AMD PDU segment format.

4.3.2.5   STATUS PDU Format

The STATUS PDU (Figure 4.15) is designed to be very simple, as the RLC PDU error ratio should normally be low in LTE due to the use of HARQ in the MAC layers. Therefore, the STATUS PDU simply lists all the missing portions of AMD PDUs by means of the following fields:

- **Control PDU Type (CPT).** This 3-bit field indicates the type of the RLC Control PDU, allowing more RLC Control PDU types to be defined in a later release of the LTE specifications. (The STATUS PDU is the only type of RLC Control PDU defined in the first version of LTE.)

- **ACK_SN.** This 10-bit field indicates the SN of the first AMD PDU which is neither received nor listed in this STATUS PDU. All AMD PDUs up to but not including this AMD PDU are correctly received by the receiver except the AMD PDUs or portions of AMD PDUs listed in the NACK_SN List.

- **NACK_SN List.** This field contains a list of SNs of the AMD PDUs that have not been completely received, optionally including indicators of which bytes of the AMD PDU are missing in the case of resegmentation.

Figure 4.15 STATUS PDU format.

4.4   Medium Access Control (MAC)

The MAC layer is the lowest sub-layer in the Layer 2 architecture of the LTE radio protocol stack. The connection to the physical layer below is through transport channels, and the connection to the RLC layer above is through logical channels. The MAC layer therefore performs multiplexing and demultiplexing between logical channels and transport channels: the MAC layer in the transmitting side constructs MAC PDUs, known as transport blocks, from MAC SDUs received through logical channels, and the MAC layer in the receiving side recovers MAC SDUs from MAC PDUs received through transport channels.

4.4.1   MAC Architecture

4.4.1.1   Overall Architecture

Figure 4.16 shows a conceptual overview of the architecture of the MAC layer.

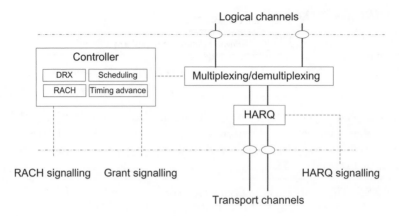

Figure 4.16 Conceptual overview of the UE-side MAC architecture.

The MAC layer can be considered to consist of a Hybrid ARQ (HARQ) entity, a multiplexing/demultiplexing entity, and a controller which performs various control functions.

The HARQ entity is responsible for the transmit and receive HARQ operations. The transmit HARQ operation includes transmission and retransmission of transport blocks, and reception and processing of ACK/NACK signalling. The receive HARQ operation includes reception of transport blocks, combining of the received data and generation of ACK/NACK signalling. In order to enable continuous transmission while previous transport blocks are being decoded, up to eight HARQ processes in parallel are used to support multiprocess 'Stop-And-Wait' (SAW) HARQ operation.

SAW operation means that upon transmission of a transport block, a transmitter stops further transmissions and awaits feedback from the receiver. When a NACK is received, or when a certain time elapses without receiving any feedback, the transmitter retransmits the transport block. Such a simple SAW HARQ operation cannot on its own utilize the transmission resources during the period between the first transmission and the retransmission. Therefore multiprocess HARQ interlaces several independent SAW processes in time so that all the transmission resources can be used by one of the processes. Each HARQ process is responsible for a separate SAW operation and manages a separate buffer.

In general, HARQ schemes can be categorized as either *synchronous* or *asynchronous*, with the retransmissions in each case being either *adaptive* or *non-adaptive*.

In a synchronous HARQ scheme, the retransmission(s) for each process occur at predefined times relative to the initial transmission. In this way, there is no need to signal information such as HARQ process number, as this can be inferred from the transmission timing. By contrast, in an asynchronous HARQ scheme, the retransmissions can occur at any time relative to the initial transmission, so additional explicit signalling is required to indicate the HARQ process number to the receiver, so that the receiver can correctly associate each retransmission with the corresponding initial transmission. In summary, synchronous HARQ schemes reduce the signalling overhead while asynchronous HARQ schemes allow more flexibility in scheduling.

In an adaptive HARQ scheme, transmission attributes such as the modulation and coding scheme, and transmission resource allocation in the frequency domain, can be changed at each retransmission in response to variations in the radio channel conditions. In a non-adaptive HARQ scheme, the retransmissions are performed without explicit signalling of new transmission attributes – either by using the same transmission attributes as those of the previous transmission, or by changing the attributes according to a predefined rule. Accordingly, adaptive schemes bring more scheduling gain at the expense of increased signalling overhead.

In LTE, asynchronous adaptive HARQ is used for the downlink, and synchronous HARQ for the uplink. In the uplink, the retransmissions may be either adaptive or non-adaptive, depending on whether new signalling of the transmission attributes is provided.

The details of the HARQ incremental redundancy schemes and timing for retransmissions are explained in Section 10.3.2.5.

In the multiplexing and demultiplexing entity, data from several logical channels can be (de)multiplexed into/from one transport channel. The multiplexing entity generates MAC PDUs from MAC SDUs when radio resources are available for a new transmission; this process includes prioritizing the data from the logical channels to decide how much data and from which logical channel(s) should be included in each MAC PDU. The demultiplexing entity reassembles the MAC SDUs from MAC PDUs and distributes them to the appropriate RLC entities. In addition, for peer-to-peer communication between the MAC layers, control messages called 'MAC Control Elements' can be included in the MAC PDU as explained in Section 4.4.2.7 below.

The controller entity is responsible for a number of functions including Discontinuous Reception (DRX), the Random Access Channel (RACH) procedure, the Data Scheduling procedure, and for maintaining the uplink timing alignment. These functions are explained in the following sections.

4.4.1.2   Logical Channels

The MAC layer provides a data transfer service for the RLC layer through logical channels. Logical channels are either Control Logical Channels which carry control data such as RRC signaling, or Traffic Logical Channels which carry user plane data. They are as follows:

Control logical channels.

- **Broadcast Control Channel (BCCH).** This is a downlink channel which is used to broadcast system information. In the RLC layer, it is associated with a TM RLC entity (see Section 4.3.1).

- **Paging Control Channel (PCCH).** This is a downlink channel which is used to notify UEs of an incoming call or a change of system information.

- **Common Control Channel (CCCH).** This channel is used to deliver control information in both uplink and downlink directions when there is no confirmed association between a UE and the eNodeB – i.e. during connection establishment. In the RLC layer, it is associated with a TM RLC entity (see Section 4.3.1).

- **Multicast Control Channel (MCCH).** This is a downlink channel which is used to transmit control information related to the reception of MBMS services (see Chapter 14). In the RLC layer, it is always associated with a UM RLC entity (see Section 4.3.1).

- **Dedicated Control Channel (DCCH).** This channel is used to transmit dedicated control information relating to a specific UE, in both uplink and downlink directions. It is used when a UE has an RRC connection with eNodeB. In the RLC layer, it is associated with an AM RLC entity[12] (see Section 4.3.1).

Traffic logical channels.

- **Dedicated Traffic Channel (DTCH).** This channel is used to transmit dedicated user data in both uplink and downlink directions. In the RLC layer, it can be associated with either a UM RLC entity or an AM RLC entity (see Section 4.3.1).

- **Multicast Traffic Channel (MTCH).** This channel is used to transmit user data for MBMS services in the downlink (see Chapter 14). In the RLC layer, it is always associated with a UM RLC entity (see Section 4.3.1).

4.4.1.3 Transport Channels

Data from the MAC layer is exchanged with the physical layer through transport channels. Data is multiplexed into transport channels depending on how it is transmitted over the air. Transport channels are classified as downlink or uplink as follows.

Downlink transport channels.

- **Broadcast Channel (BCH).** This channel is used to transport the parts of the system information which are essential for access to the DL-SCH. The transport format is fixed and the capacity is limited.

- **Downlink Shared Channel (DL-SCH).** This channel is used to transport downlink user data or control messages. In addition, the remaining parts of the system information that are not transported via the BCH are transported on the DL-SCH.

- **Paging Channel (PCH).** This channel is used to transport paging information to UEs. This channel is also used to inform UEs about updates of the system information (see Section 3.2.2).

- **Multicast Channel (MCH).** This channel is used to transport user data or control messages that require MBSFN combining (see Chapter 14).

The mapping of the downlink transport channels onto physical channels is explained in Section 6.4.

---

[12]No use case was identified for the association of a DCCH with UM RLC in the first version of the LTE specifications.

Uplink transport channels.

- **Uplink Shared Channel (UL-SCH).** This channel is used to transport uplink user data or control messages.

- **Random Access Channel (RACH).** This channel is used for access to the network when the UE does not have accurate uplink timing synchronization, or when the UE does not have any allocated uplink transmission resource (see Chapter 19).

The mapping of the uplink transport channels onto physical channels is explained in Chapter 17.

4.4.1.4 Multiplexing and Mapping between Logical Channels and Transport Channels

Figures 4.17 and 4.18 show the possible multiplexing between logical channels and transport channels in the downlink and uplink respectively.

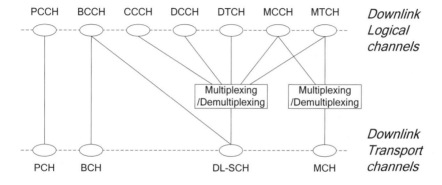

Figure 4.17 Downlink logical channel multiplexing. Reproduced by permission of © 3GPP.

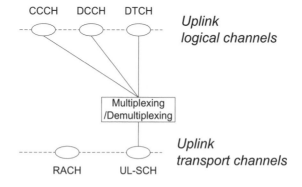

Figure 4.18 Uplink logical channel multiplexing. Reproduced by permission of © 3GPP.

Note that in the downlink, the DL-SCH carries information from all the logical channels except the PCCH. For MBMS, the MTCH and MCCH can be mapped to either the DL-SCH or the MCH depending on whether the data is transmitted from a single cell or from multiple cells respectively. This is discussed further in Chapter 14.

In the uplink, the UL-SCH carries all the information from all the logical channels.

4.4.2   MAC Functions

4.4.2.1   Scheduling

The scheduler in the eNodeB distributes the available radio resources in one cell among the UEs, and among the radio bearers of each UE. The details of the scheduling algorithm are left to the eNodeB implementation, but the signalling to support the scheduling is standardized. Some possible scheduling algorithms are discussed in Chapter 12.

In principle, the eNodeB allocates downlink or uplink radio resources to each UE based respectively on the downlink data buffered in the eNodeB and on Buffer Status Reports (BSRs) received from the UE. In this process, the eNodeB considers the QoS requirements of each configured radio bearer, and selects the size of the MAC PDU.

The usual mode of scheduling is *dynamic scheduling*, by means of downlink assignment messages for the allocation of downlink transmission resources and uplink grant messages for the allocation of uplink transmission resources; these are valid for specific single subframes.[13] They are transmitted on the Physical Downlink Control Channel (PDCCH) using a Cell Radio Network Temporary Identifier (C-RNTI) to identify the intended UE, as described in Section 9.3. This kind of scheduling is efficient for service types such as TCP or the SRBs, in which the traffic is bursty and dynamic in rate.

In addition to the dynamic scheduling, *persistent scheduling* is defined. Persistent scheduling enables radio resources to be semi-statically configured and allocated to a UE for a longer time period than one subframe, avoiding the need for specific downlink assignment messages or uplink grant messages over the PDCCH for each subframe. Persistent scheduling is useful for services such as VoIP for which the data packets are small, periodic and semi-static in size. For this kind of service the timing and amount of radio resources needed are predictable. Thus the overhead of the PDCCH is significantly reduced compared to the case of dynamic scheduling.

For the configuration or reconfiguration of a persistent schedule, RRC signalling indicates the resource allocation interval at which the radio resources are periodically assigned. Specific transmission resource allocations in the frequency domain, and transmission attributes such as the modulation and coding scheme, are signalled using the PDCCH. The actual transmission timing of the PDCCH messages is used as the reference timing to which the resource allocation interval applies. When the PDCCH is used to configure or reconfigure a persistent schedule, it is necessary to distinguish the scheduling messages which apply to a persistent schedule from those used for dynamic scheduling. For this purpose, a special identity is used, known as the Semi-Persistent Scheduling C-RNTI (SPS-C-RNTI), which for each UE is different from C-RNTI used for dynamic scheduling messages.

---

[13]The dynamic uplink transmission resource grants are valid for specific single subframes for initial transmissions, although they may also imply a resource allocation in later subframes for HARQ retransmissions.

Reconfiguration of resources used for persistent scheduling can be performed when there is a transition between a silent period and talk spurt, or when the codec rate changes. For example, when the codec rate for a VoIP service is increased, a new downlink assignment message or uplink grant message can be transmitted to configure a larger persistently-scheduled radio resource for the support of bigger VoIP packet.

4.4.2.2 Scheduling Information Transfer

Buffer Status Reports (BSRs) from the UE to the eNodeB are used to assist the eNodeB's allocation of uplink radio resources. The basic assumption underlying scheduling in LTE is that radio resources are only allocated for transmission to or from a UE if data is available to be sent or received. In the downlink direction, the scheduler in the eNodeB is obviously aware of the amount of data to be delivered to each UE; however, in the uplink direction, because the scheduling decisions are performed in the eNodeB and the buffer for the data is located in the UE, BSRs have to be sent from the UE to the eNodeB to indicate the amount of data in the UE that needs to be transmitted over the UL-SCH.[14]

Two types of BSR are defined in LTE: a long BSR and a short BSR; which one is transmitted depends on the amount of available uplink transmission resources for sending the BSR, on how many groups of logical channels have non-empty buffers, and on whether a specific event is triggered at the UE. The long BSR reports the amount of data for four logical channel groups, whereas the short BSR reports the amount of data for only one logical channel group. Although the UE might actually have more than four logical channels configured, the overhead would be large if the amount of data in the UE were to be reported for every logical channel individually. Thus, grouping the logical channels into four groups for reporting purposes represents a compromise between efficiency and accuracy.

A BSR can be triggered in the following situations:

- whenever data arrives for a logical channel which has a higher priority than the logical channels whose buffers previously contained data;

- whenever a certain time has elapsed since the last transmission of a BSR;

- whenever the serving cell changes.

If a UE is not allocated with enough UL-SCH resources to send a BSR, either a single-bit 'Scheduling Request' (SR) is sent over the Physical Uplink Control Channel (PUCCH – see Chapter 17), or the random access procedure is performed to request an allocation of an uplink radio resource for sending a BSR.

Thus LTE provides suitable signalling to ensure that the eNodeB has sufficient information about the data waiting in each UE's uplink transmission buffer to allocate corresponding uplink transmission resources in a timely manner.

---

[14]Note that, unlike HSUPA, there is no possibility in LTE for a UE to transmit autonomously in the uplink by means of a transmission grant for non-scheduled transmissions. This is because the uplink transmissions from different UEs in LTE are orthogonal in time and frequency, and therefore if an uplink resource is allocated but unused, it cannot be accessed by another UE; by contrast, in HSUPA, if a UE does not use its transmission grant for non-scheduled transmissions, the resulting reduction in uplink interference can benefit other UEs. Furthermore, the short subframe length in LTE enables uplink transmission resources to be dynamically allocated more quickly than in HSUPA.

4.4.2.3 Random Access Procedure

The random access procedure is used when a UE is not allocated with uplink radio resources but has data to transmit, or when the UE is not time-synchronized in the uplink direction. Control of the random access procedure is an important part of the MAC layer functionality in LTE. The details are explained in Chapter 19.

4.4.2.4 Uplink Timing Alignment

Uplink timing alignment maintenance is controlled by the MAC layer and is important for ensuring that a UE's uplink transmissions arrive in the eNodeB without overlapping with the transmissions from other UEs. The details of the uplink timing advance mechanism used to maintain timing alignment are explained in Section 20.2.

The timing advance mechanism utilizes MAC Control Elements (see Section 4.4.2.7) to update the uplink transmission timing. However, maintaining the uplink synchronization in this way during periods when no data is transferred wastes radio resources and adversely impacts the UE battery life. Therefore, when a UE is inactive for a certain period of time the UE is allowed to lose uplink synchronization even in RRC_CONNECTED state. The random access procedure is then used to regain uplink synchronization when the data transfer resumes in either uplink or downlink.

4.4.2.5 Discontinuous Reception (DRX)

DRX functionality can be configured for an 'RRC_CONNECTED' UE[15] so that it does not always need to monitor the downlink channels. A DRX cycle consists of an 'On Duration' during which the UE should monitor the PDCCH and a 'DRX period' during which a UE can skip reception of downlink channels for battery saving purposes.

The parameterization of the DRX cycle involves a trade-off between battery saving and latency. On the one hand, a long DRX period is beneficial for lengthening the UE's battery life. For example, in the case of a web browsing service, it is usually a waste of resources for a UE continuously to receive downlink channels while the user is reading a downloaded web page. On the other hand, a shorter DRX period is better for faster response when data transfer is resumed – for example when a user requests another web page.

To meet these conflicting requirements, two DRX cycles – a short cycle and a long cycle – can be configured for each UE. The transition between the short DRX cycle, the long DRX cycle and continuous reception is controlled either by a timer or by explicit commands from the eNodeB. In some sense, the short DRX cycle can be considered as a confirmation period in case a late packet arrives, before the UE enters the long DRX cycle – if data arrives at the eNodeB while the UE is in the short DRX cycle, the data is scheduled for transmission at the next wake-up time and the UE then resumes continuous reception. On the other hand, if no data arrives at the eNodeB during the short DRX cycle, the UE enters the long DRX cycle, assuming that the packet activity is finished for the time being.

Figure 4.19 shows an example of DRX operation. The UE checks for scheduling messages (indicated by its C-RNTI on the PDCCH) during the 'On Duration' period of either the long DRX cycle or the short DRX cycle depending on the currently active cycle. When a

---

[15]Different DRX functionality applies to UEs which are in 'RRC_IDLE'. These RRC states are discussed in Chapter 3.

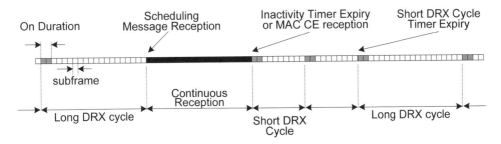

Figure 4.19 The two-level DRX procedure.

scheduling message is received during an 'On Duration', the UE starts an 'Inactivity Timer' and monitors the PDCCH in every subframe while the Inactivity Timer is running. During this period, the UE can be regarded as being in a continuous reception mode. Whenever a scheduling message is received while the Inactivity Timer is running, the UE restarts the Inactivity Timer, and when it expires the UE moves into a short DRX cycle and starts a 'Short DRX cycle timer'. The short DRX cycle may also be initiated by means of a MAC Control Element (see Section 4.4.2.7). When the short DRX cycle timer expires, the UE moves into a long DRX cycle.

In addition to this DRX behaviour, a 'HARQ Round Trip Time (RTT) timer' is defined with the aim of allowing the UE to sleep during the HARQ RTT. When decoding of a downlink transport block for one HARQ process fails, the UE can assume that the next retransmission of the transport block will occur after at least 'HARQ RTT' subframes. While the HARQ RTT timer is running, the UE does not need to monitor the PDCCH (provided that there is no other reason to be monitoring it). At the expiry of the HARQ RTT timer, the UE resumes reception of the PDCCH as normal. The HARQ RTT is illustrated in Section 10.3.2.5.

4.4.2.6 Multiplexing and Logical Channel Prioritization

Unlike the downlink, where the multiplexing and logical channel prioritization is left to the eNodeB implementation, for the uplink the process by which a UE creates a MAC PDU to transmit using the allocated radio resources is fully standardized; this is designed to ensure that the UE satisfies the QoS of each configured radio bearer in a way which is optimal and consistent between different UE implementations. Based on the uplink transmission resource grant message signalled on the PDCCH, the UE has to decide on the amount of data for each logical channel to be included in the new MAC PDU, and, if necessary, also to allocate space for a MAC Control Element.

One simple way to meet this purpose is to serve radio bearers in order of their priority. Following this principle, the data from the logical channel of the highest priority is the first to be included into the MAC PDU, followed by data from the logical channel of the next highest priority, continuing until the MAC PDU size allocated by the eNodeB is completely filled or there is no more data to transmit.

Although this kind of priority-based multiplexing is simple and favours the highest priorities, it sometimes leads to starvation of low-priority bearers. Starvation occurs when the logical channels of the lower priority cannot transmit any data because the data from higher priority logical channels always takes up all the allocated radio resources.

To avoid starvation, while still serving the logical channels according to their priorities, in LTE a Prioritized Bit Rate (PBR) is configured by the eNodeB for each logical channel. The PBR is the data rate provided to one logical channel before allocating any resource to a lower-priority logical channel.

In order to take into account both the PBR and the priority, each logical channel is served in decreasing order of priority, but the amount of data from each logical channel included into the MAC PDU is initially limited to the amount corresponding to the configured PBR. Only when all logical channels have been served up to their PBR, then if there is still room left in the MAC PDU each logical channel is served again in decreasing order of priority. In this second round, each logical channel is served only if all logical channels of higher priority have no more data for transmission.

In most cases, a MAC Control Element has higher priority than any other logical channel because it controls the operation of a MAC entity. Thus, when a MAC PDU is composed and there is a MAC Control Element to send, the MAC Control Element is included first and the remaining space is used to include data from logical channels. One exception to this rule occurs when a UE transmits the first RRC message to a target cell during a handover procedure – in this case, a MAC Control Element such as a BSR has lower priority than SRBs. This is because it is more important to complete the handover procedure as soon as possible than to inform the eNodeB of the UE's buffer status; otherwise, the data transfer interruption time would be longer and the probability of handover failure would increase due to the delayed signalling.

Figure 4.20 illustrates the LTE MAC multiplexing by way of example. First, channel 1 is served up to its PBR, channel 2 up to its PBR and then channel 3 with as much data as is available (since in this example the amount of data available is less than would be permitted by the PBR configured for that channel). After that, the remaining space in the MAC PDU is filled with data from the channel 1 which is of the highest priority until there is no further room in the MAC PDU or there is no further data from channel 1. If there is still a room after serving the channel 1, channel 2 is served in a similar way.

4.4.2.7   MAC PDU Formats

When the multiplexing is done, the MAC PDU itself can be composed. The general MAC PDU format is shown in Figure 4.21. A MAC PDU primarily consists of the MAC header and the MAC payload. The MAC header is further composed of MAC subheaders, while the MAC payload is composed of MAC Control Elements, MAC SDUs and padding.

Each MAC subheader consists of a Logical Channel ID (LCID) and a Length (L) field. The LCID indicates whether the corresponding part of the MAC payload is a MAC Control Element, and if not, to which logical channel the related MAC SDU belongs. The L field indicates the size of the related MAC SDU or MAC Control Element.

MAC Control Elements are used for MAC-level peer-to-peer signalling, including delivery of BSR information and reports of the UE's available power headroom (see Section 20.3) in the uplink, and in the downlink DRX commands and timing advance commands. For each type of MAC Control Element, one special LCID is allocated. When a MAC PDU is used to transport data from the PCCH or BCCH logical channels, the MAC PDU includes data from

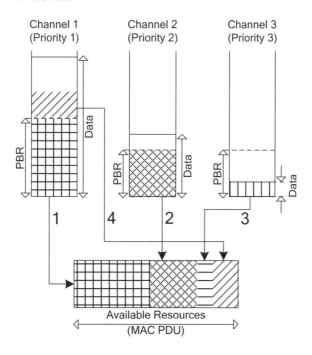

Figure 4.20 Example of MAC multiplexing.

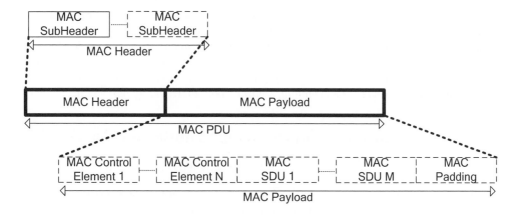

Figure 4.21 General MAC PDU format.

only one logical channel. In this case, because multiplexing is not applied, there is no need to include the LCID field in the header. In addition, if there is a one-to-one correspondence between a MAC SDU and a MAC PDU, the size of the MAC SDU can be known implicitly from the transport block size. Thus, for these cases a headerless MAC PDU format is used as a transparent MAC PDU.

4.5   Summary of the User Plane Protocols

The LTE Layer 2 protocol stack, consisting of the PDCP, RLC and MAC sublayers, acts as the interface between the radio access technology-agnostic sources of packet data traffic and the LTE physical layer. By providing functionality such as IP packet header compression, security, handover support, segmentation/concatenation, retransmission and reordering of packets, and transmission scheduling, the protocol stack enables the physical layer to be used efficiently for packet data traffic.

References[16]

[1] 3GPP Technical Specification 36.323, 'Packet Data Convergence Protocol (PDCP) Specification (Release 8)', www.3gpp.org.

[2] 3GPP Technical Specification 36.322, 'Radio Link Control (RLC) Protocol Specification (Release 8)', www.3gpp.org.

[3] 3GPP Technical Specification 36.321, 'Medium Access Control (MAC) Protocol Specification (Release 8)', www.3gpp.org.

---

[16]All web sites confirmed 18th December 2008.

Part II

Physical Layer for Downlink

5

Orthogonal Frequency Division Multiple Access (OFDMA)

Andrea Ancora, Issam Toufik, Andreas Bury and Dirk Slock

5.1 Introduction

The choice of an appropriate modulation and multiple-access technique for mobile wireless data communications is critical to achieving good system performance. In particular, typical mobile radio channels tend to be dispersive and time-variant, and this has generated interest in multicarrier modulation.

In general, multicarrier schemes subdivide the used channel bandwidth into a number of parallel subchannels as shown in Figure 5.1(a). Ideally the bandwidth of each subchannel is such that they are each non-frequency-selective (i.e. having a spectrally-flat gain); this has the advantage that the receiver can easily compensate for the subchannel gains individually in the frequency domain.

Orthogonal Frequency Division Multiplexing (OFDM) is a special case of multicarrier transmission which is highly attractive for implementation. In OFDM, the non-frequency-selective narrowband subchannels into which the frequency-selective wideband channel is divided are overlapping but orthogonal, as shown in Figure 5.1(b). This avoids the need to separate the carriers by means of guard-bands, and therefore makes OFDM highly spectrally efficient. The spacing between the subchannels in OFDM is such they can be perfectly separated at the receiver. This allows for a low-complexity receiver implementation, which makes OFDM attractive for high-rate mobile data transmission such as the LTE downlink.

It is worth noting that the advantage of separating the transmission into multiple narrowband subchannels cannot itself translate into robustness against time-variant channels if no channel coding is employed. The LTE downlink combines OFDM with channel coding

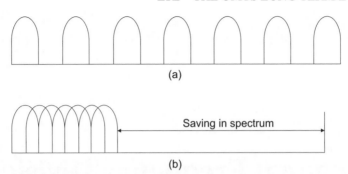

Figure 5.1 Spectral efficiency of OFDM compared to classical multicarrier modulation: (a) classical multicarrier system spectrum; (b) OFDM system spectrum.

and Hybrid Automatic Repeat reQuest (HARQ) to overcome the deep fading which may be encountered on the individual subchannels. These aspects are considered in Chapter 10 and lead to the LTE downlink falling under the category of system often referred to as 'Coded OFDM' (COFDM).

5.1.1 History of OFDM Development

Multicarrier communication systems were first introduced in the 1960s [1, 2], with the first OFDM patent being filed at Bell Labs in 1966. Initially only analogue design was proposed, using banks of sinusoidal signal generators and demodulators to process the signal for the multiple subchannels. In 1971, the Discrete Fourier Transform (DFT) was proposed [3], which made OFDM implementation cost-effective. Further complexity reductions were realized in 1980 by the application of the Winograd Fourier Transform (WFT) or the Fast Fourier Transform (FFT) [4].

OFDM then became the modulation of choice for many applications for both wired systems (such as Asymmetric Digital Subscriber Line (ADSL)) and wireless systems. Wireless applications of OFDM tended to focus on broadcast systems, such as Digital Video Broadcasting (DVB) and Digital Audio Broadcasting (DAB), and relatively low-power systems such as Wireless Local Area Networks (WLANs). Such applications benefit from the low complexity of the OFDM receiver, while not requiring a high-power transmitter in the consumer terminals. This avoids one of the main disadvantages of OFDM, namely that the transmitters in high-power applications tend to be more expensive because of the high Peak to Average Power Ratio (PAPR); this aspect is discussed in Section 5.2.2.

The first cellular mobile radio system based on OFDM was proposed in [5]. Since then, the processing power of modern digital signal processors has increased remarkably, paving the way for OFDM, after much research and development, to find its way into the LTE downlink. Here, the key benefits of OFDM which come to the fore are not only the low-complexity receiver but also the ability of OFDM to be adapted in a straightforward manner to operate in different channel bandwidths according to spectrum availability.

5.2   OFDM

5.2.1   Orthogonal Multiplexing Principle

A high-rate data stream typically faces a problem in having a symbol period T_s much smaller than the channel delay spread T_d if it is transmitted serially. This generates Inter-symbol Interference (ISI) which can only be undone by means of a complex equalization procedure. In general, the equalization complexity grows with the square of the channel impulse response length.

In OFDM, the high-rate stream of data symbols is first serial-to-parallel converted for modulation onto M parallel subcarriers as shown in Figure 5.2. This increases the symbol duration on each subcarrier by a factor of approximately M, such that it becomes significantly longer than the channel delay spread.

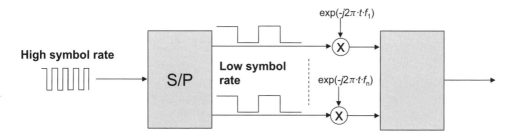

Figure 5.2 Serial-to-parallel conversion operation for OFDM.

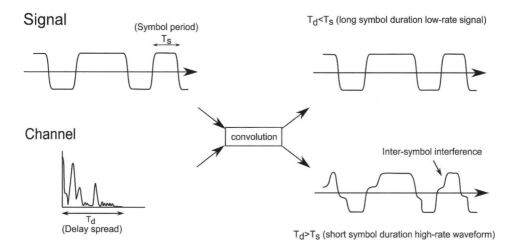

Figure 5.3 Effect of channel on signals with short and long symbol duration.

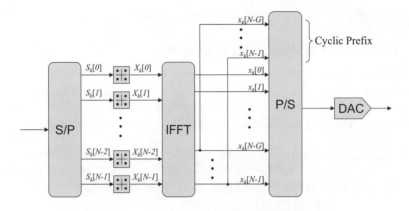

OFDM Transmitter

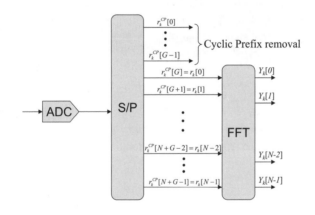

OFDM Receiver

Figure 5.4 OFDM system model: (a) transmitter; (b) receiver.

This operation has the important advantage of requiring a much less complex equalization procedure in the receiver, under the assumption that the time-varying channel impulse response remains substantially constant during the transmission of each modulated OFDM symbol. Figure 5.3 shows how the resulting long symbol duration is virtually unaffected by ISI compared to the short symbol duration, which is highly corrupted.

Figure 5.4 shows the typical block diagram of an OFDM system. The signal to be transmitted is defined in the frequency domain. A Serial to Parallel (S/P) converter collects serial data symbols into a data block $\mathbf{S}_k = [S_k[0], S_k[1], \ldots, S_k[M-1]]^T$ of dimension M, where the subscript k is the index of an OFDM symbol (spanning the M sub-carriers). The M parallel data streams are first independently modulated resulting in the complex vector $\mathbf{X}_k = [X_k[0], X_k[1], \ldots, X_k[M-1]]^T$. Note that in principle it is possible to

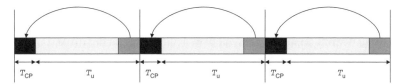

Figure 5.5 OFDM cyclic prefix insertion.

use different modulations (e.g. QPSK or 16QAM) on each sub-carrier; due to channel frequency selectivity, the channel gain may differ between sub-carriers, and thus some sub-carriers can carry higher data-rates than others. The vector of data symbols $\mathbf{X_k}$ then passes through an Inverse FFT (IFFT) resulting in a set of N complex time-domain samples $\mathbf{x_k} = [x_k[0], \ldots, x_k[N-1]]^T$. In a practical OFDM system, the number of processed sub-carriers is greater than the number of modulated sub-carriers (i.e. $N \geq M$), with the un-modulated sub-carriers being padded with zeros.

The next key operation in the generation of an OFDM signal is the creation of a guard period at the beginning of each OFDM symbol, to eliminate the remaining impact of ISI caused by multipath propagation. The guard period is obtained by adding a Cyclic Prefix (CP) at the beginning of the symbol $\mathbf{x}_k$. The CP is generated by duplicating the last G samples of the IFFT output and appending them at the beginning of $\mathbf{x}_k$. This yields the time domain OFDM symbol $[x_k[N-G], \ldots, x_k[N-1], x_k[0], \ldots, x_k[N-1]]^T$, as shown in Figure 5.5.

To avoid ISI completely, the CP length G must be chosen to be longer than the longest channel impulse response to be supported. The CP converts the linear (i.e. aperiodic) convolution of the channel into a circular (i.e. periodic) one which is suitable for DFT processing. This important feature of CP used in OFDM is explained more formally later in this section.

The output of the IFFT is then Parallel-to-Serial (P/S) converted for transmission through the frequency-selective channel.

At the receiver, the reverse operations are performed to demodulate the OFDM signal. Assuming that time- and frequency-synchronization is achieved, a number of samples corresponding to the length of the CP are removed, such that only an ISI-free block of samples is passed to the DFT. If the number of subcarriers N is designed to be a power of 2, a highly efficient FFT implementation may be used to transform the signal back to the frequency domain. Among the N parallel streams output from the FFT, the modulated subset of M subcarriers are selected and further processed by the receiver.

Let $x(t)$ be the signal symbol transmitted at time instant t. The received signal in a multipath environment is then given by

$$r(t) = x(t) * h(t) + z(t) \tag{5.1}$$

where $h(t)$ is the continuous-time impulse response of the channel, $*$ represents the convolution operation and $z(t)$ is the additive noise. Assuming that $x(t)$ is band-limited to $[-\frac{1}{2T_\text{s}}, \frac{1}{2T_\text{s}}]$, the continuous-time signal $x(t)$ can be sampled at sampling rate T_s such that the Nyquist criterion is satisfied.

As a result of the multipath propagation, several replicas of the transmitted signals arrive at the receiver at different delays.

The received discrete-time OFDM symbol k including CP, under the assumption that the channel impulse response has a length smaller than or equal to G, can be expressed as

$$
\mathbf{r^{CP}} =
\begin{bmatrix}
r_k^{CP}[0] \\
r_k^{CP}[1] \\
\vdots \\
r_k^{CP}[G-2] \\
r_k^{CP}[G-1] \\
r_k^{CP}[G] \\
\vdots \\
r_k[N+G-1]
\end{bmatrix}
= \mathbf{A} \cdot
\begin{bmatrix}
h[0] \\
h[1] \\
\vdots \\
h[G-1]
\end{bmatrix}
+
\begin{bmatrix}
z_k[0] \\
z_k[1] \\
\vdots \\
z_k[G-2] \\
z_k[G-1] \\
z_k[G] \\
\vdots \\
z_k[N+G-1]
\end{bmatrix}
\tag{5.2}
$$

where

$$
\mathbf{A} =
\begin{bmatrix}
x_k[N-G] & x_{k-1}[N-1] & x_{k-1}[N-2] & \cdots & x_{k-1}[N-G+1] \\
x_k[N-G+1] & x_k[N-G] & x_{k-1}[N-1] & \cdots & x_{k-1}[N-G+2] \\
\vdots & \vdots & \ddots & \ddots & \vdots \\
x_k[N-2] & x_k[N-3] & \ddots & x_k[N-G] & x_{k-1}[N-1] \\
x_k[N-1] & x_k[N-2] & \ddots & x_k[N-G+1] & x_k[N-G] \\
x_k[0] & x_k[N-1] & \ddots & x_k[N-G+2] & x_k[N-G+1] \\
\vdots & \cdots & \cdots & \ddots & \vdots \\
x_k[N-1] & x_k[N-2] & \cdots & \cdots & x_k[N-G]
\end{bmatrix}
$$

In general broadband transmission systems, one of the most complex operations the receiver has to handle is the equalization process to recover $x_k[n]$ (from Equation (5.2)).

Equation (5.2) can be written as the sum of intra-OFDM symbol interference (generated by the frequency-selective behaviour of the channel within an OFDM symbol) and the inter-OFDM symbol interference (between two consecutive OFDM block transmissions at time k and time $(k-1)$). This can be expressed as

$$
\mathbf{r^{CP}} = \mathbf{A}_{\text{Intra}} \cdot
\begin{bmatrix}
h[0] \\
h[1] \\
\vdots \\
h[G-1]
\end{bmatrix}
+ \mathbf{A}_{\text{Inter}} \cdot
\begin{bmatrix}
h[0] \\
h[1] \\
\vdots \\
h[G-1]
\end{bmatrix}
+
\begin{bmatrix}
z_k[0] \\
z_k[1] \\
\vdots \\
z_k[N+G-1] \\
z_k[N-G] \\
\vdots \\
z_k[N+G-1]
\end{bmatrix}
\tag{5.3}
$$

where

$$
\mathbf{A}_{\text{Intra}} =
\begin{bmatrix}
x_k[N-G] & 0 & \cdots & 0 \\
x_k[N-G+1] & x_k[N-G] & \cdots & 0 \\
\vdots & \ddots & \ddots & \vdots \\
x_k[N-1] & \cdots & \ddots & x_k[N-G] \\
x_k[0] & x_k[N-1] & \cdots & x_k[N-G+1] \\
\vdots & \ddots & \ddots & \vdots \\
x_k[N-1] & x_k[N-2] & \cdots & x_k[N-G]
\end{bmatrix}
$$

and

$$
\mathbf{A}_{\text{Inter}} =
\begin{bmatrix}
0 & x_{k-1}[N-1] & x_{k-1}[N-2] & \cdots & x_{k-1}[N-G+1] \\
0 & 0 & x_{k-1}[N-1] & \cdots & x_{k-1}[N-G+2] \\
\vdots & \ddots & \ddots & \ddots & \vdots \\
0 & 0 & \cdots & 0 & x_{k-1}[N-1] \\
0 & 0 & \cdots & \cdots & 0 \\
\vdots & \ddots & \ddots & \ddots & \vdots \\
0 & 0 & \cdots & \cdots & 0
\end{bmatrix}
$$

In order to suppress the inter-OFDM symbol interference, the first G samples of the received signal are discarded, leading to

$$
\begin{bmatrix}
r_k[0] \\
r_k[1] \\
\vdots \\
r_k[N-1]
\end{bmatrix}
=
\begin{bmatrix}
r_k^{\text{CP}}[G] \\
r_k^{\text{CP}}[G+1] \\
\vdots \\
r_k^{\text{CP}}[N+G-1]
\end{bmatrix}
$$

$$
=
\begin{bmatrix}
x_k[0] & x_k[N-1] & \cdots & x_k[N-G+1] \\
x_k[1] & x_k[0] & \cdots & x_k[N-G+2] \\
\vdots & \ddots & \ddots & \vdots \\
x_k[N-1] & x_k[N-2] & \cdots & x_k[N-G]
\end{bmatrix}
\cdot
\begin{bmatrix}
h[0] \\
h[1] \\
\vdots \\
h[G-1]
\end{bmatrix}
$$

$$
+
\begin{bmatrix}
z_k[G] \\
z_k[G+1] \\
\vdots \\
z_k[N+G-1]
\end{bmatrix}
$$

Adding zeros to the channel vector can extend the signal matrix without changing the output vector. This can be expressed as

$$
\begin{bmatrix} r_k[0] \\ r_k[1] \\ \vdots \\ r_k[N-1] \end{bmatrix} = \mathbf{B} \cdot \begin{bmatrix} h[0] \\ h[1] \\ \vdots \\ h[G-1] \\ 0 \\ \vdots \\ 0 \end{bmatrix} + \begin{bmatrix} z_k[G] \\ z_k[G+1] \\ \vdots \\ z_k[N+G-1] \end{bmatrix}
$$

where matrix $\mathbf{B}$ is given by

$$
\mathbf{B} = \begin{bmatrix} x_k[0] & x_k[N-1] & \cdots & x_k[N-G+1] & x_k[N-G] & \cdots & x_k[1] \\ x_k[1] & x_k[0] & \cdots & x_k[N-G+2] & x_k[N-G+1] & \cdots & x_k[2] \\ \vdots & \ddots & \ddots & \ddots & \ddots & \ddots & \vdots \\ x_k[N-1] & x_k[N-2] & \cdots & x_k[N-G] & x_k[N-G-1] & \cdots & x_k[0] \end{bmatrix}
$$

The matrix $\mathbf{B}$ is circulant and thus it is diagonal in the Fourier domain with diagonal elements given by the FFT of its first row [6,7]. It can then be written as $\mathbf{B} = \mathbf{F}^H \mathbf{X} \mathbf{F}$, with $\mathbf{X}$ diagonal, and the equivalent received signal can be expressed as follows:

$$
\begin{bmatrix} r_k[0] \\ r_k[N-G+1] \\ \vdots \\ r_k[N-1] \end{bmatrix} = \mathbf{F}^H \cdot \begin{bmatrix} X_k[0] & 0 & \cdots & 0 \\ 0 & X_k[1] & \cdots & 0 \\ \vdots & \cdots & \ddots & \vdots \\ 0 & 0 & \cdots & X_k[N-1] \end{bmatrix} \cdot \mathbf{F} \cdot \begin{bmatrix} h[0] \\ h[1] \\ \vdots \\ h[G-1] \\ 0 \\ \vdots \\ 0 \end{bmatrix}
$$

$$
+ \begin{bmatrix} z_k[N-G] \\ z_k[N-G+1] \\ \vdots \\ z_k[N+G-1] \end{bmatrix}
$$

where $\mathbf{F}$ is the Fourier transform matrix whose elements are

$$
(\mathbf{F})_{n,m} = \frac{1}{\sqrt{N}} \exp\left[-\frac{j2\pi}{N}(nm) \right]
$$

for $0 \leq n \leq N-1$ and $0 \leq m \leq N-1$ and N is the length of the OFDM symbol. The elements on the diagonal are the eigenvalues of the matrix $\mathbf{B}$, obtained as

$$
X_k[m] = \frac{1}{\sqrt{N}} \sum_{n=1}^{N} x_k[n] \exp\left[-2j\pi m \frac{n}{N} \right] \tag{5.4}
$$

In reverse, the time-domain signal $x_k[n]$ can be obtained by

$$x_k[n] = \frac{1}{\sqrt{N}} \sum_{m=1}^{N} X_k[m] \exp\left[2j\pi m \frac{n}{N}\right]$$ (5.5)

By applying the Fourier transform, the equivalent received signal in the frequency domain can be obtained,

$$\begin{bmatrix} R_k[0] \\ \vdots \\ R_k[N-1] \end{bmatrix} = \begin{bmatrix} X_k[0] & 0 & \cdots & 0 \\ 0 & X_k[1] & \cdots & 0 \\ \vdots & & \ddots & \vdots \\ 0 & 0 & \cdots & X_k[N-1] \end{bmatrix} \begin{bmatrix} H[0] \\ H[1] \\ \vdots \\ H[N-1] \end{bmatrix} + \begin{bmatrix} Z_k[0] \\ \vdots \\ Z_k[N-1] \end{bmatrix}$$

In summary, the CP of OFDM changes the linear convolution into a circular one. The circular convolution is very efficiently transformed by means of an FFT into a multiplicative operation in the frequency domain. Hence, the transmitted signal over a frequency-selective (i.e. multipath) channel is converted into a transmission over N parallel flat-fading channels in the frequency domain:

$$R_k[m] = X_k[m] \cdot H[m] + Z_k[m]$$ (5.6)

As a result the equalization is much simpler than for single-carrier systems and consists of just one complex multiplication per subcarrier.

5.2.2 Peak-to-Average Power Ratio and Sensitivity to Nonlinearity

In the previous section, the advantages of OFDM have been shown. By contrast, this section highlights some of the main disadvantages of OFDM. One of the major drawbacks is that the OFDM signal has a high Peak-to-Average Power Ratio (PAPR).

In the general case, the OFDM transmitter can be seen as a linear transform performed over a large block of independently identically distributed (i.i.d) QAM-modulated complex symbols (in the frequency domain). From the central limit theorem [8, 9], the time-domain OFDM symbol may be approximated as a Gaussian waveform. The amplitude variations of the OFDM modulated signal can therefore be very high. However, practical Power Amplifiers (PAs) of RF transmitters are linear only within a limited dynamic range. Thus, the OFDM signal is likely to suffer from non-linear distortion caused by clipping. This gives rise to out-of-band spurious emissions and in-band corruption of the signal. To avoid such distortion, the PAs have to operate with large power back-offs, leading to inefficient amplification and/or expensive transmitters.

The PAPR is one measure of the high dynamic range of the input amplitude (and hence a measure of the expected degradation). To analyse the PAPR mathematically, let $x[n]$ be the signal after IFFT as given by Equation (5.5) where the time index k can be dropped without loss of generality. The PAPR of an OFDM symbol is defined as the square of the peak amplitude divided by the mean power, i.e.

$$\text{PAPR} = \frac{\max_n\{|x[n]|^2\}}{E\{|x[n]|^2\}}$$ (5.7)

Under the hypothesis that the Gaussian approximation is valid, the amplitude of $x[n]$ has a Rayleigh distribution, while the power has a central chi-square distribution with two degrees of freedom. The Cumulative Distribution Function (CDF) $F_x(\alpha)$ of the normalized power is given by

$$F_x(\alpha) = \Pr\left(\frac{|x[n]|^2}{E\{|x[n]|^2\}} < \alpha\right) = 1 - e^{-\alpha} \tag{5.8}$$

If there is no oversampling, the time-domain samples are mutually uncorrelated and the probability that the PAPR[1] is above a certain threshold PAPR_0 is given by

$$\Pr(\text{PAPR} > \text{PAPR}_0) = 1 - F(\text{PAPR}_0)^N = 1 - (1 - e^{-\text{PAPR}_0})^N \tag{5.9}$$

Figure 5.6 plots the distribution of the PAPR given by Equation (5.9) for different values of the number of subcarriers N. The figure shows that a high PAPR does not occur very often. However, when it does occur, degradation due to PA non-linearities may be expected.

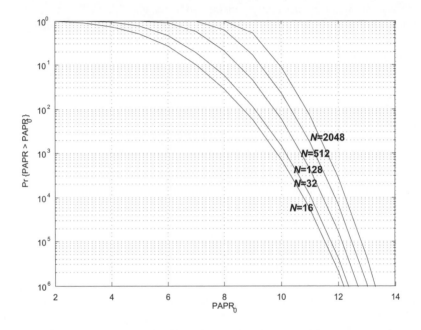

Figure 5.6 PAPR distribution for different numbers of OFDM subcarriers.

In order to evaluate the impacts of distortion on the OFDM signal reception, in Chapter 22 a useful framework for modelling non-linearities is developed.

5.2.2.1   PAPR Reduction Techniques

Many techniques have been studied for reducing the PAPR of a transmitted OFDM signal. Although no such techniques are specified for the LTE downlink signal generation, an overview of the possibilities is provided below. In general in LTE the cost and complexity

---

[1]Note that the CDF of the PAPR is $F_{\text{PAPR}}(\eta) = [F_x(\eta)]^N$ for N i.i.d. samples.

of generating the OFDM signal with acceptable Error Vector Magnitude (EVM) is left to the eNodeB implementation. As OFDM is not used for the LTE uplink (see Section 15.2), such considerations do not directly apply to the transmitter in the UE.

Techniques for PAPR reduction of OFDM signals can be broadly categorized into three main concepts:

- **Clipping and filtering [10–12].** The time-domain signal is clipped to a predefined level. This causes spectral leakage into adjacent channels, resulting in reduced spectral efficiency as well as in-band noise degrading the bit error rate performance. Out-of-band radiation caused by the clipping process can, however, be reduced by filtering. If discrete signals are clipped directly, the resulting clipping noise will all fall in band and thus cannot be reduced by filtering. To avoid this problem, one solution consists of oversampling the original signal by padding the input signal with zeros and processing it using a longer IFFT. The oversampled signal is clipped and then filtered to reduce the out-of-band radiation.

- **Selected mapping [13].** Multiple transmit signals which represent the same OFDM data symbol are generated by multiplying the OFDM symbol by different phase vectors. The representation with the lowest PAPR is selected. To recover the phase information, it is of course necessary to use separate control signalling to indicate to the receiver which phase vector was used.

- **Coding techniques [14, 15].** These techniques consist of finding the code words with the lowest PAPR from a set of codewords to map the input data. A look-up table may be used if N is small. It is shown that complementary codes have good properties to combine both PAPR and forward error correction.

The latter two concepts are not applicable in the context of LTE; selected mapping would require additional signalling, while techniques based on codeword selection are not compatible with the data scrambling used in the LTE downlink.

5.2.3 Sensitivity to Carrier Frequency Offset and Time-Varying Channels

The orthogonality of OFDM relies on the condition that transmitter and receiver operate with exactly the same frequency reference. If this is not the case, the perfect orthogonality of the subcarriers is lost, causing subcarrier leakage, also known as Inter-Carrier Interference (ICI), as can be seen in Figure 5.7.

Frequency errors typically arise from a mismatch between the reference frequencies of the transmitter and the receiver local oscillators. On the receiver side in particular, due to the importance of using low-cost components in the mobile handset, local oscillator frequency drifts are usually greater than in the eNodeB and are typically a function of parameters such as temperature changes and voltage variation. This difference between the reference frequencies is widely referred to as Carrier Frequency Offset (CFO). Phase noise in the UE receiver may also result in frequency errors.

The CFO can be several times larger than the subcarrier spacing. It is usually divided into an integer part and a fractional part. Thus the frequency error can be written as

$$f_0 = (\Gamma + \epsilon)\Delta f \tag{5.10}$$

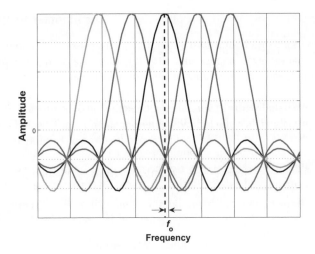

Figure 5.7 Loss of orthogonality between OFDM subcarriers due to frequency offset.

where Δf is the subcarrier spacing, Γ is an integer and $-0.5 < \epsilon < 0.5$. If $\Gamma \neq 0$, then the modulated data are in the wrong positions with respect to the subcarrier mapping performed at the transmitter. This simply results in a Bit Error Rate (BER) of 0.5 if the frequency offset is not compensated at the receiver independently of the value of ϵ. In the case of $\Gamma = 0$ and $\epsilon \neq 0$, the perfect subcarrier orthogonality is lost, resulting in ICI which can degrade the BER. Typically only synchronization errors of up to a few percent of the subcarrier spacing are tolerable in OFDM systems.

Even in an ideal case where the local oscillators are perfectly aligned, the relative speed between transmitter and receiver also generates a frequency error due to Doppler.

In the case of a single-path channel, UE mobility in a constant direction with respect to the angle of arrival of the signal results in a Doppler shift f_d, while in a scattering environment this becomes a Doppler spread with spectral density $P(f)$ as discussed further in Section 8.3.1.

It can be shown [16, 17] that, for both flat and dispersive channels, the ICI power can be computed as a function of the generic Doppler spectral density $P(f)$ as follows:

$$P_{\mathrm{ICI}} = \int_{-f_{d\mathrm{max}}}^{f_{d\mathrm{max}}} P(f)(1 - \mathrm{sinc}^2(T_s f))\, \mathrm{d}f \qquad (5.11)$$

where $f_{d\mathrm{max}}$ is the maximum Doppler frequency, and the transmitted signal power is normalized.

ICI resulting from a mismatch f_o between the transmitter and receiver oscillator frequencies can be modelled as a Doppler shift arising from single-path propagation:

$$P(f) = \delta(f - f_o) \qquad (5.12)$$

Hence, substituting (5.12) into (5.11), the ICI power in the case of a deterministic CFO is given by

$$P_{\mathrm{ICI,CFO}} = 1 - \mathrm{sinc}^2(f_o T_s) \qquad (5.13)$$

For the classical Jakes model of Doppler spread (see Section 8.3.1), the expression (5.11) can be written as

$$P_{\text{ICI,Jakes}} = 1 - 2 \int_0^1 (1 - f) J_0(2\pi f_{d_{\max}} T_s f) \, df \tag{5.14}$$

where J_0 is the zero-th order Bessel function.

When no assumptions on the shape of the Doppler spectrum can be made, an upper bound on the ICI given by Equation (5.11) can be found by applying the Cauchy–Schwartz inequality, leading to [17]

$$P_{\text{ICI}} \leq \frac{\int_0^1 [1 - \text{sinc}^2(f_d T_s f)]^2 \, df}{\int_0^1 1 - \text{sinc}^2(f_d T_s f) \, df} \tag{5.15}$$

This upper bound on P_{ICI} is valid only in the case of frequency spread and does not cover the case of a deterministic CFO.

Using Equations (5.15), (5.14) and (5.13), the SIR in the presence of ICI can be expressed as

$$SIR_{\text{ICI}} = \frac{1 - P_{\text{ICI}}}{P_{\text{ICI}}} \tag{5.16}$$

Figures 5.8 and 5.9 plot these P_{ICI} and SIR_{ICI} for the cases provided. These figures show that the highest ICI is introduced by a constant frequency offset. In the case of a Doppler spread, the ICI impairment is lower. Figure 5.9 shows that, in the absence of any other impairment such as interference ($SIR = \infty$), the SIR_{ICI} rapidly decays as a function of frequency misalignments.

The sensitivity of the BER depends on the modulation order. It is shown in reference [18] that QPSK modulation can tolerate up to $\epsilon_{\max} = 0.05$ whereas 64-QAM requires $\epsilon \leq 0.01$.

5.2.4 Timing Offset and Cyclic Prefix Dimensioning

In the case of a memoryless channel (i.e. no delay spread), OFDM is insensitive to timing synchronization errors provided that the misalignment remains within the cyclic prefix duration. In other words, if $T_o \leq T_{\text{CP}}$ (with T_o being the timing error), then orthogonality is maintained thanks to the cyclic nature of the CP. Any symbol timing delay only introduces a constant phase shift from one subcarrier to another. The received signal at the mth subcarrier is given by

$$R_k[m] = X_k[m] \exp\left(j2\pi \frac{dm}{N} \right) \tag{5.17}$$

where d is the timing offset in samples corresponding to a duration equal to T_o.

This phase shift can be recovered as part of the channel estimation operation. It is worth highlighting that the insensitivity to timing offsets would not hold for any kind of guard period other than a cyclic prefix; for example, zero-padding would not exhibit the same property, resulting in a portion of the useful signal power being lost.

In the general case of a channel with delay spread, for a given CP length the maximum tolerated timing offset without degrading the OFDM reception is reduced by an amount equal to the length of the channel impulse response: $T_o \leq T_{\text{CP}} - T_d$ as shown in Figure 5.10. For greater timing errors, ISI and ICI occur. The effect caused by an insufficient CP is discussed

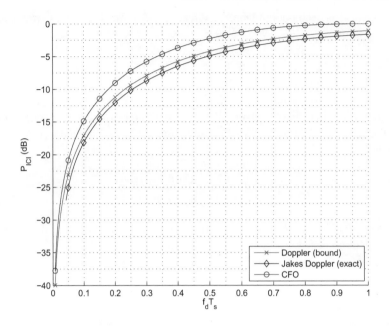

Figure 5.8 P_{ICI} for the case of a classical Doppler distribution and a deterministic CFO.

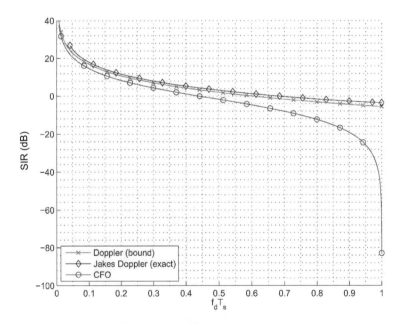

Figure 5.9 SIR_{ICI} for classical Doppler distribution and deterministic CFO.

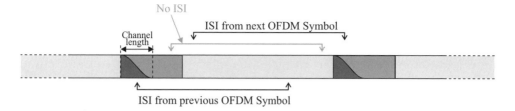

Figure 5.10 OFDM sensitivity to timing offsets.

in the following section. Timing synchronization hence becomes more critical in long-delay-spread channels.

Initial timing acquisition in LTE is normally achieved by the cell-search and synchronization procedures (see Chapter 7). Thereafter, for continuous tracking of the timing-offset, two classes of approach exist, based on either CP correlation or Reference Signals (RSs). A combination of the two is also possible. The reader is referred to [19] for a comprehensive survey on OFDM synchronization techniques.

5.2.4.1 Effect of Insufficient Cyclic Prefix Length

As already introduced before, if an OFDM system is designed with a CP of length G samples such that $L < G$ where L is the length of the channel impulse response (in number of samples), the system benefits from turning the linear convolution into a circular one to keep the subcarriers orthogonal. The condition of a sufficient CP is therefore strictly related to the orthogonality property of OFDM.

As shown in [20], for an OFDM symbol consisting of $N + G$ samples where N is the FFT size, the power of the ICI and ISI can be computed by

$$P_{\text{ICI}} = 2 \sum_{k=G}^{N+G-1} |h[k]|^2 \frac{N(k - G) - (k - G)^2}{N^2} \tag{5.18}$$

$$P_{\text{ISI}} = \sum_{k=G}^{N+G-1} |h[k]|^2 \frac{(k - G)^2}{N^2} \tag{5.19}$$

Conversely, the signal power P_S is reduced and can be written as

$$P_S = \sum_{k=0}^{G-1} |h(k)|^2 + \sum_{k=G}^{N+G-1} |h(k)|^2 \frac{(N - k + G)^2}{N^2} \tag{5.20}$$

The resulting SIR due to the CP being too short can then be written as

$$SIR_{\text{og}} = \frac{P_S}{P_{\text{ISI}} + P_{\text{ICI}}} \tag{5.21}$$

Figures 5.11 and 5.12 plot Equations (5.19) to (5.21) for the case of the normal CP length in LTE assuming a channel with a uniform and normalized power-delay profile of length $L < N + G$, where the dashed line marks the boundary of the cyclic prefix ($L = G$).

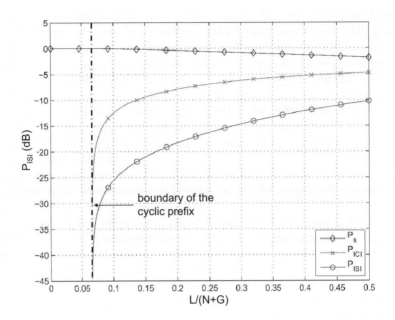

Figure 5.11 Power of signal, ICI and ISI in case of a too-short guard interval.

5.3 OFDMA

Orthogonal Frequency Division Multiple Access (OFDMA) is an extension of OFDM to the implementation of a multiuser communication system. In the discussion above, it has been assumed that a single user receives data on all the subcarriers at any given time. OFDMA distributes subcarriers to different users at the same time, so that multiple users can be scheduled to receive data simultaneously. Usually, subcarriers are allocated in contiguous groups for simplicity and to reduce the overhead of indicating which subcarriers have been allocated to each user.

OFDMA for mobile communications was first proposed in [21] based on multicarrier FDMA (Frequency Division Multiple Access), where each user is assigned to a set of randomly selected subchannels.

OFDMA enables the OFDM transmission to benefit from multiuser diversity, as discussed in Chapter 12. Based on feedback information about the frequency-selective channel conditions from each user, adaptive user-to-subcarrier assignment can be performed, enhancing considerably the total system spectral efficiency compared to single-user OFDM systems.

OFDMA can also be used in combination with Time Division Multiple Access (TDMA), such that the resources are partitioned in the time-frequency plane – i.e. groups of subcarriers for a specific time duration. In LTE, such time-frequency blocks are known as Resource Blocks (RBs), as explained in Section 6.2. Figure 5.13 depicts such an OFDMA/TDMA mixed strategy as used in LTE.

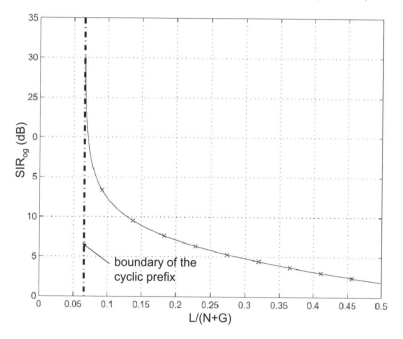

Figure 5.12 Effective SIR as a function of channel impulse response length for a given CP length.

5.3.1 Parameter Dimensioning

As highlighted in the previous sections, certain key parameters determine the performance of OFDM and OFDMA systems. Inevitably, some compromises have to be made in defining these parameters appropriately to maximize system spectral efficiency while maintaining robustness against propagation impairments.

For a given system, the main propagation characteristics which should be taken into account when designing an OFDM system are the expected delay spread T_d, the maximum Doppler frequency $f_{d_{max}}$, and, in case of cellular systems, the targeted cell size.

The propagation characteristics impose constraints on the choice of the CP length and of the subcarrier spacing.

As already mentioned, the CP should be longer than the channel impulse response in order to ensure robustness against ISI. For cellular systems, and especially for large cells, longer delay spreads may typically be experienced than those encountered, for example, in WLAN systems, implying the need for a longer CP. On the other hand, a longer CP for a given OFDM symbol duration corresponds to a larger overhead in terms of energy per transmitted bit. Out of the $N + G$ transmitted symbols, only N convey information, leading to a rate loss. This reduction in bandwidth efficiency can be expressed as a function of the CP duration $T_{CP} = GT_s$ and the OFDM symbol period $T_u = NT_s$ (where T_s is the sampling period), as follows:

$$\beta_{overhead} = \frac{T_{CP}}{T_u + T_{CP}} \tag{5.22}$$

■ User 1   ■ User 2   ■ User 3   □ User 4

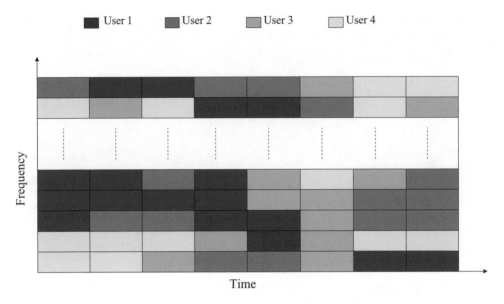

Figure 5.13 Example of resource allocation in a combined OFDMA/TDMA system.

It is clear that to maximize spectral efficiency, T_u should be chosen to be large relative to the CP duration, but small enough to ensure that the channel does not vary within one OFDM symbol.

Further, the OFDM symbol duration T_u is related to the subcarrier spacing by $\Delta f = 1/T_u$. Choosing a large T_u leads to a smaller subcarrier separation Δf, which has a direct impact on the system sensitivity to Doppler and other sources of frequency offset, as explained in Section 5.2.3.

Thus, in summary, the following three design criteria can be identified:

$$T_{CP} \geq T_d \qquad \text{to prevent ISI,}$$

$$\frac{f_{d_{max}}}{\Delta f} \ll 1 \qquad \text{to keep ICI due to Doppler sufficiently low,} \qquad (5.23)$$

$$T_{CP}\Delta f \ll 1 \qquad \text{for spectral efficiency.}$$

5.3.2 Physical Layer Parameters for LTE

LTE aims at supporting a wide range of cellular deployment scenarios, including indoor, urban, suburban and rural situations covering both low and high User Equipment (UE) mobility conditions (up to 350 or even 500 km/h). The cell sizes may range from home-networks only a few metres wide to large cells with radii of tens of kilometres.

Deployed carrier frequencies are likely to range from 400 MHz to 4 GHz, with bandwidths ranging from 1.4 to 20 MHz. All these cases imply different delay spreads and Doppler frequencies.

The 'normal' parameterization of the LTE downlink uses a $\Delta f = 15$ kHz subcarrier spacing with a CP length of approximately 5 µs. This subcarrier spacing is a compromise between the percentage overhead of the CP and the sensitivity to frequency offsets. A 15 kHz subcarrier spacing is sufficiently large to allow for high mobility and to avoid the need for closed-loop frequency adjustments.

In addition to the normal parameterization, it is possible to configure LTE with an extended CP of length approximately 17 µs.[2] This is designed to ensure that even in large suburban and rural cells, the delay spread should be contained within the CP. This, however, comes at the expense of a higher overhead from the CP as a proportion of the total system transmission resources.

LTE is also designed to support a multi-cell broadcast transmission mode known as Multimedia Broadcast Single Frequency Network (MBSFN), in which the UE receives and combines synchronized signals from multiple cells. In this case the relative timing offsets from the multiple cells must all be received at the UE's receiver within the CP duration, if ISI is to be avoided, thus requiring a significantly longer CP. In order to avoid a further overhead, the sub-carrier spacing is halved in this case, to allow the OFDM symbol length to be doubled at the expense of increasing the sensitivity to mobility and frequency errors. An extended CP of length of approximately 33 µs can therefore be provided while remaining $\frac{1}{4}$ of OFDM symbol length. The MBSFN transmission mode is discussed in more detail in Chapter 14.

These modes and their corresponding parameters are summarized in Figure 5.14. It is worth noticing that when LTE is configured with the normal CP length, the CP length for the first OFDM symbol in each 0.5 ms interval is slightly longer than that of the next six OFDM symbols. This characteristic is due to the need to accommodate an integer number of OFDM symbols, namely 7, into each 0.5 ms interval, with assumed FFT block-lengths of 2048.

The actual FFT size and sampling frequency for the LTE downlink are not specified. However, the above parameterizations are designed to be compatible with a sampling frequency of 30.72 MHz. Thus, the basic unit of time in the LTE specifications, of which all other time periods are a multiple, is defined as $T_s = 1/30.72$ µs. This is itself chosen for backward compatibility with UMTS, for which the chip rate is 3.84 MHz – exactly one eighth of the assumed LTE sampling frequency.

In the case of a 20 MHz system bandwidth, an FFT order of 2048 may be assumed for efficient implementation. However, in practice the implementer is free to use other Discrete Fourier Transform sizes.

Lower sampling frequencies (and proportionally lower FFT orders) are always possible to reduce RF and baseband processing complexity for narrower bandwidth deployments: for example, for a 5 MHz system bandwidth the FFT order and sampling frequency could be scaled down to 512 and $f_s = 7.68$ MHz respectively, while only 300 subcarriers are actually modulated with data.

For the sake of simplifying terminal implementation, the direct current (d.c.) subcarrier is left unused, in order to avoid d.c. offset errors in direct conversion receivers. Figure 5.15 graphically depicts the subcarrier allocations in a typical LTE deployment.

The OFDMA parameters used in the downlink are defined in the 3GPP Technical Specification in [22].

---

[2]The length of the extended CP is $\frac{1}{4}$ of the OFDM symbol.

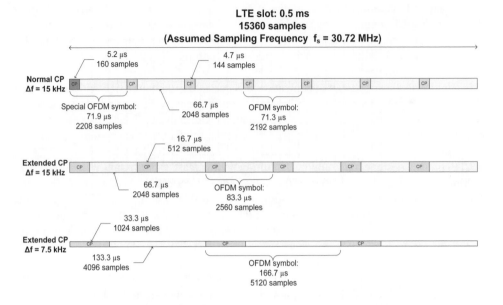

Figure 5.14 LTE OFDM symbol and cyclic prefix lengths.

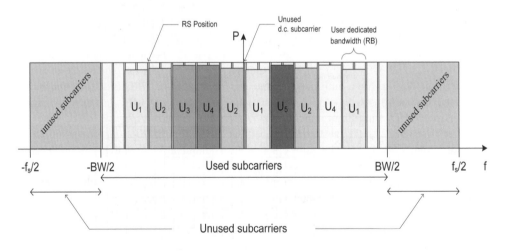

Figure 5.15 LTE OFDMA spectrum allocation.

5.4 Conclusion

In this chapter we have reviewed the key features, benefits and sensitivities of OFDM and
OFDMA systems. In summary, it can be noted that:

- OFDM is a mature technology.

- It is already widely deployed and is especially suited for broadcast or downlink applications because of the low receiver complexity while requiring a high transmitter complexity (expensive PA).

- It benefits from efficient implementation by means of the FFT.

- It achieves high transmission rates of broadband transmission, with low receiver complexity.

- It makes use of a CP to avoid ISI, enabling block-wise processing.

- It exploits orthogonal subcarriers to avoid the spectrum wastage associated with inter-subcarrier guard-bands.

- The parameterization allows the system designer to balance tolerance of Doppler and delay spread depending on the deployment scenario.

- It can be extended to a multiple-access scheme, OFDMA, in a straightforward manner.

These factors together have made OFDMA the technology of choice for the LTE downlink.

References[3]

[1] R. W. Chang, 'Synthesis of Band-limited Orthogonal Signals for Multichannel Data Transmission'. *Bell Systems Technical Journal*, Vol. 46, pp. 1775–1796, December 1966.

[2] B. R. Saltzberg, 'Performance of an Efficient Parallel Data Transmission System'. *IEEE Trans. on Communications*, Vol. 15, pp. 805–811, December 1967.

[3] S. B. Weinstein and P. M. Ebert, 'Data Transmission by Frequency-Division Multiplexing using the Discrete Fourier Transform'. *IEEE Trans. on Communications*, Vol. 19, pp. 628–634, October 1971.

[4] A. Peled and A. Ruiz, 'Frequency Domain Data Transmission using Reduced Computational Complexity Algorithms' in *Proc. IEEE International Conference on Acoustics, Speech and Signal Processing*, Vol. 5, pp. 964–967, April 1980.

[5] L. J. Cimini, 'Analysis and Simulation of Digital Mobile Channel using Orthogonal Frequency Division Multiplexing'. *IEEE Trans. on Communications*, Vol. 33, pp. 665–675, July 1985.

[6] G. H. Golub and C. F. Van Loan, *Matrix Computations*. John Hopkins University Press, 1996.

[7] R. A. Horn and C. R. Johnson, *Matrix Analysis*. Cambridge University Press, 1990.

[8] S. N. Bernstein, 'On the Work of P. L. Chebyshev in Probability Theory'. *The Scientific Legacy of P. L. Chebyshev. First Part: Mathematics*, edited by S. N. Bernstein. Academiya Nauk SSSR, Moscow-Leningrad, p. 174, 1945.

[9] T. Henk, *Understanding Probability: Chance Rules in Everyday Life*. Cambridge University Press, 2004.

[10] X. Li and L. J. Cimini, 'Effects of Clipping and Filtering on the Performance of OFDM'. *IEEE Comm. Lett.*, Vol. 2, pp. 131–133, May 1998.

---

[3] All web sites confirmed 18th December 2008.

[11] L. Wane and C. Tellambura, 'A Simplified Clipping and Filtering Technique for PAPR Reduction in OFDM Systems'. *IEEE Sig. Proc. Lett.*, Vol. 12, pp. 453–456, June 2005.

[12] J. Armstrong, 'Peak to Average Power Reduction for OFDM by Repeated Clipping and Frequency Domain Filtering'. *Electronics Letters*, Vol. 38, pp. 246–247, February 2002.

[13] A. Jayalath, 'OFDM for Wireless Broadband Communications (Peak Power Reduction, Spectrum and Coding)'. PhD thesis, School of Computer Science and Software Engineering, Monash University, 2002.

[14] A. Jones, T. Wilkinson, and S. Barton, 'Block Coding Scheme for Reduction of Peak to Mean Envelope Power Ratio of Multicarrier Transmission Schemes'. *Electronics Letters*, Vol. 30, pp. 2098–2099, December 1994.

[15] C. Tellambura, 'Use of M-sequence for OFDM Peak-to-Average Power Ratio Reduction'. *Electronics Letters*, Vol. 33, pp. 1300–1301, July 1997.

[16] Y. Li and L. J. Cimini, 'Bounds on the Interchannel Interference of OFDM in Time-Varying Impairments'. *IEEE Trans. on Communications*, Vol. 49, pp. 401–404, March 2001.

[17] X. Cai and G. B. Giannakis, 'Bounding Performance and Suppressing Intercarrier Interference in Wireless Mobile OFDM'. *IEEE Trans. on Communications*, Vol. 51, pp. 2047–2056, March 2003.

[18] J. Heiskala and J. Terry, *OFDM Wireless LANs: A Theoretical and Practical Guide*. SAMS Publishing, 2001.

[19] L. Hanzo and T. Keller, *OFDM And MC-CDMA: A Primer*. Wiley-IEEE Press, 2006.

[20] A. G. Burr, 'Irreducible BER of COFDM on IIR channel'. *Electronics Letters*, Vol. 32, pp. 175–176, February 1996.

[21] R. Nogueroles, M. Bossert, A. Donder, and V. Zyablov, 'Improved Performance of a Random OFDMA Mobile Communication System' in *Proc. IEEE Vehicular Technology Conference*, Vol. 3, pp. 2502–2506, May 1998.

[22] 3GPP Technical Specification 36.321, 'Evolved Universal Terrestrial Radio Access (E-UTRA); Physical Channels and Modulation (Release 8)', www.3gpp.org.

6

Introduction to Downlink Physical Layer Design

Matthew Baker

6.1 Introduction

The LTE downlink transmissions from the eNodeB consist of user-plane and control-plane data from the higher layers in the protocol stack (as described in Chapters 3 and 4) multiplexed together with physical layer signalling to support the data transmission. The multiplexing of these various parts of the downlink signal is facilitated by the Orthogonal Frequency Division Multiple Access (OFDMA) structure described in Chapter 5, which enables the downlink signal to be subdivided into small units of time and frequency.

This subdivided structure is introduced below, together with an outline of the general steps in forming the transmitted downlink signal in the physical layer.

6.2 Transmission Resource Structure

The downlink transmission resources in LTE possess dimensions of time, frequency and space. The spatial dimension, measured in 'layers', is accessed by means of multiple antenna transmission and reception; the techniques for exploiting multiple spatial layers are explained in Chapter 11.

The time-frequency resources are subdivided according to the following structure: the largest unit of time is the 10 ms radio frame, which is further subdivided into ten 1 ms subframes, each of which is split into two 0.5 ms slots. Each slot comprises seven OFDM symbols in the case of the normal cyclic prefix length, or six if the extended cyclic prefix is configured in the cell (see Section 5.3.2). In the frequency domain, resources are grouped in units of 12 subcarriers (thus occupying a total of 180 kHz), such that one unit of 12 subcarriers for a duration of one slot is termed a Resource Block (RB).[1]

The smallest unit of resource is the Resource Element (RE), which consists of one subcarrier for a duration of one OFDM symbol. A resource block is thus comprised of 84 resource elements in the case of the normal cyclic prefix length, and 72 resource elements in the case of the extended cyclic prefix.

The detailed resource structure is shown in Figure 6.1 for the normal cyclic prefix length.

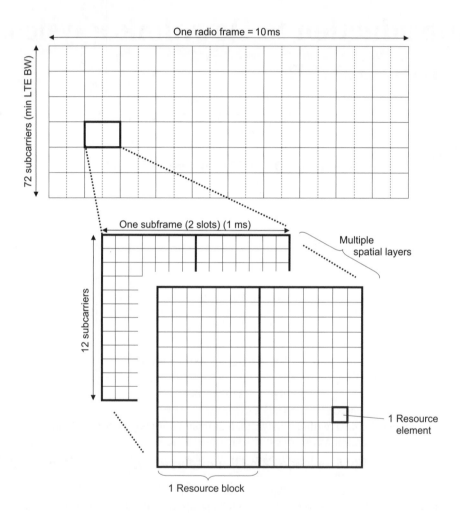

Figure 6.1 Basic time-frequency resource structure of LTE (normal cyclic prefix case).

---

[1]In the case of the 7.5 kHz subcarrier spacing which may be available for Multimedia Broadcast Multicast Service (MBMS) transmission in later releases of LTE (see Section 5.3.2), one RB consists of 24 subcarriers for a duration of one slot.

Within certain resource blocks, some resource elements are reserved for special purposes: synchronization signals (Chapter 7), reference signals (Chapter 8), control signalling and critical broadcast system information (Chapter 9). The remaining resource elements are used for data transmission, and are usually allocated in pairs of RBs (the pairing being in the time domain).

The structure shown in Figure 6.1 assumes that all subframes are available for downlink transmission. This is known as 'Frame Structure Type 1' and is applicable for Frequency Division Duplexing (FDD) in paired radio spectrum, or for a standalone downlink carrier. For Time Division Duplexing (TDD) in unpaired spectrum, the basic structure of RBs and REs remains the same, but only a subset of the subframes are available for downlink transmission; the remaining subframes are used for uplink transmission or for special subframes which allow for switching between downlink and uplink transmission. In the centre of the special subframes a guard period is provided which allows the uplink transmission timing to be advanced as described in Section 20.2. This TDD structure is known as 'Frame Structure Type 2' and is illustrated in Figure 6.2 together with details of the various possible configurations of downlink and uplink subframes. Further details of TDD operation using this frame structure are described in Chapter 23.

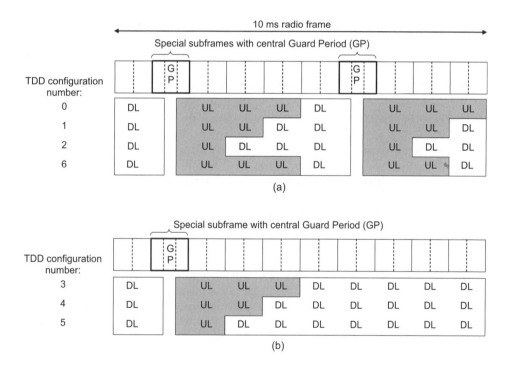

Figure 6.2 LTE subframe structure for TDD operation: (a) configurations with 5 ms periodicity of switching from downlink (DL) to uplink (UL); (b) configurations with 10 ms periodicity of switching from downlink (DL) to uplink (UL).

6.3   Signal Structure

The role of the physical layer is primarily to translate data into a reliable signal for transmission across the radio interface between the eNodeB and the User Equipment (UE). Each block of data is first protected against transmission errors, usually first with a Cyclic Redundancy Check (CRC), and then with channel coding; these aspects are explained in detail in Chapter 10. After channel coding, the steps in the formation of the downlink LTE signal are illustrated in Figure 6.3.

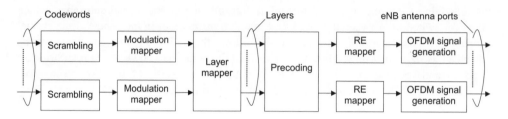

Figure 6.3 General signal structure for LTE downlink. Reproduced by permission of © 3GPP.

The initial scrambling stage is applied to all downlink physical channels, and serves the purpose of interference rejection. The scrambling sequence in all cases uses an order-31 Gold code, which can provide 2^{31} sequences which are not cyclic shifts of each other. Gold codes [1, 2] also possess the attractive feature that they can be generated with very low implementation complexity, as they can be derived from the modulo-2 addition of two maximum-length sequences (otherwise known as *M-sequences*), which can be generated from a simple shift-register.[2] A shift-register implementation of the LTE scrambling sequence generator is illustrated in Figure 6.4.

Figure 6.4 Shift-register implementation of scrambling sequence generator.

---

[2]Gold Codes were also used in WCDMA (Wideband Code Division Multiple Access), for the uplink long scrambling codes.

The scrambling sequence generator is re-initialized every subframe (except for the physical broadcast channel which is discussed in Section 9.2.1), based on the identity of the cell, the subframe number (within a radio frame) and the UE identity. This randomizes interference between cells and between UEs. In addition, in cases where multiple data streams (codewords) are transmitted via multiple layers, the identity of the codeword is also used in the initialization.

As a useful feature for avoiding unnecessary complexity, the scrambling sequence generator described here is the same as for the pseudo-random sequence used for the reference signals as described in Chapter 8, the only difference being in the method of initialization; in all cases, however, a fast-forward of 1600 places is applied at initialization, in order to ensure low cross-correlation between sequences used in adjacent cells.

Following the scrambling stage, the data bits from each channel are mapped to complex-valued modulation symbols depending on the relevant modulation scheme, then mapped to layers, precoded as explained in Chapter 11, mapped to resource elements, and finally translated into a complex-valued OFDM signal by means of an IFFT.

6.4   Introduction to Downlink Operation

In order to communicate with an eNodeB supporting one or more cells, the UE must first identify the downlink transmission from one of these cells and synchronize with it. This is achieved by means of special synchronization signals which are embedded into the OFDM structure described above. The procedure for cell search and synchronization is described in the next chapter.

The next step for the UE is to estimate the downlink radio channel in order to be able to perform coherent demodulation of the information-bearing parts of the downlink signal.

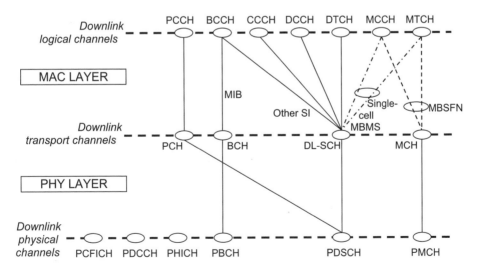

Figure 6.5 Summary of downlink physical channels and mapping to higher layers.

Some suitable techniques for this process are described in Chapter 8, based on the pilot signals (known in LTE as *reference signals*) which are inserted into the downlink signal.

Chapter 9 describes the parts of the downlink signal which carry data originating from higher protocol layers, including the Physical Broadcast Channel (PBCH), the Physical Downlink Shared Channel (PDSCH) and, in case of MBMS transmission, the Physical Multicast Channel (PMCH). In addition, the design of the downlink control signalling is explained, including its implications for the ways in which downlink transmission resources may be allocated to different users for data transmission.

The downlink physical channels described in the following chapters are summarized in Figure 6.5, together with their relationship to the higher-layer channels.

The subsequent chapters explain the key techniques which enable these channels to make efficient use of the radio spectrum: channel coding and link adaptation are explained in Chapter 10, the LTE schemes for exploiting multiple antennas are covered in Chapter 11, and techniques for effective scheduling of transmission resources to multiple users are described in Chapter 12.

References

[1] W. Huffman and V. Pless, *Handbook of Coding Theory, vol II*. Amsterdam: North-Holland, 1998.

[2] R. McEliece, *Finite Fields for Computer Scientists and Engineers*. Boston, MA: Kluwer Academic Publishers, 2003.

7

Synchronization and Cell Search

Fabrizio Tomatis and Stefania Sesia

7.1 Introduction

A UE wishing to access an LTE cell must first undertake a *cell search procedure*. This consists of a series of synchronization stages by which the UE determines time and frequency parameters that are necessary to demodulate the downlink and to transmit uplink signals with the correct timing. The UE also acquires some critical system parameters.

Three major synchronization requirements can be identified in the LTE system: the first is symbol timing acquisition, by which the correct symbol start position is determined, for example to set the FFT window position; the second is carrier frequency synchronization, which is required to reduce or eliminate the effect of frequency errors[1] arising from a mismatch of the local oscillators between the transmitter and the receiver, as well as the Doppler shift caused by any UE motion; thirdly, sampling clock synchronization is necessary.

7.2 Synchronization Sequences and Cell Search in LTE

Two relevant cell search procedures exist in LTE:

- **Initial synchronization**, whereby the UE detects an LTE cell and decodes all the information required to register to it. This would be required, for example, when the UE is switched on, or when it has lost the connection to the serving cell.

- **New cell identification**, performed when a UE is already connected to an LTE cell and is in the process of detecting a new neighbour cell. In this case, the UE reports to the serving cell measurements related to the new cell, in preparation for handover.

---

[1]Frequency offsets may arise from factors such as temperature drift, ageing and imperfect calibration.

LTE – The UMTS Long Term Evolution: from Theory to Practice   Stefania Sesia, Issam Toufik and Matthew Baker
© 2009 John Wiley & Sons, Ltd

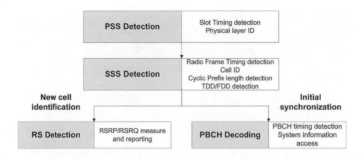

Figure 7.1 Information acquired at each step of the cell search procedure.

This cell search procedure is repeated periodically until either the serving cell quality becomes satisfactory again, or the UE moves to another serving cell as explained in Chapter 13.

In both scenarios, the synchronization procedure makes use of two specially designed physical signals which are broadcast in each cell: the Primary Synchronization Signal (PSS) and the Secondary Synchronization Signal (SSS). The detection of these two signals not only enables time and frequency synchronization, but also provides the UE with the physical layer identity of the cell and the cyclic prefix length, and informs the UE whether the cell uses Frequency Division Duplex (FDD) or Time Division Duplex (TDD).

In the case of the initial synchronization, in addition to the detection of synchronization signals, the UE proceeds to decode the Physical Broadcast CHannel (PBCH), from which critical system information is obtained (see Section 9.2.1). In the case of new cell identification, the UE does not need to decode the PBCH; it simply makes quality-level measurements based on the reference signals transmitted from the newly-detected cell and reports these to the serving cell (see Chapter 8).

The cell search and synchronization procedure is summarized in Figure 7.1, showing the information ascertained by the UE at each stage. The PSS and SSS structure is specifically designed to facilitate this acquisition of information.

The PSS and SSS structure in time is shown in Figure 7.2 for the FDD case and in Figure 7.3 for TDD: the synchronization signals are transmitted periodically, twice per 10 ms radio frame. In an FDD cell, the PSS is always located in the last OFDM (Orthogonal Frequency Division Multiplexing) symbol of the first and 11th slots of each radio frame (see Chapter 6), thus enabling the UE to acquire the slot boundary timing independently of the Cyclic Prefix (CP) length. The SSS is located in the symbol immediately preceding the PSS, a design choice enabling coherent detection of the SSS relative to the PSS, based on the assumption that the channel coherence duration is significantly longer than one OFDM symbol. In a TDD cell, the PSS is located in the third symbol of the 3rd and 13th slots, while the SSS is located three symbols earlier; coherent detection can be used under the assumption that the channel coherence time is significantly longer than four OFDM symbols.

The precise position of the SSS changes depending on the length of the CP which is chosen for the cell. At this stage of the cell detection process, the CP length is unknown

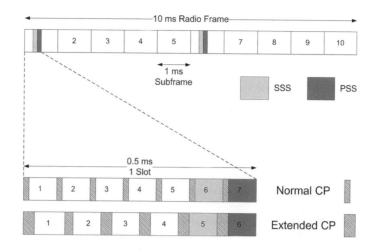

Figure 7.2 PSS and SSS frame and slot structure in time domain in the FDD case.

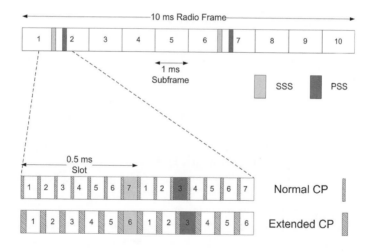

Figure 7.3 PSS and SSS frame and slot structure in time domain in the TDD case.

a priori to the UE, and it is therefore blindly detected by checking for the SSS at the two possible positions.[2]

While the PSS in a given cell is the same in every subframe in which it is transmitted, the two SSS transmissions in each radio frame change in a specific manner as described in

---

[2]Hence a total of four possible SSS positions must be checked if the UE is searching for both FDD and TDD cells.

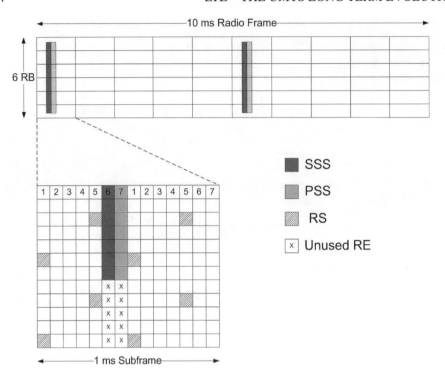

Figure 7.4 PSS and SSS frame structure in frequency and time domain for an FDD cell.

Section 7.2.3 below, thus enabling the UE to establish the position of the 10 ms radio frame boundary.

In the frequency domain, the mapping of the PSS and SSS to subcarriers is shown in Figure 7.4. The PSS and SSS are transmitted in the central six Resource Blocks[3] (RBs), enabling the frequency mapping of the synchronization signals to be invariant with respect to the system bandwidth (which can vary from 6 to 110 RBs); this allows the UE to synchronize to the network without any a priori knowledge of the allocated bandwidth. The PSS and SSS are each comprised of a sequence of length 62 symbols, mapped to the central 62 subcarriers around the d.c. subcarrier which is left unused. This means that the five resource elements at each extremity of each synchronization sequence are not used. This structure enables the UE to detect the PSS and SSS using a size-64 FFT and a lower sampling rate than would have been necessary if all 72 subcarriers were used in the central six resource blocks. The shorter length for the synchronization sequences also avoids the possibility in a TDD system of a high correlation with the uplink demodulation reference signals which use the same kind of sequence as the PSS (see Chapter 16).

In the case of multiple transmit antennas being used at the eNodeB, the PSS and SSS are always transmitted from the same antenna port[4] in any given subframe, while between

---

[3] Six Resource Blocks correspond to 72 subcarriers.
[4] The concept of an antenna port in LTE is explained in Section 8.2.

different subframes they may be transmitted from different antenna ports in order to benefit from time-switched antenna diversity.

The particular sequences which are transmitted for the PSS and SSS in a given cell are used to indicate the physical layer cell identity to the UE. There are 504 unique physical layer cell identities in LTE, grouped into 168 groups of three identities. The three identities in a group would usually be assigned to cells under the control of the same eNodeB. Three PSS sequences are used to indicate the cell identity within the group, and 168 SSS sequences are used to indicate the identity of the group.[5]

The PSS uses sequences known as *Zadoff–Chu*. This category of sequences is widely used in LTE, including for random access preambles and the uplink reference symbols in addition to the PSS. Therefore, the following section is devoted to an explanation of the fundamental principles behind these sequences, before discussing the specific constructions of the PSS and SSS sequences in the subsequent sections.

7.2.1 Zadoff–Chu Sequences

Zadoff–Chu (ZC) sequences (also known as Generalized Chirp-Like (GCL) sequences) are named after the papers [1] and [2]. They are non-binary unit-amplitude sequences [3], which satisfy a Constant Amplitude Zero Autocorrelation (CAZAC) property. CAZAC sequences are complex signals of the form $e^{j\alpha_k}$. The ZC sequence of odd-length N_{ZC} is given by

$$a_q(n) = \exp\left[-j2\pi q \frac{n(n+1)/2 + ln}{N_{ZC}}\right] \qquad (7.1)$$

where $q \in \{1, \ldots, N_{ZC} - 1\}$ is the ZC sequence root index, $n = 0, 1, \ldots, N_{ZC} - 1, l \in \mathbb{N}$ is any integer. In LTE $l = 0$ is used for simplicity.

ZC sequences have the following important properties.

Property 1. A ZC sequence has constant amplitude, and its N_{ZC}-point DFT also has constant amplitude. The constant amplitude property limits the Peak-to-Average Power Ratio (PAPR) and generates bounded and time-flat interference to other users. It also simplifies the implementation as only phases need to be computed and stored, not amplitudes.

Property 2. ZC sequences of any length have 'ideal' cyclic autocorrelation (i.e. the correlation with the circularly shifted version of itself is a delta function). The zero autocorrelation property may be formulated as:

$$r_{kk}(\sigma) = \sum_{n=0}^{N_{ZC}-1} a_k(n)a_k^*[(n+\sigma)] = \delta(\sigma) \qquad (7.2)$$

where $r_{kk}(\cdot)$ is the discrete periodic autocorrelation function of a_k at lag σ. This property is of major interest when the received signal is correlated with a reference sequence and the received reference sequences are misaligned. As an example, Figure 7.5 shows the difference between the periodic autocorrelation of a truncated Pseudo Noise (PN) sequence (as used in WCDMA [4]) and a ZC sequence. Both are 839 symbols long in this example. The ZC periodic autocorrelation is exactly zero for $\sigma \neq 0$ and it is non-zero for $\sigma = 0$, whereas the PN periodic autocorrelation shows significant peaks, some above 0.1, at non-zero lags.

---

[5]It may become possible to reserve a subset of the available physical layer cell identities for a specific use, such as to indicate that the cell is a Home eNodeB.

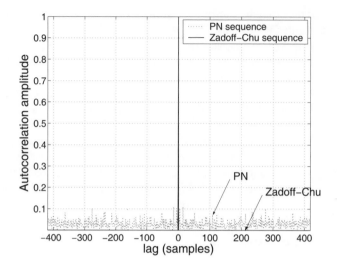

Figure 7.5 Zadoff–Chu versus PN sequence: periodic autocorrelation.

The main benefit of the CAZAC property is that it allows multiple orthogonal sequences to be generated from the same ZC sequence. Indeed, if the periodic autocorrelation of a ZC sequence provides a single peak at the zero lag, the periodic correlation of the same sequence against its cyclic shifted replica provides a peak at lag N_{CS}, where N_{CS} is the number of samples of the cyclic shift. This creates a Zero-Correlation Zone (ZCZ) between the two sequences. As a result, as long as the ZCZ is dimensioned to cope with the largest possible expected time misalignment between them, the two sequences are orthogonal for all transmissions within this time misalignment.

Property 3. The absolute value of the cyclic cross-correlation function between any two ZC sequences is constant and equal to $1/\sqrt{N_{ZC}}$, if $|q_1 - q_2|$ (where q_1 and q_2 are the sequence indices) is relatively prime with respect to N_{ZC} (a condition that can be easily guaranteed if N_{ZC} is a prime number). The cross-correlation of $1/\sqrt{N_{ZC}}$ at all lags achieves the theoretical minimum cross-correlation value for any two sequences that have ideal autocorrelation.

Selecting N_{ZC} as a prime number results in $N_{ZC} - 1$ ZC sequences which have the optimal cyclic cross-correlation between any pair. However, it is not always convenient to use sequences of prime length. In general, a sequence of non-prime length may be generated by either cyclic extension or truncation of a prime-length ZC sequence. Cyclic extension or truncation preserves both the constant amplitude property and the zero cyclic autocorrelation property for different cyclic shifts.

A further useful property of ZC sequences is that the DFT of a ZC sequence $x_u(n)$ (in Equation (7.1)) is a weighted cyclicly-shifted ZC sequence $X_w(k)$ such that $w = -1/u \bmod N_{ZC}$. This means that a ZC sequence can be generated directly in the frequency domain without the need for a DFT operation.

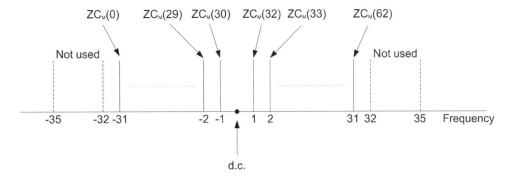

Figure 7.6 PSS sequence mapping in the frequency domain.

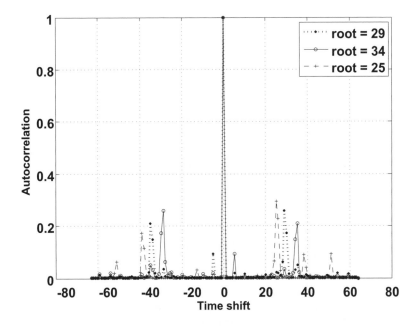

Figure 7.7 Autocorrelation profile at 7.5 kHz frequency offset for roots = 25, 29, 34.

7.2.2 Primary Synchronization Signal (PSS) Sequences

The PSS is constructed from a frequency-domain ZC sequence of length 63, with the middle
element punctured to avoid transmitting on the d.c. subcarrier.

The mapping of the PSS sequence to the subcarriers is shown in Figure 7.6.

Three PSS sequences are used in LTE, corresponding to the three physical layer
identities within each group of cells. The selected roots for the three ZC PSS sequences
are $M = 29, 34, 25$, such that the frequency domain length-63 sequence for root M is given

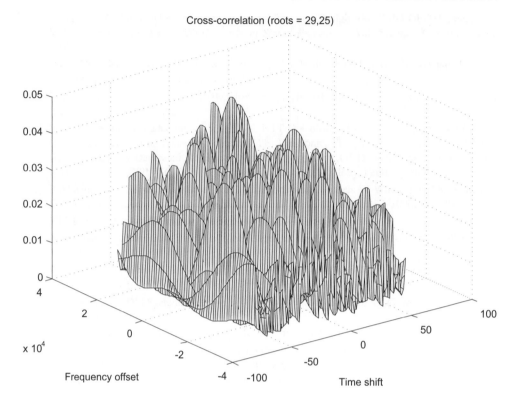

Figure 7.8 Cross-correlation of the PSS sequence pair 25 and 29.

by

$$ZC_M^{63}(n) = \exp\left[-j\frac{\pi M n(n+1)}{63}\right], \quad n = 0, \ 1, \ \ldots, 62 \qquad (7.3)$$

This set of roots for the ZC sequences was chosen for its good periodic autocorrelation and cross-correlation properties. In particular, these sequences have a low-frequency offset sensitivity, defined as the ratio of the maximum undesired autocorrelation peak in the time domain to the desired correlation peak computed at a certain frequency offset. This allows a certain robustness of the PSS detection during the initial synchronization, as shown in Figure 7.7.

Figures 7.8 and 7.9 show respectively the cross-correlation (for roots 29 and 25) as a function of timing and frequency offset and the autocorrelation as a function of timing offset (for root 29). It can be seen that the average and peak values of the cross-correlation are low relative to the autocorrelation. The residual cross-correlation signal can then be considered as white noise with low variance. Furthermore, the ZC sequences are robust against frequency drifts as shown in Figure 7.10. Thanks to the flat frequency-domain autocorrelation property and to the low frequency offset sensitivity, the PSS can be easily detected during the initial synchronization with a frequency offset up to ±7.5 kHz.

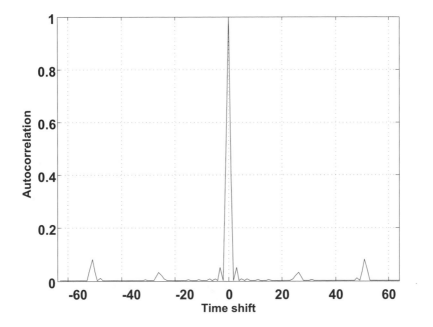

Figure 7.9 Autocorrelation of the PSS sequence 29 as a function of time offset.

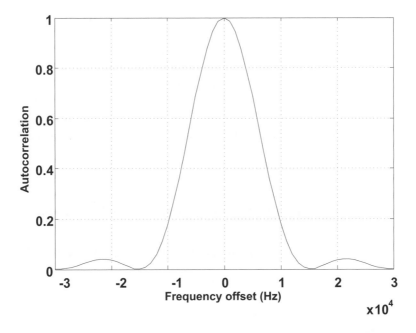

Figure 7.10 Autocorrelation of the PSS sequence as a function of frequency offset.

From the point of view of the UE, the selected root combination satisfies time-domain root-symmetry, in that sequences 29 and 34 are complex conjugates of each other and can be detected with a single correlator, thus allowing for some complexity reduction.

The UE must detect the PSS without any a priori knowledge of the channel, so non-coherent correlation is required for PSS timing detection. A maximum likelihood detector, as explained in Section 7.3, finds the timing offset m_M^* that corresponds to the maximum correlation, i.e.

$$m_M^* = \text{argmax}_m \left| \sum_{i=0}^{N-1} Y[i+m]S_M^*[i] \right|^2 \qquad (7.4)$$

where i is time index, m is the timing offset, N is the PSS time domain signal length, $Y[i]$ is the received signal at time instant i and $S_M[i]$ is the PSS with root M replica signal at time i as defined in Equation (7.3).

7.2.3   Secondary Synchronization Signal (SSS) Sequences

The SSS sequences are based on maximum length sequences, known as M-sequences, which can be created by cycling through every possible state of a shift register of length n. This results in a sequence of length $2n - 1$.

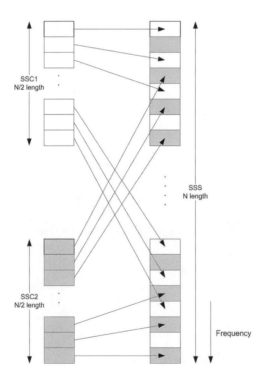

Figure 7.11 SSS sequence mapping.

Each SSS sequence is constructed by interleaving, in the frequency-domain, two length-31 BPSK-modulated secondary synchronization codes, denoted here SSC1 and SSC2, as shown in Figure 7.11.

These two codes are two different cyclic shifts of a single length-31 M-sequence. The cyclic shift indices of the M-sequences are derived from a function of the physical layer cell identity group (as given in Table 6.11.2.1-1 in [5]). The two codes are alternated between the first and second SSS transmissions in each radio frame. This enables the UE to determine the 10 ms radio frame timing from a single observation of a SSS, which is important for UEs handing over to LTE from another radio access technology. For each transmission, SSC2 is scrambled by a sequence that depends on the index of SSC1. The sequence is then scrambled by a code that depends on the PSS. The scrambling code is one-to-one mapped to the physical layer identity within the group corresponding to the target eNodeB. The sequence construction is illustrated in Figure 7.12; details of the scrambling operations are given in [5].

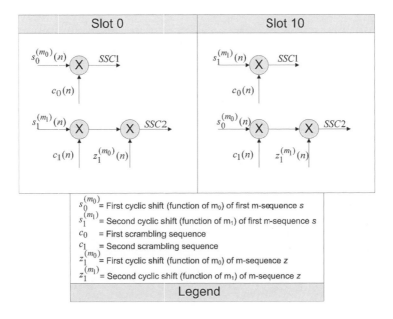

Figure 7.12 SSS sequence generation.

The SSS sequences have good frequency-domain properties, being spectrally flat as shown in Figure 7.13. As in the case of the PSS, the SSS can be detected with a frequency offset up to ± 7.5 kHz. In the time domain, the cross-correlation between any cyclic shifts of an SSS sequence is not as good as for classical M-sequences (for which the cross-correlation is known to be -1), owing to the effects of the scrambling operations. Figure 7.14 illustrates the cross-correlation in frequency domain.

From the point of view of the UE, the SSS detection is done after the PSS detection, and the channel can therefore be assumed to be known based on the PSS sequence. It follows that

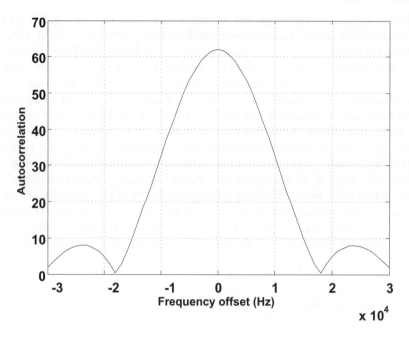

Figure 7.13 Autocorrelation of an SSS sequence as a function of frequency offset.

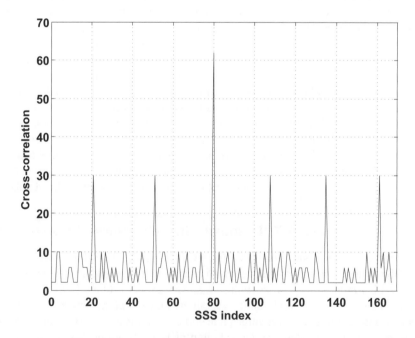

Figure 7.14 Cross-correlation of a pair of SSS sequences.

Table 7.1 Cell identification test parameters.

Parameter (e-NodeB)	Unit	Cell1	Cell2	Cell3
Relative delay of 1st path for synchronous case	ms	0	0	Half CP length
Relative delay of 1st path for asynchronous case	ms	0	1.5	3
SNR	dB	5.18	0.29	−0.75 in worst case
PSS Physical layer ID for case different PSS		PSS1	PSS2	PSS3
PSS Physical layer ID for case equal PSS		PSS1	PSS2	PSS1

a coherent detection method as described in Section 7.3 in Equation (7.8) can be applied:

$$\hat{\mathbf{S}}_m = \underset{\mathbf{S}}{\operatorname{argmin}} \left(\sum_{n=1}^{N} |\mathbf{y}[n] - \mathbf{S}[n, n]\hat{h}_n|^2 \right) \tag{7.5}$$

where the symbols $\mathbf{S}[n, n]$ represent the SSS sequences and $\hat{h}_n$ are the estimated channel coefficients.

However, in the case of synchronized neighbouring eNodeBs, the performance of a coherent detector can be degraded. This is because if an interfering eNodeB employs the same PSS as the one used by the target cell, the phase difference between the two eNodeBs can have an impact on the quality of the estimation of the channel coefficients. On the other hand, the performance of a non-coherent detector degrades if the coherence bandwidth of the channel is less than the six resource blocks occupied by the SSS.

In order to reduce the complexity of the SSS detector, the equivalence between M-sequence matrices and Walsh–Hadamard matrices can be exploited [6]. Using this property, a fast M-sequence transform is equivalent to a fast Walsh–Hadamard transform with index remapping. Thanks to this property the complexity of the SSS detector is reduced to $N \log_2 N$ where $N = 32$.

7.2.4 Cell Search Performance

A key requirement for the LTE cell search procedure is that the delay for the UE to detect a newly appearing cell and report it to the serving eNodeB should be less than a prescribed acceptable threshold. As explained in Chapter 13, the serving eNodeB uses the reported information to prepare intra- or inter-frequency handover.

The performance requirements for new cell identification specified in 3GPP are defined for real and generic deployment conditions. A multicell environment consisting of three cells is assumed, with different transmitted powers from each eNodeB. Scenarios covering both synchronized and unsynchronized eNodeBs are analysed, as summarized in Table 7.1. For the propagation channel, various multipath fading conditions with associated UE speeds are considered, and in particular ETU5 (Extended Typical Urban with UE speed 5 km/h), ETU300 (UE speed 300 km/h) and EPA5 (Extended Pedestrian A with UE speed 5 km/h).[6] At the UE, two receive antennas are assumed. Further details of the assumed scenarios can be found in [7].

---

[6]Further details of these propagation models are given in Chapter 21.

Table 7.2 Cell identification test scenarios.

Test case (synch, asynch eNodeBs)	Cell3 (Target)		Cell1 (Interference)		Cell2 (Interference)	
1,5	PSS3	SSS3a, SSS3b	PSS1	SSS1a, SSS1b	PSS2	SSS2a, SSS2b
2,6	PSS1	SSS3a, SSS3b	PSS1	SSS1a, SSS1b	PSS2	SSS2a, SSS2b
3,7	PSS1	SSS1a, SSS3b	PSS1	SSS1a, SSS1b	PSS2	SSS2a, SSS2b
4,8	PSS3	SSS1a, SSS1b	PSS1	SSS1a, SSS1b	PSS2	SSS2a, SSS2b

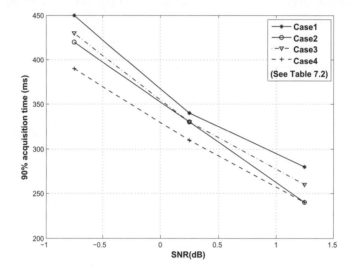

Figure 7.15 Cell search performance with synchronous eNodeBs. Reproduced by permission of © NXP Semiconductors.

The cell search performance is measured in terms of the 90-percentile cell identification delay, i.e. the maximum time required to detect the target cell 90% of the time. In the example shown in Table 7.1, the target cell for detection is 'Cell3'.

Typical performance is shown in Figures 7.15 and 7.16, without assuming any margin for non-ideal UE receiver implementation and without including any reporting delay for the Reference Signal Received Power (RSRP) measurement. Various simulation scenarios have been considered to analyse the impact on the detection performance of different PSS and SSS sequences combination as indicated in Table 7.2. More detailed performance results can be found in [8].

To arrive at performance figures (in terms of acquisition time) for the initial synchronization case (as opposed to new cell identification) the time taken for successful detection would be adapted by considering the decoding of the PBCH instead of the reporting of measurements made on the reference signals of the target cell.

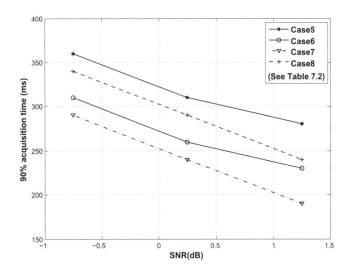

Figure 7.16 Cell search performance with asynchronous eNodeBs. Reproduced by permission of © NXP Semiconductors.

For the case of inter-frequency handover, the performance can be derived from the intra-frequency performance: timing constraints would be applied based on the inter-frequency monitoring gap pattern provided by the eNodeB for performing the necessary synchronization and measurements, as explained in Chapter 13.

7.3 Coherent Versus Non-Coherent Detection

As explained above, both coherent and non-coherent detection may play a part in the synchronization procedures: in the case of the PSS, non-coherent detection is used, while for SSS sequence detection, coherent or non-coherent techniques can be used.

This section gives some theoretical background to the difference between coherent and non-coherent detection. From a conceptual point of view, a coherent detector takes advantage of knowledge of the channel, while a non-coherent detector uses an optimization metric corresponding to the average channel statistics.

We consider a generic system model where the received sequence $\mathbf{y}_m$ at time instant m is given by Equation (7.6), where the matrix $\mathbf{S}_m = \text{diag}\,[s_{m,1}, \ldots, s_{m,N}]$ represents the transmitted symbol at time instant m and $\mathbf{h} = [h_1, \ldots, h_N]^\mathrm{T}$ is the channel vector, i.e.

$$\mathbf{y}_m = \mathbf{S}_m \mathbf{h} + \boldsymbol{v}_m \tag{7.6}$$

where $\boldsymbol{v}_m$ is zero-mean complex Gaussian noise with variance σ^2. In the following sections, expressions for coherent and non-coherent detection are derived.

7.3.1 Coherent Detection

Maximum Likelihood (ML) coherent detection of a transmitted sequence consists of finding the sequence which maximizes the probability that the sequence was transmitted, conditioned on the knowledge of the channel. Coherent detectors therefore require channel estimation to be performed first.

Since the noise is assumed to be independently identically distributed (i.i.d.) and detection is symbol-by-symbol, we focus on one particular symbol interval, neglecting the time instant without loss of generality. The problem becomes

$$\hat{\mathbf{S}} = \underset{\mathbf{S}}{\text{argmax}} \ \Pr(\mathbf{y} \mid \mathbf{S}, \mathbf{h}) = \underset{\mathbf{S}}{\text{argmax}} \ \frac{1}{(\pi N_0)^N} \exp\left[-\frac{\|\mathbf{y} - \mathbf{Sh}\|^2}{N_0}\right] \tag{7.7}$$

Maximizing Equation (7.7) is equivalent to minimizing the magnitude of the argument of the exponential, i.e.

$$\hat{\mathbf{S}} = \underset{\mathbf{S}}{\text{argmin}} \ (\|\mathbf{y} - \mathbf{Sh}\|^2)$$

$$= \underset{\mathbf{S}}{\text{argmin}} \left(\sum_{n=1}^{N} |\mathbf{y}_n - S_{n,n} h_n|^2\right) \tag{7.8}$$

Equation (7.8) is a minimum squared Euclidean distance rule where the symbols $S_{n,n}$ are weighted by the channel coefficient h_n.

7.3.2 Non-Coherent Detection

When channel knowledge is not available or cannot be exploited, non-coherent detection can be used. The trick consists of removing the dependency on a particular channel condition by averaging over the distribution of the random channel coefficients $\mathbf{h}$. The ML detection problem maximizes the following conditional probability:

$$\hat{\mathbf{S}} = \underset{\mathbf{S}}{\text{argmax}} \ \Pr(\mathbf{y} \mid \mathbf{S})$$

$$= \underset{\mathbf{S}}{\text{argmax}} \ \underset{\mathbf{h}}{\mathbb{E}}[\Pr(\mathbf{y} \mid \mathbf{S}, \mathbf{h})] = \int_{\mathbf{h}} \Pr(\mathbf{y} \mid \mathbf{S}, \mathbf{h}) \Pr(\mathbf{h}) \, d\mathbf{h}$$

$$= \int_{\mathbf{h}} \frac{1}{(\pi N_0)^N} \exp\left[-\frac{\|\mathbf{y}_m - S_m\mathbf{h}\|^2}{N_0}\right] \Pr(\mathbf{h}) \, d\mathbf{h} \tag{7.9}$$

Substituting the PDF of an AWGN channel, the ML non-coherent detector yields

$$\hat{\mathbf{S}} = \underset{\mathbf{S}}{\text{argmax}}\left\{\frac{1}{N_0^N \pi^{2N} \det(R_\mathbf{h})} \int_{\mathbf{h}} \exp\left[-\left(\frac{\|\mathbf{y} - \mathbf{Sh}\|^2}{N_0} + \mathbf{h}^H R_\mathbf{h}^{-1} \mathbf{h}\right)\right]\right\} \tag{7.10}$$

It can be shown [9] that the solution of the integral in Equation (7.10) is given by

$$\int_{\mathbf{h}} \exp\left[-\left(\frac{\|\mathbf{y} - \mathbf{Sh}\|^2}{N_0} + \mathbf{h}^H R_\mathbf{h}^{-1} \mathbf{h}\right)\right]$$

$$= \frac{\pi^N}{\det(A)} \exp\left[-\frac{1}{N_0}\mathbf{y}^H\mathbf{y} + \frac{1}{N_0}\mathbf{y}^H S (S^H S + N_0 R_\mathbf{h}^{-1})^{-1} S^H \mathbf{y}\right] \tag{7.11}$$

Note that $S^H S = I$ under the assumption of normalization by the signal energy.

The maximization in Equation (7.9) is done over the input symbols S, so all terms which do not depend on S can be discarded. Hence

$$\hat{S} = \underset{S}{\text{argmax}} \; \{\exp[y^H S(S^H S + N_0 R_h^{-1})^{-1} S^H y]\} = \underset{S}{\text{argmax}} \; \{y^H S(I + N_0 R_h^{-1})^{-1} S^H y\}$$

(7.12)

where the last equality holds because of the monotonicity of the exponential function. Depending on the form of R_h, the ML non-coherent detector can be implemented in different ways. For example, in the case of frequency non-selective channel, the channel correlation matrix can be written as $R_h = \sigma_h^2 V$ where V is the all ones matrix. It follows that

$$(I + N_0 R_h^{-1})^{-1} = \frac{\sigma_h^2}{N} V + \frac{\sigma_h^2}{N} (\alpha^{-1} - 1) I$$

where $\alpha = \sigma_h^2 / N_0 + 1$. With this hypothesis the quadratic expression in Equation (7.12) can be written as

$$y^H S(I + N_0 R_h^{-1})^{-1} S^H y = \frac{\sigma_h^2}{N N_0} \left((\alpha^{-1} - 1) \|y\|^2 + \left| \sum_{i=1}^{N} S[i, i] y[i] \right|^2 \right)$$

(7.13)

The non-coherent ML detector is thus obtained by the maximization of

$$\hat{S} = \underset{S}{\text{argmax}} \left\{ \left| \sum_{i=1}^{N} S[i, i] y[i] \right|^2 \right\}$$

(7.14)

References[7]

[1] J. D. C. Chu, 'Polyphase Codes with Good Periodic Correlation Properties'. *IEEE Trans. on Information Theory*, Vol. 18, pp. 531–532, July 1972.

[2] R. Frank, S. Zadoff and R. Heimiller, 'Phase Shift Pulse Codes With Good Periodic Correlation Properties'. *IEEE Trans. on Information Theory*, Vol. 8, pp. 381–382, October 1962.

[3] B. M. Popovic, 'Generalized Chirp-Like Polyphase Sequences with Optimum Correlation Properties'. *IEEE Trans. on Information Theory*, Vol. 38, pp. 1406–1409, July 1992.

[4] 3GPP Technical Specification 25.213, 'Spreading and modulation (FDD) (Release 8)', www.3gpp.org.

[5] 3GPP Technical Specification 36.211, 'Physical Channels and Modulation (Release 8)', www.3gpp.org.

[6] M. Cohn and A. Lempel, 'On Fast M-Sequence Transforms'. *IEEE Trans. on Information Theory*, Vol. 23, pp. 135–137, January 1977.

[7] Texas Instruments, NXP, Motorola, Ericsson, and Nokia, 'R4-072215: Simulation Assumptions for Intra-frequency Cell Identification', www.3gpp.org 3GPP TSG RAN WG4, meeting 45, Jeju, Korea, November 2007.

[8] NXP, 'R4-080691: LTE Cell Identification Performance in Multi-cell Environment', www.3gpp.org 3GPP TSG RAN WG4, meeting 46bis, Shenzen, China, February 2008.

[9] D. Reader, 'Blind Maximum Likelihood Sequence Detection over Fast Fading Communication Channels', Dissertation, University of South Australia, Australia, August 1996.

[7] All web sites confirmed 18th December 2008.

8

Reference Signals and Channel Estimation

Andrea Ancora and Stefania Sesia

8.1 Introduction to Channel Estimation and Reference Signals

A simple communication system can be generally modelled as in Figure 8.1, where the signal $\mathbf{x}$ transmitted by 'A' passes through a radio channel $\mathbf{H}$ and suffers additive noise before being received by 'B'. Mobile radio channels usually exhibit multipath fading, which causes Inter-Symbol Interference (ISI) in the received signal. In order to remove ISI, various different kinds of equalization and detection algorithms can be utilized, which may or may not exploit knowledge of the Channel Impulse Response (CIR). Orthogonal Frequency Division Multiple Access (OFDMA) is particularly robust against ISI, thanks to its structure and the use of the Cyclic Prefix (CP) which allows the receiver to perform a low-complexity single-tap scalar equalization in the frequency domain, as described in Section 5.2.1.

Figure 8.1 A simple transmission model.

As explained in Section 7.3, when the detection method exploits channel knowledge, it is generally said to be 'coherent', while otherwise it is called 'non-coherent'. Coherent detection can therefore make use of both amplitude and phase information carried by the complex signals exchanged between the eNodeBs and UEs, and not of only amplitude information as with non-coherent detection. Optimal reception by coherent detection therefore typically

LTE – The UMTS Long Term Evolution: from Theory to Practice   Stefania Sesia, Issam Toufik and Matthew Baker
© 2009 John Wiley & Sons, Ltd

requires accurate estimation of the propagation channel. This is true for both downlink and uplink, and for both Frequency Division Duplex (FDD) and Time Division Duplex (TDD).

The main advantage of coherent detection is the simplicity of the implementation compared to the more complex algorithms required by non-coherent detection for equivalent performance. However, this simplicity comes at a price, namely the overhead needed in order to be able to estimate the channel. A common and simple way to estimate the channel is to exploit known signals which do not carry any data, but which therefore cause a loss in spectral efficiency. In general, it is not an easy task to find the optimal trade-off between minimizing the spectral efficiency loss due to the overhead and providing adequate ability to track variations in the channel.

Other possible techniques for channel estimation include the use of a priori knowledge of a parametric model of the channel, exploiting the correlation properties of the channel, or using blind estimation.

Once synchronization between an eNodeB and a UE has been achieved, the main characteristic of the LTE physical processing architecture (shared with earlier systems such as GSM and UMTS) is that of being a coherent communication system, for which purpose known reference signals are inserted into the transmitted signal structure.

In general, a variety of methods can be used to embed reference signals into a transmitted signal. The reference signals can be multiplexed with the data symbols (which are unknown at the receiver) in either the frequency, time or code domains (the latter being used in the case of the common pilot channel in the UMTS downlink). A special case of time multiplexing, known as preamble-based training, involves transmitting the reference signals at the beginning of each data burst. Multiplexing-based techniques have the advantage of low receiver complexity, as the symbol detection is decoupled from the channel estimation problem. Alternatively, reference symbols may be superimposed on top of unknown data, without the two necessarily being orthogonal. Note that multiplexing reference signals in the code domain is a particular type of superposition with a constraint on orthogonality between known reference signals and the unknown data. A comprehensive analysis of the optimization of reference signal design can be found in [1, 2].

Orthogonal reference signal multiplexing is by far the most common technique. For example, to facilitate channel estimation in the UMTS downlink, two types of orthogonal reference signal are provided. The first is code-multiplexed, available to all users in a cell, and uses a specific spreading code which is orthogonal to the codes used to spread the users' data. The second type is time-multiplexed dedicated reference signals, which may in some situations be inserted into the users' data streams [3].

In the LTE downlink, the OFDM transmission can be described by a two-dimensional lattice in time and frequency, as shown in Figure 6.1 and described in Chapter 6. This structure facilitates the multiplexing of the Reference Signals (RSs), which are mapped to specific Resource Elements (REs) of the two-dimensional lattice in Figure 6.1 according to a pattern explained in Section 8.2.

In order to estimate the channel as accurately as possible, all correlations between channel coefficients in time, frequency and space should be taken into account. Since reference signals are sent only on particular OFDM REs (i.e. on particular OFDM symbols on particular subcarriers), channel estimates for the REs which do not bear reference signals have to be computed via interpolation. The optimal interpolating channel estimator in terms of mean squared error is based on a two-dimensional Wiener filter interpolation [4]. Due to the high complexity of such a filter, a trade-off between complexity and accuracy is achieved by using one-dimensional filters. In Sections 8.4, 8.5 and 8.6 the problem of channel estimation is approached from a theoretical point of view, and some possible solutions are described.

The work done in the field of channel estimation, and the corresponding literature available, is vast. Nevertheless many challenges still remain, and we refer the interested reader to [2] and [5] for a general survey of open issues in this area.

8.2   Design of Reference Signals in LTE

In the LTE downlink, three different types of reference signal are provided [6]:

- Cell-specific RSs (often referred to as 'common' RSs, as they are available to all UEs in a cell).

- UE-specific RSs, which may be embedded in the data for specific UEs.

- MBSFN-specific RSs, which are only used for Multimedia Broadcast Single Frequency Network (MBSFN) operation and are discussed further in Chapter 14.

8.2.1   Cell-Specific Reference Signals

References [7, 8] show that in an OFDM-based system an equidistant arrangement of reference symbols in the lattice structure achieves the minimum mean squared error estimate of the channel. Moreover, in the case of a uniform reference symbol grid, a 'diamond shape' in the time-frequency plane can be shown to be optimal.

In LTE, the arrangement of the symbols making up the cell-specific RSs in the time-frequency two-dimensional lattice follows these principles. Figure 8.2 illustrates the reference symbol arrangement for the normal CP length.[1]

The LTE system has been conceived to work under high-mobility assumptions, in contrast to WLAN systems which are generally optimized for pedestrian-level mobility. WLAN systems typically use a preamble-based training sequence, the frequency of which governs the degree of mobility they can support.

The required spacing in time between the reference symbols can be obtained by considering the maximum Doppler spread (highest speed) to be supported, which for LTE corresponds to 500 km/h [9]. The Doppler shift is $f_d = (f_c \, v/c)$ where f_c is the carrier frequency, v is the UE speed in metres per second, and c is the speed of light ($3 \cdot 10^8$ m/s). Considering $f_c = 2$ GHz and $v = 500$ km/h, then the Doppler shift is $f_d \simeq 950$ Hz. According to Nyquist's sampling theorem, the minimum sampling frequency needed in order to reconstruct the channel is therefore given by $T_c = 1/(2 f_d) \simeq 0.5$ ms under the above assumptions. This implies that two reference symbols per slot are needed in the time domain in order to estimate the channel correctly.

In the frequency direction there is one reference symbol every six subcarriers on each OFDM symbols which includes reference symbol, but these are staggered so that within

---

[1] In the case of the extended CP, the arrangement of the reference symbols slightly changes, but the explanations in the rest of the chapter are no less valid. The detailed arrangement of reference symbols for the extended CP can be found in [6].

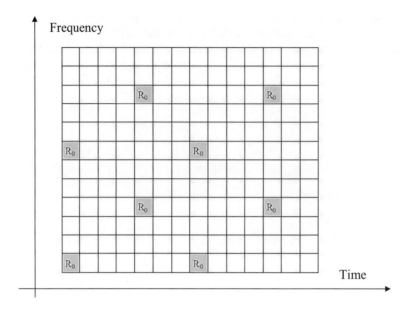

Figure 8.2 Cell-specific reference symbol arrangement in the case of normal CP length for one antenna port. Reproduced by permission of © 3GPP.

each Resource Block (RB) there is one reference symbol every 3 subcarriers, as shown in Figure 8.2. This spacing is related to the expected coherence bandwidth of the channel, which is in turn related to the channel delay spread. In particular the 90% and 50% coherence bandwidths[2] are given respectively by $B_{c,90\%} = 1/50\sigma_\tau$ and $B_{c,50\%} = 1/5\sigma_\tau$ where σ_τ is the r.m.s delay spread. In [10] the maximum r.m.s channel delay spread considered is 991 ns, corresponding to $B_{c,90\%} = 20$ kHz and $B_{c,50\%} = 200$ kHz. In LTE the spacing between two reference symbols in frequency, in one RB, is 45 kHz, thus allowing the expected frequency-domain variations of the channel to be resolved.

The LTE downlink has been specifically designed to work with multiple transmit antennas, as is discussed in detail in Chapter 11. RS patterns are therefore defined for multiple 'antenna ports' at the eNodeB. An antenna port may in practice be implemented either as a single physical transmit antenna, or as a combination of multiple physical antenna elements. In either case, the signal transmitted from each antenna port is not designed to be further deconstructed by the UE receiver: the transmitted RS corresponding to a given antenna port defines the antenna port from the point of view of the UE, and enables the UE to derive a channel estimate for that antenna port – regardless of whether it represents a single radio channel from one physical antenna or a composite channel from a multiplicity of physical antenna elements together comprising the antenna port.

Up to four cell-specific antenna ports may be used by a LTE eNodeB, thus requiring the UE to derive up to four separate channel estimates.[3] For each antenna port, a different

---

[2]$B_{c,x\%}$ is the bandwidth where the autocorrelation of the channel in the frequency domain is equal to $x\%$.
[3]Any MBSFN and UE-specific RSs, if transmitted, constitute additional independent fifth and sixth antenna ports respectively in the LTE specifications.

RS pattern is designed, with particular attention having been given to the minimization of the intra-cell interference between the multiple transmit antenna ports. In Figure 8.3 R_p indicates that the resource element is used for the transmission of an RS on antenna port p. In particular when a resource element is used to transmit an RS on one antenna port, the corresponding resource element on the other antenna ports is set to zero to limit the interference.

From Figure 8.3 it can be noticed that the density of RS for the third and fourth antenna ports is half that of the first two; this is to reduce the overhead in the system. Frequent reference symbols are useful for high-speed conditions as explained above. In cells with a high prevalence of high-speed users, the use of four antenna ports is unlikely, hence for these conditions RSs with lower density can provide sufficient channel estimation accuracy.

All the RSs (cell-specific, UE-specific or MBSFN specific) are QPSK modulated – a constant modulus modulation. This property ensures that the Peak-to-Average Power Ratio (PAPR) of the transmitted waveform is kept low. The signal can be written as

$$r_{l,n_s}(m) = \frac{1}{\sqrt{2}}[1 - 2c(2m)] + j\frac{1}{\sqrt{2}}[1 - 2c(2m + 1)] \tag{8.1}$$

where m is the index of the RS, n_s is the slot number within the radio frame and 'l' is the symbol number within the time slot. The pseudo-random sequence $c(i)$ is comprised of a length-31 Gold sequence, already introduced in Chapter 6, with different initialization values depending on the type of RSs.

The RS sequence also carries unambiguously one of the 504 different cell identities, N_{ID}^{cell}. For the cell-specific RSs, a cell-specific frequency shift is also applied, given by $N_{ID}^{cell} \bmod 6$.[4] This shift can avoid time-frequency collisions between common RS from up to six adjacent cells. Avoidance of collisions is particularly important in cases when the transmission power of the RS is boosted, as is possible in LTE up to a maximum of 6 dB relative to the surrounding data symbols. RS power boosting is designed to improve channel estimation in the cell, but if adjacent cells transmit high-power RS on the same REs the resulting inter-cell interference will prevent the benefit from being realized.

8.2.2   UE-Specific Reference Signals

UE-specific RS may be transmitted in addition to the cell-specific RSs described above. They are embedded only in the RBs to which the PDSCH is mapped for UEs which are specifically configured (by higher-layer RRC signalling) to receive their downlink data transmissions in this mode. If UE-specific RSs are used, the UE is expected to use them to derive the channel estimate for demodulating the data in the corresponding PDSCH RBs. Thus the UE-specific RS are treated as being transmitted using a distinct antenna port, with its own channel response from the eNodeB to the UE.

A typical usage of the UE-specific RSs is to enable beamforming of the data transmissions to specific UEs. For example, rather than using the physical antennas used for transmission of the other (cell-specific) antenna ports, the eNodeB may use a correlated array of physical antenna elements to generate a narrow beam in the direction of a particular UE. Such a beam will experience a different channel response between the eNodeB and UE, thus requiring the use of UE-specific RSs to enable the UE to demodulate the beamformed data coherently. The use of UE-specific beamforming is discussed in more detail in Section 11.2.

---

[4]The mod6 operation is used because RSs are spaced apart by six subcarriers in the lattice grid.

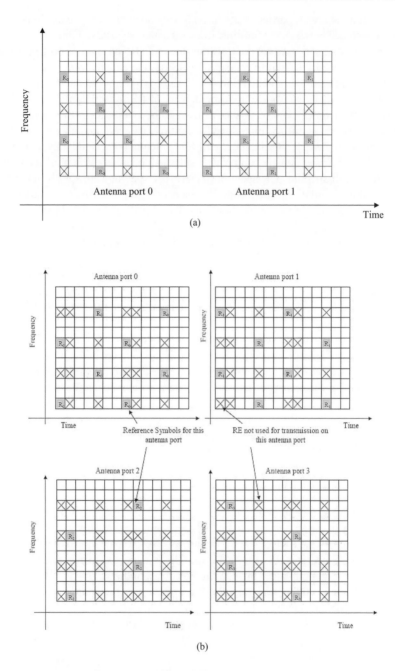

Figure 8.3 Cell-specific RS arrangement in the case of normal CP length for (a) two antenna ports, (b) four antenna ports. Reproduced by permission of © 3GPP.

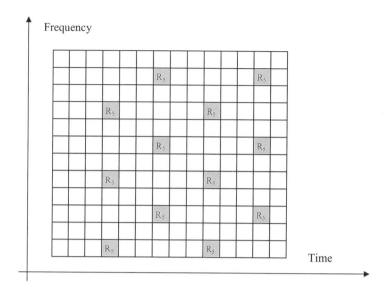

Figure 8.4   UE-specific RS arrangement with normal CP. Reproduced by permission of © 3GPP.

As identified in [11], the structure shown in Figure 8.4 (for the normal CP) has been chosen because there is no collision with the cell specific RSs, and hence the presence of a UE-specific RSs does not affect features related to the cell-specific RSs. The UE-specific RSs have a similar pattern to that of the cell-specific RSs, which allows a UE to re-use similar channel estimation algorithms. The density is half that of the cell-specific RS, hence minimizing the overhead.

The corresponding pattern for use in case of the extended CP being configured in a cell can be found in [6].

8.3   RS-Aided Channel Modelling and Estimation

The channel estimation problem is related to the channel model assumed, itself determined by the physical propagation characteristics, including the number of transmit and receive antennas, transmission bandwidth, carrier frequency, cell configuration and relative speed between eNodeB and UE receivers. In general,

- The carrier frequencies and system bandwidth mainly determine the scattering nature of the channel.

- The cell deployment governs its multipath, delay spread and spatial correlation characteristics.

- The relative speed sets the time-varying properties of the channel.

The propagation conditions characterize the channel correlation function in a three-dimensional space comprising frequency, time and spatial domains. In the general case, each MIMO (Multiple-Input Multiple-Output) multipath channel component can experience different but related spatial scattering conditions leading to a full three-dimensional correlation function across the three domains. Nevertheless, for the sake of simplicity, assuming that the multipath components of each spatial channel experience the same scattering conditions, the spatial correlation is independent from the other two domains and can be handled separately from the frequency and time domain correlations.

This framework might be suboptimal in general, but is nevertheless useful in mitigating the complexity of channel estimation as it reduces the general three-dimensional joint estimation problem into independent estimation problems.

For a comprehensive survey of MIMO channel estimation the interested reader is referred to [12].

In the following two subsections, we define a channel model and its corresponding correlation properties. These are then used as the basis for an overview of channel estimation techniques which exploit channel correlation in the context of the LTE downlink.

8.3.1 Time-Frequency Domain Correlation: The WSSUS Channel Model

The Wide-Sense Stationary Uncorrelated Scattering (WSSUS) channel model is commonly employed for the multipath channels experienced in mobile communications.

Neglecting the spatial dimension for the sake of simplicity, let $h(\tau; t)$ denote the time-varying complex baseband impulse response of a multipath channel realization at time instant t and delay τ.

When a narrowband signal $x(t)$ is transmitted, the received narrowband signal $y(t)$ can be written as

$$y(t) = \int h(\tau; t)x(t - \tau)\, d\tau \qquad (8.2)$$

Considering the channel as a random process in the time direction t, the channel is said to be delay Uncorrelated-Scattered (US) if

$$\mathbb{E}[h(\tau_a; t_1)^* h(\tau_b; t_2)] = \phi_h(\tau_a; t_1, t_2)\delta(\tau_b - \tau_a) \qquad (8.3)$$

with $\mathbb{E}[\cdot]$ being the expectation operator. According to the US assumption, two CIR components a and b at relative respective delays τ_a and τ_b are uncorrelated if $\tau_a \neq \tau_b$.

The channel is Wide-Sense Stationary (WSS)-uncorrelated if

$$\phi_h(\tau; t_1, t_2) = \phi_h(\tau; t_2 - t_1) \qquad (8.4)$$

which means that the correlation of each delay component of the CIR is only a function of the *difference* in time between each realization.

Hence, the second-order statistics of this model are completely described by its delay cross-power density $\phi_h(\tau; \Delta t)$ or by its Fourier transform, the scattering function

$$S_h(\tau; f) = \int \phi_h(\tau; \Delta t)e^{-j2\pi f \Delta t}\, d\Delta t \qquad (8.5)$$

with f being the Doppler frequency. Other related functions of interest include the *multipath intensity profile*

$$\psi_h(\tau) = \phi_h(\tau; 0) = \int S_h(\tau; f) \, df$$

the *time-correlation function*

$$\overline{\phi}_h(\tau) = \int \phi_h(\tau; \Delta t) \, d\tau$$

and the *Doppler power spectrum*

$$\overline{S}_h(f) = \int S_h(\tau; f) \, d\tau$$

A more general exposition of WSSUS models is given in [13]. Classical results were derived by Clarke [14] and Jakes [15] for the case of a mobile terminal communicating with a stationary base station in a two-dimensional propagation geometry.

These well-known results state that

$$\overline{S}_h(f) \propto \frac{1}{\sqrt{f_d^2 - f^2}} \tag{8.6}$$

for $|f| \le f_d$, $f_d = (v/c) f_c$ the maximum Doppler frequency for a mobile with relative speed v, carrier frequency f_c and propagation speed c and

$$\overline{\phi}_h(\Delta t) \propto J_0(2\pi f_d \Delta t) \tag{8.7}$$

where J_0 is the zero[th]-order Bessel function. The autocorrelation function is obtained via the inverse Fourier transform of the Power Spectral Density (PSD) (Figure 8.5 shows the PSD of the classical Doppler spectrum described by Clarke and Jakes [14, 15]). The squared magnitude of the autocorrelation function corresponding to Clarke's spectrum model is plotted in Figure 8.6 where the variable ϕ along the x-axis is effectively a spatial lag (in metres) normalized by the carrier wavelength. The Clarke and Jakes derivations are based on the assumption that the physical scattering environment is chaotic and therefore the angle of arrival of the electromagnetic wave at the receiver is a uniformly distributed random variable in the angular domain. As a consequence, the time-correlation function is strictly real-valued, the Doppler spectrum is symmetric and interestingly there is a *delay-temporal separability* property in the general bi-dimensional scattering function $S_h(\tau, \Delta t)$. In other words,

$$S_h(\tau; f) \propto \psi_h(\tau)\overline{S}_h(f) \tag{8.8}$$

or equivalently

$$\phi_h(\tau; \Delta t) \propto \psi_h(\tau)\overline{\phi}_h(\Delta t) \tag{8.9}$$

When the mobile is moving in a fixed and known direction, as for example in rural or suburban areas, the WSSUS mobile channel is in general non-separable, but can be considered to be separable when the direction of motion averages out because each multipath component is the result of omnidirectional scattering from objects surrounding the mobile, as one would expect in urban and indoor propagation scenarios. Separability is a very important assumption for reducing the complexity of channel estimation, allowing the problem to be separated into two one-dimensional operations.

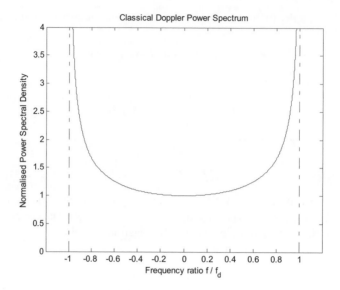

Figure 8.5 Normalized PSD for Clarke's model.

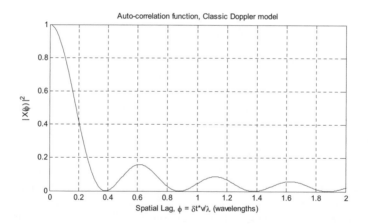

Figure 8.6 Autocorrelation function, flat Rayleigh fading, Clarke's Doppler.

8.3.2   Spatial Domain Correlation: The Kronecker Model

While the previous section addresses frequency and time correlation induced by the channel delays and Doppler spread, the spatial correlation arises from the spatial scattering conditions.

Among the possible spatial correlation models, for performance evaluation of LTE the so-called Kronecker model is generally used (see [10]). Despite its simplicity, this correlation-based analytical model is widely used for the theoretical analysis of MIMO systems and yields experimentally verifiable results when limited to two or three antennas at each end of the radio link.

Let us assume the narrowband MIMO channel $N_{\text{Rx}} \times N_{\text{Tx}}$ matrix for a system with N_{Tx} transmitting antennas and N_{Rx} receiving antennas to be

$$\mathbf{H} = \begin{bmatrix} h_{0,0} & \cdots & h_{0,N_{\text{Tx}}-1} \\ \vdots & & \vdots \\ h_{N_{\text{Rx}}-1,N_{\text{Tx}}-1} & \cdots & h_{N_{\text{Rx}}-1,N_{\text{Tx}}-1} \end{bmatrix} \qquad (8.10)$$

The narrowband assumption particularly suits OFDM systems, where each MIMO channel component $h_{n,m}$ can be seen as the complex channel coefficient of each spatial link at a given subcarrier index.

The matrix $\mathbf{H}$ is rearranged into a vector by means of the operator $\text{vec}(\mathbf{H}) = [\mathbf{v}_0^{\text{T}}, \mathbf{v}_1^{\text{T}}, \dots, \mathbf{v}_{N_{\text{Tx}}-1}^{\text{T}}]^{\text{T}}$ where $\mathbf{v}_i$ is the ith column of $\mathbf{H}$ and $\{\cdot\}^{\text{T}}$ is the transpose operation. Hence, the correlation matrix can be defined as

$$\mathbf{C}_{\text{S}} = \mathbb{E}[\text{vec}(\mathbf{H})\text{vec}(\mathbf{H})^{\text{H}}] \qquad (8.11)$$

where $\{\cdot\}^{\text{H}}$ is the Hermitian operation. The matrix in (8.11) is the full correlation matrix of the MIMO channel.

The Kronecker model assumes that the full correlation matrix results from separate spatial correlations at the transmitter and receiver, which is equivalent to writing the full correlation matrix as a Kronecker product ($\otimes$) of the transmitter and receiver correlation matrices:

$$\mathbf{C}_{\text{S}} = \mathbf{C}_{\text{Tx}} \otimes \mathbf{C}_{\text{Rx}} \qquad (8.12)$$

Typical values assumed for these correlations according to the Kronecker model are discussed in Section 21.3.7.

8.4 Frequency Domain Channel Estimation

In this section we address the channel estimation problem over one OFDMA symbol (specifically a symbol containing reference symbols) to exploit the frequency domain characteristics.

In the LTE context, as for any OFDM system with uniformly-distributed refrence symbols [16], the Channel Transfer Function (CTF) can be estimated using a maximum likelihood approach in the frequency domain at the REs containing the RSs by de-correlating the constant modulus RS. Using a matrix notation, the channel estimate $\widehat{\mathbf{H}}_p$ on reference symbol p can be written as

$$\widehat{\mathbf{H}}_p = \mathbf{H}_p + \widetilde{\mathbf{H}}_p = \mathbf{F}_p \mathbf{h} + \widetilde{\mathbf{H}}_p \qquad (8.13)$$

for $p \in (0, \dots, P)$ where P is the number of available reference symbols and $\mathbf{h}$ is the $L \times 1$ CIR vector. $\mathbf{F}_p$ is the $P \times L$ matrix obtained by selecting the rows corresponding to the reference symbol positions and the first L columns of the $N \times N$ Discrete Fourier Transform (DFT) matrix. $\widetilde{\mathbf{H}}_p$ is a $P \times 1$ zero-mean complex circular white noise vector whose $P \times P$ covariance matrix is given by $\mathbf{C}_{\widetilde{\mathbf{H}}_p}$.

The effective channel length $L \leq L_{\text{CP}}$ is assumed to be known.

8.4.1 Channel Estimation by Interpolation

8.4.1.1 Linear Interpolation Estimator

The natural approach to estimate the whole CTF is to interpolate its estimate between the reference symbol positions. In the general case, let $\mathbf{A}$ be a generic interpolation filter; then the interpolated CTF estimate at subcarrier index i can be written as

$$\widehat{\mathbf{H}}_i = \mathbf{A}\widehat{\mathbf{H}}_p \qquad (8.14)$$

Substituting Equation (8.13) into (8.14), the error of the interpolated CTF estimate is

$$\widetilde{\mathbf{H}}_i = \mathbf{H} - \widehat{\mathbf{H}}_i = (\mathbf{F}_L - \mathbf{A}\mathbf{F}_p)\mathbf{h} - \mathbf{A}\widetilde{\mathbf{H}}_p \qquad (8.15)$$

where $\mathbf{F}_L$ is the $N \times L$ matrix obtained taking the first L columns of the DFT matrix and $\mathbf{H} = \mathbf{F}_L\mathbf{h}$. In Equation (8.15), it can be seen that the channel estimation error is constituted of a bias term (itself dependent on the channel) and an error term.

The error covariance matrix is

$$\mathbf{C}_{\widetilde{\mathbf{H}}_i} = (\mathbf{F}_L - \mathbf{A}\mathbf{F}_p)\mathbf{C}_\mathbf{h}(\mathbf{F}_L - \mathbf{A}\mathbf{F}_p)^{\mathrm{H}} + \sigma_{\widetilde{\mathbf{H}}_p}^2 \mathbf{A}\mathbf{A}^{\mathrm{H}} \qquad (8.16)$$

where $\mathbf{C}_h = \mathbb{E}[\mathbf{h}\mathbf{h}^{\mathrm{H}}]$ is the channel covariance matrix.

Recalling Equation (8.14), linear interpolation would be the intuitive choice. The filter structure $\mathbf{A}$ is then given by

$$\mathbf{A} = \begin{bmatrix}
0 & 0 & \cdots & 0 & 0 \\
0 & 0 & \cdots & 0 & 0 \\
0 & 0 & \cdots & 0 & 0 \\
1 & 0 & \cdots & 0 & 0 \\
\dfrac{M-1}{M} & \dfrac{1}{M} & 0 & \cdots & 0 \\
\dfrac{M-2}{M} & \dfrac{2}{M} & 0 & \cdots & 0 \\
\vdots & \vdots & \vdots & \vdots & 0 \\
\dfrac{1}{M} & \dfrac{M-1}{M} & 0 & \cdots & 0 \\
0 & \dfrac{1}{M} & 0 & \cdots & 0 \\
0 & \dfrac{M-1}{M} & \dfrac{1}{M} & \cdots & 0 \\
0 & \dfrac{M-2}{M} & \dfrac{2}{M} & 0 & 0 \\
0 & \vdots & \vdots & \vdots & 0 \\
0 & \dfrac{1}{M} & \dfrac{M-1}{M} & 0 & 0 \\
0 & 0 & 1 & \cdots & 0 \\
\vdots & \vdots & \vdots & \vdots & 0 \\
0 & 0 & \cdots & 0 & 0 \\
0 & 0 & \cdots & 0 & 0 \\
0 & 0 & \cdots & 0 & 0
\end{bmatrix} \qquad (8.17)$$

This estimator is deterministically biased, but unbiased from the Bayesian viewpoint regardless of the structure of $\mathbf{A}$.

8.4.1.2 IFFT Estimator

As a second straightforward approach, the CTF estimate over all subcarriers can be obtained by IFFT interpolation. In this case the matrix $\mathbf{A}$ from Equation (8.14) becomes:

$$\mathbf{A}_{\text{IFFT}} = \frac{1}{P}\mathbf{F}_L\mathbf{F}_p^H \tag{8.18}$$

The error of the IFFT-interpolated CTF estimate and its covariance matrix can be obtained by substituting Equation (8.18) into (8.15) and (8.16).

With the approximation of $\mathbf{I}_L \approx (1/P)\mathbf{F}_p^H\mathbf{F}_p$, where $\mathbf{I}_L$ is the $L \times L$ identity matrix, it can immediately be seen that the bias term in Equation (8.14) would disappear, providing for better performance. The error covariance matrix reduces to

$$\mathbf{C}_{\widetilde{\mathbf{H}}_{\text{IFFT}}} \approx \frac{1}{P}\sigma_{\widetilde{\mathbf{H}}_p}^2 \mathbf{F}_L\mathbf{F}_L^H \tag{8.19}$$

Given the LTE system parameters and the RS structure, in practice $(1/P)\mathbf{F}_p^H\mathbf{F}_p$ is far from being a multiple of an identity matrix. The approximation becomes an equality when $K = N$, $N/W > L$ and N/W is an integer,[5] i.e. the system would have to be dimensioned without guard-bands and the RS would have to be positioned with a spacing which is an exact factor of the FFT order N, namely a power of two.

In view of the other factors affecting the design of the RS structure outlined above, such constraints are impractical.

8.4.2 General Approach to Linear Channel Estimation

Compared to the simplistic approaches presented in the previous section, more elaborate linear estimators derived from both deterministic and statistical viewpoints are proposed in [17–19]. Such approaches include Least Squares (LS), Regularized LS, Minimum Mean-Squared Error (MMSE) and Mismatched MMSE. These can all be expressed under the following general formulation:

$$\mathbf{A}_{\text{gen}} = \mathbf{B}(\mathbf{G}^H\mathbf{G} + \mathbf{R})^{-1}\mathbf{G}^H \tag{8.20}$$

where $\mathbf{B}$, $\mathbf{G}$ and $\mathbf{R}$ are matrices that vary according to each estimator as expressed in the following subsections.

8.4.2.1 LS Estimator

The LS estimator discussed in [17] can be described by choosing

$$\mathbf{B} = \mathbf{F}_L, \mathbf{G} = \mathbf{F}_p \text{ and } \mathbf{R} = \mathbf{0}_L \tag{8.21}$$

where $\mathbf{0}_L$ is the $L \times L$ matrix containing only zeros. The estimator becomes

$$\widehat{\mathbf{H}}_{\text{LS}} = \mathbf{F}_L(\mathbf{F}_p^H\mathbf{F}_p)^{-1}\mathbf{F}_p^H\widehat{\mathbf{H}}_p \tag{8.22}$$

---

[5]W is the spacing (in terms of number of subcarriers) between reference symbols.

Substituting Equations (8.13) and (8.21) into (8.15) and (8.16), the error reduces to

$$\widetilde{\mathbf{H}}_{\text{LS}} = -\mathbf{F}_L(\mathbf{F}_p^{\text{H}}\mathbf{F}_p)^{-1}\mathbf{F}_p^{\text{H}}\widetilde{\mathbf{H}}_p \tag{8.23}$$

showing that the LS estimator, at least theoretically, is unbiased. Thus, compared to the linear interpolation estimator given by Equation (8.14), the LS estimator can be considered to be the perfect interpolator as it sets to zero the bias term of Equation (8.15) with $\mathbf{A} = \mathbf{F}_L(\mathbf{F}_p^{\text{H}}\mathbf{F}_p)^{-1}\mathbf{F}_p^{\text{H}}$. Consequently, the error covariance matrix can be shown to be

$$\mathbf{C}_{\widetilde{\mathbf{H}}_{\text{LS}}} = \sigma_{\widetilde{\mathbf{H}}_p}^2\,\mathbf{F}_L(\mathbf{F}_p^{\text{H}}\mathbf{F}_p)^{-1}\mathbf{F}_L^{\text{H}} \tag{8.24}$$

8.4.2.2 Regularized LS Estimator

As shown in [19], the choice of LTE system parameters does not allow the LS estimator to be applied directly. The expression $(\mathbf{F}_p\mathbf{F}_p^{\text{H}})^{-1}$ is ill-conditioned due to the unused portion of the spectrum corresponding to the unmodulated subcarriers.

To counter this problem, the classical robust regularized LS estimator can be used instead to yield a better conditioning of the matrix to be inverted. The same matrices $\mathbf{B}$ and $\mathbf{G}$ can be used as for the LS estimator, but the regularization matrix $\mathbf{R} = \alpha\mathbf{I}_L$ is introduced where α is a constant (computed off-line) chosen to optimize the performance of the estimator in a given Signal-to-Noise Ratio (SNR) working range.

Its expression can be written as

$$\widehat{\mathbf{H}}_{\text{reg,LS}} = \mathbf{F}_L(\mathbf{F}_p^{\text{H}}\mathbf{F}_p + \alpha\mathbf{I}_L)^{-1}\mathbf{F}_p^{\text{H}}\widehat{\mathbf{H}}_p \tag{8.25}$$

The expressions for the error and the error covariance matrix of this estimator can be deduced directly from Equations (8.15) and (8.16) by substituting $\mathbf{B}$, $\mathbf{G}$ and $\mathbf{R}$ with their corresponding expressions into (8.20).

8.4.2.3 MMSE Estimator

By using Equation (8.20), we can formulate the MMSE estimator [17] by denoting

$$\mathbf{B} = \mathbf{F}_L,\ \mathbf{G} = \mathbf{F}_p \text{ and } \mathbf{R} = \sigma_{\widetilde{\mathbf{H}}_p}^2\,\mathbf{C_h}^{-1} \tag{8.26}$$

thus giving

$$\widehat{\mathbf{H}}_{\text{MMSE}} = \mathbf{F}_L(\mathbf{F}_p^{\text{H}}\mathbf{F}_p + \sigma_{\widetilde{\mathbf{H}}_p}^2\,\mathbf{C_h}^{-1})^{-1}\mathbf{F}_p^{\text{H}}\widehat{\mathbf{H}}_p \tag{8.27}$$

Again, applying Equations (8.20) and (8.26) in (8.15) and (8.16), the error and the error covariance matrix of the MMSE estimator can be obtained straightforwardly.

The MMSE estimator belongs to the class of statistical estimators. Unlike deterministic LS and its derivations, statistical estimators need knowledge of the second-order statistics (Power Delay Profile (PDP) and noise variance) of the channel in order to perform the estimation process, normally with much better performance compared to deterministic estimators.

However, second-order statistics vary as the propagation conditions change and therefore need appropriate re-estimation regularly. For this reason statistical estimators are in general more complex due to the additional burden of estimating the second-order statistics and computing the filter coefficients.

8.4.2.4 Mismatched MMSE Estimator

To avoid the estimation of the second-order channel statistics $\mathbf{C_h}$ and the consequent on-line inversion of an $L \times L$ matrix required in the straightforward application of the MMSE of Equation (8.27), the channel PDP can be assumed to be uniform [18]. This results in $\mathbf{C_h}$ having the structure of an identity matrix in the mismatched MMSE formulation.

With reference to the general formulation in Equation (8.20), this scheme consists of taking the same $\mathbf{B}$ and $\mathbf{G}$ as in Equation (8.26) but setting

$$\mathbf{R} = \sigma_{\widetilde{\mathbf{H}}_p}^2 / \sigma_{\mathbf{h}}^2 \cdot \mathbf{I}_L$$

This gives the following expression

$$\widehat{\mathbf{H}}_{\text{M-MMSE}} = \mathbf{F}_L (\mathbf{F}_p^H \mathbf{F}_p + \sigma_{\widetilde{\mathbf{H}}_p}^2 / \sigma_{\mathbf{h}}^2 \cdot \mathbf{I}_L)^{-1} \mathbf{F}_p^H \widehat{\mathbf{H}}_p \tag{8.28}$$

Interestingly, we notice that this estimator is in practice equivalent to the regularized LS estimator in Section 8.4.2.2, where the only difference lies in the fact that the ratio $\sigma_{\widetilde{\mathbf{H}}_p}^2 / \sigma_{\mathbf{h}}^2$ can be estimated and therefore adapted.

For a given channel length L, one practical approach to avoiding the on-line inversion of the matrix in Equation (8.28) could consist of dividing the SINR working range into subranges and storing different versions of the matrix inverted off-line for each subrange.

8.4.3 Performance Comparison

For the sake of comparison between the performance of the different classes of estimator, we introduce the Truncated Normalized Mean Squared Error (TNMSE).

For each estimator $\widehat{\mathbf{H}}$, the TNMSE is computed from its covariance matrix $\mathbf{C}_{\widehat{\mathbf{H}}}$ and the true channel $\mathbf{H}$ as follows:

$$\text{TNMSE}_{\widehat{\mathbf{H}}} = \frac{\text{Ttr}(\mathbf{C}_{\widetilde{\mathbf{H}}})}{\text{Ttr}(\mathbf{F}_L \mathbf{C_h} \mathbf{F}_L^H)} \tag{8.29}$$

where $\text{Ttr}\{\cdot\}$ denotes the truncated trace operator consisting of the truncated covariance matrix considering only the K used subcarriers.

Figure 8.7 shows the performance of a LTE FDD downlink with 10 MHz transmission bandwidth ($N = 1024$) and Spatial Channel Model-A (SCM-A).

It can be seen that the IFFT and linear interpolation methods yield the lowest performance. The regularized LS and the mismatched MMSE perform exactly equally and the curve of the latter is therefore omitted. As expected, the optimal MMSE estimator outperforms any other estimator.

The TNMSE computed over all subcarriers actually hides the behaviour of each estimator against a well-known problem of frequency-domain channel estimation techniques: the band-edge effect. This can be represented by the Gibbs [20] phenomenon in a finite-length Fourier series approximation; following this approach, Figure 8.8 shows that MMSE-based channel estimation suffers the least band-edge degradation, while all the other methods presented are highly adversely affected.

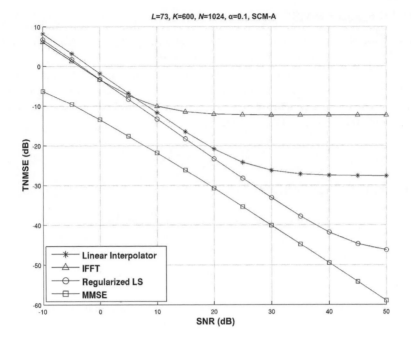

Figure 8.7 Frequency-domain channel estimation performance.

8.5   Time-Domain Channel Estimation

The main benefit of Time-Domain (TD) channel estimation is the possibility to enhance the channel estimation of one OFDM symbol containing RS by exploiting its time correlation with the channel at previous OFDM symbols containing RS.

This requires sufficient memory for buffering soft values of data over several OFDM symbols while the channel estimation is carried out.

However, the correlation between consecutive symbols decreases as the UE speed increases, as expressed by Equation (8.7). The fact that the TD correlation is inversely proportional to the UE speed sets a limit on the possibilities for TD filtering in high-mobility conditions.

TD filtering is applied to the CIR estimate, rather than to the CTF estimate in the frequency domain. The use of a number of parallel scalar filters equal to the channel length L does not imply a loss of optimality, because of the WSSUS assumption.

8.5.1   Finite and Infinite Length MMSE

The statistical TD filter which is optimal in terms of Mean Squared Error (MSE) can be approximated in the form of a finite impulse response filter [21]. The channel at the lth tap position and at time instant n is estimated as

$$\hat{\hat{h}}_{l,n} = \mathbf{w}_l^H \hat{\mathbf{h}}_{l,n}^M \tag{8.30}$$

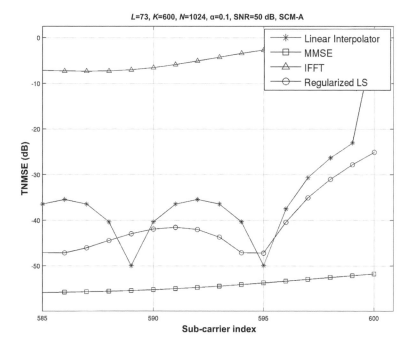

Figure 8.8 Frequency-domain channel estimation performance, band-edge behaviour.

where $\hat{h}_l$ is the smoothed CIR l^{th} tap estimate which exploits the vector $\hat{\mathbf{h}}_{l,n}^M = [\hat{h}_{l,n}, \ldots,$ $\hat{h}_{l,n-M+1}]^{\text{T}}$ of length M of the channel tap h_l across estimates at M time instants.[6] This is obtained by inverse Fourier transformation of, for example, any frequency-domain technique illustrated in previous section or even a raw estimate obtained by RS decorrelation.

The $M \times 1$ vector of Finite Impulse Response (FIR) filter coefficients $\mathbf{w}_l$ is given by

$$\mathbf{w}_l = (\mathbf{R}_h + \sigma_n^2 \mathbf{I})^{-1} \mathbf{r}_h \qquad (8.31)$$

where $\mathbf{R}_h = \mathbb{E}[\mathbf{c}_i^M (\mathbf{c}_i^M)^{\text{H}}]$ is the lth channel tap $M \times M$ correlation matrix, σ_n^2 the additive noise variance and $\mathbf{r}_h = \mathbb{E}[\mathbf{h}_l^M h_{l,n}^*]$ the $M \times 1$ correlation vector between the lth tap of the current channel realization and M previous realizations including the current one.

In practical cases, the FIR filter length M is dimensioned according to a performance-complexity trade-off as a function of UE speed.

By setting M infinite, the upper bound on performance is obtained.

The MSE performance of the finite-length estimator of a channel of length L can be analytically computed as

$$\epsilon^{(M)} = 1 - \frac{1}{\sigma_h^2} \sum_{l=0}^{L-1} \mathbf{w}_l^{\text{H}} \mathbf{r}_h \qquad (8.32)$$

---

[6]$h_{l,k}$ is the lth component of the channel vector $\mathbf{h}_k$ at time instant k. $\hat{h}_{l,k}$ is its estimate. Note that for the frequency domain treatment the time index was dropped.

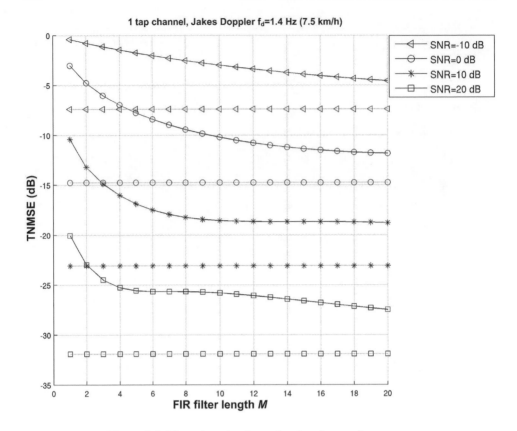

Figure 8.9 Time-domain channel estimation performance.

The upper bound given by an infinite-length estimator is therefore given by

$$\epsilon^{(\infty)} = \lim_{M \to \infty} \epsilon^{(M)} \tag{8.33}$$

From Equation (8.31) it can be observed that, unlike Frequency-Domain (FD) MMSE filtering, the size of the matrix to be inverted for a finite-length TD-MMSE estimator is independent of the channel length L but dependent on the chosen FIR order M. Similarly to the FD counterpart, the TD-MMSE estimator requires knowledge of the PDP, the UE speed and the noise variance.

Figure 8.9 shows the performance of TD-MMSE channel estimation as a function of filter length M (Equation (8.32)) for a single-tap channel with a classical Doppler spectrum for low UE speed. The performance bounds derived for an infinite-length filter in each case are also indicated.

8.5.2   Normalized Least-Mean-Square

As an alternative to TD-MMSE channel estimation, an adaptive estimation approach can be considered which does not require knowledge of second-order statistics of both channel and noise. A feasible solution is the Normalized Least-Mean-Square (NLMS) estimator.

It can be expressed exactly as in Equation (8.30) but with the $M \times 1$ vector of filter coefficients $\mathbf{w}$ updated according to

$$\mathbf{w}_{l,n} = \mathbf{w}_{l,n-1} + \mathbf{k}_{l,n-1} e_{l,n} \tag{8.34}$$

where M here denotes the NLMS filter order. The $M \times 1$ update gain vector is computed according to the well-known NLMS adaptation:

$$\mathbf{k}_{l,n} = \frac{\mu}{\|\hat{\mathbf{h}}_{l,n}\|^2} \hat{\mathbf{h}}_{l,n}^M \tag{8.35}$$

where μ is an appropriately-chosen step adaptation, $\hat{\mathbf{h}}_{l,n}^M$ is defined as for Equation (8.30) and

$$e_{l,n} = \hat{h}_{l,n} - \hat{h}_{l,n-1} \tag{8.36}$$

It can be observed that the TD-NLMS estimator requires much lower complexity compared to TD-MMSE as no matrix inversion is required, as well as not requiring any a priori statistical knowledge.

Other adaptive approaches could also be considered such as Recursive Least Squares (RLS) and Kalman-based filtering. Although more complex than NLMS, the Kalman filter is a valuable candidate and is reviewed in detail in [22].

8.6 Spatial Domain Channel Estimation

It is assumed that a LTE UE has multiple receiving antennas. Consequently, whenever the channel is correlated in the spatial domain, the correlation can be exploited to provide a further means for enhancing the channel estimate.

If it is desired to exploit spatial correlation, a natural approach is again offered by Spatial Domain (SD) MMSE filtering [23].

We consider the case of a MIMO OFDM communication system with N subcarriers, N_{Tx} transmitting antennas and N_{Rx} receiving antennas. The $N_{\text{Rx}} \times 1$ received signal vector at subcarrier k containing the RS sequence can be written as

$$\mathbf{r}(k) = \frac{1}{\sqrt{N_{\text{Tx}}}} \mathbf{W}(k) \mathbf{s}(k) + \mathbf{n}(k) \tag{8.37}$$

where $\mathbf{W}(k)$ is the $N_{\text{Rx}} \times N_{\text{Tx}}$ channel frequency response matrix at RS subcarrier k, $\mathbf{s}(k)$ is the $N_{\text{Tx}} \times 1$ known zero-mean and unit-variance transmitted RS sequence at subcarrier k, and $\mathbf{n}(k)$ is the $N_{\text{Rx}} \times 1$ complex additive white Gaussian noise vector with zero mean and variance σ_n^2.

The CTF at subcarrier k, $\mathbf{W}(k)$, is obtained by DFT from the CIR matrix at the lth tap $\mathbf{H}_l$ as

$$\mathbf{W}(k) = \sum_{l=0}^{L-1} \mathbf{H}_l \exp\left[-j2\pi \frac{lk}{N}\right] \tag{8.38}$$

The vector $\mathbf{h}$ is obtained by rearranging the elements of all $\mathbf{H}_l$ channel tap matrices as follows:

$$\mathbf{h} = [\text{vec}(\mathbf{H}_0)^{\text{T}}, \text{vec}(\mathbf{H}_1)^{\text{T}}, \ldots, \text{vec}(\mathbf{H}_{L-1})^{\text{T}}]^{\text{T}} \tag{8.39}$$

The correlation matrix of **h** is given by

$$\mathbf{C}_h = \mathbb{E}[\mathbf{hh}^H] \tag{8.40}$$

Using Equations (8.38) and (8.39), the Equation (8.37) can now be rewritten as

$$\mathbf{r}(k) = \frac{1}{\sqrt{N_{\mathrm{Tx}}}} \mathbf{G}(k)\mathbf{h} + \mathbf{n}(k) \tag{8.41}$$

where $\mathbf{G}(k) = [\mathbf{D}_0(k), \mathbf{D}_1(k), \dots, \mathbf{D}_{L-1}(k)]$, $\mathbf{D}_l(k) = \exp[-j2\pi(lk/N)]\mathbf{s}^T(k) \otimes \mathbf{I}_{N_{\mathrm{Rx}}}$.

Rearranging the received signal matrix **G** and the noise matrix at all N subcarriers into a vector as follows,

$$\mathbf{r} = [\mathbf{r}^T(0), \mathbf{r}^T(1), \dots, \mathbf{r}^T(N-1)]^T \tag{8.42}$$

Equation (8.41) can be rewritten more compactly as

$$\mathbf{r} = \frac{1}{\sqrt{N_{\mathrm{Tx}}}} \mathbf{Gh} + \mathbf{n} \tag{8.43}$$

Hence, the SD MMSE estimation of the rearranged channel impulse vector **h** can be simply obtained by

$$\hat{\mathbf{h}} = \mathbf{Qr} \tag{8.44}$$

with

$$\mathbf{Q} = \frac{1}{\sqrt{N_{\mathrm{Tx}}}} \mathbf{C}_h \mathbf{G}^H \left(\frac{1}{N_{\mathrm{Tx}}} \mathbf{GC}_h \mathbf{G}^H + \sigma_n^2 \mathbf{I}_{N \cdot N_{\mathrm{Rx}}} \right)^{-1} \tag{8.45}$$

As usual, the error covariance matrix can be computed as

$$\mathbf{C}_{\mathrm{SD\text{-}MMSE}} = \mathbb{E}[(\mathbf{h} - \hat{\mathbf{h}})(\mathbf{h} - \hat{\mathbf{h}})^H]$$

$$= \left(\sigma_w^2 \mathbf{C}_h^{-1} + \frac{1}{\sqrt{N_{\mathrm{Tx}}}} \mathbf{G}^H \mathbf{G} \right)^{-1} \tag{8.46}$$

assuming $\mathbf{C}_h$ to be invertible.

Therefore the NMSE can be computed as

$$\mathrm{NMSE}_{\mathrm{SD\text{-}MMSE}} = \frac{\mathrm{tr}(\mathbf{C}_{\mathrm{SD\text{-}MMSE}})}{\mathrm{tr}(\mathbf{C}_h)} \tag{8.47}$$

Figure 8.10 shows the SD-MMSE performance given by Equation (8.47) compared to the performance obtained by the ML channel estimation on the subcarriers which carry RS.

8.7   Advanced Techniques

The LTE specifications do not mandate any specific channel estimation technique, and there is therefore complete freedom in implementation provided that the performance requirements are met and the complexity is affordable.

Particular aspects not treated in this chapter, such as channel estimation based on the UE-specific RS, band-edge effect reduction or bursty reception, might require further improvements which go beyond the techniques described here.

Blind and semi-blind techniques are promising for some such aspects, as they try to exploit not only the a priori knowledge of the RS but also the unknown data structure. A comprehensive analysis is available in [24] and references therein.

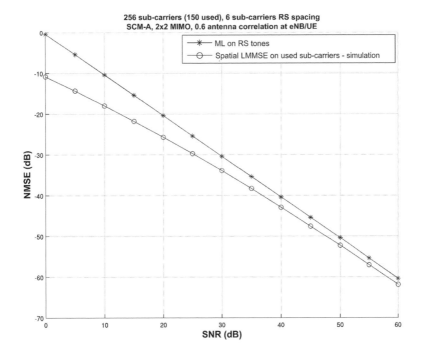

Figure 8.10 Spatial-domain channel estimation performance: CIR NMSE versus SNR.

References[7]

[1] A. Vosoughi and A. Scaglione, 'The Best Training Depends on the Receiver Architecture' in *Proc. IEEE International Conference on Acoustics, Speech, and Signal Processing (ICASSP 2004)* (Montreal, Canada), May 2004.

[2] A. Vosoughi and A. Scaglione, 'Everything You Always Wanted to Know about Training: Guidelines Derived using the Affine Precoding Framework and the CRB'. *IEEE Trans. on Signal Processing*, Vol. 54, pp. 940–954, March 2006.

[3] 3GPP Technical Specification 25.211, 'Physical Channels and Mapping of Transport Channels onto Physical Channels (FDD) (Release 8)', www.3gpp.org.

[4] N. Wiener, *Extrapolation, Interpolation, and Smoothing of Stationary Time Series*. New York: John Wiley & Sons, Ltd/Inc., 1949.

[5] D. T. M. Slock, 'Signal Processing Challenges for Wireless Communications' in *Proc. International Symposium on Control, Communications and Signal Processing (ISCCSP' 2004)* (Hammamet, Tunisia), March 2004.

[6] 3GPP Technical Specification 36.211, 'Physical Channels and Modulation (Release 8)', www.3gpp.org.

[7] R. Negi and J. Cioffi, 'Pilot Tone Selection for Channel Estimation in a Mobile OFDM System'. *IEEE Trans. on Consumer Electronics*, Vol. 44, pp. 1122–1128, August 1998.

---

[7]All web sites confirmed 18th December 2008.

[8] I. Barhumi, G. Leus and M. Moonen, 'Optimal Training Design for MIMO OFDM Systems in Mobile Wireless Channels'. *IEEE Trans. on Signal Processing*, Vol. 51, pp. 1615–1624, June 2003.

[9] 3GPP Technical Report 25.913, 'Requirements for Evolved UTRA (E-UTRA) and Evolved UTRAN (E-UTRAN) (Release 7)', www.3gpp.org.

[10] 3GPP Technical Specification 36.101, 'User Equipment (UE) Radio Transmission and Reception (Release 8)', www.3gpp.org.

[11] Motorola, Nortel, Broadcomm, Nokia, NSN, NTT DoCoMo, NEC, Mitsubishi, Alcatel-Lucent, CATT, Huawei, Sharp, Texas Instrument, ZTE, Panasonic, Philips, and Toshiba, 'R1-081108: Way Forward on Dedicated Reference Signal Design for LTE Downlink with Normal CP', www.3gpp.org, 3GPP TSG RAN WG1, meeting 52, Sorrento, Italy, February 2008.

[12] P. Almers, *et al.* 'Survey of Channel and Radio Propagation Models for Wireless MIMO Systems'. *EURASIP Journal onWireless Communications and Networking*, pp. 957–1000, July 2007.

[13] J. G. Proakis, *Digital Communications*. New York: McGraw Hill, 1995.

[14] R. H. Clarke, 'A Statistical Theory of Mobile-radio Reception'. *Bell Syst. Tech. J.*, pp. 957–1000, July 1968.

[15] W. C. Jakes, *Microwave Mobile Communications*. New York: John Wiley & Sons, Ltd/Inc., 1974.

[16] J.-J. van de Beek, O. Edfors, M. Sandell, S. K. Wilson, and P. O. Börjesson, 'On Channel Estimation in OFDM Systems' in *Proc. VTC'1995, Vehicular Technology Conference (VTC 1995)* (Chicago, USA), July 1995.

[17] M. Morelli and U. Mengali, 'A Comparison of Pilot-aided Channel Estimation Methods for OFDM Systems'. *IEEE Trans. on Signal Processing*, Vol. 49, pp. 3065–3073, December 2001.

[18] P. Hoher, S. Kaiser and P. Robertson, 'Pilot-Symbol-aided Channel Estimation in Time and Frequency' in *Proc. Communication Theory Mini-Conf. (CTMC) within IEEE Global Telecommun. Conf. (Globecom 1997)* (Phoenix, USA), July 1997.

[19] A. Ancora, C. Bona and Slock D. T. M., 'Down-Sampled Impulse Response Least-Squares Channel Estimation for LTE OFDMA' in *Proc. IEEE International Conference on Acoustics, Speech and Signal Processing (ICASSP 2007)* (Honolulu, Hawaii), April 2007.

[20] J. W. Gibbs, 'Fourier Series'. *Nature*, Vol. 59, p. 200, 1898.

[21] D. Schafhuber, G. Matz and F. Hlawatsch, 'Adaptive Wiener Filters for Time-varying Channel Estimation in Wireless OFDM Systems' in *Proc. IEEE International Conference on Acoustics, Speech, and Signal Processing (ICASSP 2003)* (Hong Kong), April 2003.

[22] J. Cai, X. Shen and J. W. Mark, 'Robust Channel Estimation for OFDM Wireless Communication Systems. An H∞ Approach'. *IEEE Trans. Wireless Communications*, Vol. 3, November 2004.

[23] H. Zhang, Y. Li, A. Reid and J. Terry, 'Channel Estimation for MIMO OFDM in Correlated Fading Channels' in *Proc. 2005 IEEE International Conference on Communications (ICC 2005)* (Soeul, South Korea), May 2005.

[24] H. Bölcskei, D. Gesbert and C. Papadias, *Space-time Wireless Systems: From Array Processing to MIMO Communications*. Cambridge: Cambridge University Press, 2006.

9

Downlink Physical Data and Control Channels

Matthew Baker and Tim Moulsley

9.1   Introduction

Chapters 7 and 8 have described the signals which enable User Equipment (UEs) to synchronize with the network and estimate the downlink radio channel in order to be able to demodulate data. In this chapter the downlink physical channels which transport the data are reviewed. This is followed by an explanation of the control-signalling channels which support the data channels by indicating the particular time-frequency transmission resources to which the data is mapped and the format in which the data itself is transmitted.

9.2   Downlink Data-Transporting Channels

9.2.1   Physical Broadcast Channel (PBCH)

In typical cellular systems the basic system information which allows the other channels in the cell to be configured and operated is carried by a Broadcast Channel (BCH). Therefore the achievable coverage for reception of the BCH is crucial to the successful operation of such cellular communication systems; LTE is no exception. As already noted in Chapter 3, the broadcast system information is divided into two categories:

LTE – The UMTS Long Term Evolution: from Theory to Practice   Stefania Sesia, Issam Toufik and Matthew Baker
© 2009 John Wiley & Sons, Ltd

- The 'Master Information Block' (MIB), which consists of a limited number of the most frequently transmitted parameters essential for initial access to the cell,[1] is carried on the Physical Broadcast Channel (PBCH).

- The other System Information Blocks (SIBs) which, at the physical layer, are multiplexed together with unicast data transmitted on the Downlink Shared Channel as discussed in Section 9.2.2.2.

This section focuses in particular on the PBCH, which has some unique design requirements:

- Detectable without prior knowledge of system bandwidth;

- Low system overhead;

- Reliable reception right to the edge of the LTE cells;

- Decodable with low latency and low impact on UE battery life.

We review here the ways in which these requirements have influenced the design selected for the PBCH in LTE, the overall structure of which is shown in Figure 9.1.

Detectability without the UE having prior knowledge of the system bandwidth is achieved by mapping the PBCH only to the central 72 subcarriers of the OFDM signal (which corresponds to the minimum possible LTE system bandwidth), regardless of the actual system bandwidth. The UE will have first identified the system centre-frequency from the synchronization signals as described in Section 7.

Low system overhead for the PBCH is achieved by deliberately keeping the amount of information carried on the PBCH to a minimum, since achieving stringent coverage requirements for a large quantity of data would result in a high system overhead. The size of the MIB is therefore just 14 bits, and, since it is repeated every 40 ms, this corresponds to a data rate on the PBCH of just 350 bps.

The main mechanisms employed to facilitate reliable reception of the PBCH in LTE are time diversity, forward error correction coding and antenna diversity.

Time diversity is exploited by spreading out the transmission of each MIB on the PBCH over a period of 40 ms. This significantly reduces the likelihood of a whole MIB being lost in a fade in the radio propagation channel, even when the mobile terminal is moving at pedestrian speeds.

The forward error correction coding for the PBCH uses a convolutional coder, as the number of information bits to be coded is small; the details of the convolutional coder are explained in Section 10.3.3. The basic code rate is $1/3$, after which a high degree of repetition of the systematic (i.e. information) bits and parity bits is used, such that each MIB is coded at a very low code-rate ($1/48$ over a 40 ms period) to give strong error protection.

Antenna diversity may be utilized at both the eNodeB and the UE. The UE performance requirements specified for LTE assume that all UEs can achieve a level of decoding performance commensurate with dual-antenna receive diversity (although it is recognized that in low-frequency deployments, such as below 1 GHz, the advantage obtained from

---

[1] The MIB information consists of the downlink system bandwidth, the PHICH structure (Physical Hybrid ARQ Indicator Channel, see Section 9.3.2.4), and the most-significant eight bits of the System Frame Number – the remaining two bits of the System Frame Number being gleaned from the 40 ms periodicity of the PBCH.

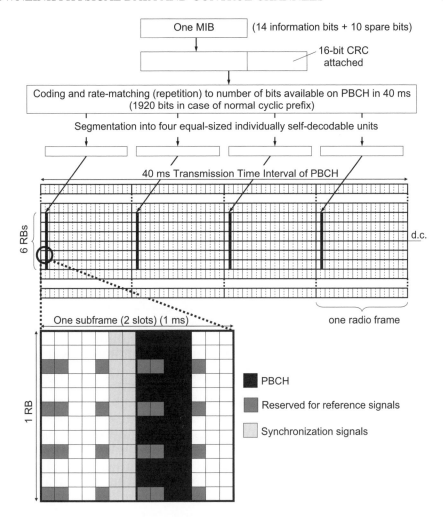

Figure 9.1 PBCH structure.

receive antenna diversity is reduced due to the correspondingly higher correlation between the antennas); this enables LTE system planners to rely on this level of performance being common to all UEs, thereby enabling wider cell coverage to be achieved with fewer cell sites than would otherwise be possible. Transmit antenna diversity may be also employed at the eNodeB to further improve coverage, depending on the capability of the eNodeB; eNodeBs with two or four transmit antenna ports transmit the PBCH using a Space-Frequency Block Code (SFBC), details of which are explained in Section 11.2.2.1.

The exact set of resource elements used by the PBCH is independent of the number of transmit antenna ports used by the eNodeB; any resource elements which may be used for reference signal transmission are avoided by the PBCH, irrespective of the actual number of transmit antenna ports deployed at the eNodeB. The number of transmit antenna ports used by the eNodeB must be determined blindly by the UE, by performing the decoding for each SFBC scheme corresponding to the different possible numbers of transmit antenna ports (namely one, two or four). This discovery of the number of transmit antenna ports is further facilitated by the fact that the Cyclic Redundancy Check (CRC) on each MIB is masked with a codeword representing the number of transmit antenna ports.

Finally, achieving low latency and a low impact on UE battery life is also facilitated by the design of the coding outlined above: the low code rate with repetition enables the full set of coded bits to be divided into four subsets, each of which is self-decodable in its own right. Each of these subsets of the coded bits is then transmitted in a different one of the four radio frames during the 40 ms transmission period, as shown in Figure 9.1. This means that if the Signal to Interference Ratio (SIR) of the radio channel is sufficiently good to allow the UE to decode the MIB correctly from the transmission in less than four radio frames, then the UE does not need to receive the other parts of the PBCH transmission in the remainder of the 40 ms period; on the other hand, if the SIR is low, the UE can receive further parts of the MIB transmission, soft-combining each part with those received already, until successful decoding is achieved.

The timing of the 40 ms transmission interval for each MIB on the PBCH is not indicated explicitly to the UE; this is determined implicitly from the scrambling and bit positions, which are re-initialized every 40 ms. The UE can therefore initially determine the 40 ms timing by performing four separate decodings of the PBCH using each of the four possible phases of the PBCH scrambling code, checking the CRC for each decoding.

When a UE initially attempts to access a cell by reading the PBCH, a variety of approaches may be taken to carry out the necessary blind decodings. A simple approach is always to perform the decoding using a soft combination of the PBCH over four radio frames, advancing a 40 ms sliding window one radio frame at a time until the window aligns with the 40 ms period of the PBCH and the decoding succeeds. However, this would result in a 40–70 ms delay before the PBCH can be decoded. A faster approach would be to attempt to decode the PBCH from the first single radio frame, which should be possible provided the SIR is sufficiently high; if the decoding fails for all four possible scrambling code phases, the PBCH from the first frame could be soft-combined with the PBCH bits received in the next frame – there is a 3-in-4 chance that the two frames contain data from the same transport block. If decoding still fails, a third radio frame could be combined, and failing that a fourth. It is evident that the latter approach may be much faster (potentially taking only 10 ms), but on the other hand requires slightly more complex logic.

9.2.2 Physical Downlink Shared Channel (PDSCH)

The Physical Downlink Shared Channel (PDSCH) is the main data-bearing downlink channel in LTE. It is used for all user data, as well as for broadcast system information which is not carried on the PBCH, and for paging messages – there is no specific physical layer paging channel in the LTE system. In this section, the use of the PDSCH for user data is explained; the use of the PDSCH for system information and paging is covered in the next section.

Data is transmitted on the PDSCH in units known as *transport blocks*, each of which corresponds to a MAC-layer Protocol Data Unit (PDU) as described in Section 4.4. Transport blocks may be passed down from the MAC layer to the physical layer once per Transmission Time Interval (TTI), where a TTI is 1 ms, corresponding to the subframe duration.

9.2.2.1 General Use of the PDSCH

When employed for user data, one or, at most, two transport blocks can be transmitted per UE per subframe, depending on the transmission mode selected for the PDSCH for each UE. The transmission mode configures the multi-antenna technique usually applied:

Transmission Mode 1: Transmission from a single eNodeB antenna port;

Transmission Mode 2: Transmit diversity (see Section 11.2.2.1);

Transmission Mode 3: Open-loop spatial multiplexing (see Section 11.2.2.3);

Transmission Mode 4: Closed-loop spatial multiplexing (see Section 11.2.2.3);

Transmission Mode 5: Multi-user Multiple-Input Multiple-Output (MIMO) (see Section 11.2.3);

Transmission Mode 6: Closed-loop rank-1 precoding (see Section 11.2.2.3);

Transmission Mode 7: Transmission using UE-specific reference signals (see Sections 11.2.2.2 and 8.2).

With the exception of transmission mode 7, the phase reference for demodulating the PDSCH is given by the cell-specific Reference Signals (RSs) described in Section 8.2, and the number of eNodeB antenna ports used for transmission of the PDSCH is the same as the number of antenna ports used in the cell for the PBCH. In transmission mode 7, UE-specific RSs (also covered in Section 8.2) provide the phase reference for the PDSCH. The configured transmission mode also affects the transmission of the associated downlink control signalling, as described in Section 9.3, and the channel quality feedback from the UE (see Section 10.2.1).

After channel coding (see Section 10.3.2) and mapping to spatial layers according to the selected transmission mode, the coded data bits are mapped to modulation symbols depending on the modulation scheme selected for the current radio channel conditions and required data rate. The modulation order may be varied between two bits per symbol (using QPSK (Quadrature Phase Shift Keying)) and six bits per symbol (using 64QAM (Quadrature Amplitude Modulation)). Support for reception of 64QAM modulation is mandatory for all classes of LTE UE and is designed to achieve the high peak downlink data rates that are required for LTE. The available modulation schemes are illustrated in Figure 9.2 by means of their constellation diagrams.

The resource elements used for the PDSCH can be any which are not reserved for other purposes (i.e. reference signals, synchronization signals, PBCH and control signalling). Thus when the control signalling informs a UE that a particular pair of resource blocks[2] in a subframe are allocated to that UE, it is only the *available* resource elements within those resource blocks which actually carry PDSCH data.

Normally the allocation of pairs of resource blocks to PDSCH transmission for a particular UE is signalled to the UE by means of dynamic control signalling transmitted at the start of

---

[2]The term 'pair of resource blocks' here means a pair of resource blocks which occupy the same set of 12 subcarriers and are contiguous in time, thus having a duration of one subframe.

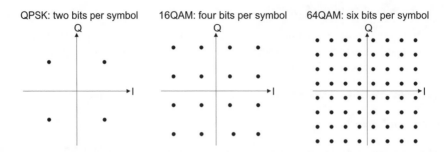

Figure 9.2 Constellations of modulation schemes applicable to PDSCH transmission.

the relevant subframe using the Physical Downlink Control Channel (PDCCH), as described in Section 9.3.

The mapping of data to physical resource blocks can be carried out in one of two ways: *localized mapping* and *distributed mapping*.

Localized resource mapping entails allocating all the available resource elements in a pair of resource blocks to the same UE. This is suitable for most scenarios, including the use of dynamic channel-dependent scheduling of resource blocks according to frequency-specific channel quality information reported by the UE (see Sections 10.2.1 and 12.4).

Distributed resource mapping entails separating in frequency the two physical resource blocks comprising each pair, as shown in Figure 9.3. This is a useful means of obtaining frequency diversity for small amounts of data which would otherwise be constrained to a narrow part of the downlink bandwidth and would therefore be more suscepti-ble to narrow-band fading. An amount of data corresponding to up to two pairs of resource blocks may be transmitted to a UE in this way. An example of a typical use for this transmission mode could be a Voice-over-IP (VoIP) service, where, in order to minimize overhead, certain frequency resources may be 'persistently-scheduled' (see Section 4.4.2.1) – in other words, certain resource blocks in the frequency domain are allocated on a periodic basis to a specific UE by RRC signalling rather than by dynamic PDCCH signalling. This means that the resources allocated are not able to benefit from dynamic channel-dependent scheduling and therefore the frequency diversity which is achieved through distributed mapping is a useful tool to improve performance. Moreover, as the amount of data to be transmitted per UE for a VoIP service is small (typi-cally sufficient to occupy only one or two pairs of resource blocks in a given sub-frame), the degree of frequency diversity obtainable via localized scheduling is very limited.

When data is mapped using the distributed mode, a frequency-hop occurs at the slot boundary in the middle of the subframe. This results in the block of data for a given UE being transmitted on one set of 12 subcarriers in the first half of the subframe and on a different set of 12 subcarriers in the second half of the subframe. This is illustrated in Figure 9.3.

The potential increase in the number of VoIP users which can be accommodated in a cell as a result of using distributed resource mapping as opposed to localized resource mapping is illustrated by way of example in Figure 9.4. In this example, the main simulation parameters are as given in Table 9.1.

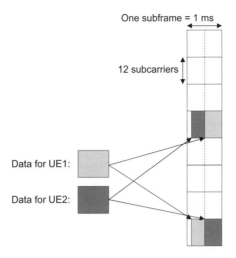

Figure 9.3 Frequency-distributed data mapping in LTE downlink.

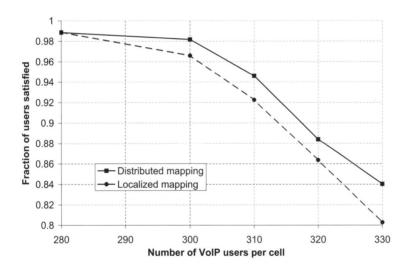

Figure 9.4 Example of increase in VoIP capacity arising from frequency-distributed resource mapping.

9.2.2.2 Special Uses of the PDSCH

As noted above, the PDSCH is used for some special purposes in addition to normal user data transmission.

One such use is sometimes referred to as the 'Dynamic BCH'. This consists of all the broadcast system information (i.e. SIBs) that is not carried on the PBCH. The resource blocks used for broadcast data of this sort are indicated by signalling messages on the

Table 9.1 Key simulation parameters for Figure 9.4.

Parameter	Value
Carrier frequency	2 GHz
Bandwidth	5 MHz
Channel model	Urban micro, 3 km/h
Total eNodeB transmit power	43 dBm
VoIP model	12.2 kbps; voice activity factor 0.43
Modulation and coding scheme	Fixed: QPSK, code rate 2/3
Satisfaction criterion	98% packets within 50 ms

PDCCH in the same way as for other PDSCH data, except that the identity indicated on the PDCCH is not the identity of a specific UE but is, rather, a designated broadcast identity known as the System Information Radio Network Temporary Identifier (SI-RNTI), which is fixed in the specifications (see Section 7.1 of [1]) and therefore known a priori to all UEs. Some constraints exist as to which subframes may be used for particular system information messages on the PDSCH; these are explained in Section 3.2.2.

Another special use of the PDSCH is for paging, as no separate physical channel is provided in LTE for this purpose. In previous systems such as WCDMA, a special 'Paging Indicator Channel' was provided, which was specially designed to enable the UE to wake up its receiver periodically for a very short period of time, in order to minimize the impact on battery life; on detecting a paging indicator (typically for a group of UEs), the UE would then keep its receiver switched on to receive a longer message indicating the exact identity of the UE being paged. By contrast, in LTE the PDCCH signalling is already very short in duration, and therefore the impact on UE battery life of monitoring the PDCCH from time to time is low. Therefore the normal PDCCH signalling can be used to carry the equivalent of a paging indicator, with the detailed paging information being carried on the PDSCH in a resource block indicated by the PDCCH. In a similar way to broadcast data, paging indicators on the PDCCH use a single fixed identifier, in this case the Paging RNTI (P-RNTI). Rather than providing different paging identifiers for different groups of UEs, different UEs monitor different subframes for their paging messages, as described in Section 3.4.

9.2.3 Physical Multicast Channel (PMCH)

Although Multimedia Broadcast and Multicast Services (MBMS) are not included in the first release of the LTE specifications, nonetheless the physical layer structure to support MBMS is defined ready for deployment in a later release. All UEs must be aware of the possible existence of MBMS transmissions at the physical layer, in order to enable such transmissions to be introduced later in a backward-compatible way.

The basic structure of the Physical Multicast Channel (PMCH) is very similar to the PDSCH. However, the PMCH is designed for 'single-frequency network' operation, whereby multiple cells transmit the same modulated symbols with very tight time-synchronization, ideally so that the signals from different cells are received within the duration of the cyclic prefix. This is known as MBSFN (MBMS Single Frequency Network) operation, and is discussed in more detail in Section 14.3. As the channel in MBSFN operation is in effect a composite channel from multiple cells, it is necessary for the UE to perform a separate channel estimate for MBSFN reception from that performed for reception of data from a single cell. Therefore, in order to avoid the need to mix normal reference symbols and reference symbols for MBSFN in the same subframe, frequency-division multiplexing of the PMCH and PDSCH is not permitted within a given subframe; instead, certain subframes may be specifically designated for MBSFN, and it is in these subframes that the PMCH would be transmitted.

The key differences from PDSCH in respect of the PMCH are as follows:

- The dynamic control signalling (PDCCH and PHICH – see Section 9.3) cannot occupy more than two OFDM symbols in an MBSFN subframe. The PDCCH is used only for uplink resource grants and not for the PMCH, as the scheduling of MBSFN data on the PMCH is carried out by higher-layer signalling.

- The pattern of reference symbols embedded in the PMCH is different from that in the PDSCH, as discussed in Chapter 8. (Note, however, that the common reference symbol pattern embedded in the OFDM symbols carrying control signalling at the start of each subframe remains the same as in the non-MBSFN subframes.)

- The extended cyclic prefix is always used. Note, however, that if the non-MBSFN subframes use the normal cyclic prefix, then the normal cyclic prefix is also used in the OFDM symbols used for the control signalling at the start of each MBSFN subframe. This results in there being some spare time samples whose usage is unspecified between the end of the last control signalling symbol and the first PMCH symbol, the PMCH remaining aligned with the end of the subframe; the eNodeB may transmit an undefined signal (e.g. a cyclic extension) during these time samples, or alternatively it may switch off its transmitter – the UE cannot assume anything about the transmitted signal during these samples.

The latter two features are designed so that a UE making measurements on a neighbouring cell does not need to know in advance the allocation of MBSFN and non-MBSFN subframes. The UE can take advantage of the fact that the first two OFDM symbols in all subframes use the same cyclic prefix duration and reference symbol pattern.

The exact pattern of MBSFN subframes in a cell is indicated in the system information carried on the part of the broadcast channel mapped to the PDSCH. The system information also indicates whether the pattern of MBSFN subframes in neighbouring cells is the same as or different from that in the current cell; however, if the pattern in the neighbouring cell is different, the UE can only ascertain the pattern by reading the system information of that cell.

Further details of multicast and broadcast operation in LTE are explained in Chapter 14.

9.3   Downlink Control Channels

9.3.1   Requirements for Control Channel Design

The control channels in LTE are provided in order to support efficient data transmission. In common with other wireless systems, the control channels convey physical layer signals or messages which cannot be carried sufficiently efficiently, quickly or conveniently by higher layers. The design of the control channels transmitted in the LTE downlink aims to balance a number of somewhat conflicting requirements, the most important of which are discussed below.

9.3.1.1 Physical Layer Signalling to Support the MAC Layer

The general requirement to support Medium Access Control (MAC) operation is very similar to that in WCDMA, but there are a number of differences of detail, mainly arising from the frequency domain resource allocation supported in the LTE multiple access schemes.

The use of the uplink transmission resources on the Physical Uplink Shared Channel (PUSCH) is determined dynamically by an uplink scheduling process in the eNodeB, and therefore physical layer signalling must be provided to indicate to UEs which time/frequency resources they have been granted permission to use.

The eNodeB also schedules downlink transmissions on the PDSCH, and therefore similar physical layer messages from the eNodeB are needed to indicate which resources in the frequency domain contain the downlink data transmissions intended for particular UEs, together with parameters such as the modulation and code rate used for the data. Explicit signalling of this kind avoids the considerable additional complexity which would arise if UEs needed to search for their data among all the possible combinations of data packet size, format and resource allocation.

In order to facilitate efficient operation of Hybrid Automatic Repeat reQuest (HARQ) and ensure that uplink transmissions are made at appropriate power levels, further physical layer signals are also needed to convey acknowledgements of uplink data packets received by the eNodeB, and power control commands to adjust the uplink transmission power (as explained in Section 20.3).

9.3.1.2 Flexibility, Overhead and Complexity

The LTE physical layer specification is intended to allow operation in any system bandwidth from six resource blocks (1.08 MHz) to 110 resource blocks (19.8 MHz). It is also designed to support a range of scenarios including, for example, just a few users in a cell each demanding high data rates, or very many users with low data rates. Considering the possibility that both uplink resource grants and downlink resource allocations could be required for every UE in each subframe, the number of control channel messages carrying resource information could be as many as a couple of hundred if every resource allocation were as small as one resource block. Since every additional control channel message implies additional overhead which will consume downlink resources, it is desirable that the control channel is designed to minimize unnecessary overhead for any given signalling load, whatever the system bandwidth.

Similar considerations apply to the signalling of HARQ acknowledgements for each uplink packet transmission.

Furthermore, as in any mobile communication system, the power consumption of the terminals is an important consideration for LTE. Therefore, the control signalling must be designed so that the necessary scaleability and flexibility is achieved without undue decoding complexity.

9.3.1.3 Coverage and Robustness

In order to achieve good coverage it must be possible to configure the system so that the control channels can be received with sufficient reliability over a substantial part of every cell. This can be seen by considering, as an example, messages indicating resource allocation. If any such messages are not received correctly, then the corresponding data transmission will also fail, with a direct and proportionate impact on throughput efficiency. Techniques such as channel coding and frequency diversity can be used to make the control channels more robust. However, in order to make good use of system resources, it is desirable to be able to adapt the transmission parameters of the control signalling for different UEs or groups of UEs, so that lower code rates and higher power levels are only applied for those UEs for which it is necessary (e.g. near the cell border, where signal levels are likely to be low and interference from other cells high).

Also, it is desirable to avoid unintended reception of control channels from other cells, for example, by applying cell-specific randomization, particularly under conditions of high inter-cell interference.

9.3.1.4 System-Related Design Aspects

Since the different parts of LTE are intended to provide a complete system, some aspects of control channel design cannot be considered in isolation.

A basic design decision in LTE is that a control channel is intended to be transmitted in a single cell to a particular UE (or in some cases a group of UEs). In addition, in order to minimize signalling latency when indicating a resource allocation, a control channel transmission should be completed within one subframe. Therefore, in order to reach multiple UEs in a cell within a subframe, it must be possible to transmit multiple control channels within the duration of a single subframe. However, in those cases where the information sent via the control channels is intended for reception by more than one UE (for example, when relating to the transmission of a SIB on the PDSCH), it is more efficient to arrange for all the UEs to receive a single transmission rather than to transmit the same information to each UE individually. This requires that both common and dedicated control channel messages should be supported.

As noted previously, some scenarios may be characterized by low average data rates to a large number of UEs, for example when supporting VoIP traffic. If the data arrives at the eNodeB on a regular basis, as is typical for VoIP, then it is possible to predict in advance when resources will need to be allocated in the downlink or granted in the uplink. In such cases the number of control channel messages which need to be sent can be dramatically reduced by means of 'persistent scheduling' as discussed in Section 4.4.2.1.

9.3.2 Control Channel Structure and Contents

In this section we describe the downlink control channel features selected for inclusion in the LTE specifications and give some additional background on the design decisions. In general, the downlink control channels can be configured to occupy the first 1, 2 or 3 OFDM symbols in a subframe, extending over the entire system bandwidth as shown in Figure 9.5. There are two special cases: in subframes containing MBSFN transmissions there may be 0, 1 or 2 symbols for control signalling, while for narrow system bandwidths (less than 10 resource

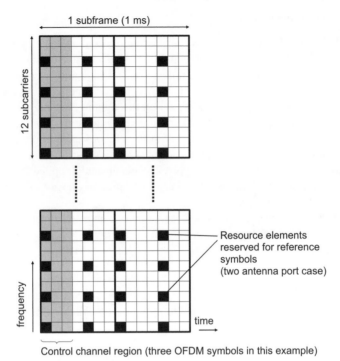

Figure 9.5 Time-frequency region used for downlink control signalling.

blocks) the number of control symbols is increased, and may be 2, 3 or 4 to ensure sufficient coverage at the cell border. This flexibility allows the control channel overhead to be adjusted according to the particular system configuration, traffic scenario and channel conditions.

9.3.2.1 Physical Control Format Indicator Channel (PCFICH)

The PCFICH carries a Control Format Indicator (CFI) which indicates the number of OFDM symbols (i.e. normally 1, 2 or 3) used for transmission of control channel information in each subframe. In principle the UE could deduce the value of the CFI without a channel such as the PCFICH, for example by multiple attempts to decode the control channels assuming each possible number of symbols, but this would result in significant additional processing load.

For carriers dedicated to MBSFN there are no physical control channels, so the PCFICH is not present in these cases.

Three different CFI values are used in the first version of LTE, and a fourth codeword is reserved for future use. In order to make the CFI sufficiently robust each codeword is 32 bits in length, as shown in Table 9.2. These 32 bits are mapped to 16 resource elements using QPSK modulation.

The PCFICH is transmitted on the same set of antenna ports as the PBCH, with transmit diversity being applied if more than one antenna port is used.

In order to achieve frequency diversity, the 16 resource elements carrying the PCFICH are distributed across the frequency domain. This is done according to a predefined pattern in the

Table 9.2 Control Format Indicator codewords.

CFI	CFI codeword $< b_0, b_1, \ldots, b_{31} >$
1	$< 0, 1, 1, 0, 1, 1, 0, 1, 1, 0, 1, 1, 0, 1, 1, 0, 1, 1, 0, 1, 1, 0, 1, 1, 0, 1, 1, 0, 1, 1, 0, 1 >$
2	$< 1, 0, 1, 1, 0, 1, 1, 0, 1, 1, 0, 1, 1, 0, 1, 1, 0, 1, 1, 0, 1, 1, 0, 1, 1, 0, 1, 1, 0, 1, 1, 0 >$
3	$< 1, 1, 0, 1, 1, 0, 1, 1, 0, 1, 1, 0, 1, 1, 0, 1, 1, 0, 1, 1, 0, 1, 1, 0, 1, 1, 0, 1, 1, 0, 1, 1 >$
4 (Reserved)	$< 0, 0, 0, 0, 0, 0, 0, 0, 0, 0, 0, 0, 0, 0, 0, 0, 0, 0, 0, 0, 0, 0, 0, 0, 0, 0, 0, 0, 0, 0, 0, 0 >$

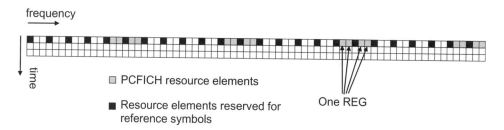

Figure 9.6 PCFICH mapping to Resource Element Groups (REGs).

first OFDM symbol in each downlink subframe (see Figure 9.6), so that the UEs can always locate the PCFICH information, which is a prerequisite to being able to decode the rest of the control signalling.

To minimize the possibility of confusion with PCFICH information from a neighbouring cell, a cell-specific frequency offset is applied to the positions of the PCFICH resource elements; this offset depends on the Physical Cell ID, which is deduced from the primary and secondary synchronization signals as explained in Section 7.2. In addition, a cell-specific scrambling sequence (again a function of the Physical Cell ID) is applied to the CFI codewords, so that the UE can preferentially receive the PCFICH from the desired cell.

9.3.2.2   Physical Downlink Control Channel (PDCCH)

A PDCCH carries a message known as Downlink Control Information (DCI), which includes resource assignments and other control information for a UE or group of UEs. In general, several PDCCHs can be transmitted in a subframe.

Each PDCCH is transmitted using one or more so-called Control Channel Elements (CCEs), where each CCE corresponds to nine sets of four physical resource elements known as Resource Element Groups (REGs). Four QPSK symbols are mapped to each REG. The resource elements occupied by reference symbols are not included within the REGs, which means that the total number of REGs in a given OFDM symbol depends on whether or not cell-specific reference signals are present. The concept of REGs (i.e. mapping in groups of four resource elements) is also used for the other downlink control channels (the PCFICH and PHICH).

Four PDCCH formats are supported, as listed in Table 9.3.

Table 9.3 PDCCH formats.

PDCCH format	Number of CCEs (n)	Number of REGs	Number of PDCCH bits
0	1	9	72
1	2	18	144
2	4	36	288
3	8	72	576

CCEs are numbered and used consecutively, and, to simplify the decoding process, a PDCCH with a format consisting of n CCEs may only start with a CCE with a number equal to a multiple of n.

The number of CCEs used for transmission of a particular PDCCH is determined by the eNodeB according to the channel conditions. For example, if the PDCCH is intended for a UE with a good downlink channel (e.g. close to the eNodeB), then one CCE is likely to be sufficient. However, for a UE with a poor channel (e.g. near the cell border) then eight CCEs may be required in order to achieve sufficient robustness. In addition, the power level of a PDCCH may be adjusted to match the channel conditions.

9.3.2.3 Formats for Downlink Control Information

The control channel messages are required to convey various pieces of information, but the useful content depends on the specific case of system deployment and operation. For example, if the infrastructure does not support MIMO, or if a UE is configured in a transmission mode which does not involve MIMO, there is no need to signal the parameters which are only required for MIMO transmissions. In order to minimize the signalling overhead it is therefore desirable that several different message formats are available, each containing the minimum payload required for a particular scenario. However, to avoid too much complexity in implementation and testing, it is desirable not to specify too many formats. The set of Downlink Control Information (DCI) message formats in Table 9.4 is specified in the first version of LTE. These are designed to cover the most useful cases. Additional formats may be defined in future.

In general the number of bits required for resource assignment depends on the system bandwidth, and therefore the message sizes also vary with the system bandwidth. Table 9.4 gives the number of bits in a PDCCH for uplink and downlink bandwidths of 50 resource blocks, corresponding to a spectrum allocation of about 10 MHz. In order to avoid additional complexity at the UE receiver, Formats 0 and 1A are designed to be always the same size. However, since these messages may have different numbers of bits, for example if the uplink and downlink bandwidths are different, leading to different numbers of bits required for indicating resource assignments, the smaller format size is extended by adding padding bits to be the same size as the larger.

The information content of the different DCI message formats is listed below.

Format 0. DCI Format 0 is used for the transmission of resource grants for the PUSCH. The following information is transmitted:

- Flag to differentiate between Format 0 and Format 1A

- Resource block grant

- Modulation and coding scheme

- HARQ information and redundancy version

- Power control command for scheduled PUSCH

- Request for transmission of an aperiodic CQI report (see Section 10.2.1).

Table 9.4 Supported DCI formats.

DCI format	Purpose	Number of bits including CRC (for a system bandwidth of 50 RBs and four antennas at eNodeB)
0	PUSCH grants	42
1	PDSCH assignments with a single codeword	47
1A	PDSCH assignments using a compact format	42
1B	PDSCH assignments for rank-1 transmission	46
1C	PDSCH assignments using a very compact format	26
1D	PDSCH assignments for multi-user MIMO	46
2	PDSCH assignments for closed-loop MIMO operation	62
2A	PDSCH assignments for open-loop MIMO operation	58
3	Transmit Power Control (TPC) commands for multiple users for PUCCH and PUSCH with 2-bit power adjustments	42
3A	Transmit Power Control (TPC) commands for multiple users for PUCCH and PUSCH with 1-bit power adjustments	42

Format 1.   DCI Format 1 is used for the transmission of resource assignments for single codeword PDSCH transmissions (transmission modes 1, 2 and 7 (see Section 9.2.2.1)). The following information is transmitted:

- Resource allocation type (see Section 9.3.3.1)

- Resource block assignment

- Modulation and coding scheme

- HARQ information

- Power control command for Physical Uplink Control Channel (PUCCH).

Format 1A. DCI Format 1A is used for compact signalling of resource assignments for single codeword PDSCH transmissions, and for allocating a dedicated preamble signature to a UE for contention-free random access (see Section 19.3.2). The following information is transmitted:

- Flag to differentiate between Format 0 and Format 1A

- Flag to indicate that the distributed mapping mode (see Section 9.2.2.1) is used for the PDSCH transmission (otherwise the allocation is a contiguous set of physical resource blocks)

- Resource block assignment

- Modulation and coding scheme

- HARQ information

- Power control command for PUCCH.

Format 1B. DCI Format 1B is used for compact signalling of resource assignments for PDSCH transmissions using closed loop precoding with rank-1 transmission (transmission mode 6). The information transmitted is the same as in Format 1A, but with the addition of an indicator of the precoding vector applied for the PDSCH transmission.

Format 1C. DCI Format 1C is used for very compact transmission of PDSCH assignments. When format 1C is used, the PDSCH transmission is constrained to using QPSK modulation. This is used, for example, for signalling paging messages and some broadcast system information messages (see Section 9.2.2.2). The following information is transmitted:

- Resource block assignment

- Modulation and coding scheme

- Redundancy version.

Format 1D. DCI Format 1D is used for compact signalling of resource assignments for PDSCH transmissions using multi-user MIMO (transmission mode 5). The information transmitted is the same as in Format 1B, but instead of one of the bits of the precoding vector indicators, there is a single bit to indicate whether a power offset is applied to the data symbols. This feature is needed to show whether or not the transmission power is shared between two UEs. Future versions of LTE may extend this to the case of power sharing between larger numbers of UEs.

Format 2. DCI Format 2 is used for the transmission of resource assignments for PDSCH for closed-loop MIMO operation (transmission mode 4). The following information is transmitted:

- Resource allocation type (see Section 9.3.3.1)

- Resource block assignment

- Power control command for PUCCH

- HARQ information

- Modulation and coding schemes for each codeword

- Number of spatial layers

- Precoding information.

Format 2A. DCI Format 2A is used for the transmission of resource assignments for PDSCH for open-loop MIMO operation (transmission mode 3). The information transmitted is the same as for Format 2, except that if the eNodeB has two transmit antenna ports, there is no precoding information, and for four antenna ports two bits are used to indicate the transmission rank.

Formats 3 and 3A. DCI Formats 3 and 3A are used for the transmission of power control commands for PUCCH and PUSCH with 2-bit or 1-bit power adjustments respectively. These DCI formats contain individual power control commands for a group of UEs.

CRC attachment. In order that the UE can identify whether it has received a PDCCH transmission correctly, error detection is provided by means of a 16-bit CRC appended to each PDCCH. Furthermore, it is necessary that the UE can identify which PDCCH(s) are intended for it. This could in theory be achieved by adding an identifier to the PDCCH payload; however, it turns out to be more efficient to scramble the CRC with the 'UE identity', which saves the additional payload but at the cost of a small increase in the probability of falsely detecting a PDCCH intended for another UE.

In addition, for UEs which support antenna selection for uplink transmissions (see Section 17.5), the requested antenna may be indicated using Format 0 by applying an antenna-specific mask to the CRC. This has the advantage that the same size of DCI message can be used, irrespective of whether antenna selection is used.

PDCCH construction. In order to provide robustness against transmission errors, the PDCCH information bits are coded as described in Section 10.3.3. The set of coded and rate-matched bits for each PDCCH are then scrambled with a cell-specific scrambling sequence; this reduces the possibility of confusion with PDCCH transmissions from neighbouring cells. The scrambled bits are mapped to blocks of four QPSK symbols (REGs). Interleaving is applied to these symbol blocks, to provide frequency diversity, followed by mapping to the available physical resource elements on the set of OFDM symbols indicated by the PCFICH.

This mapping process excludes the resource elements reserved for reference signals and the other control channels (PCFICH and PHICH).

The PDCCHs are transmitted on the same set of antenna ports as the PBCH, and transmit diversity is applied if more than one antenna port is used.

9.3.2.4   Physical Hybrid ARQ Indicator Channel (PHICH)

The PHICH carries the HARQ ACK/NACK, which indicates whether the eNodeB has correctly received a transmission on the PUSCH. The HARQ indicator is set to 0 for a positive ACKnowledgement (ACK) and 1 for a Negative ACKnowledgement (NACK). This information is repeated in each of three BPSK (Binary Phase Shift Keying) symbols.

Multiple PHICHs are mapped to the same set of resource elements. These constitute a PHICH group, where different PHICHs within the same PHICH group are separated through different complex orthogonal Walsh sequences. The sequence length is four for the normal cyclic prefix (or two in the case of the extended cyclic prefix). As the sequences are complex, the number of PHICHs in a group (i.e. the number of UEs receiving their acknowledgements on the same set of downlink resource elements) can be up to twice the sequence length. A cell-specific scrambling sequence is applied.

Factor-3 repetition coding is applied for robustness, resulting in three instances of the orthogonal Walsh code being transmitted for each ACK or NACK. The error rate on the PHICH is intended to be of the order of 10^{-2} for ACKs and as low as 10^{-4} for NACKs. The resulting PHICH construction, including repetition and orthogonal spreading, is shown in Figure 9.7.

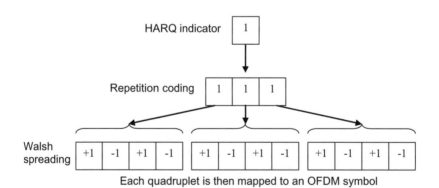

Figure 9.7   PHICH signal construction.

The PHICH duration, in terms of the number of OFDM symbols used in the time domain, is configurable (by an indication transmitted on the PBCH), normally to either one or three OFDM symbols. In some special cases[3] the three-OFDM-symbol duration is reduced to two OFDM symbols. As the PHICH cannot extend into the PDSCH transmission region, the

---

[3]The special cases when the PHICH duration is two OFDM symbols are (i) MBSFN subframes on mixed carriers supporting MBSFN and unicast data, and (ii) the second and seventh subframes in case of frame structure type 2 for Time Division Duplex (TDD) operation.

duration configured for the PHICH puts a lower limit on the size of the control channel region at the start of each subframe (as signalled by the PCFICH).

Finally, each of the three instances of the orthogonal code of a PHICH transmission is mapped to a REG on one of the first three OFDM symbols of each subframe,[4] in such a way that each PHICH is partly transmitted on each of the available OFDM symbols. This mapping is illustrated in Figure 9.8 for each possible PHICH duration.

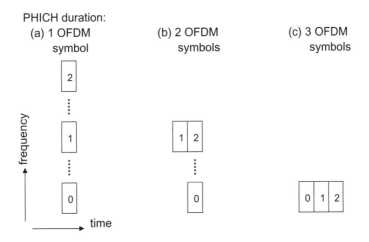

Figure 9.8 Examples of the mapping of the three instances of a PHICH orthogonal code to OFDM symbols, depending on the configured PHICH duration.

The PBCH also signals the number of PHICH groups configured in the cell, which enables the UEs to deduce to which remaining resource elements in the control region the PDCCHs are mapped.

In order to obviate the need for additional signalling to indicate which PHICH carries the ACK/NACK response for each PUSCH transmission, the PHICH index is implicitly associated with the index of the lowest uplink resource block used for the corresponding PUSCH transmission. This relationship is such that adjacent PUSCH resource blocks are associated with PHICHs in different PHICH groups, to enable some degree of load balancing. However, this mechanism alone is not sufficient to enable multiple UEs to be allocated the same resource blocks for a PUSCH transmission, as occurs in the case of uplink multi-user MIMO (see Section 17.5); in this case, different cyclic shifts of the uplink demodulation reference signals are configured for the different UEs which are allocated the same time-frequency PUSCH resources, and the same cyclic shift index is then used to shift the PHICH allocations in the downlink so that each UE will receive its ACK or NACK on a different PHICH. This mapping of the PHICH allocations is illustrated in Figure 9.9.

The PHICHs are transmitted on the same set of antenna ports as the PBCH, and transmit diversity is applied if more than one antenna port is used.

---

[4]The mapping avoids resource elements used for reference symbols or PCFICH.

PHICH index

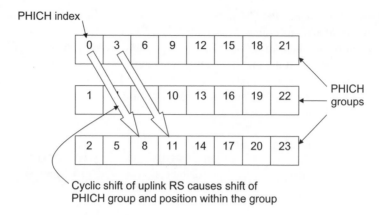

Cyclic shift of uplink RS causes shift of
PHICH group and position within the group

Figure 9.9 Indexing of PHICHs within PHICH groups, and shifting in case of cyclic shifting of the uplink demodulation reference signals.

9.3.3 Control Channel Operation

9.3.3.1 Resource Allocation

Conveying indications of physical layer resource allocation is one of the major functions provided by the PDCCHs. However, the exact use of the PDCCHs depends on the design of the algorithms implemented in the eNodeB. Nevertheless, it is possible to outline some general principles which are likely to be followed in typical systems.

In each subframe, PDCCHs indicate the frequency domain resource allocations. As discussed in Section 9.2.2.1, resource allocations are normally localized, meaning that a Physical Resource Block (PRB) in the first half of a subframe is paired with the PRB at the same frequency in the second half of the subframe. For simplicity, the explanation here is in terms of the first half subframe only.

The main design challenge for the signalling of frequency domain resource allocations (in terms of a set of resource blocks) is to find a good compromise between flexibility and signalling overhead. The most flexible, and arguably the simplest, approach is to send each UE a bitmap in which each bit indicates a particular PRB. This would work well for small system bandwidths, but for large system bandwidths (i.e. up to 110 PRBs) the bitmap would need 110 bits, which would be a prohibitive overhead – particularly for transmission of small packets, where the PDCCH message could be larger than the data packet! One possible solution would be to send a combined resource allocation message to all UEs, but this was rejected on the grounds of the high power needed to reach all UEs reliably, including those at the cell edges. A number of different approaches were considered by 3GPP, and those adopted are listed in Table 9.5.

Further details of the different resource allocation methods are given below.

Resource allocation Type 0. In resource allocations of Type 0, a bitmap indicates the Resource Block Groups (RBGs) which are allocated to the scheduled UE, where a RBG is a set of consecutive PRBs. The RBG size (P) is a function of the system bandwidth as shown

Table 9.5 Methods for indicating resource allocation.

Method	UL/DL	Description	Number of bits required (see main text for definitions)
Direct bitmap	DL	The bitmap comprises 1 bit per RB. This method is the only one applicable when the bandwidth is less than 10 resource blocks.	N_{RB}^{DL}
Bitmap: 'Type 0'	DL	The bitmap addresses Resource Block Groups (RBGs), where the group size (2, 3 or 4) depends on the system bandwidth.	$\lceil N_{RB}^{DL}/P \rceil$
Bitmap: 'Type 1'	DL	The bitmap addresses individual resource blocks in a subset of RBGs. The number of subsets (2, 3, or 4) depends on the system bandwidth. The number of bits is arranged to be the same as for Type 0, so the same DCI format can carry either type of allocation.	$\lceil N_{RB}^{DL}/P \rceil$
Contiguous allocations: 'Type 2'	DL or UL	Any possible arrangement of contiguous resource block allocations can be signalled in terms of a starting position and number of resource blocks.	$\lceil \log_2(N_{RB}^{DL}(N_{RB}^{DL}+1)) \rceil$ or $\lceil \log_2(N_{RB}^{UL}(N_{RB}^{UL}+1)) \rceil$
Distributed allocations	DL	In the downlink a limited set of resource allocations can be signalled where the resource blocks are scattered across the frequency domain and shared between two UEs. For convenience the number of bits available for using this method is the same as for contiguous allocations Type 2, and the same DCI format can carry either type of allocation.	$\lceil \log_2(N_{RB}^{DL}(N_{RB}^{DL}+1)) \rceil$

in Table 9.6. The total number of RBGs (N_{RBG}) for a downlink system bandwidth of N_{RB}^{DL} PRBs is given by $N_{RBG} = \lceil N_{RB}^{DL}/P \rceil$. An example for the case of $N_{RB}^{DL} = 25$, $N_{RBG} = 13$ and $P = 2$ is shown in Figure 9.10, where each bit in the bitmap indicates a pair of PRBs (i.e. two PRBs which are adjacent in frequency).

Resource allocation Type 1. In resource allocations of Type 1, individual PRBs can be addressed (but only within a subset of the PRBs available). The bitmap used is slightly

Table 9.6 RBG size for Type 0 resource allocation.

Downlink bandwidth N_{RB}^{DL}	RBG size (P)
$0 \leq 10$	1
11–26	2
27–63	3
64–110	4

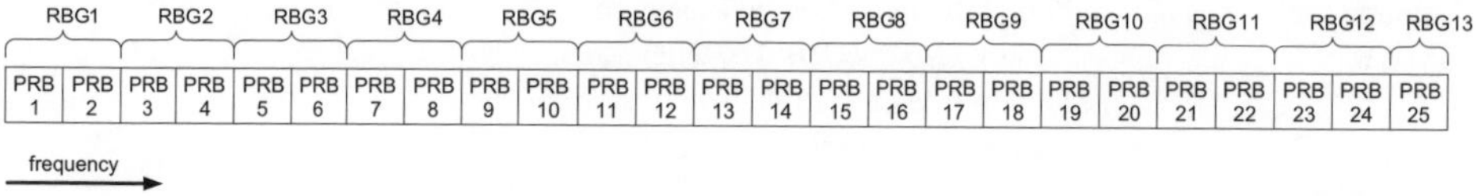

Figure 9.10 PRB addressed by a bitmap Type 0, each bit addressing a complete RBG.

smaller than for Type 0, since some bits are used to indicate the subset of the RBG which is addressed, and a shift in the position of the bitmap. The total number of bits (including these additional flags) is the same as for Type 0. An example for the case of $N_{RB}^{DL} = 25$, $N_{RBG} = 11$ and $P = 2$ is shown in Figure 9.11. One bit is used for subset selection and another bit to indicate the shift.

The motivation for providing this method of resource allocation is flexibility in spreading the resources across the frequency domain to exploit frequency diversity.

Resource allocation Type 2.   In resource allocations of Type 2, the resource allocation information indicates to a scheduled UE either:

- a set of contiguously allocated PRBs, or

- a distributed allocation comprising multiple non-consecutive PRBs (see Section 9.2.2.1).

The distinction between the two allocation methods is made by a 1-bit flag in the resource allocation message. PRB allocations may vary from a single PRB up to a maximum number of PRBs spanning the system bandwidth. A Type 2 resource allocation field consists of a Resource Indication Value (RIV) corresponding to a starting resource block (RB$_{START}$) and a length in terms of contiguously-allocated resource blocks (L_{CRBs}). The resource indication value is defined by

$$\text{if} \quad (L_{CRBs} - 1) \leq \lfloor N_{RB}^{DL}/2 \rfloor \quad \text{then} \quad RIV = N_{RB}^{DL}(L_{CRBs} - 1) + RB_{START}$$

else

$$RIV = N_{RB}^{DL}(N_{RB}^{DL} - L_{CRBs} + 1) + (N_{RB}^{DL} - 1 - RB_{START})$$

An example of a method for reversing the mapping to derive the resource allocation from the RIV can be found in [2].

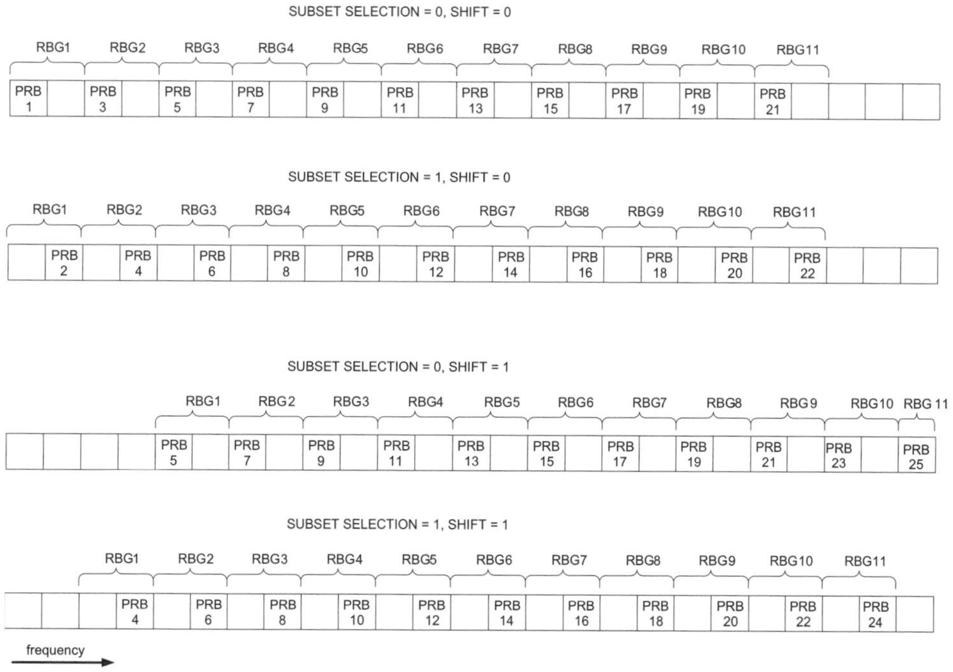

SUBSET SELECTION = 0, SHIFT = 0

RBG1	RBG2	RBG3	RBG4	RBG5	RBG6	RBG7	RBG8	RBG9	RBG10	RBG11		
PRB 1	PRB 3	PRB 5	PRB 7	PRB 9	PRB 11	PRB 13	PRB 15	PRB 17	PRB 19	PRB 21		

SUBSET SELECTION = 1, SHIFT = 0

RBG1	RBG2	RBG3	RBG4	RBG5	RBG6	RBG7	RBG8	RBG9	RBG10	RBG11		
PRB 2	PRB 4	PRB 6	PRB 8	PRB 10	PRB 12	PRB 14	PRB 16	PRB 18	PRB 20	PRB 22		

SUBSET SELECTION = 0, SHIFT = 1

		RBG1	RBG2	RBG3	RBG4	RBG5	RBG6	RBG7	RBG8	RBG9	RBG10	RBG 11
		PRB 5	PRB 7	PRB 9	PRB 11	PRB 13	PRB 15	PRB 17	PRB 19	PRB 21	PRB 23	PRB 25

SUBSET SELECTION = 1, SHIFT = 1

	RBG1	RBG2	RBG3	RBG4	RBG5	RBG6	RBG7	RBG8	RBG9	RBG10	RBG11	
	PRB 4	PRB 6	PRB 8	PRB 10	PRB 12	PRB 14	PRB 16	PRB 18	PRB 20	PRB 22	PRB 24	

frequency →

Figure 9.11 PRBs addressed by a bitmap Type 1, each bit addressing a subset of a RBG, depending on a subset selection and shift value.

9.3.3.2 PDCCH Transmission and Blind Decoding

The previous discussion has covered the structure and possible contents of an individual PDCCH message, and transmission by an eNodeB of multiple PDCCHs in a subframe. This section addresses the question of how these transmissions are organized so that a UE can locate the PDCCHs intended for it, while at the same time making efficient use of the resources allocated for PDCCH transmission.

A simple approach, at least for the eNodeB, would be to allow the eNodeB to place any PDCCH anywhere in the PDCCH resources (or CCEs) indicated by the PCFICH. In this case the UE would need to check all possible PDCCH locations, PDCCH formats and DCI formats, and act on those messages with correct CRCs (taking into account that the CRC is scrambled with a UE identity). Carrying out such a 'blind decoding' of all the possible combinations would require the UE to make many PDCCH decoding attempts in every subframe. For small system bandwidths the computational load would be reasonable, but for large system bandwidths, with a large number of possible PDCCH locations, it would become a significant burden, leading to excessive power consumption in the UE receiver. For example, blind decoding of 100 possible CCE locations for PDCCH Format 0 would be equivalent to continuously receiving a data rate of around 4 Mbps.

The alternative approach adopted for LTE is to define for each UE a limited set of CCE locations where a PDCCH may be placed. Such a constraint may lead to some limitations as to which UEs can be sent PDCCHs within the same subframe, which would thus restrict the UEs to which the eNodeB could grant resources. Therefore it is important for good system performance that the set of possible PDCCH locations available for each UE is not too small.

The set of CCE locations in which the UE may find its PDCCHs can be considered as a 'search space'. In LTE the search space is a different size for each PDCCH format. Moreover, separate *dedicated* and *common* search spaces are defined, where a dedicated search space is configured for each UE individually, while all UEs are informed of the extent of the common search space. Note that the dedicated and common search spaces may overlap for a given UE. The sizes of the common and dedicated search spaces are listed in Table 9.7.

Table 9.7 Search spaces for PDCCH formats.

PDCCH format	Number of CCEs (n)	Number of candidates in common search space	Number of candidates in dedicated search space
0	1	—	6
1	2	—	6
2	4	4	2
3	8	2	2

With such small search spaces it is quite possible in a given subframe that the eNodeB cannot find CCE resources to send PDCCHs to all the UEs that it would like to, because having assigned some CCE locations the remaining ones are not in the search space of a particular UE. To minimize the possibility of such blocking persisting into the next subframe, a UE-specific hopping sequence is applied to the starting positions of the dedicated search spaces.

In order to keep under control the computational load arising from the total number of blind decoding attempts, the UE is not required to search for all the defined DCI formats simultaneously. Typically, in the dedicated search space, the UE will always search for Formats 0 and 1A, which are both the same size and are distinguished by a flag in the message. In addition, a UE may be required to receive a further format (i.e. 1, 1B or 2, depending on the PDSCH transmission mode configured by the eNodeB).

In the common search space the UE will search for Formats 1A and 1C. In addition the UE may be configured to search for Format 3 or 3A, which have the same size as formats 0 and 1A, and may be distinguished by having the CRC scrambled by a different (common) identity, rather than a UE-specific one.

Considering the above, the UE would be required to carry out a maximum of 44 blind decodings in any subframe. This does not include checking the same message with different CRC values, which requires only a small additional computational complexity.

It is also worth noting that the PDCCH structure is adapted to avoid situations where a PDCCH CRC 'pass' might occur for multiple positions in the configured search-spaces due to repetition in the channel coding (for example, if a PDCCH was mapped to a high number of CCEs with a low code rate, then the CRC could pass for an overlapping smaller set of CCEs as well if the channel coding repetition was aligned). Such situations are avoided by adding a padding bit to any PDCCH messages having a size which could result in this problem occurring.

9.3.4 Scheduling Process from a Control Channel Viewpoint

To summarize the operation of the downlink control channels, a typical sequence of steps carried out by the eNodeB could be envisaged as follows:

1. Determine which UEs should be granted resources in the uplink, based on information such as channel quality measurements, scheduling requests and buffer status reports. Also decide on which resources should be granted.

2. Determine which UEs should be scheduled for packet transmission in the downlink, based on information such as channel quality indicator reports, and in the case of MIMO, rank indication and preferred precoding matrix.

3. Identify any common control channel messages which are required (e.g. power control commands using DCI Format 3).

4. For each message decide on the PDCCH format (i.e. 1, 2, 4 or 8 CCEs), and any power offset to be applied, in order to reach the intended UE(s) with sufficient reliability, while minimizing PDCCH overhead.

5. Determine how much PDCCH resource (in terms of CCEs) will be required, how many OFDM symbols would be needed for these PDCCHs and therefore what should be signalled on PCFICH.

6. Map each PDCCH to a CCE location within the appropriate search space.

7. If any PDCCHs cannot be mapped to a CCE location because all locations in the relevant search space have already been assigned, either:

   • continue to next step (step 8) accepting that one or more PDCCHs will not be transmitted, and not all DL-SCH and/or UL-SCH resources will be used, with a likely loss in throughput, or:

   • allocate one more OFDM symbol to support the required PDCCHs and possibly revisit step 1 and/or 2 and change UE selection and resource allocation (e.g. to fully use uplink and downlink resources).

8. Allocate the necessary resources to PCFICH and PHICH.

9. Allocate resources to PDCCHs.

10. Check that total power level per OFDM symbol does not exceed maximum allowed, and adjust if necessary.

11. Transmit downlink control channels.

Whatever approach the eNodeB implementer follows, the potentially high complexity of the scheduling process is clear, particularly bearing in mind that a cell may easily contain many hundreds of active UEs.

References[5]

[1] 3GPP Technical Specification 36.321, 'Multiplexing and Channel Coding (FDD) (Release 8)', www.3gpp.org.

[2] NEC, 'R1-072119: DL Unicast Resource Allocation Signalling', www.3gpp.org, 3GPP TSG RAN WG1, meeting 49, Kobe, Japan, May 2007.

---

[5] All web sites confirmed 18th December 2008.

10

Channel Coding and Link Adaptation

Brian Classon, Ajit Nimbalker, Stefania Sesia and
Issam Toufik

10.1 Introduction

The principle of link adaptation is fundamental to the design of a radio interface which
is efficient for packet-switched data traffic. Unlike the early versions of UMTS (Universal
Mobile Telecommunication System), which used fast closed-loop power control to support
circuit-switched services with a roughly constant data rate, link adaptation in HSPA (High
Speed Packet Access) and LTE adjusts the transmitted information data rate (modulation
scheme and channel coding rate) dynamically to match the prevailing radio channel capacity
for each user. Link adaptation is therefore very closely related to the design of the channel
coding scheme used for forward error correction.

For the downlink data transmissions in LTE, the eNodeB typically selects the modulation
scheme and code rate depending on a prediction of the downlink channel conditions. An
important input to this selection process is the Channel Quality Indicator (CQI) feedback
transmitted by the User Equipment (UE) in the uplink. CQI feedback is an indication
of the data rate which can be supported by the channel, taking into account the Signal-
to-Interference plus Noise Ratio (SINR) and the characteristics of the UE's receiver.
Section 10.2 explains the principles of link adaptation as applied in LTE; it also shows how
the eNodeB can select different CQI feedback modes to trade off the improved downlink link
adaptation enabled by CQI against the uplink overhead caused by the CQI itself.

The LTE specifications are designed to provide the signalling necessary for interoperability between the eNodeB and the UEs so that the eNodeB can optimize the link adaptation,

but the exact methods used by the eNodeB to exploit the information that is available are left to the manufacturer's choice of implementation.

In general, in response to the CQI feedback the eNodeB can select between QPSK, 16-QAM and 64-QAM schemes and a wide range of code rates. As discussed further in Section 10.2.1, the optimal switching points between the different combinations of modulation order and code rate depend on a number of factors, including the required quality of service and cell throughput.

The channel coding scheme for forward error correction, on which the code rate adaptation is based, was the subject of extensive study during the standardization of LTE. The chapter therefore continues with a review of the key theoretical aspects of the types of channel coding studied for LTE: convolutional codes, turbo codes with iterative decoding, and Low-Density Parity Check (LDPC) codes. The theory of channel coding has seen intense activity in recent decades, especially since the discovery of turbo codes offering near-Shannon limit performance, and the development of iterative processing techniques in general. 3GPP was an early adopter of these advanced channel coding techniques, with the turbo code being standardized in the first version of the UMTS as early as 1999. Later releases of UMTS, for HSPA, added more advanced channel coding features with the introduction of link adaptation, including Hybrid Automatic Repeat reQuest (HARQ), a combination of ARQ and channel coding which provides more robustness against fading; these schemes include incremental redundancy, whereby the code rate is progressively reduced by transmitting additional parity information with each retransmission. However, the underlying turbo code from the first version of UMTS remained untouched. Meanwhile, the academic and research communities were generating new insights into code design, iterative decoding and the implementation of decoders. Section 10.3.2 explains how these developments impacted the design of the channel coding for LTE, and in particular the decision to enhance the turbo code from UMTS by means of a new contention-free interleaver, rather than to adopt a new LDPC code.

For the LTE uplink transmissions, the link adaptation process is similar to that for the downlink, with the selection of modulation and coding schemes also being under the control of the eNodeB. An identical channel coding structure is used for the uplink, while the modulation scheme may be selected between QPSK and 16QAM, and, for the highest category of UE, also 64QAM. The main difference from the downlink is that instead of basing the link adaptation on CQI feedback, the eNodeB can directly make its own estimate of the supportable uplink data rate by channel sounding, for example using the Sounding Reference Signals (SRSs) which are described separately in Section 16.6.

A final important aspect of link adaptation is its use in conjunction with multi-user scheduling in time and frequency, which enables the radio transmission resources to be shared efficiently between users as the channel capacity to individual users varies. The CQI can therefore be used not only to adapt the modulation and coding rate to the channel conditions, but also for the optimization of the time/frequency selective scheduling and for inter-cell interference management. These aspects are discussed in Chapter 12.

10.2   Link Adaptation and Feedback Computation

In cellular communication systems, the quality of the signal received by a UE depends on the channel quality from the serving cell, the level of interference from other cells, and the noise level. To optimize system capacity and coverage for a given transmission power, the transmitter should try to match the information data rate for each user to the variations in received signal quality (see, for example, [1, 2] and references therein). This is commonly referred to as link adaptation and is typically based on Adaptive Modulation and Coding (AMC).

The degrees of freedom for the AMC consist of the modulation and coding schemes:

- **Modulation Scheme.** Low-order modulation (i.e. few data bits per modulated symbol, e.g. QPSK) is more robust and can tolerate higher levels of interference but provides a lower transmission bit rate. High-order modulation (i.e. more bits per modulated symbol, e.g. 64AQM) offers a higher bit rate but is more prone to errors due to its higher sensitivity to interference, noise and channel estimation errors; it is therefore useful only when the SINR is sufficiently high.

- **Code rate.** For a given modulation, the code rate can be chosen depending on the radio link conditions: a lower code rate can be used in poor channel conditions and a higher code rate in the case of high SINR. The adaptation of the code rate is achieved by applying puncturing or repetition to the output of a mother code, as explained in Section 10.3.2.4.

A key issue in the design of the AMC scheme for LTE was whether all Resource Blocks (RBs) allocated to one user in a subframe should use the same Modulation and Coding Scheme (MCS) (see, for example, [3–6]) or whether the MCS should be frequency-dependent within each subframe. It was shown that in general only a small throughput improvement arises from a frequency-dependent MCS compared to an RB-common MCS in the absence of transmission power control, and therefore the additional control signalling overhead associated with frequency-dependent MCS is not justified. Therefore in LTE the modulation and channel coding rates are constant over the allocated frequency resources for a given user, and time-domain channel-dependent scheduling and AMC is supported instead. In addition, when multiple transport blocks are transmitted to one user in a given subframe using multistream Multiple-Input Multiple-Output (MIMO) (as discussed in Chapter 11), each transport block can use an independent MCS.

In LTE the UE can be configured to report CQIs to assist the eNodeB in selecting an appropriate MCS to use for the downlink transmissions. The CQI reports are derived from the downlink received signal quality, typically based on measurements of the downlink reference signals (see Section 8.2). It is important to note that, like HSDPA, the reported CQI is not a direct indication of SINR in LTE. Instead, the UE reports the highest MCS that it can decode with a transport block error rate probability not exceeding 10%. Thus the information received by the eNodeB takes into account the characteristics of the UE's receiver, and not just the prevailing radio channel quality. Hence a UE that is designed with advanced signal processing algorithms (for example, using interference cancellation techniques) can report a higher channel quality and, depending on the characteristics of the eNodeB's scheduler, can receive a higher data rate.

A simple method by which a UE can choose an appropriate CQI value could be based on a set of Block Error Rate (BLER) thresholds, as illustrated by way of example in Figure 10.1. The UE would report the CQI value corresponding to the MCS that ensures $\text{BLER} \leq 10^{-1}$ based on the measured received signal quality.

The list of modulation schemes and code rates which can be signalled by means of a CQI value is shown in Table 10.1 (from [7]).

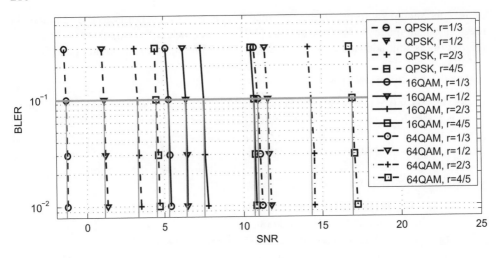

Figure 10.1 Typical BLER versus Signal-to-Noise Ratio (SNR) for different modulation and coding schemes. From left to right, the curves in this example correspond to QPSK, 16QAM and 64QAM, rates 1/3, 1/2, 2/3 and 4/5.

Table 10.1 CQI table. Reproduced by permission of © 3GPP.

CQI index	Modulation	Approximate code rate	Efficiency (information bits per symbol)
0	No transmission	—	—
1	QPSK	0.076	0.1523
2	QPSK	0.12	0.2344
3	QPSK	0.19	0.3770
4	QPSK	0.3	0.6016
5	QPSK	0.44	0.8770
6	QPSK	0.59	1.1758
7	16QAM	0.37	1.4766
8	16QAM	0.48	1.9141
9	16QAM	0.6	2.4063
10	64QAM	0.45	2.7305
11	64QAM	0.55	3.3223
12	64QAM	0.65	3.9023
13	64QAM	0.75	4.5234
14	64QAM	0.85	5.1152
15	64QAM	0.93	5.5547

AMC can exploit the UE feedback by assuming that the channel fading is sufficiently slow. This requires the channel coherence time to be at least as long as the time between the UE's measurement of the downlink reference signals and the subframe containing the correspondingly-adapted downlink transmission on the Physical Downlink Shared CHannel (PDSCH). This time is typically 7–8 ms (equivalent to a UE speed of ~16 km/h at 1.9 GHz).

However, a trade-off exists between the amount of CQI information reported by the UEs and the accuracy with which the AMC can match the prevailing conditions. Frequent reporting of the CQI in the time domain allows better matching to the channel and interference variations, while fine resolution in the frequency domain allows better exploitation of frequency-domain scheduling. However, both lead to increased feedback overhead in the uplink. Therefore, the eNodeB can configure both the time-domain update rate and the frequency-domain resolution of the CQI, as discussed in the following section.

10.2.1   CQI Feedback in LTE

The periodicity and frequency resolution to be used by a UE to report CQI are both controlled by the eNodeB. In the time domain, both periodic and aperiodic CQI reporting are supported. The Physical Uplink Control CHannel (PUCCH, see Section 17.3.1) is used for periodic CQI reporting only; the Physical Uplink Shared CHannel (PUSCH, see Section 17.2) is used for aperiodic reporting of the CQI, whereby the eNodeB specifically instructs the UE to send an individual CQI report embedded into a resource which is scheduled for uplink data transmission.

The frequency granularity of the CQI reporting is determined by defining a number of sub-bands (N), each comprised of k contiguous Physical Resource Blocks (PRBs). The value of k depends on the type of CQI report considered. In each case the number of sub-bands spans the whole system bandwidth and is given by $N = \lceil N_{\mathrm{RB}}^{\mathrm{DL}}/k \rceil$, where $N_{\mathrm{RB}}^{\mathrm{DL}}$ is the number of RBs across the system bandwidth. The CQI reporting modes can be Wideband CQI, eNodeB-configured sub-band feedback, or UE-selected sub-band feedback. These are explained in detail in the following section. In addition, in the case of multiple transmit antennas at the eNodeB, CQI value(s) may be reported for a second codeword.

For some downlink transmission modes, additional feedback signalling consisting of Precoding Matrix Indicators (PMI) and Rank Indications (RI) is also transmitted by the UE. This is explained in Section 11.2.2.4.

10.2.1.1   Aperiodic CQI Reporting

Aperiodic CQI reporting on the PUSCH is scheduled by the eNodeB by setting a CQI request bit in an uplink resource grant sent on the Physical Downlink Control CHannel (PDCCH).

The type of CQI report is configured by the eNodeB by RRC signalling. Table 10.2 summarizes the relationship between the configured downlink transmission mode (see Section 9.2.2.1) and the possible CQI reporting type. The CQI reporting type can be:

- **Wideband feedback.** The UE reports one wideband CQI value for the whole system bandwidth.

- **eNodeB-configured sub-band feedback.** The UE reports a wideband CQI value for the whole system bandwidth. In addition, the UE reports a CQI value for each sub-band, calculated assuming transmission only in the relevant sub-band. Sub-band CQI reports are encoded differentially with respect to the wideband CQI using 2-bits as

Table 10.2 Aperiodic CQI feedback types on PUSCH for each PDSCH transmission mode.

PDSCH transmission mode	Wideband only	UE-Selected sub-bands	eNodeB-configured sub-bands
Mode 1: Single antenna port		X	X
Mode 2: Transmit diversity		X	X
Mode 3: Open-loop spatial multiplexing		X	X
Mode 4: Closed-loop spatial multiplexing	X	X	X
Mode 5: Multi-user MIMO			X
Mode 6: Closed-loop rank-1 predcoding	X	X	X
Mode 7: UE-specific reference signals		X	X

Table 10.3 Sub-band size (k) versus system bandwidth for eNodeB-configured aperiodic CQI reports. Reproduced by permission of © 3GPP.

System bandwidth (RBs)	Sub-band size (k RBs)
6–7	(Wideband CQI only)
8–10	4
11–26	4
27–63	6
64–110	8

follows:

Sub-band differential CQI offset = Sub-band CQI index − Wideband CQI index

Possible sub-band differential CQI offsets are $\{\leq -1, 0, +1, \geq +2\}$. The sub-band size k is a function of system bandwidth as summarized in Table 10.3.

- **UE-selected sub-band feedback.** The UE selects a set of M preferred sub-bands of size k (where k and M are given in Table 10.4 for each system bandwidth range) within the whole system bandwidth. The UE reports one wideband CQI value and one CQI value reflecting the average quality of the M selected sub-bands. The UE also reports the positions of the M selected sub-bands using a combinatorial index r defined as

$$r = \sum_{i=0}^{M-1} \binom{N - s_i}{M - i}$$

where the set $\{s_i\}_{i=0}^{M-1}$, $1 \leq s_i \leq N$, $s_i < s_{i+1}$ contains the M sorted sub-band indices and

$$\binom{x}{y} = \begin{cases} \binom{x}{y} & \text{if } x \geq y \\ 0 & \text{if } x < y \end{cases}$$

Table 10.4 Sub-band size k and number of preferred sub-bands (M) versus downlink system bandwidth for aperiodic CQI reports for UE-selected sub-bands feedback. Reproduced by permission of © 3GPP.

System bandwidth (RBs)	Sub-band size (k RBs)	Number of preferred sub-bands (M)
6–7	(Wideband CQI only)	(Wideband CQI only)
8–10	2	1
11–26	2	3
27–63	3	5
64–110	4	6

is the extended binomial coefficient, resulting in a unique label $r \in \{0, \ldots, \binom{N}{M} - 1\}$. Some possible algorithms for deriving the combinatorial index r in the UE and extracting the information from it in the eNodeB can be found in [8] and [9] respectively.

The CQI value for the M selected sub-bands for each codeword is encoded differentially using 2-bits relative to its respective wideband CQI as defined by

Differential CQI

$\quad$ = Index for average of M preferred sub-bands − Wideband CQI index

Possible differential CQI values are $\{\leq +1, +2, +3, \geq +4\}$.

10.2.1.2 Periodic CQI Reporting

If the eNodeB wishes to receive periodic reporting of the CQI, the UE will transmit the reports using the PUCCH.[1] Only wideband and UE-selected sub-band feedback is possible for periodic CQI reporting, for all downlink (PDSCH) transmission modes. As with the aperiodic CQI reporting, the type of periodic reporting is configured by the eNodeB by RRC signalling. For the wideband periodic CQI reporting, the period can be configured to {2, 5, 10, 16, 20, 32, 40, 64, 80, 160} ms or Off (for FDD; see [7] section 7.2.2 for TDD).

While the wideband feedback mode is similar to that sent via the PUSCH, the 'UE-selected sub-band' CQI using PUCCH is different. In this case, the total number of sub-bands N is divided into J fractions called *bandwidth parts*. The value of J depends on the system bandwidth as summarized in Table 10.5. In case of periodic UE-selected sub-band CQI reporting, one CQI value is computed and reported for a single selected sub-band from each bandwidth part, along with the corresponding sub-band index.

---

[1] If PUSCH transmission resources are scheduled for the UE in one of the periodic subframes, the periodic CQI report is sent on the PUCCH instead.

Table 10.5 Periodic CQI reporting with UE-selected sub-bands: sub-band size (k) and bandwidth parts (J) versus downlink system bandwidth. Reproduced by permission of © 3GPP.

System bandwidth (RBs)	Sub-band size (k RBs)	Number of bandwidth parts (J)
6–7	(Wideband CQI only)	1
8–10	4	1
11–26	4	2
27–63	6	3
64–110	8	4

10.3 Channel Coding

Channel coding, and in particular the channel decoder, has retained its reputation for being the dominant source of complexity in the implementation of wireless communications, in spite of the relatively recent prevalence of advanced antenna techniques with their associated complexity.

Section 10.3.1 introduces the theory behind the families of channel codes of relevance to LTE. This is followed in Sections 10.3.2 and 10.3.3 by an explanation of the practical design and implementation of the channel codes used in LTE for data and control signalling respectively.

10.3.1 Theoretical Aspects of Channel Coding

The next subsection explains convolutional codes, as not only do they remain relevant for small data blocks, but also an understanding of them is a prerequisite for understanding turbo codes. The turbo coding principle and the Soft-Input Soft-Output (SISO) decoding algorithms are then discussed. The section concludes with a brief introduction to LDPC codes.

10.3.1.1 From Convolutional Codes to Turbo Codes

A convolutional encoder $\mathcal{C}(k, n, m)$ is composed of a shift register with m stages. At each time instant, k information bits enter the shift register and k bits in the last position of the shift register are dropped. The set of n output bits is a linear combination of the content of the shift register. The *rate* of the code is defined as $R_c = k/n$. Figure 10.2 shows, as example, the convolutional encoder used in LTE [10] with $m = 6$, $n = 3$, $k = 1$ and rate $R_c = 1/3$. The linear combinations are defined via n generator sequences $\mathbf{G} = [\mathbf{g}_0, \ldots, \mathbf{g}_{n-1}]$ where $\mathbf{g}_\ell = [g_{\ell,0}, \ g_{\ell,1}, \ldots, g_{\ell,m}]$.

The generator sequences used in Figure 10.2 are

$$\mathbf{g}_0 = [1\,0\,1\,1\,0\,1\,1], \quad \mathbf{g}_1 = [1\,1\,1\,1\,0\,0\,1], \quad \mathbf{g}_2 = [1\,1\,1\,0\,1\,0\,1]$$

or using octal notation

$$\mathbf{g}_0 = [133](\text{oct}), \quad \mathbf{g}_1 = [171](\text{oct}), \quad \mathbf{g}_2 = [165](\text{oct})$$

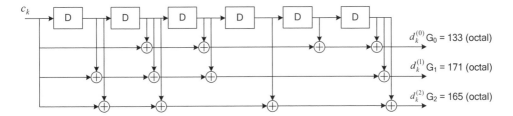

Figure 10.2 Rate $1/3$ convolutional encoder used in LTE with $m = 6$, $n = 3$, $k = 1$ [10]. Reproduced by permission of © 3GPP.

A convolutional encoder can be described by a trellis diagram [11], which is a representation of a finite state machine including the time dimension.

Consider an input block with L bits encoded with a rate $1/n$ (i.e. $k = 1$) convolutional encoder, resulting in a codeword of length $(L + m) \times n$ bits, including m trellis termination bits (or *tail bits*) which are inserted at the end of the information block to drive the shift register contents back to all zeros at the end of the encoding process. Note that using tail bits is just one possible way of terminating an input sequence. Other trellis termination methods include simple truncation (i.e. no tail bits appended) and so-called *tail-biting* [12]. In the tail-biting approach, the initial and final states of the convolutional encoder are required to be identical. Usually tail-biting for feed-forward convolutional encoders is achieved by initializing the shift register contents with the last m information bits in the input block. Tail-biting encoding facilitates uniform protection of the information bits and suffers no rate-loss owing to the tail bits. Tail-biting convolutional codes can be decoded using, for example, the *Circular Viterbi Algorithm* (CVA) [13, 14].

Let the received sequence $\mathbf{y}$ be expressed as

$$\mathbf{y} = \sqrt{E_b}\mathbf{x} + \mathbf{n} \tag{10.1}$$

where

$$\mathbf{n} = [n_0, \ n_1, \ \ldots, \ n_\ell, \ \ldots, \ n_{(L+m) \times (n-1)}]$$

and $n_\ell \sim N(0, N_0)$ is the Additive White Gaussian Noise (AWGN), E_b is the energy per bit. The transmitted codeword is

$$\mathbf{x} = [\mathbf{x}_0, \ \mathbf{x}_1, \ \ldots, \ \mathbf{x}_\ell, \ \ldots, \ \mathbf{x}_{L+m-1}]$$

where $\mathbf{x}_\ell$ is the convolutional code output sequence at time instant ℓ for the input information bit i_ℓ, given by $\mathbf{x}_\ell = [x_{\ell,0}, \ \ldots, \ x_{\ell,n-1}]$ and equivalently

$$\mathbf{y} = [\mathbf{y}_0, \ \mathbf{y}_1, \ \ldots, \ \mathbf{y}_\ell, \ \ldots, \ \mathbf{y}_{L+m-1}]$$

where $\mathbf{y}_\ell = [y_{\ell,0}, \ \ldots, \ y_{\ell,n-1}]$ is the noisy received version of $\mathbf{x}$. $(L + m)$ is the total trellis length.

10.3.1.2 Soft-Input Soft-Output (SISO) Decoders

In order to define the performance of a communication system, the codeword error probability or bit error probability can be considered. The minimization of the bit error probability is in general more complicated and requires the maximization of the a-posteriori bit probability (MAP symbol-by-symbol). The minimization of the codeword/sequence error probability is in general easier and is equivalent to the maximization of A Posteriori Probability (APP) for each codeword; this is expressed by the MAP sequence detection rule, whereby the estimate $\hat{\mathbf{x}}$ of the transmitted codeword is given by

$$\hat{\mathbf{x}} = \underset{\mathbf{x}}{\text{argmax}} \, P(\mathbf{x} \mid \mathbf{y}) \qquad (10.2)$$

When all codewords are equiprobable, the MAP criterion is equivalent to the Maximum Likelihood (ML) criterion which selects the codeword that maximizes the probability of the received sequence $\mathbf{y}$ conditioned on the estimated transmitted sequence $\mathbf{x}$, i.e.

$$\hat{\mathbf{x}} = \underset{\mathbf{x}}{\text{argmax}} \, P(\mathbf{y} \mid \mathbf{x}) \qquad (10.3)$$

Maximizing Equation (10.3) is equivalent to maximizing the logarithm of $P(\mathbf{y} \mid \mathbf{x})$, as $\log(\cdot)$ is a monotonically increasing function. This leads to simplified processing.[2] The log-likelihood function for a memoryless channel can be written as

$$\log P(\mathbf{y} \mid \mathbf{x}) = \sum_{i=0}^{L+m-1} \sum_{j=0}^{n-1} \log P(y_{i,j} \mid x_{i,j}) \qquad (10.4)$$

For an AWGN channel, the conditional probability in Equation (10.4) is $P(y_{i,j} \mid x_{i,j}) \sim N(\sqrt{E_b}x_{i,j}, N_0)$, hence

$$\log P(\mathbf{y} \mid \mathbf{x}) \propto \|\mathbf{y}_i - \sqrt{E_b}\mathbf{x}_i\|^2 \qquad (10.5)$$

The maximization of the metric in Equation (10.5) yields a codeword that is closest to the received sequence in terms of the Euclidean distance [15]. This maximization can be performed in an efficient manner by operating on the trellis.

As an example, Figure 10.3 shows a simple convolutional code with generator polynomials $\mathbf{g}_0 = [1\ 0\ 1]$ and $\mathbf{g}_1 = [1\ 1\ 1]$ and Figure 10.4 represents the corresponding trellis diagram. Each edge in the trellis corresponds to a transition from a state s to a state s', which can be obtained for a particular input information bit. In Figure 10.4 the edges are parametrized with the notation $i_\ell/x_{\ell,0}\, x_{\ell,1}$, i.e. the input/output of the convolutional encoder. The shift registers of the convolutional code are initialized to the all-0 state. In the example, m tail bits are added at the end, in order to terminate the trellis, hence the final state will again be the all-0.

Let $M(\mathbf{y}_i \mid \mathbf{x}_i) = \sum_{j=0}^{n-1} \log P(y_{i,j} \mid x_{i,j})$ denote the branch metric at the ith trellis step (i.e. the cost of choosing a branch at trellis step i), given by Equation (10.5).

At the ℓth transition in the trellis, a 'partial path metric' can be derived, which is the accumulated cost of following a specific path inside the trellis until the ℓth transition. This partial path metric can be written as

$$M_s^\ell(\mathbf{y} \mid \mathbf{x}) = \sum_{i=0}^{\ell} M(\mathbf{y}_i \mid \mathbf{x}_i) = M_{s'}^{\ell-1}(\mathbf{y} \mid \mathbf{x}) + M(\mathbf{y}_\ell \mid \mathbf{x}_\ell) \qquad (10.6)$$

---

[2]The processing is simplified because the multiplication operation can be transformed to the simpler addition operation in the logarithmic domain.

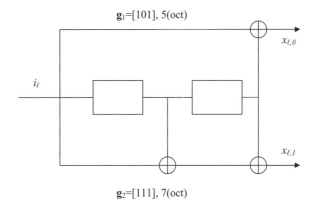

Figure 10.3 Rate $\frac{1}{2}$ convolutional encoder with $m = 2$, $m = 2$, $k = 1$, corresponding to generator polynomials $\mathbf{g}_0 = [1\ 0\ 1]$ and $\mathbf{g}_1 = [1\ 1\ 1]$.

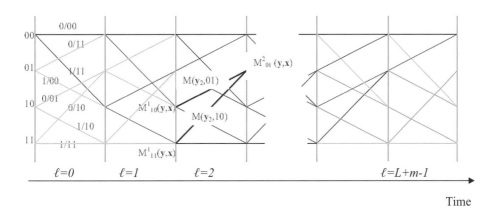

Figure 10.4 Trellis corresponding to convolutional code with generator polynomials $\mathbf{g}_0 = [1\ 0\ 1]$ and $\mathbf{g}_1 = [1\ 1\ 1]$.

The Viterbi Algorithm (VA) [16, 17] is an efficient way of using the trellis to compute the 'best' partial path metric at each trellis step by adding, comparing and selecting metrics as briefly described in the following. For each state s the VA computes the possible partial path metrics corresponding to all the edges arriving in the state s, and it selects the best partial metric. In the example in Figure 10.4, at time $\ell = 2$, there are two possible paths ending, for example, in the state $s = 01$. The VA computes

$$M^2_{01}(\mathbf{y} \mid \mathbf{x}) = \max\{M^1_{10}(\mathbf{y} \mid \mathbf{x}) + M(\mathbf{y}_2 \mid 01), M^1_{11}(\mathbf{y} \mid \mathbf{x}) + M(\mathbf{y}_2 \mid 10)\}$$

The transition that yields the best path metric is selected as the survival edge and the other one is discarded. This is carried on for each state and for $\ell = 0, \ldots, L + m - 1$. At the last stage

$\ell = L + m - 1$, the VA selects the best total metric among the different metrics computed for each state (in the case of trellis terminated convolutional code the last state is known to be the all-0 state) and it traces back the selected path in the trellis to provide the estimated input sequence.

Although the original VA outputs a hard-decision estimate of the input sequence, the VA can be modified to output soft information[3] along with the hard-decision estimate of the input sequence [18]. The reliability of coded or information bits can be obtained via the computation of the a posteriori probability.

In the following analysis, the convolutional code $\mathcal{C}(1, n, m)$ encodes the information bit i_ℓ, at time ℓ, for $\ell = 0, \ldots, L + m - 1$.

In case of a systematic convolutional code, where the uncoded information bits also appear unmodified in the coded output,[4] the codeword $\mathbf{x} \in \mathcal{C}(1, n, m)$ is characterized by $x_{\ell,0} = i_\ell$, the systematic bit at time instant ℓ.

Assuming BPSK (Binary Phase Shift Keying) modulation ($0 \rightarrow +1$, and $1 \rightarrow -1$), the Log Likelihood Ratio (LLR) of an information symbol (or bit) i_ℓ is

$$\Lambda(i_\ell) = \Lambda(x_{\ell,0}) = \log \frac{\mathrm{APP}(i_\ell = 1)}{\mathrm{APP}(i_\ell = -1)} = \log \frac{P(i_\ell = 1 \mid \mathbf{y})}{P(i_\ell = -1 \mid \mathbf{y})}$$

The decoder can make a decision by comparing $\Lambda(i_\ell)$ to zero. Thus, the sign of the LLR gives an estimate of the information bit i_ℓ (LLR $\geq 0 \rightarrow 0$, and LLR $< 0 \rightarrow 1$), and the magnitude indicates the reliability of the estimate of the bit. The LLR can be decomposed as follows:

$$\Lambda(i_\ell) \propto \mathrm{L_{ch}}(i_\ell) + \mathrm{L_{a\ priori}}(i_\ell) + \mathrm{L_{Ext}}(i_\ell) \tag{10.7}$$

where

$$\mathrm{L_{ch}}(i_\ell) = \frac{P(y_{\ell,0} \mid i_\ell = 1)}{P(y_{\ell,0} \mid i_\ell = -1)} \tag{10.8}$$

is also called the 'channel observation'. For a BPSK modulated AWGN channel with signal to noise ratio E_b/N_0 it can be shown that $\mathrm{L_{ch}}(i_\ell) = (4E_b/N_0)y_{\ell,0}$.

$$\mathrm{L_{a\ priori}}(i_\ell) = \log \frac{P(i_\ell = 1)}{P(i_\ell = -1)} \tag{10.9}$$

is the a priori information (usually equal to zero because of equiprobability of the information bits). $\mathrm{L_{Ext}}(i_\ell)$ is the extrinsic information of the bit i_ℓ, i.e. the LLR obtained from all the bits of the convolutional code except the systematic bit corresponding to i_ℓ. The a posteriori LLR $\Lambda(i_\ell)$ in Equation (10.7) can be computed via the BCJR algorithm named after its inventors, Bahl, Cocke, Jelinek and Raviv [19]. The BCJR algorithm is a SISO decoding algorithm that uses two Viterbi-like recursions going forwards and backwards in the trellis to compute Equation (10.7) efficiently. For this reason it is also referred to as a 'forward-backward' algorithm. A detailed description of the BCJR algorithm can be found in [19].

In order to explain the BCJR algorithm it is better to write the APP from Equation (10.7) in terms of the joint probability of a transition in the trellis from the state s_ℓ at time instant ℓ,

---

[3]A soft decision gives additional information about the reliability of the decision [15].

[4]The equations and rationale given here can be easily generalized to non-systematic convolutional codes.

to the state $s_{\ell+1}$ at time instant $\ell + 1$,

$$\Lambda(i_\ell) = \log\left(\frac{\sum_{S^+} P(s_\ell = s', s_{\ell+1} = s, \mathbf{y})}{\sum_{S^-} P(s_\ell = s', s_{\ell+1} = s, \mathbf{y})}\right)$$

$$= \log\left(\frac{\sum_{S^+} p(s', s, \mathbf{y})}{\sum_{S^-} p(s', s, \mathbf{y})}\right) \tag{10.10}$$

where $s, s' \in S$ are possible states of the convolutional encoder, S^+ is the set of ordered pairs (s', s) such that a transition from state s' to state s at time ℓ is caused by the input bit $i_\ell = 0$. Similarly, S^- is the set of transitions caused by $i_\ell = 1$. The probability $p(s', s, \mathbf{y})$ can be decomposed as follows:

$$p(s', s, \mathbf{y}) = p(s', \mathbf{y}_{t<\ell})p(s, \mathbf{y}_\ell \mid s')p(\mathbf{y}_{t>\ell} \mid s) \tag{10.11}$$

where $\mathbf{y}_{t<\ell} \overset{\Delta}{=} [\mathbf{y}_0, \dots, \mathbf{y}_{\ell-1}]$ and $\mathbf{y}_{t>\ell} \overset{\Delta}{=} [\mathbf{y}_\ell, \dots, \mathbf{y}_{L+m-1}]$.

By defining

$$\alpha_\ell(s') \overset{\Delta}{=} p(s', \mathbf{y}_{t<\ell}) \tag{10.12}$$

$$\beta_{\ell+1}(s') \overset{\Delta}{=} p(\mathbf{y}_{t>\ell} \mid s) \tag{10.13}$$

$$\gamma_\ell(s', s) \overset{\Delta}{=} p(s, \mathbf{y}_\ell \mid s') \tag{10.14}$$

the APP in Equation (10.10) takes the usual form [19], i.e.

$$\Lambda(i_\ell) = \log\left(\frac{\sum_{S^+} \alpha_\ell(s')\gamma_\ell(s', s)\beta_{\ell+1}(s)}{\sum_{S^-} \alpha_\ell(s')\gamma_\ell(s', s)\beta_{\ell+1}(s)}\right) \tag{10.15}$$

The probability $\alpha_\ell(s')$ is computed iteratively in a forward recursion

$$\alpha_{\ell+1}(s) = \sum_{s' \in S} \alpha_\ell(s')\gamma_\ell(s', s) \tag{10.16}$$

with initial condition $\alpha_0(0) = 1$, $\alpha_0(s) = 0$ for $s \neq 0$, assuming that the initial state of the encoder is 0. Similarly, the probability $\beta_\ell(s')$ is computed via a backward recursion

$$\beta_\ell(s') = \sum_{s \in S} \beta_{\ell+1}(s)\gamma_\ell(s', s) \tag{10.17}$$

with initial condition $\beta_{L+m-1}(0) = 1$, $\beta_{L+m-1}(s) = 0$ for $s \neq 0$ assuming trellis termination is employed using tail bits. The transition probability can be computed as

$$\gamma_\ell(s', s) = P(\mathbf{y}_\ell \mid \mathbf{x}_\ell)P(i_\ell)$$

The MAP algorithm consists of initializing $\alpha_0(s')$, $\beta_{L+m-1}(s)$, computing the branch metric $\gamma_\ell(s', s)$, and continuing the forward and backward recursion in Equations (10.16) and (10.17) to compute the updates of $\alpha_{\ell+1}(s)$ and $\beta_\ell(s')$.

The complexity of the BCJR is approximately three times that of the Viterbi decoder. In order to reduce the complexity a log-MAP decoder can be considered, where all operations

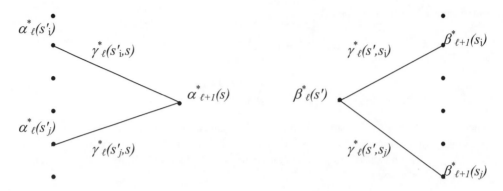

Figure 10.5 Forward and backward recursions for the BCJR decoding algorithm.

are performed in the logarithmic domain. Thus, the forward and backward recursions $\alpha_{\ell+1}(s)$, $\beta_\ell(s')$ and $\gamma_\ell(s, s')$ are replaced by $\alpha^*_{\ell+1}(s) = \log[\alpha_{\ell+1}(s)]$, $\beta^*_\ell(s') = \log[\beta_\ell(s')]$ and $\gamma^*_\ell(s, s') = \log[\gamma_\ell(s, s')]$. This gives advantages for implementation and can be shown to be numerically more stable. By defining

$$\max{}^*(z_1, z_2) \overset{\Delta}{=} \log(e^{z_1} + e^{z_2}) = \max(z_1, z_2) + \log(1 + e^{-|z_1 - z_2|}) \qquad (10.18)$$

it can be shown that the recursion in Equations (10.16) and (10.17) becomes

$$\alpha^*_{\ell+1}(s) = \log\left(\sum_{s' \in S} e^{\alpha^*_\ell(s') + \gamma^*_\ell(s', s)}\right)$$

$$= \max_{s' \in S}{}^* \{\alpha^*_\ell(s') + \gamma^*_\ell(s', s)\} \qquad (10.19)$$

with initial condition $\alpha^*_0(0) = 0$, $\alpha^*_0(s) = -\infty$ for $s \neq 0$. Similarly,

$$\beta^*_\ell(s') = \max_{s' \in S}{}^* \{\beta^*_{\ell+1}(s) + \gamma^*_\ell(s', s)\} \qquad (10.20)$$

with initial condition $\beta^*_{L+m}(0) = 0$, $\beta^*_{L+m}(s) = -\infty$ for $s \neq 0$. Figure 10.5 shows a schematic representation of the forward and backward recursion.

The APP in Equation (10.15) becomes

$$\Lambda(i_\ell) = \max_{(s', s) \in S^+}{}^* \{\alpha^*_\ell(s') + \gamma^*_\ell(s', s) + \beta^*_{\ell+1}(s)\}$$

$$- \max_{(s', s) \in S^-}{}^* \{\alpha^*_\ell(s') + \gamma^*_\ell(s', s) + \beta^*_{\ell+1}(s)\} \qquad (10.21)$$

In order to reduce the complexity further, the *max-log-MAP* approximation can be used, where the $\max{}^* = \log(e^x + e^y)$ is approximated by $\max(x, y)$. The advantage is that the algorithm is simpler and faster, with the complexity of the forward and backward passes being equivalent to a Viterbi decoder. The drawback is the loss of accuracy arising from the approximation.

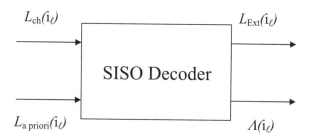

$L_{ch}(i_\ell)$ $L_{Ext}(i_\ell)$

SISO Decoder

$L_{a\ priori}(i_\ell)$ $\Lambda(i_\ell)$

Figure 10.6 Schematic view of the SISO decoder.

Figure 10.6 gives a schematic view of the input and output of a SISO decoder. The two soft inputs are the channel likelihood on coded or information bits, the channel observations $(L_{ch}(x_{i_\ell}))$ and the a priori probabilities $(L_{a\ priori}(i_\ell))$. The soft outputs are the a posteriori LLRs for the information (and/or coded) bits $(\Lambda(i_\ell))$ computed via the BCJR (full MAP, log-MAP or max-log-MAP) algorithm introduced above and the extrinsic LLRs $(L_{Ext}(i_\ell))$ obtained by using Equation (10.7).

Convolutional codes are the most widely used family of error correcting codes owing to their reasonably good performance and the possibility for extremely fast decoders based on the VA, as well as their flexibility in supporting variable input codeword sizes. However, it is well known that there remains a significant gap between the performance of convolutional codes and the theoretical limits set by Shannon.[5] In the early 1990s, an encoding and decoding algorithm based on convolutional codes was proposed [20], which exhibited performance within a few tenths of a deciBel from the Shannon limit – the turbo code family was born. Immediately after turbo codes were discovered, Low-Density Parity Check (LDPC) codes [23–25] that also provided near-Shannon limit performance were also rediscovered.

10.3.1.3 Turbo Codes

Berrou, Glavieux and Thitimajasimha introduced turbo codes and the concept of iterative decoding to achieve near-Shannon limit performance [20, 26]. Some of the reasoning regarding probabilistic processing can be found in [18, 27], as recognized by Berrou in [28].

A turbo encoder consists of a concatenation of two convolutional encoders linked by an interleaver. For instance, the turbo encoder adopted in UMTS and LTE [10] is schematically represented in Figure 10.7 with two identical convolutional codes with generator polynomial given by $\mathbf{G} = [1, \mathbf{g}_0/\mathbf{g}_1]$ where $\mathbf{g}_0 = [1011]$ and $\mathbf{g}_1 = [1101]$. Thus, a turbo code encodes the input block twice (with and without interleaving) to generate two distinct set of parity bits. Each constituent encoder may be terminated to the all zero state by using tail bits. The nominal code rate of the turbo code shown in Figure 10.7 is $1/3$.

Like convolutional codes, the optimal decoder for turbo codes would ideally be the MAP or ML decoder. However the number of states in the trellis of a turbo code is significantly larger due to the interleaver Π, thus making a true ML or MAP decoder intractable (except for

---

[5] 'Shannon's theorem', otherwise known as the 'noisy-channel coding theorem', states that if and only if the information rate is less than the channel capacity, then there exists a coding-decoding scheme with vanishing error probability when the code's block length tends to infinity [21, 22].

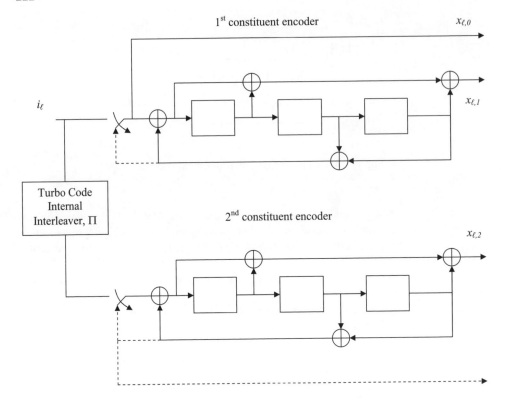

Figure 10.7 Schematic view of parallel turbo code used in LTE and UMTS [10]. Reproduced by permission of © 3GPP.

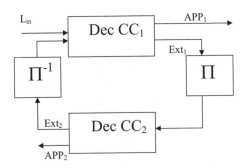

Figure 10.8 Schematic turbo decoder representation.

trivial block sizes). Therefore, Berrou *et al.* [26] proposed the principle of iterative decoding. This is based on a suboptimal approach using a separate optimal decoder for each constituent convolutional coder, with the two constituent decoders iteratively exchanging information via a (de)interleaver. The classical decoding structure is shown in Figure 10.8.

The SISO decoder for each constituent convolutional code can be implemented via the BCJR algorithm (or the MAP algorithm) which was briefly described in the previous section. The SISO decoder outputs an a posteriori LLR value for each information bit, and this can be used to obtain a hard decision estimate as well as a reliability estimate. In addition, the SISO decoder generates extrinsic LLRs for each information bit that is utilized by the other decoder as a priori information after suitable (de)interleaving. Thus the two decoders cooperate by iteratively exchanging the extrinsic information via the (de)interleaver. After a certain number of iterations, the a posteriori LLR output can be used to obtain final hard decision estimates of the information bits.

10.3.1.4 Low-Density Parity-Check (LDPC) Codes

Low-Density Parity-Check (LDPC) codes were first studied by Gallager in his doctoral thesis [23] and extensively analysed by many researchers in recent years. LDPC codes are linear parity-check codes with a parity-check equation given by $H\mathbf{c}^T = 0$, where H is the $(n - k) \times n$ parity-check matrix of the code $\mathcal{C}(k, n)$ and $\mathbf{c}$ is a length-n valid codeword belonging to the code $\mathcal{C}$. Similarly to turbo codes, ML decoding for LDPC codes becomes too complex as the block size increases. In order to approximate ML decoding, Gallager in [23] introduced an iterative decoding technique which can be considered as the forerunner of message-passing algorithms [24, 29], which lead to an efficient LDPC decoding algorithm with a complexity which is linear with respect to the block size.

The term 'low-density' refers to the fact that the parity-check matrix entries are mostly zeros – in other words, the density of ones is low. The parity-check matrix of an LDPC code can be represented graphically by a 'Tanner graph', in which two types of node, variable and check, are interconnected. The variable nodes and check nodes correspond to the codeword bits and the parity-check constraints respectively. A variable node v_j is connected to a check node c_i if the corresponding codeword bit participates in the parity-check equation, i.e. if $H(i, j) = 1$. Thus, the Tanner graph is an excellent tool by which to visualize the code constraints and to describe the decoding algorithms. Since an LDPC code has a low density of ones in H, the number of interconnections in the Tanner graph is small (and typically linear with respect to the codeword size).

Figure 10.9 shows an example of a Tanner graph of a (10, 5) code with the parity-check matrix H given in Equation (10.22),[6]

$$H = \begin{bmatrix} 1 & 1 & 1 & 1 & 0 & 1 & 1 & 0 & 0 & 0 \\ 0 & 0 & 1 & 1 & 1 & 1 & 1 & 1 & 0 & 0 \\ 0 & 1 & 0 & 1 & 0 & 1 & 0 & 1 & 1 & 1 \\ 1 & 0 & 1 & 0 & 1 & 0 & 0 & 1 & 1 & 1 \\ 1 & 1 & 0 & 0 & 1 & 0 & 1 & 0 & 1 & 1 \end{bmatrix} \tag{10.22}$$

As mentioned above, LDPC codes are decoded using message passing algorithms such as Belief Propagation (BP) or the Sum-Product Algorithm (SPA). The idea of BP is to calculate approximate marginal a posteriori LLRs by applying Bayes' rule locally and iteratively at the nodes in the Tanner graph. The variable nodes and check nodes in the Tanner graph exchange LLR messages along their interconnections in an iterative fashion, thus cooperating with each other in the decoding process. Only extrinsic messages are passed along the interconnections

---

[6]This particular matrix is not actually 'low density', but it is used for the sake of illustration.

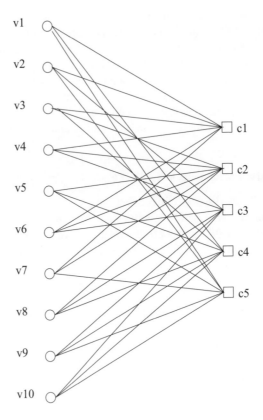

Figure 10.9 A Tanner graph representing the parity-check matrix of a (3, 6)-regular LDPC code $C(10, 5)$.

to ensure that a node does not receive repeats of information which it already possesses, analogous to the exchange of extrinsic information in turbo decoding.

It is important to note that the number of operations per iteration in the BP algorithm is linear with respect to the number of message nodes. However, the BP decoding of LDPC codes is also suboptimal, like the turbo codes, owing to the cycles in the Tanner graph.[7] Cycles increase the dependencies between the extrinsic messages being received at each node during the iterative process. However, the performance loss can be limited by avoiding excessive occurrences of short cycles. After a certain number of iterations, the a posteriori LLRs for the codeword bits are computed to obtain the final hard decision estimates. Typically, more iterations are required for LDPC codes to converge than are required for turbo codes. In practice, a variation of BP known as the Layered BP (LBP) algorithm is preferred, as it requires half as many iterations as the conventional BP algorithm [30].

So far, LDPC codes have been discussed from a decoding perspective, i.e. as the null space of a sparse matrix. A straightforward encoding algorithm based on the parity-check matrix

---

[7]A cycle of length P in a Tanner graph is a set of P connected edges that starts and ends at the same node. A cycle verifies the condition that no nodes (except the initial and final node) appear more than once.

would require a number of operations that is proportional to the square of the codeword size. However, there exist several LDPC code designs (including those used in the IEEE 802.16e, IEEE 802.11n and DVB-S2 standards) which have linear encoding complexity and extremely good performance.

In recent years, several analytical tools have been developed to understand and design LDPC codes. Density evolution (or threshold analysis) [29, 31] is a valuable tool by which the theoretical performance of an ensemble of LDPC codes using the message passing decoder can be estimated in a fast and efficient manner. Rather than focusing on a particular parity-check matrix, density evolution characterizes the performance of an ensemble (or family) of codes with similar characteristics. For instance, an LDPC ensemble is often defined by its left and right degree distributions using polynomials $\lambda(x) \triangleq \sum_{i=2}^{d_v} \lambda_i x^{i-1}$ and $\rho(x) \triangleq \sum_{i=2}^{d_c} \rho_i x^{i-1}$, where λ_i and ρ_i are the fractions of interconnections in the Tanner graph connected respectively to variable nodes and check nodes of degree i, d_v denotes the maximum variable node degree and d_c is the maximum check node degree. The code rate of the ensemble is given by

$$R = 1 - \left(\frac{\int_0^1 \rho(x)\, dx}{\int_0^1 \lambda(x)\, dx} \right)$$

It has been shown that for long code lengths ($n \to \infty$) all the randomly-constructed codes belonging to the same ensemble behave alike (*Concentration Theorem* [29]). For an LDPC code ensemble, density evolution characterizes the performance in terms of a threshold value (typically expressed as a SNR value) above which there exists at least one code in the ensemble that can achieve arbitrarily small BER [29]. For SNRs above the threshold, the BER is bounded away from zero for any number of decoder iterations. Given its similarity to the Shannon-limit, the threshold is sometimes referred to as the 'capacity' of LDPC codes under message passing decoding. Using density evolution to optimize the degree distributions, several publications have demonstrated LDPC code performance with a few tenths of a decibel of the Shannon limit [24, 32].

The structure of LDPC codes is another important aspect which greatly affects the encoding and decoding complexity. A randomly-constructed LDPC code has a random Tanner graph structure and leads to a complicated hardware design. Therefore, practical LDPC designs such as the 802.16e and 802.11n LDPC codes are often based on structured LDPC codes to facilitate efficient encoding and decoding. With structured or vectorized LDPC codes, the parity-check matrices are constructed using submatrices which are all-zero or circulant (shifted identity matrices), and such codes require substantially lower encoding/decoding complexity without sacrificing performance. Much literature exists on this subject, and the interested reader is referred, for example, to [33] and references therein.

10.3.2 Channel Coding for Data Channels in LTE

Turbo codes found a home in UMTS relatively rapidly after their publication in 1993, with the benefits of their near-Shannon limit performance outweighing the associated costs of memory and processing requirements. Once the turbo code was defined in UMTS there was little incentive to reopen the specification as long as it was functional. In fact, the effort required for the standardization of the turbo codes for UMTS was near-epic: every aspect of

Table 10.6 Major features of UMTS and LTE channel coding schemes.

Channel coding	UMTS	LTE
Constituent code	Tailed, eight states, $R = 1/3$ mother code	Same
Turbo interleaver	Row/column permutation	Contention-free quadratic permutation polynomial (QPP) interleaver
Rate matching	Performed on concatenated code blocks	Virtual Circular Buffer (CB) rate matching, performed per code block
Hybrid ARQ	Redundancy Versions (RVs) defined, Chase operation allowed	RVs defined on virtual CB, Chase operation allowed
Control channel	256-state tailed convolutional code	64-state tail-biting convolutional code, CB rate matching
Per code block operations	Turbo coding only	CRC attachment, turbo coding, rate matching, modulation mapping

the detailed design became a protracted engagement in meeting rooms around the world. The result of the effort, while not exactly beautiful, contained a certain appeal.

Therefore, although in UMTS Releases 5, 6 and 7 the core turbo code was not touched, it was enhanced by the ability to select different redundancy versions for HARQ retransmissions. However the decoder, originally working at 384 kbps, was starting to show the strain at 10 Mbps. Managing to get the UMTS decoder to run at ever higher data rates was viewed as a publication-worthy achievement [34]. As the LTE effort began in 3GPP, every aspect of system design was under scrutiny in the effort to lay out a viable roadmap for the next generations of cellular devices; with data rates of 100 Mbps to 300 Mbps in view, and possibly greater beyond Release 8, the UMTS turbo code needed to be re-examined.

Sections 10.3.2.1 and 10.3.2.2 explain the eventual decision not to use LDPC codes for LTE, while replacing the turbo interleaver with a 'contention-free' interleaver. The subsequent sections explain the specific differences between LTE channel coding and UMTS as summarized in Table 10.6.

10.3.2.1 The Lure of the LDPC Code

The draw of LDPC codes is clear: performance almost up to the Shannon limit, with promises of up to eight times less complexity than turbo codes [35]. LDPC codes had also recently been standardized as an option in IEEE 802.16-2005. The complexity angle is all-important in LTE, provided that the excellent performance of the turbo code is maintained. However, LDPC proposals put forward for LTE claimed widely varying complexity benefits, from no benefits to 2.4 times [36] up to 7.35 times [37]. In fact, it turns out that the complexity benefit is code-rate dependent, with roughly a factor of two reduction in the operations count at code rate 1/2 [38]. This comparison assumes that the LDPC code is decoded with the Layered

Belief Propagation (LBP) decoder [30] while the turbo code is decoded with a log-MAP decoder.

On operations count alone, LDPC would be the choice for LTE. However, at least three factors curbed enthusiasm for LDPC. First, LDPC decoders have significant implementation complexity for memory and routing, which makes simple operation counts unrepresentative. Second, HARQ is a key feature introduced in HSPA, and the incremental redundancy performance of LDPC is unclear. Finally, turbo codes were already standardized in UMTS, and a similar lengthy standardization process was undesired when perhaps a relatively simple enhancement to the known turbo code would suffice. It was therefore decided to keep the same turbo code constituent encoders as in the UMTS turbo code, including the tailing method, but to enhance it using a new contention-free interleaver that enables efficient decoding at the high data rates targeted for LTE. Table 10.7 gives a comparison of some of the features of turbo and LDPC codes which will be explained in the following sections.

10.3.2.2 Contention-Free Turbo Decoding

The key to high data-rate turbo decoding is to parallelize the turbo decoder efficiently while maintaining excellent performance. The classical turbo decoder has two MAP decoders (usually realized via a single hardware instantiation) that exchange extrinsic information in an iterative fashion between the two component codes, as described in Section 10.3.1. Thus, the first consideration is whether the parallelism is applied internally to the MAP decoder (single codeword parallelization) or externally by employing multiple turbo decoders (multiple codeword parallelizations with no exchange of extrinsic information between codewords).

If external parallelism is adopted, an input block may be split into X pieces, resulting in X codewords, and the increase in processing speed is obtained by operating up to X turbo decoders in parallel on the X codewords. In this case, in addition to a performance loss due to the smaller turbo interleaver size in each codeword, extra cost is incurred for memory and ancillary gate counts for forward-backward decoders. For this reason, one larger codeword with internal MAP parallelization is preferred. With one larger codeword, the MAP algorithm is run in parallel on each of the X pieces and the pieces can exchange extrinsic information, thus benefiting from the coding gains due to the large interleaver size. Connecting the pieces is most easily done with forward and backward state initialization based on the output of the previous iterations, although training within adjacent pieces during the current iteration is also possible [39]. As long as the size of each piece is large enough (e.g. 32 bits or greater), performance is essentially unaffected by the parallel processing.

While it is clear that the MAP algorithm can be parallelized, the MAP is not the entire turbo decoder. Efficient handling of the extrinsic message exchange (i.e. the (de)interleaving operation) is required as well. Since multiple MAP processors operate in parallel, multiple extrinsic values need to be read from or written to the memories concurrently. The memory accesses depend on the interleaver structure.

The existing UMTS interleaver has a problem with memory access contentions. Contentions result in having to read or write from/to the same memory at the same time (as shown in Figure 10.10).

Contention resolution is possible (see, e.g., [40] and [41]), but at a cost of extra hardware, and the resolution time (cycles) may vary for every supported interleaver size.

Table 10.7 Comparison of turbo and LDPC codes.

	Turbo codes	LDPC codes
Standards	UMTS, LTE, WiMAX	IEEE 802.16e, IEEE802.11n DVB-S2
Encoding	Linear time	Linear time with certain designs (802.16e/802.11n)
Decoding	log-MAP, and variants	Scheduling: Layered Belief Propagation, turbo-decoding Node processing: Min-Sum, normalized Min-Sum
Main decoding concern	Computationally intensive (more operations)	Memory intensive (more memory), routing network
Throughput	No inherent parallelism. Parallelism obtained via contention-free interleaver	Inherently parallelizable. Structured LDPC codes for flexible design
Performance	Comparable for information block sizes 10^3–10^4. Slightly better than LDPC at small block sizes. Four to eight iterations on average	Can be slightly better than turbo codes at large block lengths (10^4 or larger, iterative decoding threshold analysis. Very low error floor (e.g. at BLER around 10^{-7}). 10 to 15 iterations on average
HARQ	Simple for both Chase and IR (via mother code puncturing)	Simple for Chase. IR possible using model matrix extension or puncturing from a mother code
Complexity comparison (operations count)	Slightly larger operations count than LDPC at high code rates	Slightly smaller operations count than turbo at high code rates

Complex memory management may also be used as contention resolution for any arbitrary interleaver [42].

A new, Contention-Free (CF) interleaver solves the problem by ensuring that no contentions occur in the first place. An interleaver $\pi(i)$, $0 \leq i < K$, is said to be contention-free for a window size W when it satisfies the following for both interleaver $\psi = \pi$ and deinterleaver $\psi = \pi^{-1}$ [43]:

$$\left\lfloor \frac{\psi(u_1 W + v)}{W} \right\rfloor \neq \left\lfloor \frac{\psi(u_2 W + v)}{W} \right\rfloor \tag{10.23}$$

where $0 \leq vW$, $u_1 \geq 0$, $u_2 < M$ for all $u_1 \neq u_2$. The terms on both sides of Equation (10.23) are the memory bank indices that are accessed by the M processors on the vth step. Inequality

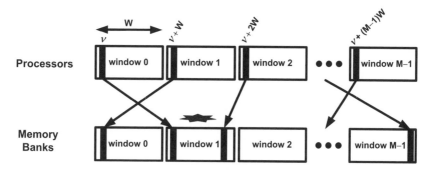

Figure 10.10 Illustration of a memory access contention in window 1 due to concurrent access by Processors 0 and 2.

(10.23) requires that for any given position v of each window, the memory banks accessed be unique between any two windows, thus eliminating access contentions. For significant hardware savings, instead of using M physically-separate memories, it is better to use a single physical memory and fetch/store M values on each cycle from a single address. This requires the CF interleaver also to satisfy a vectorized decoding property where the intra-window permutation is the same for each window:

$$\pi(uW + v) \bmod W = \pi(v) \bmod W \tag{10.24}$$

for all $1 \leq u < M$ and $0 \leq v < W$.

Performance, implementation complexity and flexibility are concerns. However, even a simple CF interleaver composed of look-up tables (for each block size) and a bit-reversal permutation [44] can be shown to have excellent performance. In terms of flexibility, a maximally contention-free interleaver can have a parallelism order (number of windows) that is any factor of the block size. A variety of possible parallelism factors provides freedom for each individual manufacturer to select the degree of parallelism based on the target data rates for different UE categories.

After consideration of performance, available flexible classes of CF interleavers and complexity benefits, a new contention-free interleaver was selected for LTE.

10.3.2.3   The LTE Contention-Free Interleaver

The main choices for the CF interleaver included Almost Regular Permutation (ARP) [45] and Quadratic Permutation Polynomial (QPP) [46] interleavers. The ARP and QPP interleavers share many similarities and they are both suitable for LTE turbo coding, offering flexible parallelism orders for each block size, low-complexity implementation, and excellent performance. A detailed overview of ARP and QPP proposals for LTE (and their comparison with the UMTS Release-99 turbo interleaver) is given in reference [47]. Of these two closely competing designs, the QPP interleaver was selected for LTE as it provides more parallelism factors M, and requires less storage, thus making it better-suited to high data rates.

For an information block size K, a QPP interleaver is defined by the following polynomial:

$$\pi(i) = (f_1 i + f_2 i^2) \bmod K \tag{10.25}$$

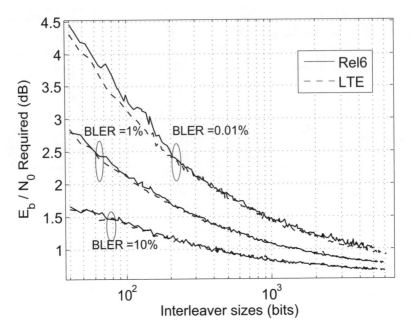

Figure 10.11 Performance of LTE QPP interleaver design versus UMTS turbo interleaver in static AWGN with QPSK modulation and eight iterations of a max-log-MAP decoder.

where i is the output index ($0 \leq i \leq K - 1$), $\pi(i)$ is the input index and f_1 and f_2 are the coefficients that define the permutation with the following conditions:

- f_1 is relatively prime to block size K;

- all prime factors of K also factor f_2.

The inverses of QPP interleavers can also be described via permutation polynomials but they are not necessarily quadratic, as such a requirement may result in inferior turbo code performance for certain block sizes. Therefore, it is better to select QPP interleavers with low-degree inverse polynomials (maximum degree of the inverse polynomial equal to four for LTE QPP interleavers) without incurring a performance penalty. In general, permutation polynomials (quadratic and non-quadratic) are implementation-friendly as they can be realized using only adders in a recursive fashion. A total of 188 interleavers are defined for LTE, of which 153 have quadratic inverses while the remaining 35 have degree-3 and degree-4 inverses. The performance of the LTE QPP interleavers is shown alongside the UMTS turbo code in Figure 10.11.

An attractive (and perhaps the most important) feature of QPP interleavers is that they are 'maximum contention-free', which means that they provide the maximum flexibility in supported parallelism, i.e. every factor of K is a supported parallelism factor. For example, for $K = 1024$, supported parallelism factors include $\{1, 2, 4, 8, 16, 32, 64, 128, 256, 512, 1024\}$, although factors that result in a window size less than 32 may not be required in practice.

The QPP interleavers also have the 'even-even' property whereby even and odd indices in the input are mapped to even and odd indices respectively in the output; this enables the encoder and decoder to process two information bits per clock cycle, facilitating radix-4 implementations (i.e. all log-MAP decoding operations can process two bits at a time, including forward/backward recursions, extrinsic LLR generation and memory read/write).

One key impact of the decision to use a CF interleaver is that not all input block sizes are natively supported. As the amounts of parallelism depend on the factorization of the block size, certain block sizes (e.g. prime sizes) are not natively supportable by the turbo code. Since the input can be any size, filler bits are used whenever necessary to pad the input to the nearest QPP interleaver size. The QPP sizes are selected such that:

- The number of interleavers is limited (fewer interleavers implies more filler bits).

- The fraction of filler bits is roughly the same as the block size increases (spacing increases as block size increases).

- Multiple parallelism values are available (block sizes are spaced an integer number of bytes apart).

For performance within approximately 0.1 dB of the baseline UMTS interleaver, as few as 45 interleaver sizes [48] are feasible, although the percentage of filler bits may be high (nearly 12%). For LTE, the following 188 byte-aligned interleaver sizes spaced in a semi-log manner were selected with approximately 3% filler bits [10]:

$$K = \begin{cases} 40 + 8t & \text{if } 0 \leq t \leq 59 & (40\text{--}512 \text{ in steps of 8 bits}) \\ 512 + 16t & \text{if } 0 < t \leq 32 & (528\text{--}1024 \text{ in steps of 16 bits}) \\ 1024 + 32t & \text{if } 0 < t \leq 32 & (1056\text{--}2048 \text{ in steps of 32 bits}) \\ 2048 + 64t & \text{if } 0 < t \leq 64 & (2112\text{--}6144 \text{ in steps of 64 bits}) \end{cases} \tag{10.26}$$

The maximum turbo interleaver size was also increased from 5114 in UMTS to 6144 in LTE, such that a 1500 byte TCP/IP packet would be segmented into only two segments rather than three, thereby minimizing potential segmentation penalty and (marginally) increasing turbo interleaver gain.

10.3.2.4 Rate-Matching

The Rate-Matching (RM) algorithm selects bits for transmission from the rate 1/3 turbo coder output via puncturing and/or repetition. Since the number of bits for transmission is determined based on the available physical resources, the RM should be capable of generating puncturing patterns for arbitrary rates. Furthermore, the RM should send as many new bits as possible in retransmissions to maximize the Incremental Redundancy (IR) HARQ gains (see Section 4.4 for further details about the HARQ protocol).

The main contenders for LTE RM were to use the same (or similar) algorithm as HSPA, or to use Circular Buffer (CB) RM as in CDMA2000 1xEV and WiMAX. The primary advantage of the HSPA RM is that while it appears complex, it has been well studied and is well understood. However, a key drawback is that there are some severe performance degradations at higher code rates, especially near code rates 0.78 and 0.88 [49].

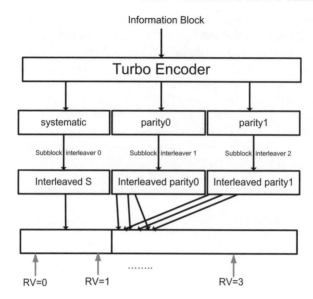

Figure 10.12 Rate-matching algorithm based on the CB. RV = 0 starts at an offset relative to the beginning of the CB to enable systematic bit puncturing on the first transmission.

Therefore, circular buffer RM was selected for LTE as it generates puncturing patterns simply and flexibly for any arbitrary code rate, with excellent performance.

In the CB approach, as shown in Figure 10.12, each of the three output streams of the turbo coder (systematic part, parity0, and parity1) is rearranged with its own interleaver (called a sub-block interleaver). In LTE the 12 tail bits are distributed equally into the three streams as well, resulting in the sub-block size $K_s = K + 4$, where K is the QPP interleaver size. Then, an output buffer is formed by concatenating the rearranged systematic bits with the interlacing of the two rearranged parity streams. For any desired code rate, the coded bits for transmission are simply read out serially from a certain starting point in the buffer, wrapping around to the beginning of the buffer if the end of the buffer is reached.

A Redundancy Version (RV) specifies a starting point in the circular buffer to start reading out bits. Different RVs are specified by defining different starting points to enable HARQ operation. Usually RV = 0 is selected for the initial transmission to send as many systematic bits as possible. The scheduler can choose different RVs on transmissions of the same packet to support both IR and Chase combining HARQ.

The turbo code tail bits are uniformly distributed into the three streams, with all streams the same size. Each sub-block interleaver is based on the traditional row-column interleaver with 32 columns (for all block size), and a simple length-32 intra-column permutation.

- The bits of each stream are written row-by-row into a matrix with 32 columns (number of rows determined by the stream size), with dummy bits padded to the front of each stream to completely fill the matrix.

- A length-32 column permutation is applied and the bits are read out column-by-column to form the output of the sub-block interleaver,

[0, 16, 8, 24, 4, 20, 12, 28, 2, 18, 10, 26, 6, 22, 14, 30, 1, 17, 9, 25, 5, 21, 13, 29, 3, 19, 11, 27, 7, 23, 15, 31]

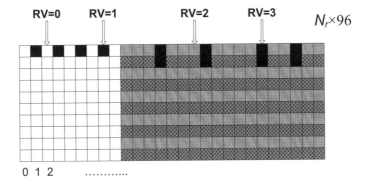

Figure 10.13 Two-dimensional visualization of the VCB. The starting points for the four RVs are the top of the selected columns.

This sub-block interleaver has the property that it naturally first puts all the even indices and then all the odd indices into the rearranged sub-block.

Given the even-even property of the QPP permutations and the above property of the sub-block interleaver, the sub-block interleaver of the second parity stream is offset by an odd value $\delta = 1$ to ensure that the odd and even input indices have equal levels of protection. Thus, for index i, if $\pi_{sys}(i)$ denotes the permutation of the systematic bit sub-block interleaver, then the permutation of the two parity sub-block interleavers are $\pi_{par0}(i) = \pi_{sys}(i)$, and $\pi_{par1}(i) = (\pi_{sys}(i) + \delta) \bmod K_s$, where K_s is the subblock size. With the offset $\delta = 1$ used in LTE, the first K_s parity bits in the interlaced parity portion of the circular buffer correspond to the K_s systematic bits, thus ensuring equal protection to all systematic bits [50]. Moreover, the offset enables the systematic bit puncturing feature whereby a small percentage of systematic bits are punctured in an initial transmission to enhance performance at high code rates. With the offset, $RV = 0$ results in partially systematic codes that are self-decodable at high coding rates, i.e. avoiding the 'catastrophic' puncturing patterns which have been shown to exist at some code rates in UMTS.

A two-dimensional interpretation of the circular buffer (with a total of 96 columns) is shown in Figure 10.13 (where different shadings indicate bits from different streams and black cells indicate dummy bits). The one-dimensional CB may be formed by reading bits out column-by-column from the two-dimensional CB. Bits are read column-by-column starting from a column top RV location, and the dummy bits are discarded during the output bit generation. Although the dummy bits can be discarded during subblock interleaving, in LTE the dummy bits are kept to allow a simpler implementation.

This leads to the foremost advantage of the LTE CB approach, in that it enables efficient HARQ operation, because the CB operation can be performed without requiring an intermediate step of forming any actual physical buffer. In other words, for any combination of the 188 stream sizes and 4 RV values, the desired codeword bits can be equivalently obtained directly from the output of the turbo encoder using simple addressing based on sub-block permutation.

Therefore the term 'Virtual Circular Buffer' (VCB) is more appropriate in LTE. The LTE VCB operation also allows Systematic Bit Puncturing (SBP) by defining $RV = 0$ to skip the

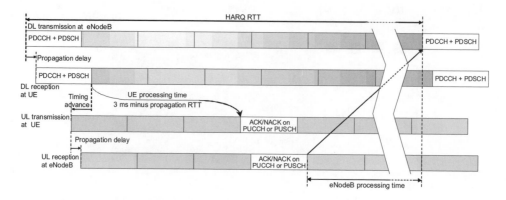

Figure 10.14 Timing diagram of the downlink HARQ (SAW) protocol.

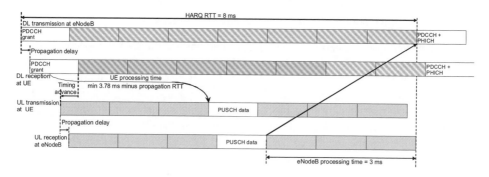

Figure 10.15 Timing diagram of the uplink HARQ (SAW) protocol.

first two systematic columns of the CB, leading to approximately 6% punctured systematic bits (with no wrap around). Thus, with systematic bit puncturing and uniform spaced RVs, the four RVs start at the top of columns 2, 26, 50 and 74.

10.3.2.5 HARQ in LTE

The physical layer in LTE supports HARQ on the physical downlink and uplink shared channels, with separate control channels to send the associated acknowledgement feedback. In Frequency Division Duplex (FDD) operation, eight Stop-And-Wait (SAW) HARQ processes are available in both uplink and downlink with a typical Round-Trip Time (RTT) of 8 ms (see Figures 10.14 and 10.15). Each HARQ process is identified with a unique three-bit HARQ process IDentifier (HARQ ID), and requires a separate soft buffer allocation in the receiver for the purpose of combining the retransmissions. There are several fields in the downlink control information to aid the HARQ operation:

- New Data Indicator (NDI): toggled whenever a new packet transmission begins;

- Redundancy Version (RV): indicates the RV selected for the transmission or retransmission;

- MCS: modulation and coding scheme.

As explained in Section 4.4.1.1, the LTE downlink HARQ is asynchronous and adaptive, and therefore every downlink transmission is accompanied by explicit signalling of control information. The uplink HARQ is synchronous, and either non-adaptive or adaptive. The uplink non-adaptive HARQ operation requires a predefined RV sequence 0, 2, 3, 1, 0, 2, 3, 1, ... for successive transmissions of a packet due to the absence of explicit control signalling. For the adaptive HARQ operation, the RV is explicitly signalled. There are also other uplink modes in which the redundancy version (or the modulation) is combined with other control information to minimize control signalling overhead. Aspects of HARQ control signalling specifically related to Time Division Duplex (TDD) operation are discussed in Section 23.4.3.

10.3.2.6 Limited Buffer Rate Matching (LBRM)

A major contributor to the UE implementation complexity is the UE HARQ soft buffer size, which is the total memory (over all the HARQ processes) required for LLR storage to support HARQ operation. The aim of Limited Buffer Rate Matching (LBRM) is therefore to reduce the required UE HARQ soft buffer sizes while maintaining the peak data rates and having minimal impact on the system performance. LBRM simply shortens the length of the virtual circular buffer of the code block segments for certain larger size of Transport Block[8] (TB), with the RV spacing being compressed accordingly. With LBRM, the effective mother code rate for a TB becomes a function of the TB size and the allocated UE soft buffer size for that TB. For example, for FDD operation and the lowest categories of UE (i.e. UE category 1 and category 2 with no spatial multiplexing – see Section 1.4), the limitation on the buffer is transparent, i.e. the LBRM does not result in any shortening of the soft buffer. For the higher UE categories (i.e. categories 3, 4 and 5), the soft buffer size is calculated assuming 8 HARQ processes and a 50% buffer reduction, which corresponds to a mother code rate of 2/3 for the largest TB. Since the eNodeB knows the soft buffer capability of the UE, it only transmits those code bits out of the VCB that can be stored in the UE's HARQ soft buffer for all (re)transmissions of a given TB.

10.3.2.7 Overall Channel Coding Chain for Data

The overall flow diagram of the turbo coded channels in LTE is shown in Figure 10.16.

The physical layer first attaches a 24-bit CRC to each TB received from the MAC layer. This is used at the receiver to verify correct reception and to generate the HARQ ACK/NACK feedback.

The TB is then segmented into 'code blocks' according to a rule which, given an arbitrary TB size, is designed to minimize the number of filler bits needed to match the available QPP sizes. This is accomplished by allowing two adjacent QPP sizes to be used when segmenting a TB, rather than being restricted to a single QPP size. The filler bits would then be placed in the first segment. However, in the first release of the LTE specifications, the set of possible

---

[8]A transport block is equivalent to a MAC PU – see Section 4.4.

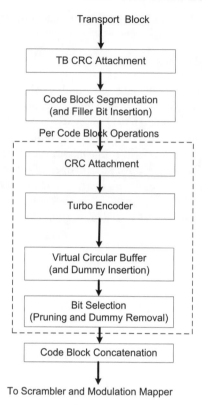

Figure 10.16 Flow diagram for turbo coded channels in LTE.

TB sizes is restricted such that the segmentation rule described above always results in a single QPP size for each segment with no filler bits. It is possible that future releases of LTE may support additional TB sizes.

Following segmentation, a further 24-bit CRC is attached to each code block if the TB was split into two or more code blocks. This code-block-level CRC can be utilized to devise rules for early termination of the turbo decoding iterations to reduce decoding complexity. It is worth noting that the polynomial used for the code-block-level CRC is different from the polynomial used for the transport block CRC. This is a deliberate design feature in order to avoid increasing the probability of failing to detect errors as a result of the use of individual CRCs per code block; if all the code block CRCs pass, the decoder should still check the transport block CRC, which, being based on a different polynomial, is likely to detect an error which was not detected by a code-block CRC.

A major difference between LTE and HSDPA is that in LTE most of the channel coding operations on the Physical Downlink Shared CHannel (PDSCH) are performed at a code-block level, whereas in HSDPA only turbo coding is performed at the code block level. Although in the LTE specifications [10] the code block concatenation for transmission is done prior to the scrambler and modulation mapper, each code block is associated with a distinct set of modulation symbols, implying that the scrambling and modulation mapping

Table 10.8 Differences between LTE and HSPA convolutional coding.

Property	LTE convolutional code	HSPA convolutional code
Number of states	64	256
Tailing method	Tail-biting	Tailed
Generators	[133, 171, 165](oct) $R = 1/3$	[561, 753](oct) $R = 1/2$ [557, 663, 711](oct) $R = 1/3$
Normalized decoding complexity	1/2 (assuming two iterations of decoding)	1
Rate matching	Circular buffer	Algorithmic calculation of rate-matching pattern

operations may in fact be done individually for each code block, which facilitates an efficient pipelined implementation.

10.3.3   Coding for Control Channels in LTE

Unlike the data, control information (such as is sent on the Physical Downlink Control CHannel (PDCCH) and Physical Broadcast CHannel (PBCH)) is coded with a convolutional code, as the code blocks are significantly smaller and the additional complexity of the turbo coding operation is therefore not worthwhile.

The PDCCH is especially critical from a decoding complexity point-of-view, since a UE must decode a large number of potential control channel locations as discussed in Section 9.3.3.2. Another relevant factor in the code design for the PDCCH and the PBCH is that they both carry a relatively small number of bits, making the tail bits (which were inserted in UMTS to reset the state of the decoder and avoid iterative decoding) a more significant overhead. Therefore, it was decided to adopt a tailbiting convolutional code for LTE, but, in order to limit the decoding complexity, using a new convolutional code with only 64 states instead of the 256-state convolutional code used in UMTS. These key differences are summarized in Table 10.8.

The LTE convolutional code, shown in Figure 10.2, offers slightly better performance for the target information block sizes, as shown in Figure 10.17. The initial value of the shift register of the encoder is set to the values corresponding to the last six information bits in the input stream so that the initial and final states of the shift register are the same. The decoder can utilize a Circular Viterbi Algorithm [14] or MAP algorithm [51], with decoding complexity approximately twice that of the Viterbi decoder with two iterations (passes through the trellis). With only a quarter the number of states, the overall complexity of the LTE convolutional code can therefore be argued to be half that of the HSPA code, provided that only two iterations are used.

The rate-matching for the convolutional code in LTE uses a similar circular buffer method as for the turbo code. A 32-column interleaver is used, with no interlacing in the circular buffer (the three parity streams are concatenated in the circular buffer). This structure gives good performance at higher code rates as well as lower code rates, and therefore LTE has no need for an additional (different) $R = 1/2$ generator polynomial as used in UMTS.

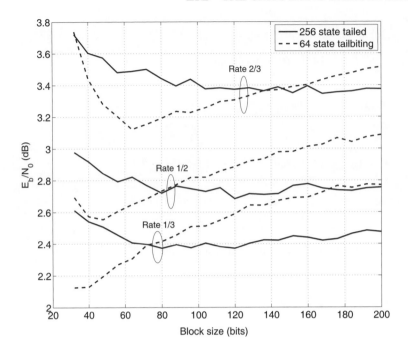

Figure 10.17 E_b/N_0 (dB) versus block size (bits) for BLER target of 1% for convolutional codes with rate $1/3, 1/2$, and $2/3$. The HSPA convolutional code has 256-state code with tail bits and the 64-state LTE code is tail-biting.

Other even smaller items of control information use block codes (for example, a Reed–Muller code for CQI, or a simple list of codewords for the Physical Control Format Indicator CHannel (PCFICH)). With small information words, block codes lend themselves well to a maximum likelihood decoding approach.

10.4   Concluding Remarks

The LTE channel coding is a versatile design that has benefited from the decades of research and development in the area of iterative processing. Although the turbo codes used in LTE and UMTS are of the same form as Berrou's original scheme, the LTE turbo code with its contention-free interleaver provides hardware designers with sufficient design flexibility to support the high data rates offered by the first release of LTE and beyond. However, with increased support for parallelism comes the cost of routing the extrinsic values to and from the memory. The routing complexity in the turbo decoder with a large number of processors (e.g. $M = 64$) may in fact become comparable to that of an LDPC code with similar processing capability. Therefore, it is possible that the cost versus performance trade-offs between turbo and LDPC codes will be reinvestigated in the future. Nevertheless, it is clear that the turbo code will continue to shine for a long time to come.

References[9]

[1] A. J. Goldsmith and S. G. Chua, 'Adaptive Coded Modulation for Fading Channels'. *IEEE Trans. on Communications*, Vol. 46, pp. 595–602, May 1998.

[2] J. Hayes, 'Adaptive feedback communications'. *IEEE Trans. on Communication Technologies*, Vol. 16, pp. 15–22, February 1968.

[3] NTT DoCoMo, Fujitsu, Mitsubishi Electric Corporation, NEC, QUALCOMM Europe, Sharp, and Toshiba Corporation, 'R1-060039: Adaptive Modulation and Channel Coding Rate Control for Single-antenna Transmission in Frequency Domain Scheduling in E-UTRA Downlink', www.3gpp.org, 3GPP TSG RAN WG1 LTE ad Hoc, Helsinki, Finland, January 2006.

[4] LG Electronics, 'R1-060051: Link Adaptation in E-UTRA Downlink', www.3gpp.org, 3GPP TSG RAN WG1 LTE ad Hoc, Helsinki, Finland, January 2006.

[5] NTT DoCoMo, Fujitsu, Mitsubishi Electric Corporation, NEC, Sharp, and Toshiba Corporation, 'R1-060040: Adaptive Modulation and Channel Coding Rate Control for MIMO Transmission with Frequency Domain Scheduling in E-UTRA Downlink', www.3gpp.org, 3GPP TSG RAN WG1 LTE ad Hoc, Helsinki, Finland, January 2006.

[6] Samsung, 'R1-060076: Adaptive Modulation and Channel Coding Rate', www.3gpp.org, 3GPP TSG RAN WG1 LTE ad Hoc, Helsinki, Finland, January 2006.

[7] 3GPP Technical Specification 36.213, 'Physical layer procedures (Release 8)', www.3gpp.org.

[8] Huawei, 'R1-080555: Labelling Complexity of UE Selected Subbands on PUSCH', www.3gpp.org, 3GPP TSG RAN WG1, meeting 51bis, Sevilla, Spain, January 2008.

[9] Huawei, 'R1-080182: Labelling of UE-selected Subbands on PUSCH', www.3gpp.org, 3GPP TSG RAN WG1, meeting 51bis, Sevilla, Spain, January 2008.

[10] 3GPP Technical Specification 36.212, 'Multiplexing and Channel Coding (FDD) (Release 8)', www.3gpp.org.

[11] J. G. Proakis, *Digital Communications*. New York: McGraw-Hill, 1995.

[12] J. H. Ma and J. W. Wolf, 'On Tail-Biting Convolutional Codes'. *IEEE Trans. on Communications*, Vol. 34, pp. 104–111, February 1986.

[13] R. V. Cox and C. V. Sundberg, 'An Efficient Adaptive Circular Viterbi Algorithm for Decoding Generalized Tailbiting Convolutional Codes'. *IEEE Trans. on Vehicular Technology*, Vol. 43, pp. 57–68, February 1994.

[14] H. Ma and W. J. Wolf, 'On Tail-biting Convolutional Codes'. *IEEE Trans. on Communications*, Vol. 34, pp. 104–111, February 1986.

[15] S. Lin and D. J. Costello, *Error and Control Coding: Fundamentals and Applications*. Second Edition, Prentice Hall: Englewood Cliffs, NJ, 2004.

[16] A. J. Viterbi, 'Error Bounds for Convolutional Codes and an Asymptotically Optimum Decoding Algorithm'. *IEEE Trans. on Information Theory*, Vol. 13, pp. 260–269, April 1967.

[17] G. D. Forney Jr., 'The Viterbi Algorithm', in *Proc. of the IEEE*, March 1973.

[18] J. Haguenauer and P. Hoeher, 'A Viterbi algorithm with soft-decision outputs and its applications', in *Proc. of IEEE Globecom*, Dallas, TX, 1989.

[19] L. R. Bahl, J. Cocke, F. Jelinek and J. Raviv, 'Optimal Decoding of Linear Codes for Minimizing Symbol Error Rate'. *IEEE Trans. on Information Theory*, Vol. 20, pp. 284–287, March 1974.

[20] C. Berrou and A. Glavieux, 'Near Optimum Error Correcting Coding and Decoding: Turbo-codes'. *IEEE Trans. on Communications*, Vol. 44, pp. 1261–1271, October 1996.

---

[9]All web sites confirmed 18th December 2008.

[21] C. E. Shannon, 'Communication in the Presence of Noise', in *Proc. IRE*, Vol. 37, pp. 10–21, January 1949.

[22] C. E. Shannon, *A Mathematical Theory of Communication*. Urbana, IL: University of Illinois Press, 1998.

[23] R. G. Gallager, 'Low-Density Parity-Check Codes'. Ph.D. Thesis, Cambridge, MA: MIT Press, 1963.

[24] S. Y. Chung, 'On the Construction of Some Capacity-Approaching Coding Scheme'. Ph.D. Thesis, Massachusetts Institute of Technology, September 2000.

[25] D. J. C. MacKay and R. M. Neal, 'Near Shannon Limit Performance of Low Density Parity Check Codes'. *IEEE Electronics Letters*, Vol. 32, pp. 1645–1646, August 1996.

[26] C. Berrou, P. Glavieux and A. Thitimajshima, 'Near Shannon Limit Error-correcting Coding and Decoding', in *Proc. IEEE International Conference on Communications*, Geneva, Switzerland, 1993.

[27] G. Battail, 'Coding for the Gaussian Channel: the Promise of Weighted-Output Decoding'. *International Journal of Satellite Communications*, Vol. 7, pp. 183–192, 1989.

[28] C. Berrou, 'The Ten-Year Old Turbo Codes are Entering into Service'. *IEEE Communication Magazine*, Vol. 41, no. 8, 2003.

[29] T. J. Richardson and R. L Urbanke, 'The Capacity of Low-density Parity-check Codes Under Message-passing Decoding'. *IEEE Trans. on Information Theory*, Vol. 47, pp. 599–618, February 2001.

[30] M. M. Mansour and N. Shanbhag, 'High-throughput LDPC Decoders'. *IEEE Trans. on Very Large Scale Integration (VLSI) Systems*, Vol. 11, pp. 976–996, October 2003.

[31] T. J. Richardson, M. A Shokrollahi and R. L Urbanke, 'Design of Capacity-approaching Irregular Low-density Parity-check Codes'. *IEEE Trans. on Information Theory*, Vol. 47, pp. 619–637, February 2001.

[32] S-Y. Chung, G. D., Jr. Forney, T. J. Richardson and R. Urbanke, 'On the Design of Low-density Parity-check Codes within 0.0045 dB of the Shannon Limit'. *IEEE Communication Letters*, Vol. 5, pp. 58–60, February 2001.

[33] J. Chen, R. M. Tanner, J. Zhang and M. P. C Fossorier, 'Construction of Irregular LDPC Codes by Quasi-Cyclic Extension'. *IEEE Trans. on Information Theory*, Vol. 53, pp. 1479–1483, 2007.

[34] M. Bickerstaff, L. Davis, C. Thomas, D. Garrett and C. Nicol, 'A 24 Mb/s Radix-4 LogMAP Turbo Decoder for 3GPP-HSDPA Mobile Wireless', in *Proc. of 2003 IEEE International Solid State Circuits Conference*, San Francisco, CA, 2003.

[35] Flarion Technologies Inc, 'Vector LDPC codes for mobile broadband communications', http://www.flarion.com/viewpoint/whitepapers/vector_LDPC.pdf, 2003.

[36] Motorola, 'R1-060384: LDPC Codes for EUTRA', www.3gpp.org, 3GPP TSG RAN WG1 meeting 44, Denver, USA, February 2006.

[37] Samsung, 'R1-060334: LTE Channel Coding', www.3gpp.org, 3GPP TSG RAN WG1 meeting 44, Denver, USA, February 2006.

[38] Intel, ITRI, LG, Mitsubishi, Motorola, Samsung, and ZTE, 'R1-060874: Complexity Comparison of LDPC Codes and Turbo Codes', www.3gpp.org, 3GPP TSG RAN WG1 meeting 44bis, Athens, Greece, March 2006.

[39] T. K. Blankenship, B. Classon and V. Desai, 'High-Throughput Turbo Decoding Techniques for 4G', in *Proc. of International Conference 3G and Beyond*, pp. 137–142, 2002.

[40] Panasonic, 'R1-073357: Interleaver for LTE turbo code', www.3gpp.org, 3GPP TSG RAN WG1, meeting 47, Riga, Latvia, November 2006.

[41] Nortel and Samsung, 'R1-073265: Parallel Decoding Method for the Current 3GPP Turbo Interleaver', www.3gpp.org, 3GPP TSG RAN WG1, meeting 47, Riga, Latvia, November 2006.

[42] A. Tarable, Benedetto S. and G. Montorsi, 'Mapping Interleaving Laws to Parallel Turbo and LDPC Decoder Architectures'. *IEEE Trans. on Information Theory*, Vol. 50, pp. 2002–2009, September 2004.

[43] A. Nimbalker, T. K. Blankenship, B. Classon, T. Fuja, and D. J. Costello Jr, 'Contention-free Interleavers', in *Proc. IEEE International Symposium on Information Theory*, Chicago, IL, 2004.

[44] A. Nimbalker, T. K. Blankenship, B. Classon, T. Fuja and D. J. Costello Jr, 'Inter-window Shuffle Interleavers for High Throughput Turbo Decoding' in *Proc. of 3rd International Symposium on Turbo Codes and Related Topics*, Brest, France, 2003.

[45] C. Berrou, Y. Saouter, C. Douillard, S. Kerouedan and M. Jezequel, 'Designing Good Permutations for Turbo Codes: Towards a Single Model', in *Proc. IEEE International Conference on Communications*, Paris, France, 2004.

[46] O. Y. Takeshita, 'On Maximum Contention-free Interleavers and Permutation Polynomials over Integer Rings'. *IEEE Trans. on Information Theory*, Vol. 52, pp. 1249–1253, March 2006.

[47] A. Nimbalker, Y Blankenship, B. Classon, T. Fuja and T. K. Blankenship, 'ARP and QPP Interleavers for LTE Turbo Coding', in *Proc. IEEE Wireless Communications and Networking Conference*, 2008.

[48] Motorola, 'R1-063061: A Contention-free Interleaver Design for LTE Turbo Codes', www.3gpp.org, 3GPP TSG RAN WG1 meeting 47, Riga, Latvia, November 2006.

[49] Siemens, 'R1-030421: Turbo Code Irregularities in HSDPA', www.3gpp.org, 3GPP TSG RAN WG1, meeting 32, Marne La Vallée, Paris, May 2003.

[50] Motorola, 'R1-071795: Parameters for Turbo Rate-Matching', www.3gpp.org, 3GPP TSG RAN WG1, meeting 48bis, Malta, March 2007.

[51] R. Y. Shao, S. Lin, and M. Fossorier, 'An Iterative Bidirectional Decoding Algorithm for Tail Biting Codes', in *Proc. of IEEE Information Theory Workshop (ITW)*, Kruger National Park, South Africa, 1999.

11

Multiple Antenna Techniques

David Gesbert, Cornelius van Rensburg, Filippo Tosato and Florian Kaltenberger

11.1 Fundamentals of Multiple Antenna Theory

11.1.1 Overview

The value of multiple antenna systems as a means to improve communications was recognized in the very early ages of wireless transmission. However, most of the scientific progress in understanding their fundamental capabilities has occurred only in the last 20 years, driven by efforts in signal and information theory, with a key milestone being achieved with the invention of so-called Multiple-Input Multiple-Output (MIMO) systems in the mid-1990s.

Although early applications of beamforming concepts can be traced back as far as 60 years in military applications, serious attention has been paid to the utilization of multiple antenna techniques in mass-market commercial wireless networks only since around 2000. The first such attempts used only the simplest forms of space-time processing algorithms. Today, the key role which MIMO technology plays in the latest wireless communication standards for Personal, Wide and Metropolitan Area Networks (PANs, WANs and MANs) testifies to its anticipated importance. Aided by rapid progress in the areas of computation and circuit integration, this trend culminated in the adoption of MIMO for the first time in a cellular mobile network standard in the Release 7 version of HSDPA (High Speed Downlink Packet Access); soon after, the development of LTE broke new ground in being the first global mobile cellular system to be designed with MIMO as a key component from the start.

In this chapter, we first provide the reader with the theoretical background necessary for a good understanding of the role and advantages promised by multiple antenna techniques in wireless communications in general. We focus on the intuition behind the main technical results and show how key progress in information theory yields practical lessons in algorithm and system design for cellular communications. As can be expected, there is still a gap between the theoretical predictions and the performance achieved by schemes that must meet the complexity constraints imposed by commercial considerations.

We distinguish between single-user MIMO and multi-user MIMO theory and techniques (see below for a definition), although a common set of concepts captures the essential MIMO benefits in both cases. Single-user MIMO techniques dominate the algorithms selected for LTE, with multi-user MIMO not being used to the maximum extent in the first version of LTE, despite its potential.

Following an introduction of the key elements of MIMO theory, in both the single-user and the multi-user cases, we proceed to describe the actual methods adopted for LTE, paying particular attention to the factors leading to these choices. However, the main goal of this section is not to provide exhaustive tutorial information on MIMO systems (for which the reader may refer, for example, to [1–3]) but rather to explain the combination of underlying theoretical principles and system design constraints which influenced specific choices for LTE.

While traditional wireless communications (Single-Input Single-Output (SISO)) exploit time- or frequency-domain pre-processing and decoding of the transmitted and received data respectively, the use of additional antenna elements at either the base station (eNodeB) or User Equipment (UE) side (on the downlink and/or uplink) opens an extra spatial dimension to signal precoding and detection. So-called space-time processing methods exploit this dimension with the aim of improving the link's performance in terms of one or more possible metrics, such as the error rate, communication data rate, coverage area and spectral efficiency (bps/Hz/cell).

Depending on the availability of multiple antennas at the transmitter and/or the receiver, such techniques are classified as Single-Input Multiple-Output (SIMO), Multiple-Input Single-Output (MISO) or MIMO. Thus in the scenario of a multi-antenna enabled base station communicating with a single antenna UE, the uplink and downlink are referred to as SIMO and MISO respectively. When a (high-end) multi-antenna terminal is involved, a full MIMO link may be obtained, although the term MIMO is sometimes also used in its widest sense, thus including SIMO and MISO as special cases. While a point-to-point multiple-antenna link between a base station and one UE is referred to as Single-User MIMO (SU-MIMO), Multi-User MIMO (MU-MIMO) features several UEs communicating simultaneously with a common base station using the same frequency- and time-domain resources.[1] By extension, considering a multicell context, neighbouring base stations sharing their antennas in virtual MIMO fashion to communicate with the same set of UEs in different cells will be termed multicell multi-user MIMO (although this latter scenario is not supported in the first version of LTE, and is therefore addressed only in outline in the context of future versions in Chapter 24). The overall evolution of MIMO concepts, from the simplest diversity setup to the futuristic multicell multi-user MIMO, is illustrated in Figure 11.1.

Despite their variety and sometimes perceived complexity, single-user and multi-user MIMO techniques tend to revolve around just a few fundamental principles, which aim at

---

[1]Note that in LTE a single eNodeB may in practice control multiple cells; in such a case, we consider each cell as an independent base station for the purpose of explaining the MIMO techniques; the simultaneous transmissions in the different cells address different UEs and are typically achieved using different fixed directional physical antennas; they are therefore not classified as multi-user MIMO.

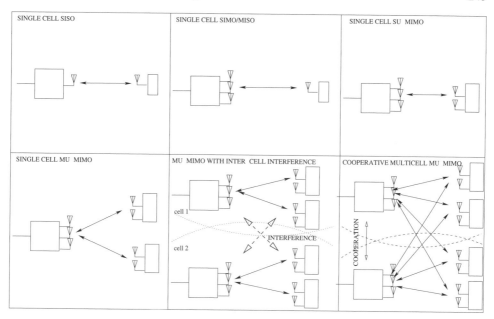

Figure 11.1 The evolution of MIMO technology, from traditional single antenna communication, to multi-user MIMO scenarios, to the possible multicell MIMO networks of the future.

leveraging some key properties of multi-antenna radio propagation channels. As introduced in Section 1.3, there are basically three advantages associated with such channels (over their SISO counterparts):

- Diversity gain.

- Array gain.

- Spatial multiplexing gain.

Diversity gain corresponds to the mitigation of the effect of multipath fading, by means of transmitting or receiving over multiple antennas at which the fading is sufficiently decorrelated. It is typically expressed in terms of an *order*, referring either to the number of effective independent diversity branches or to the slope of the bit error rate curve as a function of the Signal-to-Noise Ratio (SNR) (or possibly in terms of an SNR gain in the system's link budget).

While diversity gain is fundamentally related to improvement of the *statistics* of instantaneous SNR in a fading channel, array gain and multiplexing gain are of a different nature, rather being related to geometry and the theory of vector spaces. Array gain corresponds to a spatial version of the well-known matched-filter gain in time-domain receivers, while multiplexing gain refers to the ability to send multiple data streams in parallel and to separate them on the basis of their spatial signature. The latter is much akin to the multiplexing of users separated by orthogonal spreading codes, timeslots or frequency assignments, with the great advantage that, unlike Code, Time or Frequency Division Multiple Access (CDMA, TDMA or FDMA), MIMO multiplexing does not come at the cost of bandwidth expansion; it does, however, suffer the expense of added antennas and signal processing complexity.

These aspects are analysed further by introducing a common signal model and notation for the main families of MIMO techniques. The model is valid for single-user MIMO, yet it is sufficiently general to capture all the key principles mentioned above, as well as being easily extensible to the multi-user MIMO case (see Section 11.2.3). The model is first presented in a general way, covering theoretically optimal transmission schemes, and then particularized to popular MIMO approaches. We consider models for both uplink and downlink, or when possible a generic formulation which includes both possibilities. LTE-related schemes, specifically for the downlink, are addressed subsequently. We focus on the Frequency Division Duplex (FDD) case. Discussion of some aspects of MIMO which are specific to Time Division Duplex (TDD) operation can be found in Section 23.5.

11.1.2   MIMO Signal Model

Let $\mathbf{Y}$ be a matrix of size $N \times T$ denoting the set of (possibly precoded) signals being transmitted from N distinct antennas over T symbol durations (or, in the case of some frequency-domain systems, T subcarriers), where T is a parameter of the MIMO algorithm (defined below). Thus the nth row of $\mathbf{Y}$ corresponds to the signal emitted from the nth transmit antenna. Let $\mathbf{H}$ denote the $M \times N$ channel matrix modelling the propagation effects from each of the N transmit antennas to any one of M receive antennas, over an arbitrary subcarrier whose index is omitted here for simplicity. We assume $\mathbf{H}$ to be invariant over T symbol durations. The matrix channel is represented by way of example in Figure 11.2. Then the $M \times T$ signal $\mathbf{R}$ received over T symbol durations over this subcarrier can be conveniently written as

$$\mathbf{R} = \mathbf{HY} + \mathbf{N} \qquad\qquad (11.1)$$

where $\mathbf{N}$ is the additive noise matrix of dimension $M \times T$ over all M receiving antennas. We will use $\mathbf{h}_i$ to denote the ith column of $\mathbf{H}$, which will be referred to as the *receive spatial signature* of (i.e. corresponding to) the ith transmitting antenna. Likewise, the jth row of $\mathbf{H}$ can be termed the *transmit spatial signature* of the jth receiving antenna.

Mapping the symbols to the transmitted signal

Let $\mathbf{X} = (x_1, x_2, \ldots, x_P)$ be a group of P QAM symbols to be sent to the receiver over the T symbol durations. Thus these symbols must be *mapped* to the transmitted signal $\mathbf{Y}$ before launching into the air. The choice of this mapping function $\mathbf{X} \rightarrow \mathbf{Y}(\mathbf{X})$ determines which one out of several possible MIMO transmission methods results, each yielding a different combination of the diversity, array and multiplexing gains. Meanwhile, the so-called *spatial rate* of the chosen MIMO transmission method is given by the ratio P/T.

Note that, in the most general case, the considered transmit (or receive) antennas may be attached to a single transmitting (or receiving) device (base station or UE), or distributed over different devices. The symbols in $(x_1, x_2, \ldots, x_P)$ may also correspond to the data of one or possibly multiple users, giving rise to the so-called single-user MIMO or multi-user MIMO models.

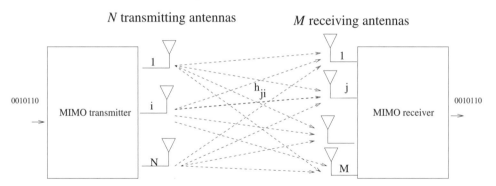

Figure 11.2 Simplified transmission model for a MIMO system with N transmit antennas, M receive antennas, giving rise to a $M \times N$ channel matrix, with MN links.

In the following sections, we explain classical MIMO techniques to illustrate the basic principles of this technology. We first assume a base station to single-user communication. The techniques are then generalized to multi-user MIMO situations.

11.1.3   Single-User MIMO Techniques

Several classes of SU-MIMO transmission methods are discussed below, both optimal and suboptimal.

11.1.3.1   Optimal Transmission over MIMO Systems

The optimal way of communicating over the MIMO channel involves a channel-dependent precoder, which fulfils the roles of both transmit beamforming and power allocation across the transmitted streams, and a matching receive beamforming structure. Full channel knowledge is therefore required at the transmit side for this method to be applicable. Consider a set of $P = NT$ symbols to be sent over the channel. The symbols are separated into N streams (or layers) of T symbols each. Stream i consists of symbols $[x_{i,1}, x_{i,2}, \ldots, x_{i,T}]$. Note that in an ideal setting, each stream may adopt a distinct code rate and modulation. This is clarified below. The transmitted signal can now be written as

$$\mathbf{Y(X)} = \mathbf{VP\bar{X}} \tag{11.2}$$

where

$$\bar{\mathbf{X}} = \begin{pmatrix} x_{1,1} & x_{1,2} & \cdots & x_{1,T} \\ \vdots & \vdots & & \vdots \\ x_{N,1} & x_{N,2} & \cdots & x_{N,T} \end{pmatrix} \tag{11.3}$$

and where $\mathbf{V}$ is an $N \times N$ transmit beamforming matrix, and $\mathbf{P}$ is a $N \times N$ diagonal power-allocation matrix with $\sqrt{p_i}$ as its ith diagonal element, where p_i is the power allocated to the ith stream. Of course, the power levels must be chosen so as not to exceed the available transmit power, which can often be conveniently expressed as a constraint on the total normalized transmit power P_t.[2] Under this model, the information-theoretic capacity of the MIMO channel in bps/Hz can be obtained as [3]

$$C_{\text{MIMO}} = \log_2 \det(I + \rho \mathbf{HVP}^2 \mathbf{V}^H \mathbf{H}^H) \tag{11.4}$$

where $\{\cdot\}^H$ denotes the Hermitian operator for a matrix or vector and ρ is the so-called transmit SNR, given by the ratio of the transmit power over the noise power.

The optimal (capacity-maximizing) precoder (**VP**) in Equation (11.4) is obtained by the concatenation of *singular vector beamforming* and the so-called *waterfilling power allocation*.

Singular vector beamforming means that **V** should be a unitary matrix (i.e. $\mathbf{V}^H \mathbf{V}$ is the identity matrix of size N) chosen such that $\mathbf{H} = \mathbf{U \Sigma V}^H$ is the Singular-Value Decomposition[3] (SVD) of the channel matrix **H**. Thus the ith right singular vector of **H**, given by the ith column of **V**, is used as a transmit beamforming vector for the ith stream. At the receiver side, the optimal beamformer for the ith stream is the ith left singular vector of **H**, obtained as the ith row of $\mathbf{U}^H$:

$$\mathbf{u}_i^H \mathbf{R} = \lambda_i \sqrt{p_i}[x_{i,1}, x_{i,2}, \ldots, x_{i,T}] + \mathbf{u}_i^H \mathbf{N} \tag{11.5}$$

where λ_i is the ith singular value of **H**.

Waterfilling power allocation is the optimal power allocation and is given by

$$p_i = [\mu - 1/(\rho \lambda_i^2)]_+ \tag{11.6}$$

where $[x]_+$ is equal to x if x is positive and zero otherwise. μ is the so-called 'water level', a positive real variable which is set such that the total power constraint is satisfied.

Thus the optimal SU-MIMO multiplexing scheme uses SVD-based transmit and receive beamforming to decompose the MIMO channel into a number of parallel non-interfering subchannels, dubbed 'eigen-channels', each one with an SNR being a function of the corresponding singular value λ_i and chosen power level p_i.

Contrary to what would perhaps be expected, the philosophy of optimal power allocation across the eigen-channels is *not* to equalize the SNRs, but rather to render them more unequal, by 'pouring' more power into the better eigen-channels, while allocating little power (or even none at all) to the weaker ones because they are seen as not contributing enough to the total capacity. This waterfilling principle is illustrated in Figure 11.3.

The underlying information-theoretic assumption here is that the information rate on each stream can be adjusted finely to match the eigen-channel's SNR. In practice this is done by selecting a suitable Modulation and Coding Scheme (MCS) for each stream.

11.1.3.2 Beamforming with Single Antenna Transmitter or Receiver

In the case where either the receiver or the transmitter is equipped with only a single antenna, the MIMO channel exhibits only one active eigen-channel, and hence multiplexing of more than one data stream is not possible.

In *receive* beamforming, $N = 1$ and $M > 1$ (assuming a single-stream). In this case, one symbol is transmitted at a time, such that the symbol-to-transmit-signal mapping function is

---

[2] In practice there may be a limit on the maximum transmission power from each antenna.
[3] The reader is referred to [4] for an explanation of generic matrix operations and terminology.

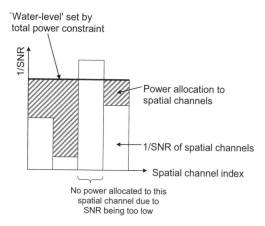

Figure 11.3 The waterfilling principle for optimal power allocation.

characterized by $P = T = 1$, and $\mathbf{Y}(\mathbf{X}) = \mathbf{X} = x$, where x is the one QAM symbol to be sent. The received signal vector is given by

$$\mathbf{R} = \mathbf{H}x + \mathbf{N} \qquad (11.7)$$

The receiver combines the signals from its M antennas through the use of weights $\mathbf{w} = [w_1, \ldots, w_M]$. Thus the received signal after antenna combining can be expressed as

$$z = \mathbf{wR} = \mathbf{wH}s + \mathbf{wN} \qquad (11.8)$$

After the receiver has acquired a channel estimate (as discussed in Chapter 8), it can set the beamforming vector $\mathbf{w}$ to its optimal value to maximize the received SNR. This is done by aligning the beamforming vector with the UE's channel, via the so-called Maximum Ratio Combining (MRC) $\mathbf{w} = \mathbf{H}^H$, which can be viewed as a spatial version of the well-known matched filter. Note that cancellation of an interfering signal can also be achieved, by selecting the beamforming vector to be orthogonal to the channel from the interference source. These simple concepts are illustrated vectorially in Figure 11.4.

The maximum ratio combiner provides a factor of M improvement in the received SNR compared to the $M = N = 1$ case – i.e. an array gain of $10 \log_{10}(M)$ dB in the link budget.

In *transmit* beamforming, $M = 1$ and $N > 1$. The symbol-to-transmit-signal mapping function is characterized by $P = T = 1$, and $\mathbf{Y}(\mathbf{X}) = \mathbf{w}x$, where x is the one QAM symbol to be sent. $\mathbf{w}$ is the transmit beamforming vector of size $N \times 1$, computed based on channel knowledge, which is itself often obtained via a receiver-to-transmitter feedback link.[4] Assuming perfect channel knowledge at the transmitter side, the SNR-maximizing solution is given by the transmit MRC, which can be seen as a matched prefilter:

$$\mathbf{w} = \frac{\mathbf{H}^H}{\|\mathbf{H}\|} \qquad (11.9)$$

[4]In some situations other techniques such as receive/transmit channel reciprocity may be used, as discussed in Section 23.5.2.

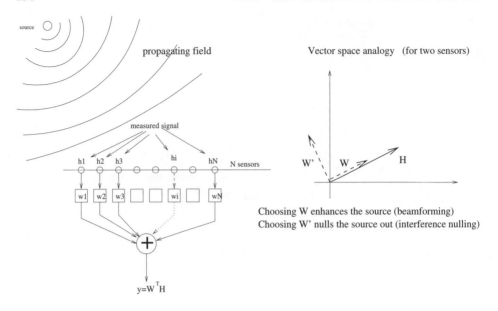

Figure 11.4 The beamforming and interference cancelling concepts.

where the normalization by $\|\mathbf{H}\|$ enforces a total power constraint across the transmit antennas. The transmit MRC pre-filter provides a similar gain as its receive counterpart, namely $10 \log_{10}(N)$ dB in average SNR improvement.

11.1.3.3 Spatial Multiplexing without Channel Knowledge at the Transmitter

When $N > 1$ and $M > 1$, multiplexing of up to $\min(M, N)$ streams is theoretically possible even without transmit channel knowledge. Assume for instance that $M \geq N$. In this case one considers N streams, each transmitted using one different transmitted antenna. As the transmitter does not have knowledge of the matrix $\mathbf{H}$, the design of the spatial multiplexing scheme cannot be improved by the use of a channel-dependent precoder. Thus the precoder is simply the identity matrix. In this case, the symbol-to-transmit-signal mapping function is characterized by $P = NT$ and by

$$\mathbf{Y}(\mathbf{X}) = \bar{\mathbf{X}} \tag{11.10}$$

At the receiver, a variety of linear and non-linear detection techniques may be implemented to recover the symbol matrix $\bar{\mathbf{X}}$. A low-complexity solution is offered by the linear case, whereby the receiver superposes N beamformers $\mathbf{w}_1, \mathbf{w}_2, \ldots, \mathbf{w}_N$.

The detection of stream $[x_{i,1}, x_{i,2}, \ldots, x_{i,T}]$ is achieved by applying $\mathbf{w}_i$ as follows:

$$\mathbf{w}_i \mathbf{R} = \mathbf{w}_i \mathbf{H} \bar{\mathbf{X}} + \mathbf{w}_i \mathbf{N} \tag{11.11}$$

The design criterion for the beamformer $\mathbf{w}_i$ can be interpreted as a compromise between single-stream beamforming and cancelling of interference (created by the other $N - 1$ streams). Inter-stream interference is fully cancelled by selecting the Zero-Forcing (ZF) receiver given by

$$\mathbf{W} = \begin{pmatrix} \mathbf{w}_1 \\ \mathbf{w}_2 \\ \vdots \\ \mathbf{w}_N \end{pmatrix} = (\mathbf{H}^{\mathrm{H}}\mathbf{H})^{-1}\mathbf{H}^{\mathrm{H}} \tag{11.12}$$

However, for optimal performance, $\mathbf{w}_i$ should strike a balance between alignment with respect to $\mathbf{h}_i$ and orthogonality with respect to all other signatures $\mathbf{h}_k$, $k \neq i$. Such a balance is achieved by, for example, a Minimum Mean-Squared Error (MMSE) receiver.

Beyond classical linear detection structures such as the ZF or MMSE receivers, more advanced but nonlinear detectors can be exploited which provide a better error rate performance at the chosen SNR operating point, at the cost of extra complexity. Examples of such detectors include the Successive Interference Cancellation (SIC) detector and the Maximum Likelihood Detector (MLD). The principle of the SIC detector is to treat individual streams, which are channel-encoded, like layers which are peeled off one by one by a processing sequence consisting of linear detection, decoding, remodulating, re-encoding and subtraction from the total received signal $\mathbf{R}$. On the other hand, the MLD attempts to select the most likely set of all streams, simultaneously, from $\mathbf{R}$, by an exhaustive search procedure or a lower-complexity equivalent such as the sphere-decoding technique [3].

Multiplexing gain

The multiplexing gain corresponds to the multiplicative factor by which the spectral efficiency is increased by a given scheme. Perhaps the single most important requirement for MIMO multiplexing gain to be achieved is for the various transmit and receive antennas to experience a sufficiently different channel response. This translates into the condition that the spatial signatures of the various transmitters (the $\mathbf{h}_i$'s) (or receivers) be sufficiently decorrelated and linearly independent to allow for the channel matrix $\mathbf{H}$ to be invertible (or more generally, well-conditioned). An immediate consequence of this condition is the limitation to $\min(M, N)$ of the number of independent streams which may be multiplexed into the MIMO channel, or more generally to rank($\mathbf{H}$) streams. As an example, single-user MIMO communication between a four-antenna base station and a dual antenna UE can, at best, support multiplexing of two data streams, and thus a doubling of the UE's data rate compared with a single stream.

11.1.3.4 Diversity

Unlike the basic multiplexing scenario in Equation (11.10), where the design of the transmitted signal matrix $\mathbf{Y}$ exhibits no redundancy between its entries, a diversity-oriented design will feature some level of repetition between the entries of $\mathbf{Y}$. For 'full diversity', each transmitted symbol $x_1, x_2, \ldots, x_P$ must be assigned to each of the transmit antennas at least once in the course of the T symbol durations. The resulting symbol-to-transmit-signal mapping function is called a Space-Time Block Code (STBC). Although many designs of STBC exist, additional properties such as the orthogonality of matrix $\mathbf{Y}$ allow improved performance and easy decoding at the receiver. Such properties are realized by the so-called Alamouti space-time code [5], explained later in this chapter. The total diversity order which can be realized in the N to M MIMO channel is MN when entries of the MIMO channel matrix are statistically uncorrelated. The intuition behind this is that $MN - 1$ represents the number of SISO links simultaneously in a state of severe fading which the system can sustain while still being able to convey the information to the receiver. The diversity order is equal to this number plus one. As in the previous simple multiplexing scheme, an advantage of diversity-oriented transmission is that the transmitter does not need knowledge of the channel **H**, and therefore no feedback of this parameter is necessary.

Diversity versus multiplexing trade-off

A fundamental aspect of the benefits of MIMO lies in the fact that any given multiple antenna configuration has a limited number of degrees of freedom. Thus there exists a compromise between reaching full beamforming gain in the detection of a desired stream of data and the perfect cancelling of undesired, interfering streams. Similarly, there exists a trade-off between the number of streams that may be multiplexed across the MIMO channel and the amount of diversity that each one of them will enjoy. Such a trade-off can be formulated from an information-theoretic point of view [6]. In the particular case of spatial multiplexing of N streams over a N to M antenna channel, with $M \geq N$, and using a linear detector, it can be shown that each stream will enjoy a diversity order of $M - N + 1$.

To some extent, increasing the spatial load of MIMO systems (i.e. the number of spatially-multiplexed streams) is akin to increasing the user load in CDMA systems. This correspondence extends to the fact that an optimal load level exists for a given target error rate in both systems.

11.1.4 Multi-User Techniques

11.1.4.1 Comparing Single-User and Multi-User MIMO

The set of MIMO techniques featuring data streams being communicated to (or from) antennas located on distinct UEs in the model is referred to as Multi-User MIMO (MU-MIMO). Although this situation is just as well described by our model in Equation (11.1), the MU-MIMO scenario differs in a number of crucial ways from its single-user counterpart. We first explain these differences qualitatively, and then present a brief survey of the most important MU-MIMO transmission techniques.

In MU-MIMO, K UEs are selected for simultaneous communication over the same time-frequency resource, from a set of U active UEs in the cell. Typically, K is much smaller than U. Each UE is assumed to be equipped with J antennas, so the selected UEs together form a set of $M = KJ$ UE-side antennas. Since the number of streams that may be communicated over an N to M MIMO channel is limited to $\min(M, N)$ (if complete interference suppression is intended using linear combining of the antennas), the upper bound on the number of streams in MU-MIMO is typically dictated by the number of base station antennas N. The number of streams which may be allocated to each UE is limited by the number of antennas J at that UE. For instance, with single-antenna UEs, up to N streams can be multiplexed, with a distinct stream being allocated to each UE. This is in contrast to SU-MIMO, where the transmission of N streams necessitates that the UE be equipped with at least N antennas. Therefore a great advantage of MU-MIMO over SU-MIMO is that the MIMO multiplexing benefits are preserved even in the case of low-cost UEs with a small number of antennas. As a result, it is generally assumed that in MU-MIMO it is the base station which bears the burden of spatially separating the UEs, be it on the uplink or the downlink. Thus the base station performs receive beamforming from several UEs on the uplink and transmit beamforming towards several UEs on the downlink.

Another fundamental contrast between SU-MIMO and MU-MIMO comes from the difference in the underlying channel model. While in SU-MIMO the decorrelation between the spatial signatures of the antennas requires rich multipath propagation or the use of orthogonal polarizations, in MU-MIMO the decorrelation between the signatures of the different UEs occurs naturally due to fact that the separation between such UEs is typically large relative to the wavelength.

11.1.4.2 Techniques for Single-Antenna UEs

In considering the case of MU-MIMO for single-antenna UEs, it is worth noting that the number of antennas available to a UE for transmission is typically less than the number available for reception. We therefore examine first the uplink scenario, followed by the downlink.

With a single antenna at each UE, the MU-MIMO uplink scenario is very similar to the one described by Equation (11.10): because the UEs in mobile communication systems such as LTE typically cannot cooperate and do not have knowledge of the uplink channel coefficients, no precoding can be applied and each UE simply transmits an independent message. Thus, if K UEs are selected for transmission in the same time-frequency resource, each UE k transmitting symbol s_k, the received signal at the base station, over a single $T = 1$ symbol period, is written

$$\mathbf{R} = \mathbf{H}\bar{\mathbf{X}} + \mathbf{N} \tag{11.13}$$

where

$$\bar{\mathbf{X}} = \begin{pmatrix} x_1 \\ \vdots \\ x_K \end{pmatrix} \tag{11.14}$$

In this case, the columns of $\mathbf{H}$ correspond to the receive spatial signatures of the different UEs. The base station can recover the transmitted symbol information by applying beam-forming filters, for example using MMSE or ZF solutions (as in Equation (11.12)). Note that no more than N UEs can be served (i.e. $K \leq N$) if inter-user interference is to be suppressed fully.

MU-MIMO in the uplink is sometimes referred to as 'Virtual MIMO', as from the point of view of a given UE there is no knowledge of the simultaneous transmissions of the other UEs. This transmission mode and its implications for LTE are discussed in Section 17.5.2.

On the downlink, which is illustrated in Figure 11.5, the base station must resort to transmit beamforming in order to separate the data streams intended for the various UEs. Over a single $T = 1$ symbol period, the signal received by UEs 1 to K can be written compactly as

$$\mathbf{R} = \begin{pmatrix} r_1 \\ \vdots \\ r_K \end{pmatrix} = \mathbf{HVP}\bar{\mathbf{X}} + \mathbf{N} \tag{11.15}$$

This time, the *rows* of $\mathbf{H}$ correspond to the transmit spatial signatures of the various UEs. $\mathbf{V}$ is the transmit beamforming matrix and $\mathbf{P}$ is the (diagonal) power allocation matrix selected

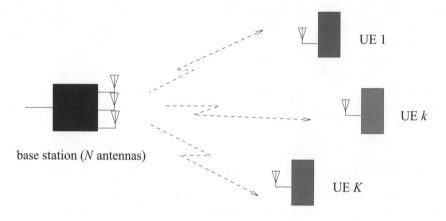

K users (UEs have 1 antenna each)

Figure 11.5 A MU-MIMO scenario in the downlink with single-antenna UEs: the base station transmits to K selected UEs simultaneously. Their contributions are separated by multiple-antenna precoding at the base station side, based on channel knowledge.

such that it fulfils the total normalized transmit power constraint P_t. To cancel out fully the inter-user interference when $K \leq N$, a transmit ZF beamforming solution may be employed (although this is not optimal due to the fact that it may require a high transmit power if the channel is ill-conditioned). Such a solution would be given by

$$\mathbf{V} = \mathbf{H}^{\mathrm{H}}(\mathbf{H}\mathbf{H}^{\mathrm{H}})^{-1} \qquad (11.16)$$

Note that regardless of the channel realization, the power allocation must be chosen to satisfy any power constraints at the base station, for example such that $\mathrm{trace}(\mathbf{V}\mathbf{P}\mathbf{P}^{\mathrm{H}}\mathbf{V}^{\mathrm{H}}) = P_t$.

11.1.4.3   Techniques for Multiple-Antenna UEs

The ideas presented above for single-antenna UEs can be generalized to the case of multiple-antenna UEs. There could, in theory, be essentially two ways of exploiting the additional antennas at the UE side. In the first approach, the multiple antennas are simply treated as multiple *virtual* UEs, allowing high-capability terminals to receive or transmit more than one stream, while at the same time spatially sharing the channel with other UEs. For instance, a four-antenna base station could theoretically communicate in a MU-MIMO fashion with two UEs equipped with two antennas each, allowing two streams per UE, resulting in a total multiplexing gain of four. Another example would be that of two single-antenna UEs, receiving one stream each, and sharing access with another two-antenna UE, the latter receiving two streams. Again, the overall multiplexing factor remains limited to the number of base-station antennas.

   The second approach for making use of additional UE antennas is to treat them as extra degrees of freedom for the purpose of strengthening the link between the UE and the base station. Multiple antennas at the UE may then be combined in MRC fashion in the case of the downlink, or in the case of the uplink space-time coding could be used. Antenna selection is another way of extracting more diversity out of the channel, as discussed in Section 17.5.

11.1.4.4 Comparing Single-User and Multi-User capacity

To illustrate the gains of multi-user multiplexing over single-user transmission, we compare the sum-rate achieved by both types of system from an information theoretic standpoint, for single antenna UEs. We compare the Shannon capacity in single-user and multi-user scenarios both for an idealized synthetic channel and for a channel obtained from real measurement data.

The idealized channel model assumes that the entries of the channel matrix **H** in Equation (11.13) are independently and identically distributed (i.i.d.) Rayleigh fading. For the measured channel case, a channel sounder was used[5] to perform real-time wideband channel measurements synchronously for two UEs moving at vehicular speed in an outdoor semi-urban hilly environment with Line-Of-Sight (LOS) propagation predominantly present. The most important parameters of the platform are summarized in Table 11.1.

Table 11.1 Parameters of the measured channel for SU-MIMO/MU-MIMO comparison. More details can be found in references [7, 8].

Parameter	Value
Centre frequency	1917.6 MHz
Bandwidth	4.8 MHz
Base station transmit power	30 dBm
Number of antennas at base station	4 (2 cross polarized)
Number of UEs	2
Number of antennas at UE	1
Number of subcarriers	160

The sum-rate capacity of a two-UE MU-MIMO system (calculated assuming a zero-forcing precoder as described in Section 11.1.4) is compared with the capacity of an equivalent MISO system serving a single UE at a time (i.e. in TDMA), employing beamforming (see Section 11.1.3.2). The base station has four antennas and the UE has a single antenna. Full Channel State Information at the Transmitter (CSIT) is assumed in both cases.

Figure 11.6 shows the ergodic (mean) sum-rate of both schemes in both channels. The mean is taken over all frames and all subcarriers and subsequently normalized to bps/Hz. It can be seen that in both the ideal and the measured channels, MU-MIMO yields a higher sum-rate than SU-MISO in general. In fact, at high SNR, the multiplexing gain of the MU-MIMO system is two while it is limited to one for the SU-MISO case.

However, for low SNR, the SU-MISO TDMA and MU-MIMO schemes perform very similarly. This is because a sufficiently high SNR is required to excite more than one MIMO transmission mode. Interestingly, the performance of both SU-MISO TDMA and MU-MIMO

---

[5]The Eurecom MIMO OpenAir Sounder (EMOS) [7].

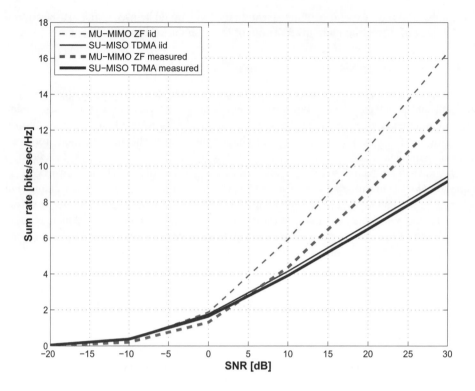

Figure 11.6 Ergodic sum-rate capacity of SU-MISO TDMA and MU-MIMO with two UEs, for an i.i.d. Rayleigh fading channel and for a measured channel.

is slightly worse in the measured channels than in the idealized i.i.d. channels. This can be attributed to the correlation of the measured channel in time (due to the relatively slow movement of the users), in frequency (due to the LOS propagation), and in space (due to the transmit antenna correlation). In the MU-MIMO case the difference between the i.i.d. and the measured channel is much higher than in the single-user TDMA case, since these correlation effects result in a rank-deficient channel matrix.

11.2   MIMO Schemes in LTE

Building on the theoretical background of the previous section, the MIMO schemes adopted for LTE are reviewed and explained. These schemes relate to the downlink unless otherwise mentioned.

11.2.1   Practical Considerations

First, a few important practical constraints are briefly reviewed which affect the real-life performance of the theoretical MIMO systems considered above, and which often are decisive when selecting a particular transmission strategy in a given propagation and system setting.

It was argued above that the full MIMO benefits (array gain, diversity gain and multiplexing gain) assume ideally decorrelated antennas and full-rank MIMO channel matrices. In this regard, the propagation environment and the antenna design (e.g. the spacing) play a significant role. In the single-user case, the antennas at both the base station and the UE are typically separated by between half a wavelength and a couple of wavelengths at most. This distance is very short in relation to the distance from base station to UE. In a LOS situation, this will cause a strong correlation between the spatial signatures, limiting the use of multiplexing schemes. However, an exception to this can be created from the use of antennas whose design itself provides the necessary orthogonality properties even in LOS situations. An example is the use of two antennas (at both transmitter and receiver) that operate on orthogonal polarizations (e.g. horizontal and vertical polarizations, or better, so-called $+45°$ and $-45°$ polarizations, which give a twofold multiplexing capability even in LOS). However, the use of orthogonal polarizations at the UEs may not always be recommended as it results in non-omnidirectional beam patterns. Such exceptions aside, in single-user MIMO the condition of spatial signature independence can only be satisfied with the help of rich random multipath propagation. In diversity-oriented schemes, the invertibility of the channel matrix is not required, yet the entries of the channel matrix should be statistically decorrelated. Although a greater LOS to non-LOS energy ratio will tend to correlate the fading coefficients on the various antennas, this effect will be compensated by the reduction in fading delivered by the LOS component.

Another source of discrepancy between theoretical MIMO gains and practically achieved performance lies in the (in-)ability of the receiver, and whenever needed the transmitter, to estimate the channel coefficients perfectly. At the receiver, channel estimation is typically performed over a finite sample of Reference Signals (RSs), as discussed in Chapter 8. In the case of transmit beamforming and MIMO SVD-based precoding, the transmitter then has to acquire this channel knowledge (or directly the precoder knowledge) from the receiver usually through a limited feedback link, which causes further degradation to the available CSIT.

With MU-MIMO, the principle advantages over SU-MIMO are clear: robustness with respect to the propagation environment and spatial multiplexing gain preserved even in the case of UEs with small numbers of antennas. However, such advantages come at a price. In the downlink, MU-MIMO relies on the ability of the base station to compute the required transmit beamformer, which in turn requires CSIT. The fundamental role of CSIT in the MU-MIMO downlink can be emphasized as follows: in the extreme case of no CSIT being available and identical fading statistics for all the UEs, the MU-MIMO gains disappear and the SU-MIMO strategy becomes optimal [9].

As a consequence, one of the most difficult challenges in making MU-MIMO practical for cellular applications, and particularly for an FDD system, is devising mechanisms that allow for accurate Channel State Information (CSI) to be delivered by the UE to the base station in a resource-efficient manner. This requires the use of appropriate *codebooks* for quantization. These aspects are developed later in this chapter. A recent account of the literature on this subject may also be found in reference [10].

Another issue which arises for practical implementations of MIMO schemes is the interaction between the physical layer and the scheduling protocol. As noted in Section 11.1.4.1, in both uplink and downlink cases the number of UEs which can be served in a MU-MIMO fashion is typically limited to $K = N$, assuming linear combining structures. Often one may even decide to limit K to a value strictly less than N to preserve some degrees of freedom for per-user diversity. As the number of active users U will typically exceed K, a selection algorithm must be implemented to identify which set of users will be scheduled for simultaneous transmission over a particular time-frequency Resource Block (RB). This algorithm is not specified in LTE and various approaches are possible; as discussed in Chapter 12, a combination of rate maximization and QoS constraints will typically be considered. It is important to note that the choice of UEs that will maximize the sum-rate (the sum over the K individual rates for a given subframe) is one that favours UEs exhibiting not only good instantaneous SNR but also spatial separability among their signatures.

11.2.2 Single-User Schemes

In this section, we examine the solutions adopted in LTE for SU-MIMO. We consider first the diversity schemes used on the transmit side, then beamforming schemes, and finally we look at the spatial multiplexing mode of transmission.

11.2.2.1 Transmit Diversity Schemes

The theoretical aspects of transmit diversity were discussed in Section 11.1. Here we discuss the two main transmit diversity techniques defined in LTE. In LTE, transmit diversity is only defined for 2 and 4 transmit antennas, and one data stream, referred to in LTE as one *codeword* since one transport block CRC is used per data stream. To maximize diversity gain the antennas typically need to be uncorrelated, so they need to be well separated relative to the wavelength or have different polarization. Transmit diversity still has its value in a number of scenarios, including low SNR, low mobility (no time diversity), or for applications with low delay tolerance. Diversity schemes are also desirable for channels for which no uplink feedback signalling is available (e.g. Multimedia Broadcast/Multicast Services (MBMS) described in Chapter 14, Physical Broadcast CHannel (PBCH) in Chapter 9 and Synchronization Signals in Chapter 7).

In LTE the MIMO scheme is independently assigned for the control channels and the data channels, and is also assigned independently per UE in the case of the data channels (Physical Downlink Shared CHannel – PDSCH).

In this section we will discuss in more detail the transmit diversity techniques of Space-Frequency Block Codes (SFBCs) and Frequency Switched Transmit Diversity (FSTD), as well as the combination of these schemes as used in LTE. These transmit diversity schemes may be used in LTE for the PBCH and Physical Downlink Control CHannel (PDCCH), and also for the PDSCH if it is configured in transmit diversity mode[6] for a UE.

Another transmit diversity technique which is commonly associated with OFDM is Cyclic Delay Diversity (CDD). CDD is not used in LTE as a diversity scheme in its own right but rather as a precoding scheme for spatial multiplexing on the PDSCH; we therefore introduce it later in Section 11.2.2.3 in the context of spatial multiplexing.

---

[6]PDSCH transmission mode 2 – see Section 9.2.2.1.

Space-Frequency Block Codes (SFBCs)

If a physical channel in LTE is configured for transmit diversity operation using two eNodeB antennas, pure SFBC is used. SFBC is a frequency-domain version of the well-known Space-Time Block Codes (STBCs), also known as Alamouti codes [5]. This family of codes is designed so that the transmitted diversity streams are orthogonal and achieve the optimal SNR with a linear receiver. Such orthogonal codes only exist for the case of two transmit antennas.

STBC is used in UMTS, but in LTE the number of available OFDM symbols in a subframe is often odd while STBC operates on pairs of adjacent symbols in the time domain. The application of STBC is therefore not straightforward for LTE, while the multiple subcarriers of OFDM lend themselves well to the application of SFBC.

For SFBC transmission in LTE, the symbols transmitted from the two eNodeB antenna ports on each pair of adjacent subcarriers are defined as follows:

$$\begin{bmatrix} y^{(0)}(1) & y^{(0)}(2) \\ y^{(1)}(1) & y^{(1)}(2) \end{bmatrix} = \begin{bmatrix} x_1 & x_2 \\ -x_2^* & x_1^* \end{bmatrix} \tag{11.17}$$

where $y^{(p)}(k)$ denotes the symbols transmitted from antenna port p on the kth subcarrier.

Since no orthogonal codes exist for antenna configurations beyond 2×2, SFBC has to be modified in order to apply it to the case of 4 transmit antennas. In LTE, this is achieved by combining SFBC with FSTD.

Frequency Switched Transmit Diversity (FSTD) and its Combination with SFBC

General FSTD schemes transmit symbols from each antenna on a different set of subcarriers. For example, an FSTD transmission from 4 transmit antennas on four subcarriers might appear as follows:

$$\begin{bmatrix} y^{(0)}(1) & y^{(0)}(2) & y^{(0)}(3) & y^{(0)}(4) \\ y^{(1)}(1) & y^{(1)}(2) & y^{(1)}(3) & y^{(1)}(4) \\ y^{(2)}(1) & y^{(2)}(2) & y^{(2)}(3) & y^{(2)}(4) \\ y^{(3)}(1) & y^{(3)}(2) & y^{(3)}(3) & y^{(3)}(4) \end{bmatrix} = \begin{bmatrix} x_1 & 0 & 0 & 0 \\ 0 & x_2 & 0 & 0 \\ 0 & 0 & x_3 & 0 \\ 0 & 0 & 0 & x_4 \end{bmatrix} \tag{11.18}$$

where, as previously, $y^{(p)}(k)$ denotes the symbols transmitted from antenna port p on the kth subcarrier. In practice in LTE, FSTD is only used in combination with SFBC for the case of 4 transmit antennas, in order to provide a suitable transmit diversity scheme where no orthogonal rate-1 block codes exists. The LTE scheme is in fact a combination of two 2×2 SFBC schemes mapped to independent subcarriers as follows:

$$\begin{bmatrix} y^{(0)}(1) & y^{(0)}(2) & y^{(0)}(3) & y^{(0)}(4) \\ y^{(1)}(1) & y^{(1)}(2) & y^{(1)}(3) & y^{(1)}(4) \\ y^{(2)}(1) & y^{(2)}(2) & y^{(2)}(3) & y^{(2)}(4) \\ y^{(3)}(1) & y^{(3)}(2) & y^{(3)}(3) & y^{(3)}(4) \end{bmatrix} = \begin{bmatrix} x_1 & x_2 & 0 & 0 \\ 0 & 0 & x_3 & x_4 \\ -x_2^* & x_1^* & 0 & 0 \\ 0 & 0 & -x_4^* & x_3^* \end{bmatrix} \tag{11.19}$$

Note that mapping of symbols to antenna ports is different in the 4 transmit-antenna case compared to the 2 transmit-antenna SFBC scheme. This is because the RS density on the third and fourth antenna ports is half that of the first and second antenna ports (see Section 8.2), and hence the channel estimation accuracy may be lower on the third and fourth antenna ports. Thus, this design of the transmit diversity scheme avoids concentrating the channel estimation losses in just one of the SFBC codes, resulting in a slight coding gain.

11.2.2.2 Beamforming Schemes

The theoretical aspects of beamforming were described in Section 11.1. This section explains how it is implemented in LTE.

LTE differentiates between two transmission modes which may support beamforming for the PDSCH:

- **Closed-loop rank 1 precoding.**[7] Although this amounts to beamforming, it can also be seen as a special case of SU-MIMO spatial multiplexing and is therefore covered by this discussion in Section 11.2.2.3. In this mode the UE feeds channel information back to the eNodeB to indicate suitable precoding to apply for the beamforming operation.

- **UE-specific RSs.**[8] In this mode the UE does not feed back any precoding-related information. The eNodeB instead tries to deduce this information, for example using Direction Of Arrival (DOA) estimations from the uplink, in which case it is worth noting that calibration of the eNodeB RF paths may be necessary, as discussed in Section 23.5.2.

In this section we focus on the latter case. This mode is primarily a mechanism to extend cell coverage by concentrating the eNodeB power in the direction in which the UE is located. It typically has the following properties:

- It can conveniently be implemented by an array of closely-spaced antenna elements for creating directional transmissions. The signals from the different antenna elements are phased appropriately so that they all add up constructively at the location where the UE is situated.

- The eNodeB is responsible for ensuring that the beam is correctly directed, as the UE does not explicitly indicate a preference regarding the direction/selection of the beam.

- Other than being directed to use the UE-specific RS as the phase reference, a UE would not really be aware that it is receiving a directional transmission rather than a cell-wide transmission. To the UE, the phased array of antenna ports 'appears' as just one antenna.

One side-effect of using beamforming based on UE-specific RS is that channel quality experienced by the UE will typically be different (hopefully better) than that of any of the cell-specific RS. However, as the UE-specific RS are only provided in the specific RBs for which the beamforming transmission mode is applied, the eNodeB cannot rely on the UE being able to derive Channel Quality Indicator (CQI) feedback from the UE-specific RS. For this reason, it is specified in LTE that CQI feedback from a UE configured with UE-specific RS is derived using the cell-specific RSs (assuming transmit diversity if more than one common antenna port exists). This suggests a deployment scenario whereby at least

---

[7]PDSCH transmission mode 6 – see Section 9.2.2.1.

[8]PDSCH transmission mode 7 – see Section 9.2.2.1.

one of the common antenna ports actually uses one of the elements of the phased array. The eNodeB could then, over time, establish a suitable offset to apply to the CQI reports received from the UE to adapt them to the actual quality of the beamformed signal. Such an offset might, for example, be derived from the proportion of transport blocks positively acknowledged by the UE. An eNodeB antenna configuration of this kind also allows the possibility to use beamforming for UEs near the edge of the cell, while other antenna ports may be used for SU-MIMO spatial multiplexing to deliver high data rates to UEs closer to the eNodeB.

Another factor to consider when deploying beamforming in LTE is that it can only be applied to the PDSCH and not to the control channels. Typically the range of the PDSCH can therefore be extended by beamforming, but the overall cell range may still be limited by the range of the control channels unless other measures are taken. One approach could be to reduce the code rate used for the control channels when beamforming is applied to the PDSCH.

The effect of beamforming on neighbouring cells should also be taken into account. If the beamforming is intermittent, it can result in a problem often known as the 'flash light effect', where strong intermittent interference may disturb the accuracy of the UEs' CQI reporting in adjacent cells. This effect was shown to have the potential to cause throughput reductions in HSDPA [11], but in LTE the possibility for frequency-domain scheduling in OFDMA (as discussed in Section 12.5) provides an additional degree of freedom to avoid such issues.

11.2.2.3 Spatial Multiplexing Schemes

Introduction

We begin by introducing some terminology used to describe spatial multiplexing in LTE:

- A spatial *layer* is the term used in LTE for the different streams generated by spatial multiplexing as described in Section 11.1. A layer can be described as a mapping of symbols onto the transmit antenna ports. Each layer is identified by a (precoding) vector of size equal to the number of transmit antenna ports and can be associated with a radiation pattern.

- The *rank* of the transmission is the number of layers transmitted.

- A *codeword* is an independently encoded data block, corresponding to a single Transport Block (TB) delivered from the Medium Access Control (MAC) layer in the transmitter to the physical layer, and protected with a CRC.

For ranks greater than 1, two codewords can be transmitted. Note that the number of codewords is always less than or equal to the number of layers, which in turn is always less than or equal to the number of antenna ports.

In principle, a SU-MIMO spatial multiplexing scheme can either use a single codeword mapped to all the available layers, or multiple codewords each mapped to one or more different layers.

The main benefit of using only one codeword is a reduction in the amount of control signalling required, both for CQI reporting, where only a single value would be needed for all layers, and for HARQ ACK/NACK feedback, where only one ACK/NACK would have

Table 11.2 Codeword-to-layer mapping in LTE.

	Codeword 1	Codeword 2
Rank 1	Layer 1	
Rank 2	Layer 1	Layer 2
Rank 3	Layer 1	Layer 2 and Layer 3
Rank 4	Layer 1 and Layer 2	Layer 3 and Layer 4

to be signalled per subframe per UE. In such a case, the MLD receiver is optimal in terms of minimizing the bit error rate.

At the opposite extreme, a separate codeword could be mapped to each of the layers. The advantage of this type of scheme is that significant gains are possible by using Successive Interference Cancellation (SIC), albeit at the expense of more signalling being required. An MMSE-SIC receiver can be shown to approach the Shannon capacity [6]. Note that an MMSE receiver is viable for both transmitter structures. For LTE, a middle-way was adopted whereby at most two codewords are used, even if four layers are transmitted. The codeword-to-layer mapping is static, since only minimal gains were shown for a dynamic mapping method. The mappings are shown in Table 11.2. Note that in LTE all RBs belonging to the same codeword use the same MCS, even if a codeword is mapped to multiple layers.

Precoding

The PDSCH transmission modes for open-loop spatial multiplexing[9] and closed-loop spatial multiplexing[10] use precoding from a defined 'codebook' to form the transmitted layers. Each codebook consists of a set of predefined precoding matrices, with the size of the set being a trade-off between the number of signalling bits required to indicate a particular matrix in the codebook and the suitability of the resulting transmitted beam direction.

In the case of closed-loop spatial multiplexing, a UE feeds back to the eNodeB the most desirable entry from a predefined codebook. The preferred precoder is the matrix which would maximize the capacity based on the receiver capabilities. In a single-cell, interference-free environment the UE will typically indicate the precoder that would result in a transmission with an effective SNR following most closely the largest singular values of its estimated channel matrix.

Some important properties of the LTE codebooks are as follows:

- **Constant modulus property.** LTE uses precoders which mostly comprise pure phase corrections – that is, with no amplitude changes. This is to ensure that the Power Amplifier (PA) connected to each antenna is loaded equally. The one exception (which still maintains the constant modulus property) is using an identity matrix as the precoder. However, although the identity precoder may completely switch off one antenna on one layer, since each layer is still connected to one antenna at constant power the net effect across the layers is still constant modulus to the PA.

---

[9]PDSCH transmission mode 3 – see Section 9.2.2.1.
[10]PDSCH transmission modes 4 and 6 – see Section 9.2.2.1.

- **Nested property.** The nested property is a method of arranging the codebooks of different ranks so that the lower rank codebook is comprised of a subset of the higher rank codebook vectors. This property simplifies the CQI calculation across different ranks. It ensures that the precoded transmission for a lower rank is a subset of the precoded transmission for a higher rank, thereby reducing the number of calculations required for the UE to generate the feedback. For example, if a specific index in the codebook corresponds to the columns 1, 2 and 3 from the precoder **W** in the case of a rank 3 transmission, then the same index in the case of rank 2 transmission must consist of either columns 1 and 2 or columns 1 and 3 from **W**.

- **Minimal 'complex' multiplications.** The 2-antenna codebook consists entirely of a QPSK alphabet, which eliminates the need for any complex multiplications since all codebook multiplications use only ± 1 and $\pm j$. The 4-antenna codebook does contain some QPSK entries which require a $\sqrt{2}$ magnitude scaling as well; it was considered that the performance gain of including these precoders justified the added complexity.

The 2-antenna codebook in LTE is comprised of one 2×2 identity matrix and two DFT (Discrete Fourier Transform) matrices:

$$\begin{bmatrix} 1 & 0 \\ 0 & 1 \end{bmatrix}, \quad \begin{bmatrix} 1 & 1 \\ 1 & -1 \end{bmatrix} \quad \text{and} \quad \begin{bmatrix} 1 & 1 \\ j & -j \end{bmatrix} \tag{11.20}$$

Here the columns of the matrices correspond to the layers.

The 4 transmit antenna codebook in LTE uses a Householder generating function:

$$\mathbf{W}_H = \mathbf{I} - 2\mathbf{u}\mathbf{u}^H / \mathbf{u}^H \mathbf{u}, \tag{11.21}$$

which generates unitary matrices from input vectors **u**, which are defined in reference [12]. The advantage of using precoders generated with this equation is that it simplifies the CQI calculation (which has to be carried out for each individual precoder in order to determine the preferred precoder) by reducing the number of matrix inversions. This structure also reduces the amount of control signalling required, since the optimum rank 1 version of $\mathbf{W}_H$ is always the first column of $\mathbf{W}_H$; therefore the UE only needs to indicate the preferred precoding matrix, and not also the individual vector within it which would be optimal for the case of rank 1 transmission.

Note that both DFT and Householder matrices are unitary.

Cyclic Delay Diversity (CDD)

In the case of *open*-loop spatial multiplexing,[11] the feedback from the UE indicates only the *rank* of the channel, and not a preferred precoding matrix. In this mode, if the rank used for PDSCH transmission is greater than 1 (i.e. more than one layer is transmitted), LTE uses Cyclic Delay Diversity (CDD) [13]. CDD involves transmitting the same set of OFDM symbols on the same set of OFDM subcarriers from multiple transmit antennas, with a different delay on each antenna. The delay is applied before the Cyclic Prefix (CP) is added, thereby guaranteeing that the delay is cyclic over the Fast Fourier Transform (FFT) size. This gives CDD its name.

---

[11] PDSCH transmission mode 3 – see Section 9.2.2.1.

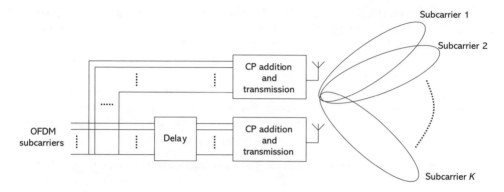

Figure 11.7 Principle of Cyclic Delay Diversity.

Adding a time delay is identical to applying a phase shift in the frequency domain. As the same time delay is applied to all subcarriers, the phase shift will increase linearly across the subcarriers with increasing subcarrier frequency. Each subcarrier will therefore experience a different beamforming pattern as the non-delayed subcarrier from one antenna interferes constructively or destructively with the delayed version from another antenna. The diversity effect of CDD therefore arises from the fact that different subcarriers will pick out different spatial paths in the propagation channel, thus increasing the frequency-selectivity of the channel. The channel coding, which is applied to a whole transport block across the subcarriers, ensures that the whole transport block benefits from the diversity of spatial paths.

Although this approach does not optimally exploit the channel in the way that ideal precoding would (by matching the precoding to the eigenvectors of the channel), it does help to ensure that any destructive fading is constrained to individual subcarriers rather than affecting a whole transport block. This can be particularly beneficial if the channel information at the transmitter is unreliable, for example due to the feedback being limited or the UE velocity being high.

The general principle of the CDD technique is illustrated in Figure 11.7.

The fact that the delay is added before the CP means that any delay value can be used without increasing the overall delay spread of the channel. By contrast, if the delay had been added after the addition of the CP, then the usable delays would have had to be kept small in order to ensure that the delay spread of the delayed symbol is no more than the maximum channel delay spread, in order to obviate any need to increase the CP length.

The time-delay/phase-shift equivalence means that the CDD operation can be implemented as a frequency-domain precoder for the affected antenna(s), where the precoder phase changes on a per-subcarrier basis according to a fixed linear function. In general the implementation designer can choose whether to implement CDD in the time domain or the frequency domain. However, one advantage of a frequency-domain implementation is that it is not limited to delays corresponding to an integer number of samples.

As an example of CDD for a case with 2 transmit antenna ports, we can express mathematically the received symbol r_k on the kth subcarrier as

$$r_k = h_{1k}x_k + h_{2k}e^{j\phi k}x_k \tag{11.22}$$

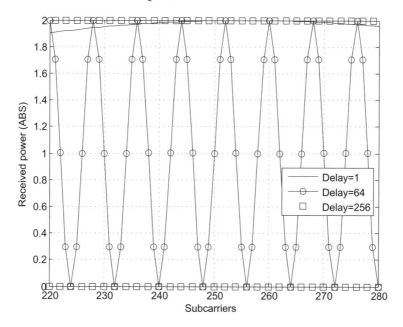

Figure 11.8 Received power for 2-transmit-antenna CDD with delays of 1, 64 and 256 samples.

where h_{pk} is the channel from the pth transmit antenna, and $e^{j\phi k}$ is the phase shift on the kth subcarrier due to the delay operation. We can see clearly that on some subcarriers the symbols from the second transmit antenna will add constructively, while on other subcarriers they will add destructively. Here, $\phi = 2\pi d_{\text{cdd}}/N$, where N is the FFT size and d_{cdd} is the delay in samples.

The number of resulting peaks and troughs in the received signal spectrum across the subcarriers therefore depends on the delay parameter d_{cdd}: as d_{cdd} is increased, the number of peaks and troughs in the spectrum also increases. Figure 11.8 shows some examples of the received spectrum for delays of $d_{\text{cdd}} = 1$, 64 and 256 with 2 transmit antennas. This helps to illustrate how CDD enhances the channel coding gain by introducing frequency selectivity into a possibly flat-fading channel.

In LTE, the values used for ϕ are π, $2\pi/3$ and $\pi/2$ for 2, 3 and 4 layer transmission respectively. For a size-2048 FFT, for example, these values correspond to $d_{\text{cdd}} =$ 1024, 682.7, 512 samples respectively.

For the application of CDD in LTE, the eNodeB transmitter combines CDD delay-based phase shifts with additional precoding using fixed unitary DFT-based precoding matrices.

The application of precoding in this way is useful when the channel coefficients of the antenna ports are correlated, since then *virtual* antennas (formed by fixed, non-channel-dependent precoding) will typically be uncorrelated. On its own, the benefit of CDD is reduced by antenna correlation. This can be illustrated by an extreme example with full correlation.

Assume that the two physical channels are identical, and the CDD delay is chosen such that at each alternate subcarrier frequency the net combined channel is the sum of the two channels, and at each other alternate frequency the net combined channel is the difference between the two physical channels:

$$r_{2k} = (h_1 + h_2)x_{2k} \tag{11.23}$$

$$r_{2k+1} = (h_1 - h_2)x_{2k+1} \tag{11.24}$$

$$h_1 = h_2 \tag{11.25}$$

Clearly, at the odd frequencies where the channels subtract, no signal will be received, and effectively the bandwidth is halved.

The use of uncorrelated virtual antennas created by fixed precoding can avoid this problem in correlated channels, while not degrading the performance if the individual antenna ports are uncorrelated.

For ease of explanation, the discussion of CDD so far has been in terms of a rank-1 transmission – i.e. with a single layer. However, in practice, CDD is only applied in LTE when the rank used for PDSCH transmission is greater than 1. In such a case, each layer benefits independently from CDD in the same way as for a single layer. For example, for a rank-2 transmission, the transmission on the second antenna port is delayed relative to the first antenna port for each layer. This means that symbols transmitted on both layers will experience the delay and hence the increased frequency selectivity.

For multilayer CDD operation, the mapping of the layers to antenna ports is carried out using precoding matrices selected from the spatial multiplexing codebooks described earlier. As the UE does not indicate a preferred precoding matrix in the open loop spatial multiplexing transmission mode in which CDD is used, the particular spatial multiplexing matrices selected from the spatial multiplexing codebooks in this case are predetermined.

In the case of 2 transmit antenna ports, the predetermined spatial multiplexing precoding matrix $\mathbf{W}$ is always the same (the first entry in the 2 transmit antenna port codebook, which is the identity matrix). Thus, the transmitted signal can be expressed as follows:

$$\begin{bmatrix} y^{(0)}(k) \\ y^{(1)}(k) \end{bmatrix} = \mathbf{W}\mathbf{D}_2\mathbf{U}_2\mathbf{x} = \frac{1}{\sqrt{2}}\begin{bmatrix} 1 & 0 \\ 0 & 1 \end{bmatrix} \frac{1}{\sqrt{2}}\begin{bmatrix} 1 & 0 \\ 0 & e^{j\phi_1 k} \end{bmatrix}\begin{bmatrix} 1 & 1 \\ 1 & -1 \end{bmatrix}\begin{bmatrix} x^{(0)}(i) \\ x^{(1)}(i) \end{bmatrix} \tag{11.26}$$

In the case of 4 transmit antenna ports, ν different precoding matrices are used from the 4 transmit antenna port codebook where ν is the transmission rank. These ν precoding matrices are applied in turn across groups of ν subcarriers in order to provide additional decorrelation between the spatial streams.

11.2.2.4   Feedback Computation and Signalling

In addition to CQI reporting as discussed in Section 10.2.1, to support MIMO operation the UE can be configured to report Precoding Matrix Indicators (PMIs) and Rank Indicators (RIs).

The precoding described above is applied relative to the phase of the common RSs for each antenna port. Thus if the UE knows the precoding matrices that could be applicable (as defined in the configured codebook), and it knows the transfer function of the channels from the different antenna ports (by making measurements on the RSs), it can determine which **W** is most suitable under the current radio conditions and signal this to the eNodeB. The preferred **W**, whose index constitutes the PMI report, is the precoder that maximizes the aggregate number of data bits which could be received across all layers.

The UE can also be configured to report the channel rank via a RI, which is calculated to maximize the capacity over the entire bandwidth, jointly selecting the preferred precoder per subband to maximize its capacity on the assumption of the selected rank. The UE also reports CQI values corresponding to the preferred rank and precoders, to enable the eNodeB to perform link adaptation and multi-user scheduling as discussed in Sections 10.2 and 12.1. The number of CQI values reported normally corresponds to the number of codewords supported by the preferred rank. Further, the CQI values themselves will depend on the assumed rank: for example, the precoding matrix for layer 1 will usually be different depending on whether or not the UE is assuming the presence of a second layer.

The eNodeB is not bound to use the precoder requested by the UE, but clearly if the eNodeB chooses another precoder then the reported CQI cannot be assumed to be valid. In addition, the eNodeB may restrict the set of precoders which the UE may evaluate and report. This is known as *codebook subset restriction*. It enables the eNodeB to prevent the UE from reporting precoders which are not useful, for example in some eNodeB antenna configurations or in correlated fading scenarios. In the case of open-loop spatial multiplexing, codebook subset restriction amounts simply to a restriction on the rank which the UE may report. For each PDSCH transmission to a UE, the eNodeB indicates via the PDCCH whether it is applying the UE's preferred precoder, and if not, which precoder is used. This enables the UE to derive the correct phase reference relative to the common RS in order to demodulate the PDSCH data.

The number of RBs to which each PMI corresponds in the frequency domain is configurable by the eNodeB. A typical value is likely to be five RBs (900 kHz).

Although the UE indicates the rank which would maximize the downlink data rate, the eNodeB can also indicate to the UE that a different rank is being used for a PDSCH transmission. This gives flexibility to the eNodeB, since the UE does not know the amount of data in the downlink buffer; if the amount of data in the buffer is small, the eNodeB may prefer to use a lower rank with higher reliability.

Thus, all three types of feedback – CQI, PMI and RI – are fundamental components of making satisfactory practical use of the available SU-MIMO transmission techniques in the LTE downlink.

11.2.3 Multi-User Schemes

MU-MIMO is one of the the most recent developments of MIMO technology and was the subject of much interest during the development of LTE. However, MU-MIMO typically entails a fairly significant change of perspective compared to other more familiar MIMO techniques. As pointed out in Section 11.2.1, the availability of accurate CSIT is the main challenge in making MU-MIMO schemes attractive for cellular applications. As a result, the main focus for MIMO in the first release of LTE is on achieving transmit antenna diversity or single-user multiplexing gain, neither of which requires such a sophisticated CSI feedback mechanism.

Thus, the support for MU-MIMO in the first version of LTE is rather limited. However, it is very likely that more advanced MU-MIMO techniques will be a part of future versions of

LTE, and that more sophisticated and resource-efficient solutions for channel estimation and feedback will play a crucial role in making these techniques a practical success.

Many variations of MU-MIMO may be considered, and the choice of a particular algorithm reflects a compromise between complexity, performance and the amount of feedback necessary to convey the CSIT to the eNodeB. In this Section we describe the main aspects of MU-MIMO techniques which emerged during the development of LTE, with the aim of highlighting the available technical solutions and the challenges which remain.

11.2.3.1 Precoding Strategies and Supporting Signalling

MU-MIMO is particularly beneficial for increasing total cell throughput in the downlink: when the eNodeB has N transmit antennas and $U \geq N$ UEs are present in the cell, it is well known that the full multiplexing gain N can be achieved even when each UE has only a single antenna, by using Spatial Division Multiple Access (SDMA) schemes based on linear precoding for transmit beamforming. Moreover, when the number of active UEs in the cell is large, a significant portion of the MU-MIMO throughput gain can be secured by exploiting multi-user diversity through relatively simple UE-selection mechanisms. All these benefits, however, depend on the level of CSIT that the eNodeB receives from each UE.

In the case of SU-MIMO, it has been shown that even a small number of feedback bits per antenna can be very beneficial in steering the transmitted energy more accurately towards the UE's antenna(s) [14–17]. More precisely, in SU-MIMO channels the accuracy of CSIT only causes an SNR offset, but does not affect the slope of the capacity-versus-SNR curve (i.e. the multiplexing gain). Yet for the MU-MIMO downlink, the level of CSI available at the transmitter does affect the multiplexing gain, because a MU-MIMO system with finite-rate feedback is essentially interference-limited, where the crucial interference rejection processing is carried out by the transmitter. Hence, providing accurate channel feedback is considerably more important for MU-MIMO than for SU-MIMO.

On the other hand, in a system with a large number of UEs, if all the UEs are to report very accurate channel measurements, the total amount of uplink resources required for CSIT feedback may soon outweigh the increase in system throughput provided by MU-MIMO techniques. Recent published results show that, given a total feedback budget, considerably higher throughput is achieved by receiving very accurate channel feedback from a relatively small fraction of the UEs, selected according to some appropriate criterion, rather than coarse feedback from many UEs [18].

One general solution to the problem of limited CSIT feedback in MU-MIMO schemes is that of utilizing a codebook of $N_q = 2^B$ N-dimensional vectors and sending to the base station a B-bit index from the codebook, selected according to some minimum-distance criterion. The codebook indices are then used by the base station to construct the precoding matrix. Typically, a real-valued CQI is also sent along with the codebook index, which can be used by the base station for MCS selection as well as user selection [17, 19–22].

Two main precoding techniques emerged as MU-MIMO candidates for LTE, both relying on a codebook-based limited feedback concept.

1. **Codebook of Unitary Precoding (UP) matrices.** The codebook contains a set of $L = N_q/N$ predefined and fixed unitary beamforming matrices of size $N \times N$. For each beamforming matrix in the codebook, each UE computes an SINR for each of the N beamforming vectors in the matrix, assuming that the other $N - 1$ vectors are

used for interfering transmissions to other UEs. Overall, the UE computes N_q SINRs and signals back to the eNodeB the codebook index corresponding to the best SINR and the value of this SINR. The eNodeB then makes use of this information to select the beamforming matrix and schedule the UEs for transmission (along with a suitable MCS for each UE), in order to maximize the cell throughput. In this scheme the eNodeB has a very limited set of unitary matrices from which to choose for precoding, and the multiplexing gain is at its maximum when enough UEs are in the cell with (approximately) orthogonal channel signatures matching the vectors in one of the codebook matrices. This happens with high probability only for a large density of UEs. On the other hand, limiting the set of precoding matrices enables efficient signalling of the selected precoder back to the UEs to enable data demodulation. This operation requires just $\log_2(L)$ bits to indicate the selected precoding matrix, or $\log_2(N_q)$ bits per UE.

2. **Codebook for Channel Vector Quantization (CVQ).** In this case the codebook contains $N_q = 2^B$ unit-norm quantization vectors and is used by each UE to quantize an N-dimensional vector of channel measurements. Before quantization, the channel vector is normalized by its amplitude, such that the quantization index captures information regarding only the 'direction' of the channel vector. The UE then feeds back this quantization index along with a real number representing an estimate of its SINR, which depends on the amplitude of the channel and the directional quantization error. In this case the UE does not know the set of possible beamforming matrices in advance. The eNodeB utilizes this feedback information collected from the UEs to select the UEs for transmission and to construct a suitable beamforming matrix, for example according to a zero-forcing beamforming criterion. One simple yet effective option is to combine UE-selection with naive zero-forcing precoding. In such a scheme the eNodeB has the flexibility to design the precoding matrix making use of the CSIT provided by UEs; however, a signalling mechanism has to be devised to send back to the UEs enough information about the designed precoder to allow successful data demodulation. One effective means for such signalling is the use of at least one precoded (or UE-specific) reference symbol for each precoding vector being used.

Note that in the limit of a large UE population, a zero-forcing beamforming solution converges to a unitary precoder because, with high probability, there will be N UEs with good channel conditions reporting orthogonal channel signatures. On the other hand, one clear issue with unitary precoding in this context is that for large codebooks it achieves a multiplexing gain upper-bounded by one (i.e. the same as time-division multiplexing between UEs). This is because, if $p = 1/L = N/2^B$ is the probability that a UE selects a given beamforming matrix in the codebook, then the probability of l out of U UEs selecting the same matrix is a binomial random variable with parameters (p, U) and mean value $\bar{l} = Up$. Hence the average number of UEs selecting the same beamforming matrix decreases exponentially with the codebook size in bits. Eventually, for large B, if U is kept constant, only a single UE will be allocated per subframe.

As for many other physical layer building blocks, not every detail of a MU-MIMO scheme needs to be specified in a standard. Only the aspects essential for interoperability are necessary, including: (1) the format and procedure for the feedback information from the UEs, which may or may not include details of the method of calculating the quantities to feed back; (2) the codebooks for feedback calculation by the UEs and precoder selection by the eNodeB; (3) the method and procedure for signalling the precoding weights used for transmission to the relevant UEs. These three aspects are described in more detail in the following.

11.2.3.2   Calculation of Precoding Vector Indicator (PVI) and CQI

As MU-MIMO in LTE supports only rank-1 transmission, i.e. one codeword, to each of the selected UEs, we refer to the precoding feedback as a Precoding Vector Indicator (PVI) for the purposes of this discussion (although in the LTE specifications the term PMI is used for both SU-MIMO and MU-MIMO).

We focus on two different assumptions which a UE could theoretically make as to the nature of the interfering precoding vectors when calculating the feedback information. Each assumption corresponds to one of the MU-MIMO precoding techniques described above. We also show that the two sets of assumptions lead to similar results, and indeed the feedback values reported by the UE are equivalent. Under certain assumptions on the codebook structure and number of interferers, the two limited feedback concepts introduced above provide identical information to the transmitter; the main difference between the two approaches in fact lies in the way the precoder is selected based on the information collected by the eNodeB.

Precoder and feedback calculation methods

For the purpose of analysis, consider a codebook for feedback from the UEs, $\mathbf{C} = \{\mathbf{C}^{(0)}, \ldots, \mathbf{C}^{(L-1)}\}$, consisting of L unitary matrices of size $N \times N$, N being the number of transmit antennas, such that the overall codebook size in number of vectors is $N_q = LN$. The MU-MIMO channel in a given subframe with linear precoding applied at the transmitter is given by

$$\mathbf{r} = \mathbf{HWx} + \mathbf{n} \tag{11.27}$$

where $\mathbf{x}$, with $\mathbb{E}[\mathbf{x}\mathbf{x}^H] = \mathbf{I}$, is the vector of independent data symbols transmitted in parallel by the N transmit antennas, $\mathbf{r} = (r_0, \ldots, r_{K-1})^T$ is the vector of signals individually received by the K users and $\mathbf{n} \sim \mathcal{CN}(\mathbf{0}, \mathbf{I})$ is an i.i.d. complex Gaussian noise vector.[12] The matrix $\mathbf{H} = [\mathbf{h}_0^T, \ldots, \mathbf{h}_{K-1}^T]^T$ contains the channel coefficients from the N antennas to the K UEs, where the *row vector* $\mathbf{h}_k$ is the channel of UE k.[13] The transmission is power-constrained such that

$$\text{trace}(\mathbf{W}^H\mathbf{W}) \leq P_t \tag{11.28}$$

where P_t denotes the total downlink transmitted power (energy per subframe). The precoding matrix, $\mathbf{W}$, can be rewritten in the form

$$\mathbf{W} = \mathbf{F} \, \text{diag} \, (\mathbf{p})^{1/2} \tag{11.29}$$

---

[12]Here the noise variance, which includes the thermal noise and possibly other interference sources modelled as Gaussian, is normalized to one without loss of generality. In fact, the power constraint P introduced next can be regarded as an SNR value.

[13]For the moment it is assumed that each UE uses one receiving antenna.

where the matrix $\mathbf{F}$ is designed without a power constraint, and the kth coefficient of vector $\mathbf{p}$, denoted by p_k, is the power level for UE k. Let $\mathcal{S}$ denote the set of UEs selected for transmission. We also assume equal power allocation across these UEs.

Therefore, if the precoding matrix is unitary, according to MU-MIMO precoding technique 1 above, it follows that

$$\mathbf{F}(\mathcal{S}) \in \{\mathbf{C}^{(0)}(\mathcal{S}), \ldots, \mathbf{C}^{(L-1)}(\mathcal{S})\} \quad \text{and} \quad p_k = P_t / |\mathcal{S}| \tag{11.30}$$

where the notation $\mathbf{C}^{(i)}(\mathcal{S})$ indicates that the precoding matrix can contain only a subset of vectors of codebook matrix $\mathbf{C}^{(i)}$. As mentioned in Section 11.2.2.4, this is referred to as 'codebook subset selection'.

If $\mathbf{F}$ is the zero-forcing solution, according to MU-MIMO precoding technique 2 above, then the precoder can be expressed as

$$\mathbf{F}(\mathcal{S}) = \hat{\mathbf{H}}(\mathcal{S})^{\text{H}} (\hat{\mathbf{H}}(\mathcal{S})\hat{\mathbf{H}}(\mathcal{S})^{\text{H}})^{-1} \quad \text{and} \quad p_k = P_t / (|\mathcal{S}| \cdot \|\mathbf{f}_k\|^2) \tag{11.31}$$

where $\mathbf{f}_k$ denotes the kth column of $\mathbf{F}$ and $\hat{\mathbf{H}}(\mathcal{S}) = [\hat{\mathbf{h}}_1^{\text{T}}, \ldots, \hat{\mathbf{h}}_{|\mathcal{S}|}^{\text{T}}]^{\text{T}}$, contains the quantized channel vectors of the selected UEs.

We also introduce the cross-talk matrix $[\mathbf{\Phi}]_{i,j} = |\mathbf{h}_i \mathbf{f}_j|^2$, such that the received SINR for user k is given by

$$\gamma_k = \frac{p_k [\mathbf{\Phi}]_{k,k}}{1 + \sum_{i \neq k} p_i [\mathbf{\Phi}]_{k,i}} \tag{11.32}$$

Let us consider the following two methods of calculating an index (the PVI) from the codebook and a real-value quantity signifying an SINR estimate (the CQI).

1. In MU-MIMO precoding technique 1 above (codebook of UP matrices), the following two operations are carried out by each UE:

   - The SINR is computed using Equation (11.32) for each vector in the codebook taken as the UE's own precoding vector. The interfering precoding vectors are assumed to be the other $N - 1$ vectors in the codebook which are orthogonal to the UE's own.

   - The largest such SINR is used to report the CQI value, while the index of the corresponding useful precoding vector represents the PVI.

2. In MU-MIMO precoding technique 2 above (codebook for CVQ), the following operations are performed instead:

   - The measured channel vector[14] is quantized to the nearest vector in the codebook in terms of *chordal distance* (see, for example, reference [23]). The quantization index represents the PVI.

   - An expected SINR is estimated under the following assumptions:

     - There are $N - 1$ interfering precoding vectors and they are isotropically distributed on the hyperplane orthogonal to the PVI vector (i.e. the UE assumes that zero-forcing beamforming will be used).

---

[14]Still assuming that each UE uses one receiving antenna.

 – The useful precoding vector is only slightly offset from the PVI vector – i.e. the angle between the two vectors is small.

It is easy to show that the operations for MU-MIMO precoding technique 1 equate to performing vector quantization on the channel vector with minimum chordal distance. We also show below that the PVI/CQI values for the two methods are totally equivalent, under the assumptions stated above.

Equivalence between channel vector quantization and PVI calculation

Let us rewrite the SINR for a generic UE k, when the selected precoding matrix is $\mathbf{C}^{(i)} = [\mathbf{c}_0^{(i)}, \ldots, \mathbf{c}_{N-1}^{(i)}]$, and the reference UE's own precoding vector is $\mathbf{c}_j^{(i)}$. We denote this SINR as $\gamma_k[\mathbf{c}_j^{(i)}]$ and define the angle $\theta_{k,j}^{(i)} \in [0, \pi/2]$ between vectors $\tilde{\mathbf{h}}_k^H$ and $\mathbf{c}_j^{(i)}$, where $\tilde{\mathbf{h}}_k = \mathbf{h}_k/\|\mathbf{h}_k\|$, as

$$\cos \theta_{k,j}^{(i)} = |\langle \tilde{\mathbf{h}}_k^H, \mathbf{c}_j^{(i)} \rangle| = |\mathbf{c}_j^{(i)H} \tilde{\mathbf{h}}_k^H| \qquad (11.33)$$

where $\langle \cdot \rangle$ denotes the inner product. As $\mathbf{C}^{(i)}$ is a generator matrix for $\mathbb{C}^N$, we can decompose $\tilde{\mathbf{h}}_k^H$ as

$$\tilde{\mathbf{h}}_k^H = \alpha_j \mathbf{c}_j^{(i)} + \sum_{l \neq j} \alpha_l \mathbf{c}_l^{(i)} \qquad (11.34)$$

where $\alpha_l = \langle \tilde{\mathbf{h}}_k^H, \mathbf{c}_l^{(i)} \rangle$ for $l = 0, \ldots, N - 1$. It follows that $\|\tilde{\mathbf{h}}_k\|^2 = \sum_{l=0}^{N-1} \alpha_l^2 = 1$. Therefore, if the precoding matrix is $\mathbf{F} = \mathbf{C}^{(i)}$, the power is equally distributed across the N precoding vectors, and UE k has reported vector $\mathbf{c}_j^{(i)}$ as PVI, by substituting Equation (11.34) into (11.32) we obtain

$$\gamma_k(\mathbf{c}_j^{(i)}) = \frac{(P/N)\|\mathbf{h}_k\|^2 \cos^2 \theta_{k,j}^{(i)}}{1 + (P/N)\|\mathbf{h}_k\|^2 \sin^2 \theta_{k,j}^{(i)}} \qquad (11.35)$$

The reported PVI can be interpreted as an approximate representation of the conjugate channel vector, $Q(\tilde{\mathbf{h}}_k^H)$, where $Q(\cdot)$ denotes a vector-quantization operation. More precisely, the reported PVI corresponds to the vector in the codebook with minimum chordal distance from $\tilde{\mathbf{h}}_k^H$, i.e.

$$Q(\tilde{\mathbf{h}}_k^H) = \underset{\mathbf{c}_j^{(i)} \in \mathbf{C}}{\text{argmin}} \sin \theta_{k,j}^{(i)} = \underset{\mathbf{c}_j^{(i)} \in \mathbf{C}}{\text{argmax}} \gamma_k[\mathbf{c}_j^{(i)}] \qquad (11.36)$$

In Figure 11.9 θ_1 is used to denote this minimum quantization angle for the channel vector of UE 1. Figure 11.9 shows a graphical representation of the channel vector, the selected PVI and the precoding vectors for MU-MIMO precoding technique 1. Note that the precoding vectors $\{\mathbf{f}_1, \mathbf{f}_2, \mathbf{f}_3\}$ are the columns of the selected unitary matrix from the codebook.

Let us now consider MU-MIMO precoding technique 2 illustrated in Figure 11.10. In this case the UE directly seeks to represent the channel vector $\tilde{\mathbf{h}}_k^H$ as accurately as possible by one of the vectors in the codebook.[15]

---

[15]It is clearly equivalent to represent vector $\tilde{\mathbf{h}}_k^T$ by an element of the codebook $\mathbf{C}^*$, obtained from $\mathbf{C}$ by taking the conjugate of its vector components. In the case of a DFT-based codebook, depending on whether $\tilde{\mathbf{h}}_k^H$ or $\tilde{\mathbf{h}}_k^T$ is quantized, we can use the inverse DFT and DFT matrices respectively (or vice versa) as base matrices for the generation of the two codebooks, as $\mathbf{C}_{\text{DFT}} = \mathbf{C}_{\text{IDFT}}^*$. The resulting quantization vectors are the conjugate of each other.

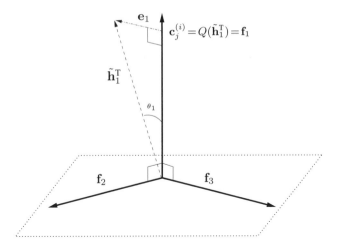

Figure 11.9 Representation of precoding vectors, channel vector and PVI vector for MU-MIMO precoding technique 1. $\|\tilde{\mathbf{h}}_1\| = \|\hat{\mathbf{h}}_1\| = 1$.

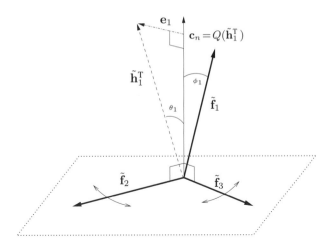

Figure 11.10 Representation of precoding vectors, channel vector and PVI vector for MU-MIMO precoding technique 2. $\tilde{\mathbf{f}}_k = \mathbf{f}_k/\|\mathbf{f}_k\|$. $\|\tilde{\mathbf{h}}_1\| = \|\hat{\mathbf{h}}_1\| = 1$.

We still use the definition in Equation (11.33) for the angle between the vectors $\tilde{\mathbf{h}}_k^H$ and $\mathbf{c}_j^{(i)}$. In this case the feedback codebook is a collection of N_q vectors, and therefore we introduce the following change of variables for ease of notation: $n = Lj + i$, with $j = 0, \ldots, N - 1, i = 0, \ldots, L - 1$ and $n = 0, \ldots, N_q - 1$. The vector quantization operation yields the vector in the codebook with minimum chordal distance from $\tilde{\mathbf{h}}_k^H$, which is given by Equation (11.36), and can be rewritten as

$$Q(\tilde{\mathbf{h}}_k^H) = \underset{\mathbf{c}_n \in \mathbf{C}}{\arg\min} \sin \theta_{k,n} = \underset{\mathbf{c}_n \in \mathbf{C}}{\arg\max} \gamma_k(\mathbf{c}_n) \tag{11.37}$$

When channel vector quantization and zero-forcing precoding are used, the UEs know only the vector they are assuming for their own transmissions when deriving the PVI to report; they do not know in advance the other vectors which will eventually be used to make up the full precoding matrix, and therefore they cannot predict the exact SINR value. However, they can derive a good approximation of the SINR averaged over the unknown components of the precoding matrix. The final expression, which is an approximate lower-bound to the average SINR at the UE, is very similar to Equation (11.35) and is expressed as follows:[16]

$$E[\gamma_k] \gtrsim \frac{p_k \|\mathbf{h}_k\|^2 \cos^2 \theta_k}{1 + (P/N)\|\mathbf{h}_k\|^2 \sin^2 \theta_k} \tag{11.38}$$

As p_k is known to the eNodeB but unknown to the UE, the CQI sent by UE k, according to MU-MIMO precoding technique 2 is the same as Equation (11.35), i.e.

$$\text{CQI}_k = \frac{(P/N)\|\mathbf{h}_k\|^2 \cos^2 \theta_k}{1 + (P/N)\|\mathbf{h}_k\|^2 \sin^2 \theta_k} \tag{11.39}$$

The eNodeB can then estimate the SINR at UE k by a simple scaling operation: $\gamma_k = (p_k/(P/N))\text{CQI}_k$. Note that the ratio $(p_k/(P/N)) = 1$ for every k if the precoding matrix is unitary.

From the analysis so far we can conclude that the only substantial difference between the two MU-MIMO precoding techniques of Section 11.2.3.1 is in the use of the same PVI and CQI information by the eNodeB to construct the precoding matrix. In MU-MIMO precoding technique 1 the choice is restricted to a precoder matched to the selected PVIs, which have to be orthogonal. In MU-MIMO precoding technique 2, the PVIs of the selected UEs need not be exactly orthogonal and the eNodeB would be allowed to adjust the precoder in order to minimize any residual interference among UEs.

In practice, only unitary codebooks are provided in the first version of LTE, and PDSCH transmission in MU-MIMO mode[17] is based on selection of precoding matrices from the same codebooks as are used for SU-MIMO (as summarized in Section 11.2.2.3 above).

In the light of the above discussion on MU-MIMO techniques, it is instructive to observe that the LTE 2-antenna codebook, and also the first eight entries in the 4-antenna codebook, are in fact DFT-based codebooks – in other words derived from a discrete Fourier transformation matrix. The main reasons for this choice are as follows:

- Simplicity in the PVI/CQI calculation, which can be done in some cases without complex multiplications.

- A DFT matrix nicely captures the characteristics of a highly correlated MISO channel. In fact, it is not difficult to see that, if the first N rows are taken from a DFT matrix of size N_q, each of the N_q column vectors that is obtained contains the phases of (i.e. the

---

[16]The derivation of this formula can be found, for example, in [20].

[17]PDSCH transmission mode 5 – see Section 9.2.2.1.

baseband representation of) a line-of-sight propagation channel from a uniform linear array with N elements to a point in space located at a given angle with respect to the antenna array boresight (see, for example, reference [6]).

- It is also easy to verify that the vectors in such a codebook can be grouped into $L = N_q/N$ unitary matrices.

11.2.3.3 User Selection Mechanism

As is clear from the discussion above, one important aspect of MU-MIMO precoding is the selection of UEs for transmission, as the throughput gain of naive linear precoding techniques very much depends on exploiting the multi-user diversity available in the system. Although the method for selecting UEs is not specified in LTE or other similar systems, we briefly review two very simple strategies which could be appropriate for implementation in conjunction with the precoding methods that we have considered.

One simple way of selecting UEs, according to MU-MIMO precoding technique 1 of Section 11.2.3.1, is to group UEs that report orthogonal PVIs and select the group providing the highest total throughput. This can be simply estimated, for example from the capacity formula $R = \sum_k \log(1 + \text{SINR}_k)$, where the summation is over the UEs in the group and the SINR value is given by the CQI.

Another more general and very simple greedy strategy, which would be appropriate for MU-MIMO precoding technique 2, consists of adding one UE at a time, as long as the additional UE increases the overall throughput, estimated according to a criterion such as the achievable sum-rate. This simple strategy can be described as follows (see, for example reference [24]). Let $R(\mathcal{S})$ denote the achievable sum-rate when the set of UEs $\mathcal{S}$ is selected for transmission.

User selection algorithm

Initialize $\mathcal{S} = \emptyset$, $R(\mathcal{S}) = 0$.
while $|\mathcal{S}| \leq N$ *do*

1. $\bar{k} = \underset{k \notin \mathcal{S}}{\text{argmax}} \ \mathcal{R}(\mathcal{S} \cup \{k\})$

2. *if* $R(\mathcal{S} \cup \{\bar{k}\}) > R(\mathcal{S})$ update $\mathcal{S} = \mathcal{S} \cup \{\bar{k}\}$, *else* exit.

11.2.3.4 Receiver Spatial Equalizers

Since in LTE each UE is assumed to support two receiving antennas, it is important to consider the possible receive beamforming structures for MU-MIMO transmission, which may affect the PVI and CQI calculation.[18]

We consider the possibilities for receiver beamforming in terms of a generic UE k with J antenna elements, and we use $\mathbf{H}_k = [\mathbf{h}_1^T, \ldots, \mathbf{h}_J^T]^T$ to denote a $J \times N$ channel for that UE.

---

[18]Note that receiver algorithms are not defined in the specifications; instead, the required demodulation performance is specified.

We also use $\bar{\mathbf{H}}_k = \mathbf{H}_k\mathbf{F} = [\bar{\mathbf{h}}_{k,0}, \ldots, \bar{\mathbf{h}}_{k,N-1}]$ to denote the product matrix between the reference UE's channel and the precoder, as defined in Equation (11.29).

Using this terminology, we describe three possible spatial equalizers, which all attempt to maximize the received SINR, with different degrees of knowledge at the UE side.

1. **MMSE receiver.** If the precoding vector for the reference UE k corresponds to the jth column of $\mathbf{F}$, then the $1 \times J$ vector of receiver combining coefficients, $\mathbf{w}_{k,j}$, is given by[19]

$$\mathbf{w}_{k,j} = \bar{\mathbf{h}}_{k,j}^{\text{H}}(\bar{\mathbf{H}}_k\bar{\mathbf{H}}_k^{\text{H}} + \mathbf{I}_J)^{-1} \tag{11.40}$$

   This receiver requires that the user estimates the N vectors of size J in the product matrix $\bar{\mathbf{H}}_k$. This knowledge can be acquired by estimating the channel matrix from the cell-specific RS and being told by explicit signalling which precoding matrix is used. Alternatively, if precoded RS were to be provided, the product matrix could be directly estimated from them. It goes without saying that the former method is best suited for a MU-MIMO scheme with a small set of possible precoding matrices, while the latter allows for full flexibility in the construction of the precoding matrix.

2. **Minimum Quantization Error (MQE) receiver.** In this case, for a given vector of the codebook, $\mathbf{c}_n$, the coefficients of the linear combination at the receiver k are given by

$$\mathbf{w}_{k,n} = \alpha\mathbf{h}_{k,n}^{(\text{eff})}\mathbf{H}_k^{\dagger} \tag{11.41}$$

   where $\mathbf{H}_k^{\dagger} = \mathbf{H}_k^{\text{H}}(\mathbf{H}_k\mathbf{H}_k^{\text{H}})^{-1}$ is the right pseudo-inverse of $\mathbf{H}_k$, α is a normalization factor, such that $\|\mathbf{w}_{k,n}\| = 1$, and (the row vector) $\mathbf{h}_{k,n}^{(\text{eff})}$ is the projection of $\mathbf{c}_n^{\text{H}}$ onto span($\mathbf{H}_k^{\text{T}}$). To express this algebraically, let $\mathbf{Q}_k^{\text{T}} = [\mathbf{q}_{k,1}^{\text{T}}, \ldots, \mathbf{q}_{k,J}^{\text{T}}]$ be a generator matrix for span($\mathbf{H}_k^{\text{T}}$), then

$$\mathbf{h}_{k,n}^{(\text{eff})} = \mathbf{c}_n^{\text{H}}\mathbf{Q}_k^{\text{H}}\mathbf{Q}_k \tag{11.42}$$

   For the derivation of this receiver we refer the interested reader to [25]. Note that the structure of Equation (11.41) is very similar to (11.40), but calculation of (11.41) does not require knowledge of the precoding matrix. This is because it minimizes the vector quantization error independently of the transmit beamformer, and by doing so it also maximizes the CQI value in Equation (11.39).

3. **SVD of the channel.** This naive receive beamformer depends only on the channel matrix $\mathbf{H}_k$ and is independent of any CSI feedback. Let $\mathbf{H}_k = \mathbf{U}_k\mathbf{\Lambda}_k\mathbf{T}_k^{\text{H}}$ be the SVD of the channel, with singular values $\lambda_{k,1} \geq \lambda_{k,2} \geq \cdots \geq \lambda_{k,J}$ $(J \leq N)$, and $\mathbf{U}_k = [\mathbf{u}_{k,1}, \ldots, \mathbf{u}_{k,J}]$. Then

$$\mathbf{w}_k = \mathbf{u}_{k,1}^{\text{H}} \tag{11.43}$$

11.2.4 Physical-Layer MIMO Performance

In this section we present some link-level and system-level simulation results to illustrate the physical layer performance of various MIMO schemes, starting with SU-MIMO and then followed by MU-MIMO schemes. Some of these results were influential in deciding on the MIMO schemes for LTE.

---

[19]Note that the noise variance is normalized to 1 as it is included in the transmit SNR, P.

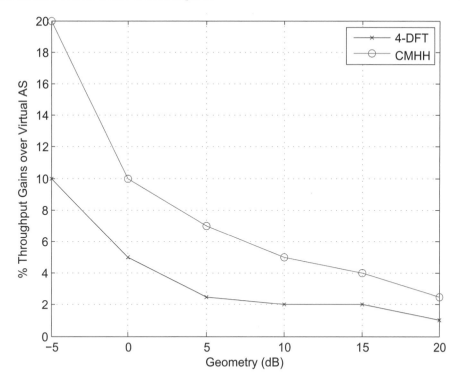

Figure 11.11 Throughput gain achieved by precoding with two different codebooks, compared to antenna selection: 4×2. Reproduced by permission of © Texas Instruments.

11.2.4.1 Precoding Performance

To demonstrate the benefits of precoding, link-level simulation results [26] in Figures 11.11 and 11.12 show the percentage throughput gains versus geometry (average SINR) when compared to antenna selection. The channel model here is the SCM-D (see Section 21.3.5) with a uniform linear array with 4λ (uncorrelated frequency selective fading) antenna spacing. Clearly the precoding gains are greater in the 4×2 case than in the 4×4 case, as expected. Similar gains compared to antenna selection are shown in reference [27]. The legends in Figures 11.11 and 11.12 refer to different codebook designs as follows:

- CMHH: Constant Modulus HouseHolder;

- 4-DFT: codebook based on four DFT matrices. This is also referred to as the unoptimized or regular DFT codebook, since the different matrices are not optimized to maximize the minimum chordal distance.

The 4-antenna LTE codebook performs similarly to the CMHH codebook.

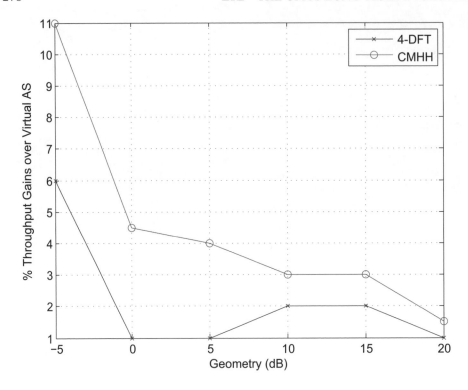

Figure 11.12   Throughput gain achieved by precoding with two different codebooks, compared to antenna selection: 4×4. Reproduced by permission of © Texas Instruments.

11.2.4.2   Multi-User MIMO performance

This section considers the two MU-MIMO precoding techniques discussed in Section 11.2.3.1, namely unitary precoding and zero-forcing precoding with equal power distribution across the selected UEs. We consider the case with $N = 4$ transmit antennas at the eNodeB and $M = 1, 2$ receive antennas per UE.

The main parameters used for the link-level simulations are listed in Table 11.3.

In Figure 11.13 the throughput is plotted versus average SNR for unitary precoding and zero-forcing precoding. As the codebook size increases, unitary precoding shows degradation due to the scheduling problem discussed earlier, namely that too few UEs select the same precoding matrix, thus making full-rank transmission less likely. As a consequence, the multiplexing gain for large codebooks falls to one, as for TDMA. The gap between the TDMA curve and the unitary precoding curves for large codebook sizes is due to the fact that for TDMA we assume that the eNodeB always matches the transmit beamformer to the reported channel vector. It is possible to devise various techniques to mitigate the scheduling problem of unitary precoding, for example by overruling the UE's selected matrix according to some criterion.

For a large number of UEs available for transmission, unitary precoding and zero-forcing will perform the same, as it becomes increasingly likely that a set of N UEs can be found with

Table 11.3 Link-level simulation parameters corresponding to Figure 11.13.

Parameter	Value
Number of transmit antennas	4
Number of receive antennas	1
Transmit antenna spacing	0.5λ
Type of transmit precoding	Unitary, Zero-Forcing
Number of UEs	20
UE speed	3 km/h
Transmission bandwidth	5 MHz
Centre frequency	2 GHz
DFT size	512
Feedback codebook type	DFT
Feedback codebook size	4, 8, 12 bits
Channel model	SCM urban micro
Number of paths	10
Feedback granularity	1 per RB
Delay between feedback and data detection	1.5 ms (three time slots)
Modulation schemes	QPSK, 16-QAM, 64-QAM
Turbo coding rates	1/3, 1/2, 2/3, 3/4, 4/5
Target BLER	10%

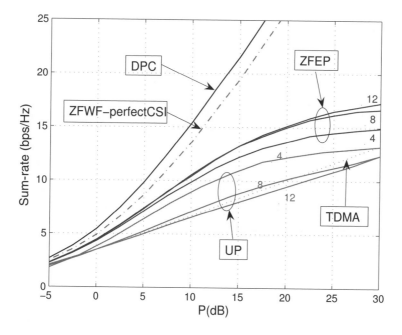

Figure 11.13 Link-level throughput for the considered MU-MIMO precoding techniques and different DFT codebook sizes: 4, 8 and 12 bits.

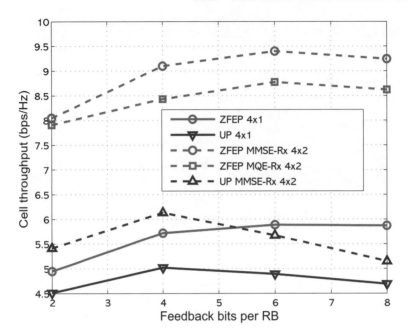

Figure 11.14 System level simulation results for two different precoding strategies and two different linear spatial equalizers (with 2 Rx antennas).

almost orthogonal channel signatures. For comparison, Figure 11.13 also shows the optimal channel sum-rate capacity (achieved by 'Dirty-Paper Coding' (DPC) [28]) and the achievable throughput for zero-forcing precoding with perfect CSIT and water-filling power allocation.

Figure 11.14 shows some system-level results for the two precoding schemes and $M = 1, 2$ receiving antennas. The feedback generation is done under the assumption of an MMSE receiver for unitary precoding, by using the MMSE solution in Equation (11.40); the strategy for zero-forcing is quantization error minimization according to Equation (11.41). The suffix 'MMSE' denotes the curves obtained by applying the MMSE spatial filter in Equation (11.40), and 'MQE' indicates the zero-forcing curve where the spatial equalizer in Equation (11.41) is used for both feedback calculation and data detection. The system-level simulation parameters are as in Table 11.3 except where shown in Table 11.4.

It can be observed that the cell throughput with Zero-Forcing Equal Power (ZFEP) slightly decreases with the granularity of the channel quantization for a number of reported bits per index larger than six. The explanation for this is the suboptimality of the UE selection mechanism. In fact, for a small codebook size UEs with very similar channel signatures fall into the same quantization bin, and hence two such UEs are never selected simultaneously (otherwise the resulting MIMO channel would be singular). Conversely, for larger codebooks two UEs with similar separation of their channel signatures may fall into separate bins, and they may be selected for transmission as a result of the proportional fair scheduling mechanism, thus causing slight degradation in throughput.

Table 11.4 System level simulation parameters corresponding to Figure 11.14.

Parameter	Value
Number of receive antennas	1,2
Transmit antenna spacing	10λ
Receive antenna spacing	0.5λ
Type of Rx beamforming	MMSE, MQE
Cellular layout	hexagonal grid, 19 cell sites, 3 sectors per site
Inter-site distance	2 km
Thermal noise spectral density	-174 dBm/Hz
Shadowing	log-normal (8 dB)
Transmit power	20 W
Front-to-back ratio of eNodeB antennas	20 dB
Path loss exponent	4
Frequency reuse factor	1
Feedback codebook size	2–8 bits
Traffic model	full queue traffic
Scheduling algorithm	proportional fair scheduling
Proportional fair factor	1
Hybrid ARQ	Chase combining
Max. number of retransmissions	4

For unitary precoding, the same degradation in throughput can be observed for large codebook sizes as in Figure 11.13.

It is worth noting, however, that the limited simulation results presented in Figures 11.13 and 11.14 do not show the performance degradation of zero-forcing beamforming compared to unitary precoding which may occur if the reported channel feedback information is very inaccurate. In general, zero-forcing beamforming can be less robust against inaccuracies in the channel representation compared to unitary precoding. This would be the case, for example, if the feedback were to be reported with a coarser frequency granularity than the coherence bandwidth of the channel, or if the codebook size for the channel feedback is small. A similar situation occurs when the channels are spatially highly uncorrelated: in such a case the 'direction' of a channel signature may change significantly between groups of adjacent subcarriers and it becomes harder to provide an accurate channel representation consistently across frequency, compared to the case of spatially correlated channels. In these cases, where it is not possible to provide sufficiently accurate channel information to the transmitter, the unitary precoding adopted in the first version of LTE can prove more robust and than a zero-forcing approach.

11.3 Concluding Remarks

We have reviewed the predominant families of MIMO techniques (both Single-User and Multi-User) of relevance to the first and future versions of LTE. LTE breaks new ground in drawing on such approaches to harness the power of MIMO systems not just for boosting the peak per-user data rate but also for improving overall system capacity and spectral efficiency.

Single-User MIMO techniques are well-developed in the first version of LTE, providing the possibility to benefit from precoding white avoiding a high control signalling overhead. However, it is clear that the early techniques leave substantial scope for further development and enhancement in later releases of the LTE specifications.

References[20]

[1] D. Gesbert, M. Shafi, P. Smith, D. Shiu and A. Naguib, 'From Theory to Practice: An Overview of MIMO Space-time Coded Wireless Systems'. *IEEE Journal on Selected Areas in Communications*. Special Issue on MIMO systems, guest edited by the authors, April 2003.

[2] H. Bölcskei, D. Gesbert, C. Papadias and A. J. van der Veen (eds), *Space-Time Wireless Systems: From Array Processing to MIMO Communications*. Cambridge: Cambridge University Press, 2006.

[3] A. Goldsmith, *Wireless Communications*. Cambridge: Cambridge University Press, 2005.

[4] G. Golub and C. van Loan, *Matrix Computations*. Balitmore, MA: John Hopkins University Press, 1996.

[5] S. M. Alamouti, 'A Simple Transmitter Diversity Technique for Wireless Communications'. *IEEE Journal on Selected Areas in Communications*, Vol. 16, pp. 1451–1458, October 1998.

[6] D. Tse and P. Viswanath, *Fundamentals of Wireless Communication*. Cambridge: Cambridge University Press, 2004.

[7] F. Kaltenberger, L. S. Cardoso, M. Kountouris, R. Knopp and D. Gesbert, 'Capacity of Linear Multi-user MIMO Precoding Schemes with Measured Channel Data', in *Proc. IEEE Workshop on Sign. Proc. Adv. in Wireless Comm. (SPAWC)*, Recife, Brazil, July 2008.

[8] F. Kaltenberger, M. Kountouris, D. Gesbert and R. Knopp, 'Correlation and Capacity of Measured Multi-user MIMO Channels', in *Proc. IEEE International Symposium on Personal, Indoor and Mobile Radio Communications*, Cannes, France, September 2008.

[9] Caire, G. and Shamai, S. (Shitz), 'On the Achievable Throughput of a Multi-antenna Gaussian Broadcast Channel'. *IEEE Trans. on Information Theory*, Vol. 49, pp. 1691–1706, July 2003.

[10] R. Heath, V. Lau, D. Love, D. Gesbert, B. Rao and A. Andrews, (eds), 'Exploiting Limited Feedback in Tomorrow's Wireless Communications Networks'. Special issue of *IEEE Journal on Selected Areas in Communications*, October 2008.

[11] A. Osseiran and A. Logothetis, 'Closed Loop Transmit Diversity in WCDMA HS-DSCH', in *Proc. IEEE Vehicular Technology Conference*, May 2005.

[12] 3GPP Technical Specification 36.211, 'E-UTRA physical channels and modulation (Release 8)', www.3gpp.org.

[13] Y. Li, J. Chuang and N. Sollenberger, 'Transmitter Diversity for OFDM Systems and Its Impact on High-Rate Data Wireless Networks'. *IEEE Journal on Selected Areas in Communications*, Vol. 17, pp. 1233–1243, July 1999.

[14] A. Narula, M. J. Lopez, M. D. Trott and G. W. Wornell, 'Efficient Use of Side Information in Multiple Antenna Data Transmission over Fading Channels'. *IEEE Journal on Selected Areas in Communications*, Vol. 16, pp. 1223–1436, October 1998.

[15] D. Love, R. Heath Jr and T. Strohmer, 'Grassmannian Beamforming for Multiple-input Multiple-output Wireless Systems'. *IEEE Trans. on Information Theory*, Vol. 49, pp. 2735–2747, October 2003.

---

[20]All web sites confirmed 18th December 2008.

[16] D. Love, R. Heath Jr, W. Santipach, and M. Honig, 'What is the Value of Limited Feedback for MIMO Channels?'. *IEEE Communication Magazine*, Vol. 42, pp. 54–59, October 2003.

[17] N. Jindal, 'MIMO Broadcast Channels with Finite Rate Feedback'. *IEEE Trans. Information Theory*, Vol. 52, pp. 5045–5059, November 2006.

[18] R. Zakhour and D. Gesbert, 'Adaptive Feedback Rate Control in MIMO Broadcast Systems', in *Proc. of IEEE Information Theory Workshop (ITW)*, San Diego, CA, January 2008.

[19] N. Jindal, 'MIMO Broadcast Channels with Finite Rate Feedback', in *Proc. IEEE Global Telecommunications Conference*, pp. 1520–1524, 2005.

[20] T. Yoo, N. Jindal and A. Goldsmith, 'Multi-antenna Downlink Channels with Limited Feedback and User Selection'. *IEEE Journal on Selected Areas in Communications*, Vol. 25, pp. 1478–1491, September 2007.

[21] T. Yoo and A. Goldsmith, 'On the Optimality of Multiantenna Broadcast Scheduling using Zero-forcing Beamforming'. *IEEE Journal on Selected Areas in Communications*, Vol. 24, pp. 528–541, March 2006.

[22] P. Ding, D. J. Love and M. D. Zoltowski, 'Multiple Antenna Broadcast Channels with Shape Feedback and Limited Feedback'. *IEEE Trans. on Signal Processing*, Vol. 55, pp. 3417–3428, July 2007.

[23] J. H. Conway, R. H. Hardin and N. J. A. Sloane, *Packing Lines, Planes, etc.: Packings in Grassmannian Frames*. AT&T Bell Laboratories, April 1996. Unpublished report available online.

[24] G. Dimic and N. Sidiropoulos, 'On Downlink Beamforming with Greedy User Selection: Performance Analysis and Simple New Algorithm'. *IEEE Trans. on Signal Processing*, Vol. 53, pp. 3857–3868, October 2005.

[25] N. Jindal, 'A Feedback Reduction Technique for MIMO Broadcast Channels', in *Proc. IEEE International Symposium on Information Theory*, Seattle, WA, July 2006.

[26] Texas Instruments, 'R1-072844: Link Level Evaluation of 4-TX Codebook for SU-MIMO', www.3gpp.org, 3GPP TSG RAN 1, Meeting 49bis Orlando, FL, June 2007.

[27] Samsung, 'R1-073181:Way-forward on Codebook for SU-MIMO Precoding', www.3gpp.org, 3GPP TSG RAN 1, Meeting 49bis Orlando, FL, June 2007.

[28] M. Costa, 'Writing on Dirty Paper'. *IEEE Trans. on Information Theory*, Vol. 29, pp. 439–441, May 1983.

12

Multi-User Scheduling and Interference Coordination

Issam Toufik and Raymond Knopp

12.1   Introduction

The eNodeB in an LTE system is responsible, among other functions, for managing resource scheduling for both uplink and downlink channels. The ultimate aim of this function is typically to fulfil the expectations of as many users of the system as possible, taking into account the Quality-of-Service (QoS) requirements of their respective applications.

A typical single-cell cellular radio system is shown in Figure 12.1, comprising K User Equipments (UEs) communicating with one eNodeB over a fixed total bandwidth B. Each UE has several data queues corresponding to different uplink logical channel groups, each with different delay and rate constraints. In the same way in the downlink, the eNodeB may maintain several buffers per UE containing dedicated data traffic, in addition to queues for broadcast services. The different traffic queues in the eNodeB would typically have different QoS constraints. In LTE, the total system bandwidth B is divided into M Resource Blocks (RBs) in the frequency domain. Data is normally split into blocks of duration $T = 1$ subframe (1 ms), otherwise known as the Transmission Time Interval (TTI). For the purpose of analysing suitable scheduling algorithms, T is assumed to be shorter than the coherence time of the channel, so that the channel can be assumed stationary for the duration of each subframe, but may vary from subframe to subframe. We also assume that the channel is constant over the subcarriers in one RB, but that the channel gain of a user may change from one RB to another in the frequency domain (i.e. B/M, which is 180 kHz, is smaller than the channel coherence bandwidth).

The goal of a resource scheduling algorithm in the eNodeB is to allocate the resource blocks and transmission powers for each subframe in order to optimize a function of a set of performance metrics, for example maximum/minimum/average throughput, maximum/minimum/average delay, total/per-user spectral efficiency or outage probability. In the

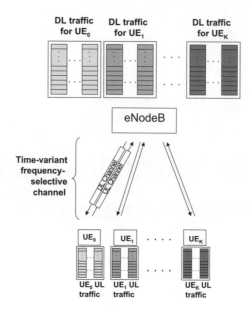

Figure 12.1 A typical single-cell cellular radio system.

downlink the resource allocation strategy is constrained by the total transmission power of the eNodeB, while in the uplink the main constraint on transmission power in different RBs arises from a multicell view of inter-cell interference.

Resource allocation algorithms must also take into account practical constraints, such as the orthogonal design of the LTE uplink and downlink multiple access schemes, which in general dictate that only one user is allocated a particular RB in any subframe.[1]

In this chapter we provide an overview of some of the key families of scheduling algorithms which are relevant for an LTE system and highlight some of the factors that an eNodeB can advantageously take into account.

12.2   General Considerations for Resource Allocation Strategies

The generic function of a resource scheduler, as shown for the downlink case in Figure 12.2, is to schedule data to a set of UEs on a shared set of physical resources. In general, scheduling algorithms can make use of two types of measurement information to inform the scheduling decisions, namely channel-state information and traffic measurements (volume and priority). These are obtained either by direct measurements at the eNodeB or via feedback signalling channels, or a combination of both. The amount of feedback used is an important consideration, since the availability of accurate channel state and traffic information helps

---

[1]Note, however, that multi-user MIMO schemes may result in more than one UE being allocated to a given RB, as discussed in Sections 11.2.3 and 17.5.2 for the downlink and uplink respectively.

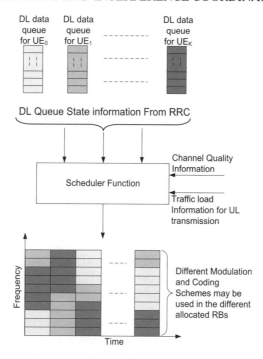

Figure 12.2 Generic view of a wideband resource scheduler.

to maximize the data rate in one direction at the expense of more overhead in the other. This fundamental trade-off, which is common to all feedback-based resource scheduling schemes, is particularly important in Frequency Division Duplex (FDD) operation where uplink-downlink reciprocity of the radio channels cannot be assumed. For Time Division Duplex (TDD) systems, amplitude coherence between the uplink and downlink channels may be used to assist the scheduling algorithm, as discussed in Section 23.5.

It is worth noting that the algorithm used by the resource scheduler is also tightly coupled with the adaptive coding and modulation scheme and the retransmission protocol (Hybrid Automatic Repeat reQuest, HARQ, see Sections 4.4 and 10.3.2.5). This is due firstly to the fact that, in addition to dynamic physical resource allocation, the channel measurements are also used to adapt the coding and modulation format (i.e. transmission spectral-efficiency) as explained in Section 10.2. Secondly, the queue dynamics, which impact the throughput and delay characteristics of the link seen by the application, depend heavily on the HARQ protocol and transport block sizes. Moreover, the combination of channel coding and retransmissions provided by HARQ enables the spectral efficiency of an individual transmission in one subframe to be traded off against the number of subframes in which retransmissions take place. Well-designed practical scheduling algorithms will necessarily consider all these aspects.

Based on the available measurement information, the eNodeB resource scheduler must manage the differing requirements of all the UEs in the cells under its control to ensure that sufficient radio transmission resources are allocated to each UE within acceptable latencies to meet their QoS requirements in a spectrally-efficient way. The details of this process are not standardized as it is largely internal to the eNodeB, allowing for vendor-specific algorithms to be developed which can be optimized for specific scenarios in conjunction with network operators. However, the key inputs available to the resource scheduling process are common, and in general some typical approaches can be identified.

Two extremes of scheduling algorithm may be identified: *opportunistic scheduling* and *fair scheduling*. The former is typically designed to maximize the sum of the transmitted data rates to all users by exploiting the fact that different users experience different channel gains and hence will experience good channel conditions at different times and frequencies. A fundamental characteristic of mobile radio channels is the fading effects arising from the mobility of the UEs in a multipath propagation environment, and from variations in the surrounding environment itself (see Chapter 21). In [1–3] it is shown that, for a multi-user system, significantly more information can be transmitted across a fading channel than a non-fading channel for the same average signal power at the receiver. This principle is known as *multi-user diversity*. With proper dynamic scheduling, allocating the channel at each given time instant only to the user with the best channel condition in a particular part of the spectrum can yield a considerable increase in the total throughput as the number of active users becomes large. In general this result relies on being able to adapt the power dynamically according to the channel state, but it has also been shown that the greater part of the performance gains can be achieved by an 'on–off' power allocation between RBs (i.e. equal power in those RBs in which a transmission takes place). This not only allows some simplification of the scheduling algorithm but also is well suited to a downlink scenario where the frequency-domain dynamic range of the transmitted power in a given subframe is limited by constraints such as the dynamic range of the UE receivers and the need to transmit wideband reference signals for channel estimation.

The main issue arising from opportunistic resource allocation schemes is the difficulty of ensuring fairness and QoS. Users' data cannot always wait until the channel conditions are sufficiently favourable for transmission, especially in slowly varying channels. Furthermore, as explained in Chapter 1, it is important that network operators can provide reliable wide area coverage, including to stationary users near the cell edge – not just to the users which happen to experience good channel conditions by virtue of their proximity to the eNodeB.

The second extreme of scheduling algorithm, fair scheduling, therefore pays more attention to latency for each user than to the total data rate achieved. This is particularly important for real-time applications such as Voice-over-IP (VoIP) or video-conferencing, where a certain minimum rate must be guaranteed independently of the channel state.

In practice, most scheduling algorithms fall between the two extremes outlined above, including elements of each to deliver the required mix of QoS. When considering the degree of fairness provided by a scheduling algorithm, a metric based on the Cumulative Density Function (CDF) of the throughput of all users is often used. A typical example is to ensure that the CDF of the throughput lies to the right-hand side of a particular line, such as that shown in Figure 12.3. The objective of such a metric is to avoid heavily penalizing the cell-edge users in order to give high throughputs to the users with good channel conditions. Other factors also need to be taken into account, especially the fact that, in a coordinated deployment, individual cells cannot be considered in isolation – nor even the individual set of cells controlled by a single eNodeB. The eNodeBs should take into account the interference generated by co-channel cells, which can be a severe limiting factor, especially for cell-edge users.

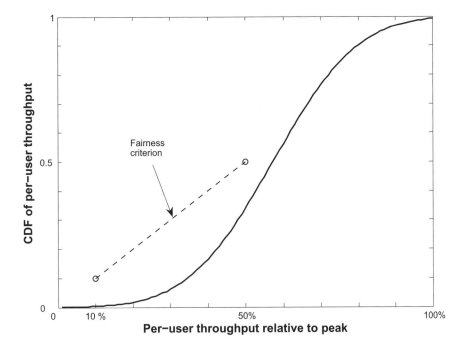

Figure 12.3 An example of metric based on the throughput CDF over all users for scheduler fairness evaluation (10–50 metric).

Similarly, the performance of the system as a whole can be enhanced if each eNodeB also takes into account the impact of the transmissions of its own cells on the neighbouring cells. These aspects, and the corresponding inter-eNodeB signalling mechanisms provided in LTE, are discussed in detail in Section 12.5.

12.3 Scheduling Algorithms

Multi-user scheduling finds its basis at the interface between information theory and queueing theory, in the theory of capacity-maximizing resource allocation. Before establishing an algorithm, a capacity-related metric is first formulated and then optimized across all possible resource allocation solutions satisfying a set of predetermined constraints. Such constraints may be physical (e.g. bandwidth and total power) or QoS-related.

Information theory offers a range of possible capacity metrics which are relevant in different system operation scenarios. Two prominent examples are explained here, namely the so-called *ergodic capacity* and *delay limited capacity*, corresponding respectively to soft and hard forms of rate guarantee for the user.

12.3.1 Ergodic Capacity

The ergodic capacity (also known as the Shannon capacity) is defined as the maximum data rate which can be sent over the channel with asymptotically small error probability, averaged over the fading process. When Channel State Information is available at the Transmitter (CSIT), the transmit power and mutual information[2] between the transmitter and the receiver can be varied depending on the fading state in order to maximize the average rates. The ergodic capacity metric considers the long-term average data rate which can be delivered to a user when the user does not have any latency constraints.

12.3.1.1 Maximum Rate Scheduling

It has been shown in [3,4] that the maximum total ergodic sum rate $\sum_{k=1}^{K} R_k$, where R_k is the total rate allocated to user k, is achieved with a transmission power given by

$$
P_k(m, f) = \begin{cases} \left[\dfrac{1}{\lambda_k} - \dfrac{N_0}{|H_k(m, f)|^2} \right]^+ & \text{if } |H_k(m, f)|^2 \geq \dfrac{\lambda_k}{\lambda_{k'}} |H_{k'}(m, f)|^2 \\ 0 & \text{otherwise} \end{cases} \tag{12.1}
$$

where $[x]^+ = \max(0, x)$, $H_k(m, f)$ is the channel gain of user k in RB m of subframe f and λ_k are constants which are chosen in order to satisfy an average per-user power constraint.

This approach is known in the literature as *maximum rate* scheduling. The result in Equation (12.1) shows that the maximum sum rate is achieved by orthogonal multiplexing where in each subchannel (i.e. each RB in LTE) only the user with the best channel gain is scheduled. This orthogonal scheduling property is in line with, and thus justifies, the philosophy of the LTE multiple-access schemes. The input power spectra given by Equation (12.1) are water-filling formulae in both frequency and time (i.e. allocating more power to a scheduled user when his channel gain is high and less power when it is low).[3]

A variant of this resource allocation strategy with no power control (relevant for cases where the power control dynamic range is limited or zero) is considered in [2]. This allocation strategy is called 'maximum-rate constant-power' scheduling, where only the user with the best channel gain is scheduled in each RB, but with no adaptation of the transmit power. It is shown in [2] that most of the performance gains offered by the maximum rate allocation in Equation (12.1) are due to multi-user diversity and not to power control, so an on–off power allocation can achieve comparable performance to maximum-rate scheduling.

12.3.1.2 Proportional Fair Scheduling

The ergodic sum rate corresponds to the optimal rate for traffic which has no delay constraint. This results in an unfair sharing of the channel resources. When the QoS required by the application includes latency constraints, such a scheduling strategy is not suitable and other fairer approaches need to be considered. One such approach is the well-known Proportional Fair Scheduling (PFS) algorithm. PFS schedules a user when its instantaneous channel

---

[2]The mutual information of two random variables is a quantity that measures the mutual dependence of the two variables.

[3]Water-filling strategies are discussed in more detail in the context of MIMO in Section 11.1.3.

quality is high relative to its own average channel condition over time. It allocates user k in RB m in any given subframe f if $k = \hat{k}_m$, where [5]

$$\hat{k}_m = \underset{k'=1,\ldots,K}{\operatorname{argmax}} \frac{R_{k'}(m, f)}{T_{k'}(f)} \tag{12.2}$$

where $T_k(f)$ denotes the long-term average throughput of user k computed in subframe f and $R_k(m, f) = \log[1 + \text{SNR}_k(m, f)]$ is the achievable rate by user k in RB m and subframe f. The long-term average user throughputs are recursively computed by

$$T_k(f) = \left(1 - \frac{1}{t_c}\right) T_k(f-1) + \frac{1}{t_c} \sum_{m=1}^{M} R_k(m, f) \mathcal{I}\{\hat{k}_m = k\} \tag{12.3}$$

where t_c is the time window over which fairness is imposed and $\mathcal{I}\{\cdot\}$ is the indicator function equal to one if $\hat{k}_m = k$ and zero otherwise. A large time window t_c tends to maximize the total average throughput; in fact, in the limit of a very long time window, PFS and maximum-rate constant-power scheduling result in the same allocation of resources. For small t_c, the PFS tends towards a round-robin[4] scheduling of users in the system [5].

Several studies of PFS have been conducted in the case of WCDMA/HSDPA, yielding insights which can be applied to LTE. In [6] the link level system performance of PFS is studied, taking into account issues such as link adaptation dynamic range, power and code resources, convergence settings, signalling overhead and code multiplexing. [7] analyses the performance of PFS in the case of VoIP, including a comparison to round-robin scheduling for different delay budgets and packet-scheduling settings. A study of PFS performance in the case of video streaming in HSDPA can be found in [8].

12.3.2 Delay-Limited Capacity

Even though PFS introduces some fairness into the system, this form of fairness may not be sufficient for applications which have a very tight latency constraint. For these cases, a different capacity metric is needed; one such example is referred to as the 'delay-limited capacity'. The delay-limited capacity (also known as zero-outage capacity) is defined as the transmission rate which can be guaranteed in all fading states under finite long-term power constraints. In contrast to the ergodic capacity, where mutual information between the transmitter and the receiver vary with the channel, the powers in delay-limited capacity are coordinated between users and RBs with the objective of maintaining constant mutual information independently of fading states. The delay-limited capacity is relevant to traffic classes where a given data rate must be guaranteed throughout the connection time, regardless of the fading dips.

The delay-limited capacity for a flat-fading multiple-access channel is characterized in [9]. The wideband case is analysed in [5, 10].

It is shown in [5] that guaranteeing a delay-limited rate incurs only a small throughput loss in high Signal-to-Noise Ratio (SNR) conditions, provided that the number of users is large. However, the solution to achieve the delay-limited capacity requires non-orthogonal scheduling of the users, with successive decoding in each RB. This makes this approach basically unsuitable for LTE.

---

[4]A round-robin approach schedules each user in turn, irrespective of their prevailing channel conditions.

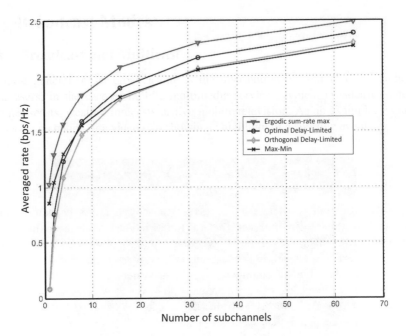

Figure 12.4 Comparison of scheduling algorithm performance. (Average SINR $= 0$ dB for all users.)

It is, however, possible to combine orthogonal multiple access with hard QoS requirements.

One such scheme is known as 'orthogonal delay-limited rate allocation'. The objective of this strategy, as for delay-limited capacity, is to find the allocation of users to RBs which maximizes the number of served users for a given total transmit power while achieving a target rate-tuple $\mathbf{R}^* = (R_1^*, R_2^*, \ldots, R_K^*)$ through an orthogonal multiplexing of the users. The optimal solution to this problem uses power adaptation across the RBs.

Another possibility is the *Max-Min* allocation. This policy guarantees that, at any given time instant, the minimum channel gain of any of the allocated users is the highest possible among all possible allocations and thus maximizes the minimum allocated rate when an equal and fixed power is used for all users. This is particularly relevant for systems where no power control (or a limited amount of power control, as in the LTE downlink) can be used.

Scheduling algorithms satisfying the two above-mentioned allocation strategies can be found in [10, 11], where the performance is compared to both optimal ergodic sum-rate and non-orthogonal optimal delay-limited rate allocations.

12.3.3 Performance of Scheduling Strategies

Figure 12.4 shows the system spectral efficiency for each of the different allocation strategies presented above, as a function of the number of RBs (under the assumption that the number of UEs is always at least equal to the number of RBs).

The curves 'Optimal Delay-Limited' and 'Orthogonal Delay-Limited' represent the maximum achievable common rate (equal rate for all users) under optimal and orthogonal multiple access respectively. The results in Figure 12.4 are obtained for an average Signal to Interference plus Noise Ratio (SINR) of 0 dB for all UEs, and Rayleigh fading with uncorrelated channel gains between RBs. Although unrealistic, this provides an idea of the achievable rates, as a function of the approximate number of degrees of freedom allowed by the system bandwidth and the propagation environment.

In all cases, it can be seen from Figure 12.4 that the per-user average throughput increases with the number of users, which proves that, even under delay-limited requirements, high multi-user diversity gains can still be achieved. It can be seen that even under hard fairness constraints it is possible to achieve performance which is very close to the optimal unfair policy; thus hard fairness constraints do not necessarily introduce a significant throughput degradation, even with orthogonal resource allocation, provided that the number of users and the system bandwidth are large. This may be a relevant scenario for a deployment targeting VoIP users, where densities of several hundred users per cell are foreseen (as mentioned in Section 1.2), with a latency constraint typically requiring each packet to be successfully delivered within 50 ms. In the case of users at different distances from the eNodeB (and hence with different SINR statistics), it has been shown in [5] that, for a high SINR scenario, PFS does not provide any significant gain and may even perform worse than the optimal non-orthogonal delay-limited scheduling; this is despite the fact that the imposed fairness constraint is less stringent. However, for low to moderate SINR, the stricter hard-fairness constraint incurs a large throughput penalty for delay-limited scheduling with respect to PFS.

The summary of scheduling strategies presented above assumes that all users have an equal and infinite queue length – often referred to as a full-buffer traffic model. In practice this is not the case, especially for real-time services, and information on users' queue lengths is necessary to guarantee system stability. If a scheduling algorithm can be found which keeps the average queue length bounded, the system is said to be stabilized [12]. One approach to ensuring that the queue length remains bounded is to use the queue length to set the priority order in the allocation of RBs. This generally works for lightly-loaded systems. References [13] and [14] show that in wideband frequency-selective channels, low average packet delay can be achieved even if the fading is very slow.

12.4 Considerations for Resource Scheduling in LTE

In LTE, each logical channel has a corresponding QoS description which should influence the behaviour of the eNodeB resource scheduling algorithm. Based on the evolution of the radio and traffic conditions, this QoS description could potentially be updated for each service in a long-term fashion. It is likely that the mapping between the QoS descriptions of different services and the resource scheduling algorithm in the eNodeB will be a key differentiating factor between radio network equipment manufacturers.

An important constraint for the eNodeB scheduling algorithm is the availability and accuracy of the channel quality information for the active UEs in the cell. The manner in which such information is provided to the scheduler in LTE differs between uplink and downlink transmissions. In practice, for the downlink this information is provided through the feedback of Channel Quality Indicators (CQIs) by UEs as described in Chapter 10, while for the uplink the eNodeB may use Sounding Reference Signals (SRSs) or other signals transmitted by the UEs to estimate the channel quality, as discussed in Chapter 16. The frequency with which CQI reports and SRS are transmitted is configurable by the eNodeB, allowing for a trade-off between the signalling overhead and the availability of up-to-date channel information. If the most recent CQI report or SRS was received a significant time before the scheduling decision is taken, the performance of the scheduling algorithm can be significantly degraded. This is due to the potential for a large change in the channel quality between the time the scheduler receives the information and the time the UE is scheduled.

In order to perform frequency-domain scheduling, the information about the radio channel needs to be frequency-specific. For this purpose, the eNodeB may configure the CQI reports to relate to specific subbands to assist the downlink scheduling, as explained in Section 10.2.1. Uplink frequency-domain scheduling can be facilitated by configuring the SRS to be transmitted over a large bandwidth. However, for cell-edge UEs the wider the transmitted bandwidth the lower the available power per RB; this means that accurate frequency-domain scheduling may be more difficult for UEs near the cell edge. Limiting the SRS to a subset of the system bandwidth will improve the channel quality estimation on these RBs but restrict the ability of the scheduler to find an optimal scheduling solution for all users. In general, provided that the bandwidth over which the channel can be estimated for scheduling purposes is greater than the intended scheduling bandwidth for data transmission by a sufficient factor, a useful element of multi-user diversity gain may still be achievable.

As noted above, in order to support QoS- and queue-aware scheduling, it is necessary for the scheduler to have not only information about the channel quality, but also information on the queue status. In the LTE downlink, knowledge of the amount of buffered data awaiting transmission to each UE is inherently available in the eNodeB; for the uplink, Section 4.4.2.2 explains the buffer status reporting mechanisms available to transfer such information to the eNodeB.

12.5 Interference Coordination and Frequency Reuse

One limiting aspect for system throughput performance in cellular networks is inter-cell interference, especially for cell edge users. Careful management of inter-cell interference is particularly important in systems such as LTE which are designed to operate with a frequency reuse factor of one.

The scheduling strategy of the eNodeB may therefore include an element of inter-cell interference coordination, whereby interference from and to the adjacent cells is taken into account in order to increase the data rates which can be provided for users at the cell edge. This implies for example imposing restrictions on what resources in time and/or frequency are available to the scheduler, or what transmit power may be used in certain time/frequency resources.

The impact of interference on the achievable data rate for a given user can be expressed analytically. If a user k is experiencing no interference, then its achievable rate in a RB m of subframe f can be expressed as

$$R_{k,\text{no-Int}}(m, f) = \frac{B}{M} \log\left[1 + \frac{P^s(m, f)|H_k^s(m, f)|^2}{N_0}\right] \qquad (12.4)$$

where $H_k^s(m, f)$ is the channel gain from the serving cell s to user k, $P^s(m, f)$ is the transmit power from cell s and N_0 is the noise power. If neighbouring cells are transmitting in the same time-frequency resources, then the achievable rate for user k reduces to

$$R_{k,\text{Int}}(m, f) = \frac{B}{M} \log\left[1 + \frac{P^s(m, f)|H_k^s(m, f)|^2}{N_0 + \sum_{i \neq s} P^i(m, f)|H_k^i(m, f)|^2}\right] \qquad (12.5)$$

where the indices i denote interfering cells.

The rate loss for user k can then be expressed as

$$R_{k,\text{loss}}(m, f) = R_{k,\text{no-Int}}(m, f) - R_{k,\text{Int}}(m, f)$$

$$= \frac{B}{M} \log\left\{ \frac{1 + \text{SNR}}{1 + \left[\frac{1}{\text{SNR}} + \frac{\sum_{i \neq s} P^i(m,f)|H_k^i(m,f)|^2}{P^s(m,f)|H_k^s(m,f)|^2}\right]^{-1}} \right\} \qquad (12.6)$$

Figure 12.5 plots the rate loss for user k as a function of the total inter-cell interference to signal ratio $\alpha = [(\sum_{i \neq s} P^i(m, f)|H_k^i(m, f)|^2)/(P^s(m, f)|H_k^s(m, f)|^2)]$, with SNR = 0 dB. It can easily be seen that for a level of interference equal to the desired signal level (i.e. $\alpha \approx 0$ dB), user k experiences a rate loss of approximately 40%.

In order to demonstrate further the significance of interference and power allocation depending on the system configuration we consider two examples of a cellular system with two cells (s_1 and s_2) and one active user per cell (k_1 and k_2 respectively). Each user receives the wanted signal from its serving cell, while the inter-cell interference comes from the other cell.

In the first example, each user is located near its respective eNodeB (see Figure 12.6(a)). The channel gain from the interfering cell is small compared to the channel gain from the serving cell ($|H_{k_1}^{s_1}(m, f)| \gg |H_{k_1}^{s_2}(m, f)|$ and $|H_{k_2}^{s_2}(m, f)| \gg |H_{k_2}^{s_1}(m, f)|$). In the second example (see Figure 12.6(b)), we consider the same scenario but with the users now located close to the edge of their respective cells. In this case the channel gain from the serving cell and the interfering cell are comparable ($|H_{k_1}^{s_1}(m, f)| \approx |H_{k_1}^{s_2}(m, f)|$ and $|H_{k_2}^{s_2}(m, f)| \approx |H_{k_2}^{s_1}(m, f)|$).

The capacity of the system with two eNodeBs and two users can be written as

$$R_{\text{Tot}} = \frac{B}{M}\left(\log\left(1 + \frac{P^{s_1}|H_{k_1}^{s_1}(m, f)|^2}{N_0 + P^{s_2}|H_{k_1}^{s_2}(m, f)|^2}\right) + \log\left(1 + \frac{P^{s_2}|H_{k_2}^{s_2}(m, f)|^2}{N_0 + P^{s_1}|H_{k_2}^{s_1}(m, f)|^2}\right)\right)$$

$$(12.7)$$

From this equation, it can be noted that the optimal transmit power operating point in terms of maximum achievable throughput is different for the two considered cases. In the first scenario, the maximum throughput is achieved when both eNodeBs transmit at maximum power, while in the second the maximum capacity is reached by allowing only one eNodeB to transmit. It can in fact be shown that the optimal power allocation for maximum capacity for this situation with two base stations is binary in the general case; this means that either both base stations should be operating at maximum power in a given RB, or one of them should be turned off completely in that RB [15].

From a practical point of view, this result can be exploited in the eNodeB scheduler by treating users in different ways depending on whether they are cell-centre or cell-edge users.

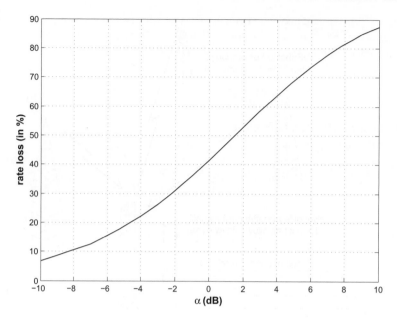

Figure 12.5 User's rate loss due to interference.

Each cell can then be divided into two parts – inner and outer. In the inner part, where users experience a low level of interference and also require less power to communicate with the serving cell, a frequency reuse factor of 1 can be adopted. For the outer part, scheduling restrictions are applied: when the cell schedules a user in a given part of band, the system capacity is optimized if the neighbouring cells do not transmit at all; alternatively, they may transmit only at low power (probably to users in the inner parts of the neighbouring cells) to avoid creating strong interference to the scheduled user in the first cell. This effectively results in a higher frequency reuse factor at the cell-edge; it is often known as 'partial frequency reuse', and is illustrated in Figure 12.7.

In order to coordinate the scheduling in different cells in such a way, communication between neighbouring cells is required. If the neighbouring cells are managed by the same eNodeB, a coordinated scheduling strategy can be followed without the need for standardized signalling. However, where neighbouring cells are controlled by different eNodeBs, standardized signalling is important, especially in multivendor networks. Inter-Cell Interference Coordination (ICIC) in LTE is normally assumed to be managed in the frequency domain rather than the time domain (as time domain coordination would interfere with the operation of the HARQ processes, especially in the uplink where synchronous HARQ is used), and the inter-eNodeB signalling is designed to support this.

In relation to the downlink transmissions, a bitmap termed the Relative Narrowband Transmit Power (RNTP[5]) indicator can be exchanged between eNodeBs over the X2 interface. Each bit of the RNTP indicator corresponds to one RB in the frequency domain and is used to inform the neighbouring eNodeBs if a cell is planning to keep the transmit

---

[5]RNTP is defined in reference [16], Section 5.2.1.

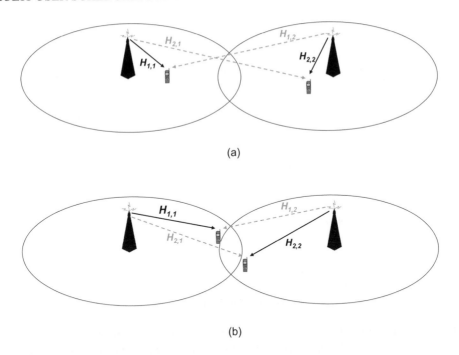

Figure 12.6 System configuration: (a) users close to eNBs; (b) users at the cell edge.

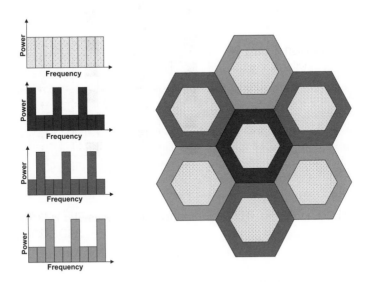

Figure 12.7 Partial frequency reuse.

power for the RB below a certain upper limit or not. The value of this upper limit, and the period for which the indicator is valid into the future, are configurable. This enables the neighbouring cells to take into account the expected level of interference in each RB when scheduling UEs in their own cells. The reaction of the eNodeB in case of receiving an indication of high transmit power in an RB in a neighbouring cell is not standardized (thus allowing some freedom of implementation for the scheduling algorithm); however, a typical response could be to avoid scheduling cell-edge UEs in such RBs. In the definition of the RNTP indicator, the transmit power per antenna port is normalized by the maximum output power of a base station or cell. The reason for this is that a cell with a smaller maximum output power, corresponding to smaller cell size, can create as much interference as a cell with a larger maximum output power corresponding to a larger cell size.

For the uplink transmissions, two messages may be exchanged between eNodeBs to facilitate some coordination of their transmit powers and scheduling of users:

- A reactive indicator, known as the 'Overload Indicator' (OI), can be exchanged over the X2 interface to indicate physical layer measurements of the average uplink interference plus thermal noise for each RB. The OI can take three values, expressing low, medium, and high levels of interference plus noise. In order to avoid excessive signalling load, it cannot be updated more often than every 20 ms.

- A proactive indicator, known as the 'High Interference Indicator' (HII), can also be sent by an eNodeB to its neighbouring eNodeBs to inform them that it will, in the near future, schedule uplink transmissions by one or more cell-edge UEs in certain parts of the bandwidth, and therefore that high interference might occur in those frequency regions. Neighbouring cells may then take this information into consideration in scheduling their own users to limit the interference impact. This can be achieved either by deciding not to schedule their own cell-edge UEs in that part of the bandwidth and only considering the allocation of those resources for cell-centre users requiring less transmission power, or by not scheduling any user at all in the relevant RBs. The HII is comprised of a bitmap with one bit per RB, and, like the OI, is not sent more often than every 20 ms. The HII bitmap is addressed to specific neighbour eNodeBs.

In addition to frequency-domain scheduling in the uplink, the eNodeB also controls the degree to which each UE compensates for the path-loss when setting its uplink transmission power. This enables the eNodeB to trade off fairness for cell-edge UEs against inter-cell interference generated towards other cells, and can also be used to maximize system capacity. This is discussed in more detail in Section 20.3.2.

In general, ICIC may be static or semi-static, with different levels of associated communication required between eNodeBs.

- For *static interference coordination*, the coordination is associated with cell planning, and reconfigurations are rare. This largely avoids signalling on the X2 interface, but it may result in some performance limitation since it cannot adaptively take into account variations in cell loading and user distributions.

- *Semi-static interference coordination* typically refers to reconfigurations carried out on a time-scale of the order of seconds or longer. The inter-eNodeB communication methods over the X2 interface can be used as discussed above. Other types of information such as traffic load information may also be used, as discussed in Section 2.6.4. Semi-static interference coordination may be more beneficial in cases of non-uniform load distributions in eNodeBs and varying cell sizes across the network.

12.6 Concluding Remarks

In summary, it can be observed that a variety of resource scheduling algorithms may be applied by the eNodeB depending on the optimization criteria required. The prioritization of data will typically take into account the corresponding traffic classes, especially in regard to balancing throughput maximization for delay-tolerant applications against QoS for delay-limited applications in a fair way.

It can be seen that multi-user diversity is an important factor in all cases, and especially if the user density is high, in which case the multi-user diversity gain enables the scheduler to achieve a high capacity even with tight delay constraints. Finally, it is important to note that individual cells, and even individual eNodeBs, cannot be considered in isolation. System optimization requires some degree of coordination between cells and eNodeBs, in order to avoid inter-cell interference becoming the limiting factor. Considering the system as a whole, the best results are in many cases realized by simple 'on–off' allocation of resource blocks, whereby some eNodeBs avoid scheduling transmissions in certain resource blocks which are used by neighbouring eNodeBs for cell-edge users.

References[6]

[1] K. Knopp and P. A. Humblet, 'Information Capacity and Power Control in Single-cell Multiuser Communications', in *Proc. IEEE International Conference on Communications*, Seattle, WA, 1995.

[2] R. Knopp and P. A. Humblet, 'Multiple-Accessing over Frequency-selective Fading Channels', in *Proc. IEEE International Symposium on Personal, Indoor and Mobile Radio Communications*, September 1995.

[3] D. N. C. Tse and S. D. Hanly, 'Multiaccess Fading Channels. I: Polymatroid Structure, Optimal Resource Allocation and Throughput Capacities'. *IEEE Trans. on Information Theory*, Vol. 44, pp. 2796–2815, November 1998.

[4] R. Knopp and P. A. Hamblet, 'Multiuser Diversity', Technical Report, Eurecom Institute, Sophia Antipolis, France, 2002.

[5] G. Caire, R. Müller, and R. Knopp, 'Hard Fairness Versus Proportional Fairness in Wireless Communications: The Single-cell Case'. *IEEE Trans. on Information Theory*, Vol. 53, pp. 1366–1385, April 2007.

[6] T. E. Kolding, 'Link and System Performance Aspects of Proportional Fair Scheduling in WCDMA/HSDPA', in *Proc. IEEE Vehicular Technology Conference*, Orlando, FL, October 2003.

[7] B. Wang, K. I. Pedersen, P. E. Kolding and T. E. Mogensen, 'Performance of VoIP on HSDPA', in *Proc. IEEE Vehicular Technology Conference*, Stockholm, May 2005.

[8] V. Vukadinovic and G. Karlsson, 'Video Streaming in 3.5G: On Throughput-Delay Performance of Proportional Fair Scheduling', in *IEEE International Symposium on Modeling, Analysis, and Simulation of Computer and Telecommunication Systems*, Monterey, CA, September 2006.

---

[6]All web sites confirmed 18th December 2008.

 [9] S. D. Hanly and D. N. C. Tse, 'Multiaccess Fading Channels. II: Delay-limited Capacities'. *IEEE Trans. on Information Theory*, Vol. 44, pp. 2816–2831, November 1998.

[10] I. Toufik, 'Wideband Resource Allocation for Future Cellular Networks'. Ph.D. Thesis, Eurecom Institute, 2006.

[11] I. Toufik and R. Knopp, 'Channel Allocation Algorithms for Multi-carrier Systems', in *Proc. IEEE Vehicular Technology Conference*, Los Angeles, CA, September 2004.

[12] M. J. Neely, E. Modiano and C. E. Rohrs, 'Power Allocation and Routing in Multi-beam Satellites with Time Varying Channels'. *IEEE Trans. on Networking*, Vol. 11, pp. 138–152, February 2003.

[13] M. Realp, A. I. Pérez-Neira and R. Knopp, 'Delay Bounds for Resource Allocation in Wideband Wireless Systems', in *Proc. IEEE International Conference on Communications*, Istanbul, June 2006.

[14] M. Realp, R. Knopp and A. I. Pérez-Neira, 'Resource Allocation in Wideband Wireless Systems' in *Proc. IEEE International Symposium on Personal, Indoor and Mobile Radio Communications*, Berlin, September 2005.

[15] A. Gjendemsjoe, D. Gesbert, G. Oien and S. Kiani, 'Binary Power Control for Sum Rate Maximization over Multiple Interfering Links'. *IEEE Trans. Wireless Communications*, August 2008.

[16] 3GPP Technical Specification 36.213, 'Evolved Universal Terrestrial Radio Access (E-UTRA); Physical Layer Procedures (Release 8)', www.3gpp.org.

13

Radio Resource Management

Francesc Boixadera

13.1 Introduction

Radio Resource Management (RRM) in cellular wireless systems aims to provide the user with a mobility experience whereby the User Equipment (UE) and the network take care of managing mobility seamlessly, without the need for any significant user intervention. Unseen to the user, the provision of this functionality often involves additional complexity, both in the system as a whole and in the UE and network equipment. During the design of cellular wireless communication standards there is an ever-present trade-off between additional UE complexity (e.g. cost, power consumption, processing power), network complexity (e.g. radio interface resource consumption, network topology) and achievable performance.

The main procedures followed by an LTE UE in order to provide support for seamless mobility are cell search, measurements, handover and cell reselection. Chapter 3 has already introduced the protocols for handling mobility, while the procedures for synchronization and cell search have been addressed in Chapter 7 (with particular reference to the design of the physical synchronization signals transmitted from each LTE cell). The present chapter gives an overview from the UE perspective of the functions and procedures followed by a UE to handle mobility with other LTE cells or cells belonging to other Radio Access Technologies (RATs).

In some areas the level of detail provided here is limited, since the definition of mobility procedures and performance requirements typically take longer to develop than the core system specifications for systems such as LTE.

13.2   Overview of UE Mobility Activities

In order to maintain service continuity as a user moves, UEs must not only be connected to a serving cell but importantly also monitor their neighbour cells. This monitoring is an ongoing activity, since propagation conditions to different eNodeBs can change rapidly at any point in time, as can interference levels. Typically the efforts by the UE and the network will always be directed towards providing service continuity on a preferred RAT according to a given preference criterion which may for example include Quality of Service (QoS), cost or network operator.

Generally while a UE has service with its preferred network, it will be requested to perform mobility decisions (handover or cell reselection) towards other base stations from the same RAT. Whenever the preferred network is not available because of poor coverage, congestion or other reasons, the network and the UE must cooperate in order to identify fallback options to other networks or RATs to preserve service continuity. A fallback RAT may in some cases only be able to provide a subset of more basic services (e.g. voice communications or low data rate packet data), thus resulting in some degradation in terms of the services provided, but at least basic service continuity can be preserved.

Mobility procedures have to be specified for a significant number of different possible scenarios, and these are described in detail in the following sections. In all cases the UE has to meet specified minimum performance requirements which are scenario-specific, such as for cell search (see Chapter 7), measurement accuracy, measurement periodicity and handover execution delay. UE power consumption and cost are also important factors, and for this reason the procedures and requirements are generally specified so as to minimize the effort required from the UE. These requirements are mainly specified in [1]. The performance requirements for mobility are usually specific to a particular scenario, consisting of:

- The current serving RAT.

- The Layer 3 state of the current serving RAT, e.g. for LTE the main Radio Resource Control (RRC) states are RRC_CONNECTED and RRC_IDLE (see Section 3.2.1).

- The RAT being monitored. When camped on an LTE cell a UE needs to monitor other cells on the same LTE frequency (LTE intra-frequency cells) and cells on other LTE frequencies (LTE inter-frequency cells). In addition it may be required to monitor one or more other RATs, such as GSM, UMTS, WiMAX, TD-SCDMA, CDMA High Rate Packet Data (HRPD), or CDMA2000. The performance requirements in each case depend on the RAT being monitored.

In the case of the 3GPP RATs (i.e. LTE, UMTS and GERAN[1]), the minimum-effort approach for the UE typically consists of a number of common stages which are broadly the same regardless of the RAT involved. Each of these common stages constitutes a decision point, either in the UE or in the network. These stages are:

- **Serving cell quality monitoring and evaluation.** Serving cell quality is monitored and evaluated on a periodic basis. If the serving cell quality is satisfactory (i.e. above a threshold which is configurable by the network), then no further action is required.

---

[1] GSM EDGE Radio Access Network.

However, if the serving cell quality is below the configured threshold, the next step is executed.

- **Initiate periodic cell search activity for candidate neighbour cells.** Candidate neighbour cells can be any combination of intra-frequency, inter-frequency and inter-RAT cells, and typically the search is performed in a defined order of priority. Cell search needs to be repeated periodically since new cells may appear and disappear at any time. Therefore, even if a UE has successfully identified a neighbour cell it will continue performing cell search activity until either the serving cell quality becomes satisfactory again, or the UE moves to another serving cell through handover, cell reselection or a cell change order. If some neighbour cells are successfully identified, then the following step is performed.

- **Neighbour cell measurement.** In this step the signal strength for the neighbour cells identified in the previous step is measured periodically (since the signal strength and quality may vary) until either the serving cell quality becomes satisfactory again, or the UE moves to another serving cell through handover, cell reselection or a cell change order.

In order to avoid the effect of short-term fluctuations due to fast fading in the radio channel, any measurement for mobility purposes is required to be obtained by performing the average over a number of evenly spaced measurement samples within a measurement period. The measurement period is specified by the performance requirements for the relevant scenario; for example, intra-frequency LTE Reference Signal Received Power (RSRP) measurements have a measurement period of 200 ms (see Section 13.4.1.1). As soon as measurements become available the next step is performed.

- **Mobility evaluation.** In this step a decision is made on whether or not the UE should move to another serving cell. Mobility evaluation can be performed within a network entity (in LTE the eNodeB decides on handover and UE redirection/cell change orders), or within the UE in the case of cell reselection. If the criteria to trigger UE mobility are fulfilled then a mobility procedure to move towards a better cell is executed, where 'better' is judged by a criterion which is scenario-specific. The destination cell may be within the same RAT (for intra- and inter-frequency handover and cell reselection) or in a different RAT (for inter-RAT handover and cell reselection).

13.3 Cell Search

The term 'cell search' refers to the UE's activity to identify the presence of neighbour cells. These neighbour cells may belong to LTE (intra-frequency and inter-frequency cell search) or to other RATs (inter-RAT cell search).

13.3.1 LTE Cell Search

The cell search process for LTE is described in full in Chapter 7. In summary, it consists of:

- Primary Synchronization Signal (PSS) detection to obtain the physical layer cell ID (within a group of three) and slot synchronization.

- Secondary Synchronization Signal (SSS) detection to obtain the Cyclic Prefix (CP) length, the physical layer cell group ID and the frame synchronization.

- Physical Broadcast CHannel (PBCH) decoding to obtain the critical system information and the 40 ms period of the system information. For new cell identification, PBCH decoding is not required but Reference Signal (RS) decoding may be performed to measure the RSRP and the Reference Signal Received Quality (RSRQ)[2] of the new cell to be reported to the network.

13.3.2 UMTS Cell Search

UMTS synchronization codes are defined in [2] and an overview is provided in [3]. A UE is synchronized to a UMTS cell when it knows the timing of the cell's frame boundaries and the identity of the cell's primary scrambling code which distinguishes the cell's transmissions from those of other cells. The most common UMTS synchronization process consists of the following stages:

- **Primary-Synchronization CHannel (P-SCH) search.** Only one UMTS P-SCH code exists, and it is repeated on the first 256 chips of every slot (0.666 ms) in all UMTS cells. During this stage the UE performs matched filter correlation between the received signal and the P-SCH sequence. These correlations are performed for all possible timing offsets within one slot. As a result a curve containing P-SCH correlation power as a function of the time offset can be obtained and correlation peaks can be observed in those locations where a P-SCH sequence is present. This gives the slot boundary timing for each detected P-SCH. For one or more of the strongest detected peaks, the following step is performed.

- **Secondary-SCH decoding.** The UTRA S-SCH code sequence in a cell is one of 15 codewords present on the first 256 chips of every slot (i.e. at the same time as the P-SCH). One S-SCH code sequence is defined for all cells belonging to the same 'code group'. Each S-SCH code sequence identifies uniquely a code group and 10 ms frame boundary position. In good signal conditions the information contained within three slots is sufficient to identify uniquely both the frame timing and the code group, but in order to reliably decode the S-SCH in difficult reception conditions longer decoding periods are required.

- **Primary scrambling code identification.** The code group identified from the S-SCH indicates a group of eight primary scrambling codes. A given cell uses one code from the indicated group as the scrambling code for all downlink channels, including the Primary Common PIlot CHannel (P-CPICH). In order to determine which code is being used, the UE performs a simple correlation against the eight possible scrambling sequences in the code group, looking for the known CPICH sequence (which is the same in all UMTS cells).

---

[2]Refer to Section 13.4 for more details on RSRP and RSRQ measurements.

- **UMTS System Frame Number (SFN) detection.** The UMTS Primary Common Control Physical CHannel (P-CCPCH) carries the Broadcast CHannel (BCH), which is encoded over a 20 ms Transmission Time Interval (TTI). However, the earlier synchronization stages only provide timing information up to a 10 ms period (i.e. one UMTS frame). The location of the even frame boundaries can be found, for example, by trial and error by performing decoding attempts on the two possible TTI boundaries. Only the correct boundary will lead to successful channel decoding and return a correct Cyclic Redundancy Check (CRC). The BCH payload carries the SFN. This step is only required in some inter-RAT scenarios: on cell reselection from LTE to UMTS and on handover to UMTS (only after handover initiation).

It is worth noting that once a UE has camped on an LTE cell it will receive a UMTS neighbour cell list containing up to 32 primary scrambling codes per UMTS carrier. This additional side information may be used by the UE to speed up the UMTS cell search process.

13.3.3   GSM Cell Search

Typically the synchronization process to a GSM cell consists of: GSM Received Signal Strength Indicator (RSSI) measurements, initial Base Station Identification Code (BSIC) identification and BSIC reconfirmation.

These synchronization stages are briefly described in the subsections below. A more extensive description of the GSM common channels is provided in [4]. When camped on an LTE cell the UE will be provided with a Neighbour Cell List (NCL) containing at least 32 GSM carrier numbers (i.e. Absolute Radio Frequency Channel Numbers (ARFCNs)) indicating the frequencies of neighbouring cells, and optionally an associated BSIC for each GSM carrier in the NCL.

13.3.3.1   GSM RSSI Measurements

When GSM monitoring is required, a UE will measure GSM RSSI for all the carriers in the NCL in every measurement period configured by the network. Once measurements for all the cells in the list have become available, the results are sorted in order of decreasing power, and the strongest N cells are passed to the following step. The GSM RSSI measurement period is 480 ms, and any GSM RSSI measurement must be the average of at least three measurement samples as evenly spaced as possible.

13.3.3.2   Initial BSIC Identification

The BSIC is contained within the GSM Synchronization Burst (SB), which carries the GSM Synchronization CHannel (SCH). In addition to the BSIC, the GSM SCH also carries information related to the GSM cell SFN. The BSIC allows a UE to distinguish two different GSM cells which share the same beacon frequency. It is a 6-bit field composed of two 3-bit fields: the Base station Colour Code (BCC) and the Network Colour Code (NCC). The BCC is also used to identify the Training Sequence Code (TSC) to be used when reading the Broadcast Control CHannel (BCCH). The NCC is used to differentiate between operators utilizing the same frequencies, e.g. on an international border when both operators have been allocated the same frequency or frequencies.

Initial BSIC identification is performed in order of decreasing power for the $N = 8$ strongest GSM carriers. Typically, initial BSIC Identification in a UE consists of the following steps:

- **GSM Frequency Control Channel (FCCH) detection.** In order to detect a Frequency Burst (FB) carried by the GSM FCCH, the UE must tune its receiver to a GSM beacon carrier frequency and perform a continuous correlation against the signal contained within the FB. The FB is transmitted on timeslot number 0 of frames 0, 10, 20, 30 and 40 of the 51-frame GSM control multiframe, as depicted in Figure 13.1. When a correlation peak is detected, coarse frame timing and coarse frequency synchronization can be acquired. Note that if continuous correlation is performed, the GSM FCCH is guaranteed to be detected under good reception conditions in no more than 11 GSM frames. One GSM frame has a duration equal to $60/13 = 4.61$ ms.

- **GSM SCH detection.** The GSM SB is carried by the GSM SCH. There is always one SB exactly one frame after the FB. GSM SCH decoding is reasonably straightforward when it is possible to receive the SB exactly one frame after FB detection. If decoding the SB immediately after the FB is not an option, for example because initial BSIC identification is being performed within a gap in the LTE signal specially created for inter-RAT monitoring and this gap is too short, then the GSM SCH can be decoded later. In that case the task of decoding the GSM SCH becomes more complicated because of the presence of the GSM idle frame in the 51-frame control multiframe. The idle frame introduces a $N/(N + 1)$ frame ambiguity which forces the UE to perform decoding attempts at two adjacent locations after FB detection separated by one GSM frame: 10/11 frames, 20/21 frames, etc. as can be seen from Figure 13.1. Given that the data carried in the SCH contains a CRC, the outcome of the CRC check can be used to determine which of the two options is the correct outcome. Once the SCH has been successfully decoded, both the BSIC and the GSM frame number are obtained, and the position of any GSM SB can be predicted without any further ambiguity.

Figure 13.1 GSM common control channel structure. Each square represents the contents of slot 0 for each GSM frame within the 51-frame control multiframe.

It is also possible to detect directly the GSM SCH without prior FCCH detection. However, this involves additional complexity, and the additional benefits in term of detection performance (delay and sensitivity) may not be sufficiently significant to justify this. Initial BSIC identification requirements from UMTS in [5] allow for the implementation of the 2-steps approach. It is very likely that the same requirements will be adopted in [1] for LTE.

13.3.3.3 BSIC Reconfirmation

BSIC Reconfirmation consists of decoding the SB periodically on GSM carriers for which the BSIC has already been detected by initial BSIC identification. In this situation the SB position within a cell can be predicted exactly because the neighbour cell SFN has been acquired by the UE in the previous step. The UE needs to check periodically that the carrier for which the BSIC was previously identified still has the same BSIC because GSM beacon channels can be reused. In addition, 3GPP base stations are not synchronized, and BSIC reconfirmation attempts can also be used to perform minor adjustments to the GSM cell timing stored in the UE, to compensate for cell timing drift relative to the serving cell to which the UE is locked. After a number of unsuccessful BSIC reconfirmation attempts a GSM carrier must be moved back to initial BSIC identification.

13.4 Measurements when Camped on LTE

A UE camped on an LTE cell must be able to perform the measurements described in this section (see [6] for further details).

The majority of UE physical layer measurements require some degree of coherent demodulation[3] and processing. Therefore, these measurements can only be performed after the UE has achieved synchronization with the target cell and knows the relevant physical layer parameters which are required in order to perform coherent processing; this includes the slot timing, frame timing and scrambling codes. There are, however, a few carrier-specific signal strength measurements which can be performed non-coherently without having achieved synchronization; these are LTE carrier RSSI, UMTS carrier RSSI and GSM RSSI (see Sections 13.4.1.2, 13.4.2.2 and 13.4.4.1 respectively).

All measurements reported from the UE by RRC signalling must be obtained by averaging multiple measurement samples which are as uniformly distributed as possible over the measurement period in the time domain, and over the measurement bandwidth in the frequency domain, [1].

The actual implementation of the measurement algorithms is not specified. Any implementation is possible as long as the associated performance requirements in [1] are met.

The measurement model adopted for LTE has been inherited from the UMTS specifications [5, 7]. The model contains four different reference points as shown in Figure 13.2.

- **Reference point 'A'** represents a measurement internal to the physical layer ('Layer 1') in support of the measurements to be provided to higher layers. An example could be a single measurement sample of LTE RSRP corresponding to a very short integration time (e.g. 1 ms). The actual measurement implementation is not constrained by the LTE specifications, but the expected value measurements obtained must correspond to the measurement definition in reference [6].

- **Reference point 'B'** represents measurements after Layer 1 filtering being reported to RRC ('Layer 3'). The actual measurement is not constrained by the specifications – i.e. the model does not state a specific sampling rate, nor even whether the sampling is periodic or not. The LTE specifications simply define the performance objective,

---

[3]See Section 7.3 for an explanation and theoretical analysis of coherent and non-coherent demodulation.

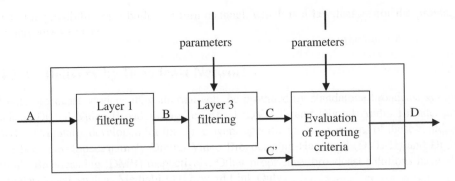

Figure 13.2 E-UTRAN measurement model. Reproduced by permission of © 3GPP.

measurement bandwidth and measurement period. The reporting rate at point B is also not specified, but it needs to be sufficient to meet the specified performance objectives [1]. Measurement performance at point 'B' can be verified by defining tests with Layer 3 filtering disabled.

- **Reference point 'C'** represents a measurement after processing in the Layer 3 filter in RRC. The reporting rate is identical to the reporting rate at point B and is therefore also dependent on the measurement type. Although this is not shown in Figure 13.2, one measurement can be used by a multiplicity of evaluations of reporting criteria in RRC, as explained in Section 3.2.5. The behaviour of the Layer 3 filters is standardized and their configuration is provided by RRC signalling. Each filtered result at point C corresponds to a filtered version of the samples available at point B.

- **Reference point 'D'** contains measurement reports sent by the UE to the eNodeB by RRC signalling. The evaluation of reporting criteria consists of checking whether actual measurement reporting is necessary at point D, i.e. whether a message needs to be sent over the radio interface. The evaluation can be based on more than one flow of measurements (C, C') at reference point C, for example to compare between different measurements. The UE evaluates the reporting criteria at least every time a new measurement result is reported. The reporting criteria are standardized and their configuration provided by RRC signalling [9] (see Chapter 3).

13.4.1 LTE Measurements

13.4.1.1 LTE Reference Signal Received Power (RSRP)

The RSRP measurement provides a cell-specific signal strength metric. This measurement is used mainly to rank different LTE candidate cells according to their signal strength and is used as an input for handover and cell reselection decisions. RSRP is defined for a specific cell as the linear average over the power contributions (in Watts) of the Resource Elements (REs) which carry cell-specific RS within the considered measurement frequency bandwidth. Normally the RS transmitted on the first antenna port are used for RSRP determination, but the RS on the second antenna port can also be used if the UE can determine that they are being transmitted. Details of the RS design can be found in Chapter 8. If receive diversity is in use by the UE, the reported value is the linear average of the power values of all diversity branches.

13.4.1.2 LTE Carrier Received Signal Strength Indicator (RSSI)

The LTE carrier RSSI is defined as the total received wideband power observed by the UE from all sources, including co-channel serving and nonserving cells, adjacent channel interference and thermal noise within the measurement bandwidth specified in [1]. LTE carrier RSSI is not reported as a measurement in its own right, but is used as an input to the LTE RSRQ measurement described below.

13.4.1.3 LTE Reference Signal Received Quality (RSRQ)

This measurement is intended to provide a cell-specific signal quality metric. Similarly to RSRP, this metric is used mainly to rank different LTE candidate cells according to their signal quality. This measurement is used as an input for handover and cell reselection decisions, for example in scenarios for which RSRP measurements do not provide sufficient information to perform reliable mobility decisions. The RSRQ is defined as the ratio $N \cdot \text{RSRP}/(\text{LTE carrier RSSI})$, where N is the number of Resource Blocks (RBs) of the LTE carrier RSSI measurement bandwidth. The measurements in the numerator and denominator are made over the same set of resource blocks. While RSRP is an indicator of the wanted signal strength, RSRQ additionally takes the interference level into account due to the inclusion of RSSI. RSRQ therefore enables the combined effect of signal strength and interference to be reported in an efficient way [10].

13.4.2 UMTS FDD Measurements

13.4.2.1 UMTS FDD CPICH Received Signal Code Power (RSCP)

UTRA FDD CPICH RSCP is a UMTS measurement equivalent to LTE RSRP. This measurement is used mainly to rank different UMTS FDD candidate cells according to their signal strength and is used as an input for decisions on handover and cell reselection to UMTS. It is defined as the received power measured on the P-CPICH [11]. If transmit diversity is applied on the P-CPICH, the received code power from each antenna is measured separately and summed together in Watts to a total received code power.

13.4.2.2 UMTS FDD Carrier RSSI

This measurement is defined as the received wideband power including thermal noise and noise generated in the receiver, for the UMTS FDD carrier, within the bandwidth defined by the receiver pulse shaping filter [11].

13.4.2.3 UMTS FDD CPICH E_c/N_0

The UTRA FDD CPICH E_c/N_0 measurement is defined as the received energy per chip (E_c) on the P-CPICH of a given cell divided by the total noise power density (N_0) on the

UMTS carrier [11]. CPICH E_c/N_0 is used mainly to rank different UMTS FDD candidate cells according to their signal quality and is used as an input for handover and cell reselection decisions. If receive diversity is not in use by the UE, the CPICH E_c/N_0 is identical to CPICH RSCP divided by UMTS Carrier RSSI. If transmit diversity is applied on the P-CPICH the received E_c from each antenna must be separately measured and summed together (in Watts) to a total received energy per chip on the P-CPICH, before calculating the E_c/N_0.

13.4.3   UMTS TDD Measurements

13.4.3.1   UMTS TDD Carrier RSSI

This measurement is defined as the received wideband power, including thermal noise and noise generated in the receiver, within the bandwidth defined by the receiver pulse shaping filter, for UMTS TDD within a specified timeslot [12].

13.4.3.2   UMTS TDD P-CCPCH RSCP

P-CCPCH RSCP is defined as the received power on the P-CCPCH of a neighbour UMTS TDD cell. This measurement is used mainly to rank different UMTS TDD candidate cells according to their signal strength and is used as an input for handover and cell reselection decisions [12].

13.4.4   GSM Measurements

13.4.4.1   GSM Carrier RSSI

GSM RSSI is the wideband received power within the GSM channel bandwidth. This measurement is performed on a GSM BCCH carrier (i.e. a beacon carrier frequency).

13.4.5   CDMA2000 Measurements

13.4.5.1   CDMA2000 1x RTT Pilot Strength

This measurement is equivalent to LTE RSRP. It is therefore used mainly to rank different CDMA2000 1x candidate cells according to their signal strength and is used as an input for handover and cell reselection decisions. A detailed definition can be found in [13].

13.4.5.2   CDMA2000 HRPD Pilot Strength

This measurement is equivalent to LTE RSRP. Therefore, it is used mainly to rank different CDMA2000 HRPD candidate cells according to their signal strength and is used as an input for handover and cell reselection decisions. The CDMA2000 HRPD pilot strength measurement is defined in [14].

13.5 LTE Mobility in RRC_IDLE – Neighbour Cell Monitoring and Cell Reselection

From the protocol point of view, the handling of mobility in RRC_IDLE is explained in Section 3.3. This section focuses on the measurement and performance aspects.

All the LTE mobility procedures in RRC_IDLE state are performed autonomously within the UE and extreme care is taken when they are defined so that they enable sufficient UE mobility performance without unduly constraining power-efficient UE implementations. The detailed implementation of cell search and measurements in RRC_IDLE is therefore deliberately not specified in significant detail in order to enable different manufacturers to optimize their own implementations subject to meeting the specified minimum performance requirements. There are two main drivers for the UE behaviour being autonomous. The first is the minimization of transmission resource usage by inactive UEs. In fact, over-the-air signalling is only required for RRC_IDLE mobility if a tracking area update is needed (see Section 13.5.1). The second driver for autonomous UE behaviour is UE power saving.

The main objective in RRC_IDLE is to camp on a serving cell where UE paging reception is sufficiently reliable that a UE can be paged in the event of an incoming call; UE operations are restricted to periodic paging reception and serving cell monitoring. These are the only activities which a UE must perform while the serving cell quality is good enough. In order to maximize UE battery life, the UE is not required to perform any frequent neighbour cell-monitoring activity (cell search and measurements) unless the serving quality drops below a specified threshold. RRC_IDLE procedures are described in detail in Chapter 3 and [15], and associated performance requirements in [1].

13.5.1 Priority-Based Cell Reselection

Priority-based cell reselection is a novel principle adopted in LTE to improve the performance of cell reselection in the presence of multiple RATs which may coexist in the same geographical location. This approach reduces the need for a UE to monitor all the available intra-system frequency layers and inter-RAT carriers by monitoring them according to a set of priority rules provided to the UE. These rules are defined in [15]. The following steps are carried out every time a UE performs cell reselection and camps on a new cell while in RRC_IDLE:

- **Decode broadcast information.** A UE decides to camp on a cell if the cell suitability criteria (see Section 3.3) defined in [15] are met. When the UE camps on a cell it will receive all the relevant system information broadcast on the BCH [9]. This information includes parameters such as cell-specific configuration information (e.g. paging and random access parameters, cell bandwidth) and configurations for RRC_IDLE measurements and cell reselection (e.g. the serving cell minimum quality threshold, i.e. S-threshold, UMTS neighbour cell list, GSM neighbour cell list). Neighbour cell priority information is also provided to the UE through dedicated signalling (not broadcast).

- **Tracking Area Update.** If the UE has moved to a cell which belongs to a different tracking area,[4] it will establish a brief signalling connection with the eNodeB to inform

---

[4]A Tracking Area is a group of contiguous cells which helps to track UEs.

it about its new location within the new tracking area before it enters the paging reception stage.

- **Paging reception.** The UE determines the paging Discontinuous Reception (DRX) cycle and other relevant paging parameters by processing the serving cell paging parameters broadcast by the serving cell. From this point, and unless a UE is being paged or cell reselection occurs, the UE will periodically wake up on every paging occasion to check for paging messages; at the same time, the UE can measure its serving cell quality. Paging reception is explained in Section 3.4.

When the LTE serving cell signal quality is poor (e.g. below a certain RSRP threshold) the UE risks losing service on the serving cell, and it must attempt to identify and reselect a new suitable cell. Therefore, all possible networks need to be searched and measured regardless of their relative priority, and the UE must camp on the highest priority network which can be detected and meets the suitability criteria. Given that the serving cell quality is poor and the risk of losing service is high, the search rate will be frequent (i.e. a small multiple of the paging DRX period) [1].

If the serving cell quality is good enough, according to the S-criterion (see Section 3.3.3), then searching for lower priority layers does not need to be performed. The UE still needs to search for cells in higher priority layers at a reduced rate compared to the previous case and the UE must reselect a higher priority cell if it meets the cell reselection criteria. Given that the quality of the received signal from the current serving cell is good and the UE can be reached, the search rate can be far less frequent in order to reduce UE power consumption (typical values under consideration are in the order of 60 s).

- **Cell reselection evaluation.** On every paging occasion the cell reselection criterion is evaluated. If a neighbour cell currently being measured meets the cell reselection criterion then cell reselection towards that cell is initiated and the UE restarts the process named *cell-specific configuration* on the new cell.

13.5.2   Measurements in Idle Mode

The UE neighbour cell search, measurement rates and reselection evaluation rates are a function of the paging DRX cycle which has been configured, of the layer being measured (cell search and measurement rates tend to be higher for LTE intra-frequency cells, followed by LTE inter-frequency cells and inter-RAT cells) and a function of whether or not the serving cell quality is above a configured threshold. If the serving cell quality is poor, measurement of neighbour cells is relatively frequent (typically a small multiple of a DRX cycle). If the serving cell received quality is good then measurement rates can be far lower.

13.6   LTE Mobility in RRC_CONNECTED – Handover

During RRC_CONNECTED a UE is actively communicating with the network to transmit and receive user data. In this situation every effort must be made by the UE and the network to maintain the radio link, and therefore UE neighbour-cell monitoring performance is given

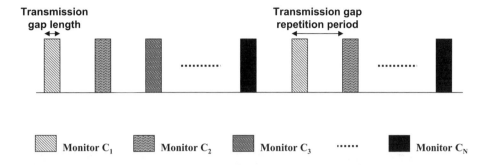

Figure 13.3 Monitoring gap pattern example (C_i is an E-UTRA carrier or a RAT).

priority over any power saving. As opposed to RRC_IDLE, in RRC_CONNECTED the UE cell search and measurements activity is always controlled and configured by the eNodeB. From the network perspective, this behaviour is explained in detail in Section 3.2.3.4.

When a better serving cell than the current one has been identified, the eNodeB will trigger handovers to other cells. UE handovers can be requested by the eNodeB to other cells on the same LTE carrier (intra-frequency handover), to LTE cells on other carriers (inter-frequency handover) and to other cells of a different RAT (inter-RAT handover).

The UE neighbour cell monitoring performance requirements and handover performance requirements during RRC_CONNECTED are defined in [1].

13.6.1 Monitoring Gap Pattern Characteristics

During RRC_CONNECTED mode, if the eNodeB decides that the UE needs to perform LTE inter-frequency and inter-RAT monitoring activities, it will provide the UE with a measurement configuration which includes a monitoring gap pattern sequence (an example of a monitoring gap pattern is given in Figure 13.3). Similar mechanisms exist in UMTS (known as 'Compressed Mode gaps' and 'FACH Measurement Occasions' depending on the state and capabilities of the UE) and in GSM (known as GSM Idle frames in GSM Dedicated and Packet Transfer Mode states). During the monitoring gaps, UE reception and transmission activities with the serving cell are interrupted. The main reasons for using monitoring gap patterns are as follows:

- The same LTE receiver can be used both to perform both intra-frequency monitoring and to receive data when there is no transmission gap.

- The presence of monitoring gaps allows the design of UEs with a single, reconfigurable receiver. A reconfigurable receiver can be used to receive data and to perform inter-RAT activity, but typically not simultaneously.

- Even if a UE has multiple receivers to perform inter-RAT monitoring activity (e.g. one LTE receiver, one UMTS receiver and one GSM receiver) there are some band configurations for which monitoring gaps are still required in the uplink direction. In particular these are useful when the uplink carrier used for transmission is immediately

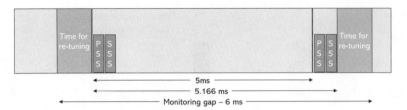

Figure 13.4 E-UTRAN inter-frequency and inter-RAT monitoring gaps, for the worst-case relative alignment between the E-UTRA target cell to be identified and the UTRA monitoring gap.

adjacent to the frequency band which the UE needs to monitor. There is always a significant power difference between the inter-RAT signal to be measured and the signal transmitted by the UE. The amount of receive filtering which can be provided, within the cost and size limitations of a UE, is not sufficient to filter out the transmitted signal at the input of the receiver front end, so the transmit signal leaks into the receiver band creating interference which saturates the radio front end stages. This interference desensitizes the radio receiver which is being used to detect inter-RAT cells.[5] Rather than address each scenario (i.e. each pair of frequency bands) with a specific solution, uplink gaps in LTE are configured in the same way for all scenarios.

LTE monitoring gap patterns contain gaps every N LTE frames[6] (i.e. the gap periodicity is a multiple of 10 ms) and these gaps have a 6 ms duration, as described in [1]. A single monitoring gap pattern is used to monitor all possible RATs (inter-frequency LTE FDD and TDD, UMTS FDD, GSM, TD-SCDMA, CDMA2000 1x and CDMA2000 HRPD).

Different gap periodicities are used to trade off between UE inter-frequency and inter-RAT monitoring performance, UE data throughput and efficient utilization of transmission resources. In general, cell identification performance increases as the monitoring gap density increases, while the ability of the UE to transmit and receive data decreases as the monitoring gap density increases.

Most RATs (LTE, UMTS FDD, TD-SCDMA, CDMA2000) broadcast sufficient pilot and synchronization information to enable a UE to synchronize and perform measurements within a useful period slightly in excess of 5 ms. This is because most RATs transmit downlink synchronization signals with a periodicity no lower than 5 ms.

For example, in LTE the PSS and SSS symbols are transmitted every 5 ms. Therefore a 6 ms gap provides sufficient additional headroom to retune the receiver to the inter-frequency LTE carrier and back to the serving LTE carrier and still to cope with the worst-case relative alignment between the gap and the cell to be identified. The worst-case relative alignment is depicted in Figure 13.4, where the LTE PSS and SSS symbols are time-aligned with the gap edges.

However, GSM requires special treatment because synchronization information is organized differently in the time domain. More detail is provided in Section 13.6.1.3.

---

[5] Receiver sensitivity considerations are described in detail in Chapter 22.

[6] Two gain pattern periods are defined in [1]. Further patterns may be defined in order to fulfil all required inter-RAT monitoring purposes.

13.6.1.1 LTE Intra-Frequency Monitoring

LTE intra-frequency monitoring aims to perform measurements both on the serving cell and on neighbouring cells which use the same carrier frequency as the serving cell. In order to be able to perform RSRP and RSRQ measurements the UE must first synchronize to and determine the cell ID of the neighbour cells. Contrary to previous 3GPP systems, if UE has to be able to perform the search without an explicit neighbour cell list being provided.

The intra-frequency measurement period is defined to be 200 ms.

Note that even when monitoring gap patterns are activated the vast majority of time (i.e. no less than 85–90% of the time) is available to perform intra-frequency monitoring. However, when DRX activity is enabled the UE must be able to take advantage of the opportunities to save power between subsequent DRX 'On periods' (see Section 4.4.2.5). Intra-frequency monitoring performance relaxations will only be defined for those cases where the periodicity of the 'On period' is larger than 40 ms. The amount of acceptable relaxation is still under discussion.

13.6.1.2 LTE Inter-Frequency Monitoring

LTE inter-frequency monitoring is very similar to intra-frequency monitoring apart from the fact that its achievable performance is constrained by the availability of monitoring gaps. For a 6 ms gap pattern only 5 ms is available for inter-frequency monitoring once the switching time has been removed. If the monitoring gaps repeat every 40 ms only $5/40 = 12.5\%$ is available for inter-frequency monitoring. For this reason LTE inter-frequency maximum cell identification time and measurement periods need to be longer than for the intra-frequency case.

As already discussed in Section 13.6.1, within one monitoring gap the presence of the PSS and SSS symbols is guaranteed since they repeat every 5 ms, and there are also sufficient RSs to perform power accumulation and obtain a measurement sample for RSRP calculation. There is also sufficient signal to perform an LTE carrier RSSI measurement to derive RSRQ.

The normal measurement bandwidth corresponds to the 6 central RBs of an LTE carrier (i.e. 1.08 MHz), which include the synchronization signals. An optional 50 RB configuration is also defined.

Similarly to the intra-frequency case, if the UE is in DRX mode some performance relaxation is required to ensure that the UE can take advantage of the DRX period to save power.

13.6.1.3 GSM Monitoring from LTE

As stated earlier in Section 13.6.1, GSM is the only RAT which does not provide sufficient information to synchronize to a cell with a single 6 ms monitoring gap.

This problem has been successfully addressed in the past for measuring and synchronizing to GSM cells from UMTS [5]. A monitoring gap pattern used for GSM monitoring must allocate time for three parallel activities: GSM RSSI measurements, initial BSIC identification and BSIC reconfirmation. This can be achieved by, for example, allocating every third monitoring gap to each one of the three activities.

Careful selection of the gap repetition period to a period which is a factor of 240 ms (i.e. 30, 40, 80, 120 and 240 ms) enables the detection of the GSM synchronization burst containing the BSIC and the SFN within a guaranteed maximum amount of time. The duration of a GSM control channel multiframe is 51 frames (each of which lasts 60/13 ms), and this is equal to 240 ms minus the duration of one frame:

$$51 \cdot \frac{60}{13} \text{ ms} = 240 - \frac{60}{13} \text{ ms}$$

Some simple analysis can be used to explain why a gap pattern containing 6 ms gaps repeating every 240 ms guarantees a maximum initial BSIC identification and BSIC reconfirmation time under good reception conditions. The maximum time always involves multiple gaps. In particular, any gap duration exceeding nine GSM timeslots (i.e. $9 \cdot (60/(13 \cdot 8))$ ms $= 5.19$ ms where 8 is the number of time slots in one GSM frame) is guaranteed to contain a GSM timeslot 0 regardless of the GSM frame boundary alignment relative to the monitoring gap. Timeslot 0 contains FB or SB, as shown in Figure 13.1. Once the receive switching overhead is added, a 6 ms gap is sufficiently large. Moreover, a gap pattern repeating every 240 ms will be guaranteed to observe the GSM FB or the GSM SB in at most 11 consecutive gaps because there is a shift of one GSM frame with respect to the GSM 51-frame control multiframe between two adjacent monitoring gaps. Thus, a single-step BSIC reconfirmation is guaranteed not to require more than 11 consecutive gaps since all it requires is decoding the SB. For the same reason, both the FB and the SB can be observed in no more than 12 consecutive gaps (i.e. $12 \cdot 240 = 2880$ ms). Therefore, the two-step initial BSIC identification single attempt is guaranteed not to require more than 2880 ms. Single-step initial BSIC reconfirmation requires the same time as BSIC reconfirmation.

More complicated analysis must be performed in order to determine the worst-case initial BSIC identification and BSIC reconfirmation times when using other monitoring gap periodicities (i.e. 40, 80 and 120 ms).

13.6.1.4 UMTS Monitoring from LTE

UMTS monitoring is performed within the available monitoring gaps. The three relevant UMTS physical channels required to perform UMTS cell identification and measurements are P-SCH, S-SCH and CPICH and are guaranteed to be present within a 6 ms gap. The CPICH is used for both the last stage of UMTS cell identification and also to perform CPICH-based measurements (RSCP and E_c/N_0).

13.6.2 Measurement Reporting

Two different types of measurement reporting are specified in LTE [9]:

- **Periodic reporting.** Measurement reports are configured to be reported periodically according to the measurement configuration parameters.

- **Event-triggered measurement reporting.** In order to limit the amount of signalling to be sent back to the eNodeB the measurement reporting activity can be configured to trigger measurement reports given that some conditions are met by the measurements performed by the UE. Typically, the reporting conditions relate to criteria used by the network to start neighbour cell measurements or trigger the execution of handovers due to poor cell coverage or poor service quality. Once an event has been met, the UE can be configured to report additional measurement information which is unrelated to the event condition. This information is used by the eNodeB RRM algorithms to determine the most adequate handover command. This is explained in more detail in Section 3.2.5.2.

13.6.3 Handover to LTE

When the RRC entity in the eNodeB decides to initiate a handover it sends a 'MOBILITY FROM E-UTRA' RRC message to the UE. This message contains the target RAT and frequency and the relevant parameters required for the UE to establish a radio link with the target handover cell, as described in Section 3.2.4.1. Handovers can be initiated by the eNodeB for several different reasons:

- **Quality-based handovers.** Typically these handovers are initiated as a result of a UE measurement report indicating that the UE can communicate with a neighbour cell with a better channel quality than that of the current serving cell.

- **Coverage-based handovers.** These handovers move the connection to another RAT because the UE is losing coverage for the current RAT. For example, a UE could be moving away from an urban area and losing LTE coverage. As a result the network hands over the connection to the second preferred RAT which the UE has detected, such as UMTS or GSM.

- **Load-based handovers.** These handovers are performed by the network in order to spread the load more evenly between different RATs belonging to the same operator when a given cell is overloaded. For example, if an LTE cell is congested then some users may be moved to nearby LTE cells or nearby UMTS cells.

Figure 13.5 [16] illustrates the major steps involved in executing a handover between two LTE cells (see also Section 3.2.3.4). The steps are as follows:

- The UE generates and transmits a measurement report to the eNodeB. Within this measurement report there is a measurement for one target cell with higher RSRP level than the current serving cell.

- The source eNodeB decides that a handover is necessary, identifies a suitable target cell (assumed in the following to be an LTE cell) and requests access to the target eNodeB controlling the target cell.

- The target eNodeB accepts the handover request and provides the source eNodeB with the parameters required for the UE to access the target cell once the handover has been executed, including the cell ID, carrier frequency and allocated uplink and downlink resources.

- The source eNodeB sends a 'MOBILITY FROM E-UTRA' RRC message to the UE.

- The UE receives the message, interrupts the radio link with the source eNodeB and initiates establishment of the new radio link with the target eNodeB. During this period data transmission is interrupted. There are a number of steps involved, including the following:

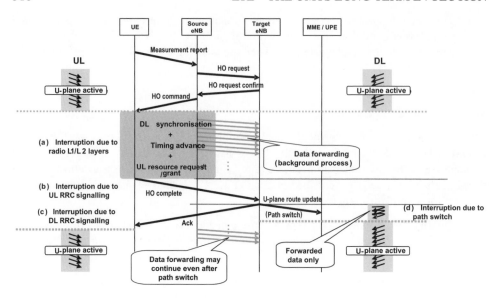

Figure 13.5 E-UTRAN handover. Reproduced by permission of © 3GPP.

– **Downlink synchronization establishment.** If the UE has been unable to synchronize to the target cell prior to receiving the 'MOBILITY FROM E-UTRA' RRC message (blind handover) then the UE will have to perform the relevant LTE synchronization steps. Even in the case when the UE had previously synchronized with the target cell, downlink synchronization is only considered established when the downlink RS quality is sufficiently good. From this point onwards data reception in the downlink from the target eNodeB may take place.

– **Timing advance.** This is provided by the target eNodeB based on the received delay measured on the Physical Random Access CHannel (PRACH) (see Section 20.2).

– **Data transmission.** The UE starts transmitting uplink data towards the target eNodeB.

In the meantime the source eNodeB may forward the UE data to the destination eNodeB (this depends on the quality of service provided).

• As soon as uplink activity has been established with the target eNodeB, an RRC message is sent to the target eNodeB to notify it that handover has been completed.

• The target eNodeB notifies the Mobility Management Entity (MME) that the handover is successful and the MME reroutes the downlink data to the target eNodeB and acknowledges the 'MOBILITY FROM E-UTRA' command back to the UE.

• When the UE receives the handover command acknowledgement, the handover is considered finished.

The above steps represent an overview of a best-case scenario when all the steps in the procedure occur as desired. There are a number of possible failure cases which can necessitate the initiation of recovery procedures; for example, the preferred target eNodeB may have no spare resources to grant access to the UE, downlink synchronization might fail, or uplink synchronization might fail because there is not sufficient uplink power. A detailed description of all these possible scenarios can be found in the RRC protocol specification in [9].

13.6.3.1 Differences between LTE Intra- and Inter-Frequency Handover

Both intra-frequency and inter-frequency handovers between LTE cells are hard handovers. Because of this, the steps involved for both handover types are very similar.

The handover interruption time is defined as the time between the end of the last TTI in which the UE has received the handover command on the Physical Downlink Control CHannel (PDCCH) and the time at which the UE is ready to start a PRACH transmission to the new uplink. The interruption time can be broken down into the following contributions:

$$T_{\text{interrupt}} = T_{\text{search}} + T_{\text{IU}} + 20 \text{ ms}$$

where T_{search} is the time required to find the target cell when the target cell is not already known when the handover command is received by the UE. If the target cell is known, then $T_{\text{search}} = 0$ ms. T_{IU} is the interruption uncertainty to locate the first available PRACH occasion in the new cell. 20 ms is added to allow for UE processing time required to execute the handover.

13.6.4 Handover to UMTS

Handover to UMTS can be decomposed into two separate stages, each of which relates to an identical stage in a single-mode handover procedure:

- **Handover initiation.** From a UE perspective this stage is no different to handover initiation in the case of LTE intra- or inter-frequency handover.

- **Radio link establishment to the target UMTS cell.** This stage is identical to UMTS inter-frequency handover. For this reason, handover execution delay requirements in [1] are very similar to intra-UMTS hard handover requirements in [5].

As in the handover to LTE case, handover to UMTS may be blind or guided depending on whether or not the UE has been able to synchronize to the target cell prior to receiving the 'MOBILITY FROM E-UTRA' RRC message.

13.6.5 Handover to GSM

Handover to GSM can also be decomposed into two separate stages, each of which relates to an identical stage in a single-mode handover procedure:

- **Handover initiation.** From a UE perspective this stage is no different to the handover initiation in the case of LTE intra- or inter-frequency handover.

- **Radio link establishment to the target GSM cell.** This stage is identical to a GSM intra-system unsynchronized handover.

The handover to GSM interruption time can be decomposed into three main contributions:

$$T_{\text{interrupt GSM}} = T_{\text{processing time}} + T_{\text{sync, blind}} + T_{\text{interruption, guided}}$$

where $T_{\text{processing time}} = 50$ ms allows different implementations to reconfigure the modem for GSM operation. $T_{\text{sync, blind}}$ is the additional time allowance to enable a UE to synchronize to the target GSM cell during a blind handover to GSM (100 ms). For a guided handover this component does not apply (i.e. equals 0 ms). $T_{\text{interruption, guided}} = 40$ ms is the worst-case interruption time for a guided handover to GSM where the UE has been able to synchronize to the target GSM cell before receiving the 'MOBILITY FROM E-UTRA' RRC message.

13.7 Concluding Remarks

The provision of seamless mobility is a key differentiator for cellular systems such as LTE, in order to deliver an uninterrupted mobile user experience. The support for such functionality in LTE not only covers LTE cells on the same frequency, but also extends to cells on different frequencies and cells using a wide range of other radio access technologies. A wide range of measurements and signalling are defined in LTE to support these different handover scenarios. These aspects are carefully designed taking into account their potential impact on other aspects of the user's experience, such as UE power consumption and data interruption.

References[6]

[1] 3GPP Technical Specification 36.133, 'Evolved Universal Terrestrial Radio Access (E-UTRA); Requirements for Support of Radio Resource Management(Release 8)', www.3gpp.org.

[2] 3GPP Technical Specification 25.213, 'Technical Specification Group Radio Access Network; Spreading and Modulation (FDD)', www.3gpp.org.

[3] H. Holma and A. Toskala, *WCDMA for UMTS*, 2nd edition. John Wiley & Sons, Ltd/Inc. 2002.

[4] M. Mouly and M.-B. Pautet, *The GSM System for Mobile Communications*. Cell & Sys, 1992.

[5] 3GPP Technical Specification 25.133, 'Technical Specification Group Radio Access Network; Requirements for Support of Radio Resource Management (FDD)', www.3gpp.org.

[6] 3GPP Technical Specification 36.214, 'Evolved Universal Terrestrial Radio Access (E-UTRA); Physical Layer – Measurements (Release 8)', www.3gpp.org.

[7] 3GPP Technical Specification 25.302, 'Technical Specification Group Radio Access Network; Services Provided by the Physical Layer', www.3gpp.org.

[8] 3GPP Technical Specification 36.213, 'Evolved Universal Terrestrial Radio Access (E-UTRA); Physical Layer Procedures (Release 8)', www.3gpp.org.

[9] 3GPP Technical Specification 36.331, 'Evolved Universal Terrestrial Radio Access (E-UTRA); Radio Resource Control (RRC); Protocol Specification', www.3gpp.org.

[10] Ericsson, 'R1-073041: Reference Signal Received Quality, RSRQ, Measurement', www.3gpp.org, 3GPP TSG RAN WG1, meeting 49bis, Orlando, USA, June 2007.

[11] 3GPP Technical Specification 25.215, 'Technical Specification Group Radio Access Network; Physical Layer – Measurements (FDD)', www.3gpp.org.

---

[6]All web sites confirmed 18th December 2008.

[12] 3GPP Technical Specification 25.225, 'Technical Specification Group Radio Access Network; Physical Layer – Measurements (TDD)', www.3gpp.org.

[13] 3GPP2 CS.0005-D, 'Upper Layer (Layer 3) Signaling Standard for CDMA2000 Spread Spectrum Systems Release D', www.3gpp2.org.

[14] 3GPP2 CS.0024-A, 'cdma2000 High Rate Packet Data Air Interface Specification', www.3gpp2.org.

[15] 3GPP Technical Specification 36.304, 'Evolved Universal Terrestrial Radio Access (E-UTRA); User Equipment (UE) Procedures in Idle Mode', www.3gpp.org.

[16] 3GPP Technical Report 25.912, 'Feasibility study for evolved Universal Terrestrial Radio Access (UTRA) and Universal Terrestrial Radio Access Network (UTRAN)', www.3gpp.org.

14

Broadcast Operation

Olivier Hus and Matthew Baker

14.1 Introduction

It was an early requirement of the Long-term Evolution (LTE) specifications [1] to support an enhanced version of the *Multimedia Broadcast/Multicast Service* (MBMS) already available in UTRAN[1] and GERAN.[2] MBMS aims to provide an efficient mode of delivery for both broadcast and multicast services over the core network.

We first give a general overview of MBMS as a broadcast and multicast delivery system, before reviewing in detail the enhancements to MBMS which can be supported by LTE compared to UMTS Release 6. In particular, we describe how LTE is able in principle to benefit from the new OFDM downlink radio interface to achieve radically improved transmission efficiency and coverage by means of multicell 'Single Frequency Network' operation.

While most of these aspects are already well-developed in the first release of the LTE specifications (Release 8), it was in fact necessary in the specification process in 3GPP to prioritize some features over others, in order to expedite the commercial deployment of LTE. The MBMS specifications for LTE are therefore not included in the first release of LTE. Nevertheless the LTE physical layer has already been designed to support MBMS, and essential components are implemented in Release 8 to ensure forward-compatibility for full MBMS operation when the feature is finalized in later releases.

---

[1]UTMS Terrestrial Radio Access Network.
[2]GSM Edge Radio Access Network.

14.2   Broadcast Modes

14.2.1   Broadcast and Multicast

In the most general sense, broadcasting is the distribution of content to an audience of multiple users; in the case of mobile multimedia services, typical broadcast content can include newscasts, weather forecasts or live mobile television. As an example, Figure 14.1 illustrates the reception of a video clip showing a highlight of a sporting event.

Figure 14.1 Receiving football game highlights as a mobile broadcast service. Adapted from image provided courtesy of Philips 'Living Memory' project.

The distribution of mobile multimedia services requires an efficient transmission system for the simultaneous delivery of content to large groups of mobile users.

There are three possible types of transmission to reach multiple users:

- **Unicast:** A bidirectional point-to-point transmission between the network and each of the multiple users; the network provides a dedicated connection to each terminal, and the same content is transmitted multiple times to individual users that request it.

- **Broadcast:** A downlink-only point-to-multipoint connection from the network to multiple terminals; the content is only transmitted once to all terminals in a geographical area, and users are free to choose whether or not to receive it.

- **Multicast:** A downlink-only point-to-multipoint connection from the network to a managed group of terminals; the content is only transmitted once to the whole group, and only users belonging to the managed user group can receive it.

The delivery of identical content to multiple users via unicast connections (point-to-point) is usually a very inefficient method of transmission, both because of the waste of radio resources (especially for high-bandwidth for multimedia applications) and because of the signalling burden in the core network. Point-to-multipoint transmission modes lessen the load on network resources by decreasing the total amount of data transmitted and reducing signalling procedures significantly, particularly in the uplink.

At the physical layer, there is no difference between broadcast and multicast downlink data transmissions. The difference between the two modes lies in the subscriber base: broadcast services are available to all users without the need for subscriptions to particular services, whereas availability of multicast services is restricted so that individual users need to subscribe in order to be able to receive selected services. Multicasting can thus be seen as 'broadcast via subscription', with the possibility of charging for the subscription.

14.2.2   UMTS Release 6 MBMS Service and Delivery System

In UMTS, MBMS was first specified in Release 6 for the delivery of content in a point-to-multipoint mode using the 3GPP core network and the UMTS Terrestrial Radio Access Network (UTRAN).

Figure 14.2 shows the architectural model for the full Release 6 MBMS 'content chain' – i.e. the path from the content provider to the MBMS receiver at the User Equipment (UE).

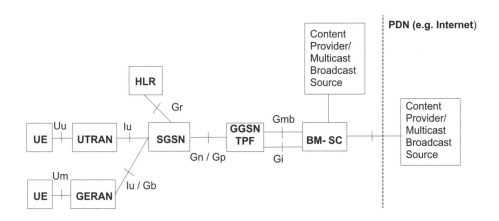

Figure 14.2   UMTS Release 6 MBMS Network Architecture Model [2]. Reproduced by permission of © 3GPP.

The full MBMS service and delivery system can be described as being made up of three subsystems, outlined in the following sections.

14.2.2.1   Content Provision

Sources of MBMS content vary, but are normally assumed to be external to the core network; one example of external content providers would be television broadcasters for mobile television. The MBMS content is generally assumed to be IP based, and by design the MBMS system is integrated with the IP Multimedia Subsystem (IMS) [3] service infrastructure based on the Internet Engineering Task Force (IETF) Session Initiation Protocol (SIP) [4].

The interface between the external content providers and the core network is the Broadcast-Multicast Service Centre (BM-SC). The main functions provided by the BM-SC are:

- Reception of MBMS content from external content providers;

- Providing application and media servers for the mobile network operator;

- Scheduling of MBMS services and transport of MBMS data into the core network;

- Management of user group membership, subscription and charging for sessions.

14.2.2.2 Core Network

In UMTS, the Gateway GPRS Support Node (GGSN) is the entry point into the core network for MBMS multicast or broadcast traffic. This is connected to the Serving GPRS Support Node (SGSN), which is the entry point into the Radio Access Network (RAN), either UTRAN or GERAN, and supports mobility procedures. The SGSN also performs control functions for each individual user: storage of user-specific information for each activated MBMS service and generation of charging data for each user (for handling by the BM-SC).

14.2.2.3 Radio Access Network – UTRAN/GERAN and UE

The RAN, whether UTRAN or GERAN, is responsible for delivering MBMS data efficiently to the designated MBMS service area. It performs the following functions:

- Choice of appropriate radio bearer based on the number of users within a cell;

- Transmission of MBMS service announcements;

- Initiation and termination of MBMS transmissions;

- Support of user mobility.

In UMTS Release 6 the transmission of MBMS is performed by single cells operating with different scrambling codes, although signals from neighbouring cells can often be combined to increase gain at reception. In Release 7, UMTS MBMS is enhanced to allow multiple cells transmitting the same MBMS service to do so using the same scrambling code in a synchronized manner; this allows joint equalization of the multicell transmissions in the UE's receiver in a manner which is to some extent analogous to the Single Frequency Network method by which MBMS transmission occurs in LTE, as described in Section 14.3.

The final element in the MBMS content and service chain is the MBMS receiver itself: the UE.

14.3   MBMS in LTE

An initial LTE design requirement was to support an enhanced version of MBMS compared to UMTS Release 6. The targets included a cell edge spectrum efficiency in an urban or suburban environment of 1 bps/Hz – equivalent to the support of at least 16 Mobile TV channels at around 300 kbps per channel in a 5 MHz carrier. This is only achievable by exploiting the special features of the LTE OFDM air interface in a Single Frequency Network mode, as described in Section 14.3.1.

It was also recognized that the user experience is not purely determined by the data rate achieved, but also by other factors such as the interruption time when switching channels. This has implications for the design of the MBMS control signalling, which is also being extensively redesigned for LTE.

14.3.1   Single Frequency Network for MBMS

A key new feature of LTE is the possibility to exploit the OFDM radio interface to transmit multicast or broadcast data as a multicell transmission over a synchronized Single Frequency Network: this is known as *Multimedia Broadcast Single Frequency Network (MBSFN)* operation.

14.3.1.1   Transmission of Data in MBSFN Operation

In MBSFN operation, MBMS data is transmitted simultaneously over the air from multiple tightly time-synchronized cells. A UE receiver will therefore observe multiple versions of the signal with different delays due to the multicell transmission. Provided that the transmissions from the multiple cells are sufficiently tightly synchronized for each to arrive at the UE within the cyclic prefix at the start of the symbol, there will be no InterSymbol Interference (ISI). In effect, this makes the MBSFN transmission appear to a UE as a transmission from a single large cell, and the UE receiver may treat the multicell transmissions in the same way as multipath components of a single-cell transmission without incurring any additional complexity. This is illustrated in Figure 14.3. The UE does not even need to know how many cells are transmitting the signal.

This Single Frequency Network reception leads to significant improvements in spectral efficiency compared to UMTS Release 6 MBMS, as the MBSFN transmission greatly enhances the SINR. This is especially true at the cell edge, where transmissions which would otherwise have constituted inter-cell interference are translated into useful signal energy – hence the received signal power is increased at the same time as the interference power being largely removed.

An example of the improvement in performance achievable using MBSFN transmission compared to single-cell point-to-multipoint transmission is shown in Figure 14.4. In this example, the probability of achieving a randomly-located UE not being in outage (defined as MBMS packet loss rate < 1%) is plotted against spectral efficiency of the MBMS data transmissions (a measure of MBMS data rate in a given bandwidth). A hexagonal cell-layout is assumed, with the MBSFN area comprising 1, 2 or 3 rings around a central cell for which the performance is evaluated. It can be seen that the achievable data rates increase markedly as the size of the MBSFN area is increased and hence the surrounding inter-cell interference

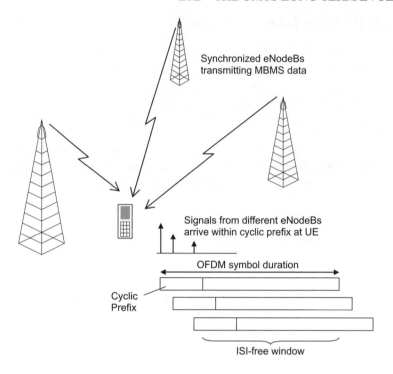

Figure 14.3 ISI-free operation with MBSFN transmission.

is reduced. A 1 km cell radius is assumed, with 46 dBm eNodeB transmission power, 15 m eNodeB antenna height and 2 GHz carrier frequency.

MBSFN data transmission takes place via the Multicast Channel (MCH) transport channel, which is mapped to the Physical Multicast Channel (PMCH) introduced in Section 9.2.3. In addition to some specific aspects of the corresponding control channel design (discussed in Section 9.2.3), the key features of the PMCH to support MBSFN transmission are:

- The extended Cyclic Prefix (CP) is used ($\sim$17 μs instead of $\sim$5 μs). As the differences in propagation delay from multiple cells will typically be considerably greater than the delay spread in a single cell, the longer CP helps to ensure that the signals remain within the CP at the UE receivers, thereby reducing the likelihood of ISI. This avoids introducing the complexity of an equalizer in the UE receiver, at the expense of a small loss in peak data rate due to the additional overhead of the longer cyclic prefix.

- The Reference Signal (RS) pattern is modified compared to non-MBSFN data transmission, as shown in Figure 14.5. The reference symbols are spaced more closely in the frequency domain than for non-MBSFN transmission, reducing the separation to every other subcarrier instead of every sixth subcarrier. This improves the accuracy of the channel estimate which can be achieved for the longer delay spreads. The channel estimate obtained by the UE from the MBSFN RS is in fact a composite channel

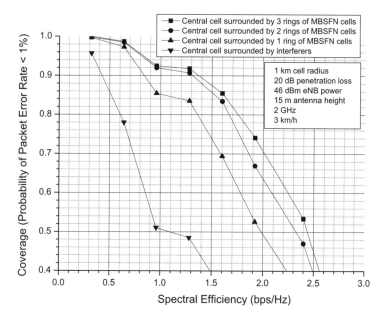

Figure 14.4   Reduction in total downlink resource usage achievable using MBSFN transmission [5]. Reproduced by permission of © 2007 Motorola.

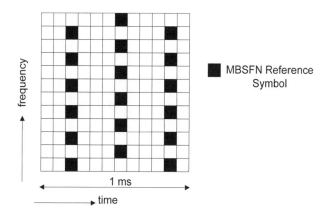

Figure 14.5 MBSFN reference symbol pattern for 15 kHz subcarrier spacing. Reproduced by permission of © 3GPP.

estimate, representing the composite channel from the set of cells transmitting the MBSFN data.

In addition to these enhancements for MBSFN transmission, a second OFDM parameterization is provided in LTE specifically for downlink-only multicast/broadcast transmissions.

This has an even longer CP, double the length of the extended CP, resulting in approximately 33 µs. This is designed to cater for deployments with very large differences in propagation delay between the signals from different cells (typically up to 10 km). This is most likely to occur for deployments at low carrier frequencies and large inter-site distances.

In order to avoid further increasing the overhead arising from the CP in this case, the number of subcarriers per unit bandwidth is also doubled, giving a subcarrier spacing of 7.5 kHz. The cost of this is an increase in inter-carrier interference, especially in high-mobility scenarios with a large Doppler spread. In choosing whether to use the 7.5 kHz subcarrier spacing, there is therefore a trade-off between support for wide-area coverage and support for high mobile velocities. It should be noted, however, that the maximum Doppler shift is lower at the low carrier frequencies which are likely to be used in the typical deployment scenario for the 7.5 kHz subcarrier spacing. The absolute frequency spacing of the reference symbols for the 7.5 kHz is the same as for the 15 kHz subcarrier spacing MBSFN pattern, resulting in a RS on every fourth subcarrier.

14.3.1.2   Time-synchronization in MBSFN Operation

We use the term *MBSFN area* to refer to the area within which cells transmit the same content for MBSFN combining at the UEs.

The transmissions from the multiple cells (eNodeBs) in an MBSFN area must be tightly time-synchronized with an accuracy of a few µs to achieve symbol-level alignment within the CP. The method of achieving symbol-level synchronization is not defined in the LTE specifications; this is left to the implementation of the eNodeB. Typical implementations are likely to use satellite-based solutions (e.g. GPS) or possibly synchronized backhaul protocols (e.g. [6]).

In addition, there is also a requirement for the content of the time-synchronized MBSFN physical resource blocks to be aligned; this requires synchronization at the service provision level, to avoid resource blocks containing different data causing interference at the receiver. A synchronization mechanism for MBMS content is therefore also necessary: this is described in Section 14.3.3.1.

Physical resources are allocated to a specific MCH by specifying a pattern of subframes: the MCH Subframe Allocation Pattern (MSAP). One MCH contains data belonging to only one MBSFN Area. MCHs from different MBSFN areas are time multiplexed onto different subframes.

14.3.2   MBMS Deployment

The deployment of MBMS in LTE is supported in various configurations of geographical cell distribution, carrier frequency allocation (known as 'frequency layers') and transmission modes.

14.3.2.1   Areas Related to MBMS

A geographical area of the network where MBMS can be transmitted is called an *MBMS Service Area*.

A geographical area where all eNodeBs can be synchronized and can perform MBSFN transmissions is called an *MBSFN Synchronization Area*.

Within an MBSFN Synchronization Area, a group of cells that are coordinated for an MBSFN transmission is called an *MBSFN Area*. An MBSFN Synchronization Area may support several MBSFN Areas and a cell within an MBSFN Synchronization Area may form part of multiple MBSFN Areas, each characterized by different transmitted content and a different set of participating cells.

MBSFN Synchronization Areas are quite independent from the definition of MBMS Service Areas.

Figure 14.6 illustrates some possible different scenarios for the deployment of MBMS services and MBSFN transmissions, with an MBMS Service Area that includes:

- An MBSFN Synchronization Area supporting one independent MBSFN Area (A) and two overlapping MBSFN Areas (B and C).

- Independent cells for non-MBSFN MBMS transmission.

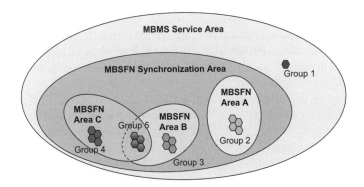

Figure 14.6 Example of deployment scenarios for MBMS Service Areas and MBSFN areas.

Different groups of cells (that transmit MBMS services) are labelled in Figure 14.6 to illustrate some examples of different modes of operation:

- Group 1: an individual cell transmitting MBMS services in an MBMS Service Area; this cell does not belong to an MBSFN Synchronization Area and cannot therefore operate in MBSFN transmission mode. It is worth noting that although there are significant benefits of MBSFN operation as explained in Section 14.3.1 above, it is intended that single-cell MBMS transmissions may still be used in LTE. Single-cell MBMS transmissions do not require the synchronization mechanism necessary for MBSFN operation and therefore provide an alternative deployment possibility for cases when tight synchronization cannot be achieved. Single-cell MBMS transmissions use the normal downlink shared channel. The selection between MBSFN transmission and single-cell transmission is performed by the MultiCell/Multicast Coordination Entity (MCE) at the time of set-up (or network reconfiguration) of the MBMS Service Area or individual cells for the transmission of MBMS services – see Section 14.3.3.1.

- Groups 2–5 belong to an MBSFN Synchronization Area and may operate in MBSFN transmission mode within their respective MBSFN Areas:

  – Groups 2, 3 and 4 may perform MBSFN transmissions in their respective MBSFN Areas A, B and C;

  – Group 5 is formed of cells that belong to overlapping MBSFN Areas A and B: the signalling of the availability of MBMS services and the management of physical resources for MBSFN transmissions in this group requires a more complex configuration than for groups 2, 3 and 4.

Overlapping MBSFN areas. Different MBMS services may have different service areas, since MBMS services may target different geographical areas or different user distributions; this has implications on the configuration and resource management of cells in MBSFN operation.

When multiple MBSFN Areas overlap, the overlapped area (illustrated by cells in Group 5 in Figure 14.6) would require the allocation of separate resources and signalling to support the different MBMS services transmitted in the different areas simultaneously in the overlapping area. This may restrict the resources available for other services in each MBSFN Area, as the non-overlapping parts of an MBSFN area cannot reuse resources which are used by a different MBSFN area in the overlap region.

One possibility for the finalization of LTE MBMS in a future release of the specifications could be to facilitate coordination of the transmission of services from different overlapping MBSFN Areas, by multiplexing joint services in the overlapped area; in effect, this would be equivalent to assigning an MBSFN Area to the smallest common denominator area. This could, however, result in a profusion of smaller MBSFN Areas and could add complexity to the assignment of resource blocks and to the combining of multicell transmissions by a receiver.

Configuration of MBSFN Areas. When MBMS in LTE is finalized, it will be necessary to support the configuration of which cells comprise each MBSFN area.

Such configuration may be static, such that the mapping of MBSFN services, areas and cells is fixed at the time of set-up of MBSFN services.

Alternatively, semi-static configuration using RRC signalling may be used, giving flexibility to the network operators for managing MBSFN areas. The allocation of MBSFN areas would typically be modified at the start and stop of MBMS sessions, with minimal changes being possible during the lifetime of the service.

More dynamic configuration of MBSFN areas may also be considered, enabling the set of cells belonging to an MBSFN area to be configured more frequently depending on the needs of particular MBMS services to be transmitted. Although a more dynamic MBSFN configuration would allow the tailoring of radio resource efficiency to the number of users in specific cells, in practice the usefulness of more frequent reconfigurations is likely to be limited, as it would require the gathering of information on the number of users for MBSFN Area cells through a counting procedure across the MBSFN Area, in order to determine in which cell MBSFN transmission should be turned on or off or whether to trigger the addition of neighbouring cells. This would require significant resource management effort in a live network.

14.3.2.2 Dedicated and Mixed Carriers

LTE is designed to support MBMS either on a dedicated carrier in which all subframes are used for MBSFN transmission or on a mixed MBSFN/unicast carrier which is shared between MBMS and unicast services. In the latter case, the MBSFN and unicast services are time-multiplexed using different subframes. Certain subframes are not allowed to be used for MBSFN transmission: subframes 0, 4 and 5 in each 10 ms radio frame are reserved for unicast transmission in order to avoid disrupting the synchronization signals, and to ensure that sufficient common RS are available for decoding the broadcast system information.

A dedicated MBMS carrier is for downlink transmissions only, and cannot support uplink connections (as there is no Physical Downlink Control Channel (PDCCH) to carry the control signalling for the uplink).

14.3.2.3 Handover

Appropriate handover procedures need to be provided to maintain MBMS service continuity as a UE moves across cell boundaries and MBMS Service Areas.

Four different scenarios exist for handover (i.e. cell reselection) in the context of MBMS service continuity:

- Intra-frequency handover within an MBSFN area on a dedicated MBMS carrier: by design, the service continuity is seamless as the terminal sees the whole service area as a single cell;

- Inter-frequency handover between MBSFN operation on a dedicated MBMS carrier and MBMS transmission on a mixed MBMS/unicast carrier: this is a difficult scenario as the lack of an uplink channel for the dedicated MBMS carrier will prevent the usual handover communication with UEs;

- Intra- or inter-frequency handover between MBSFN operation on a mixed MBMS/ unicast carrier and MBMS single-cell transmission on a mixed MBMS/unicast carrier: handover would be possible through the unicast connection; however, this may be an unrealistic solution when the number of receivers of MBSFN transmissions is high and a mechanism would be needed to ensure that resources are not overloaded;

- Intra- or inter-frequency handover between MBMS single-cell transmissions on a mixed MBMS/unicast carrier: this is a relatively simple scenario as handover procedures would be very similar to normal unicast handover.

In summary, handover procedures to exit MBSFN mode on a dedicated layer are likely to be tricky. Inter- or intra-frequency handover procedures from a mixed MBMS/unicast cell would follow the traditional handover procedures.

14.3.2.4 Notification for Non-MBMS Services

A requirement for the design of LTE [1] was that it supports concurrent MBMS and non-MBMS services for a smooth user experience: a user receiving one or several MBMS services should be notified when a dedicated service (voice, data) is incoming and should be able to take the call or get the data.

The notification mechanism considered for MBMS is paging. On a mixed MBMS/unicast carrier, normal paging procedures apply as described in Section 9.2.2.2. On a dedicated MBSFN carrier, notification is difficult as the location of the UE is not known since there is no uplink channel, and the paging area may be very large. In general this is a significant advantage of using mixed MBMS/unicast carriers, in that paging and simultaneous support of non-MBMS services is possible without necessarily requiring a dual receiver to be implemented in the UE. This is discussed further in Section 14.4.1.

14.3.3   MBMS Architecture and Protocols

This section describes the architectural principles and protocols specific to MBMS, beyond those already covered in Chapters 2–4.

14.3.3.1   MBMS Logical Architecture and User Structure

The management of both MBMS content and resources will be performed through a Multicell/Multicast Coordination Entity (MCE), as shown in Figure 14.7. This is a new node designed to coordinate the transmissions from multiple cells, which would otherwise be difficult to achieve in the flat architecture of LTE. The role of the MCE includes allocating the time/frequency radio resources used by all eNodeBs in the MBSFN area, ensuring that the same resource blocks are used across the whole MBSFN area for a given service, and deciding the radio configuration (modulation and coding scheme). Thus for MBMS the radio scheduling and configuration roles which are normally the responsibility of the eNodeBs are instead centralized.

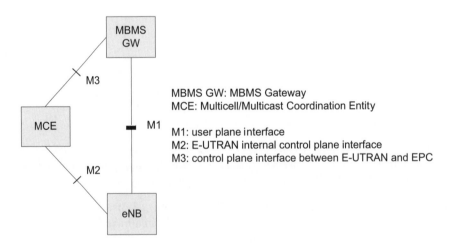

Figure 14.7   Network architecture to support MBMS in LTE. Reproduced by permission of © 3GPP.

An additional logical entity called the MBMS Gateway (MBMS GW) is also defined. It receives user-plane MBMS traffic from the Broadcast/Multicast Service Centre (BM-SC), and, in contrast to non-MBMS traffic, hosts the PDCP[3] layer of the user plane for header compression for MBMS data packets for both multicell and single-cell transmission. The MBMS GW then forwards the user-plane traffic to the eNodeBs.

Each eNodeB hosts the RLC and MAC layers of the protocol stack, including the segmentation and reassembly functions. The other functions which remain to be hosted by the eNodeB for MBMS mainly relate to single-cell MBMS operation.

The interfaces between the MBMS logical network entities are as follows:

- **M1 interface (MBMS GW – eNodeB):** A pure user plane interface; no control plane application part is defined for this interface; IP multicast is used for point-to-multipoint delivery of user packets for both single cell and multicell transmission. The SYNC protocol [7] is used to ensure that content is synchronized for multicell MBSFN transmission. This carries additional information which enables eNodeBs to identify the radio frame timing and to detect packet loss.[4] The stringent timing requirements of the SYNC protocol would not apply to eNodeBs which only transmit MBMS in single-cell mode. The integration of the SYNC protocol layer into the user-plane architecture for MBMS content synchronization is shown in Figure 14.8.

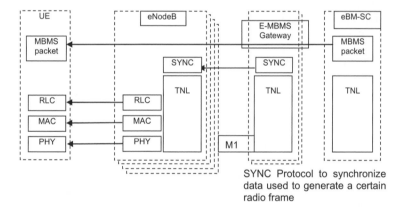

Figure 14.8   User plane architecture including SYNC protocol for MBMS content synchronization. Reproduced by permission of © 3GPP.

- **M2 interface (MCE – eNodeB):** A pure control-plane interface which conveys the session control signalling to the eNodeB, adding the necessary radio configuration data for multicell MBSFN transmission. This additional radio configuration data ensures that the RLC/MAC entities located in the eNodeB are configured appropriately and consistently in order to deliver synchronized content. The SCTP protocol [8] is used over the M3 interface to carry the application part.

---

[3]Packet Data Convergence Protocol – see Section 4.2.
[4]The details of the SYNC protocol remain to be defined in a later release of LTE, but it is expected to be based on timestamps.

- **M3 interface (MCE – MBMS GW):** A pure control plane-interface that carries the session control signalling on the SAE[5] bearer level, including MBMS 'Session Start' and 'Session Stop' messages. The Session Start message provides the information necessary for the service (including the service area over which to deliver the broadcast transmissions, and relevant QoS parameters). The SCTP protocol is used to carry the application part signalling.

The MCE may be deployed as a separate node; alternatively the M3 interface may be terminated in eNodeBs, in which case the MCE would be considered to be part of an eNodeB and the M2 interface would not exist.

14.3.3.2   MBMS Logical Channels and Multiplexing

As shown in Figure 4.17, the MBMS logical channels are integrated into the LTE Layer 2 (see Section 4.4.1.4) and consist of:

- A traffic channel, the Multicast Traffic CHannel (MTCH): a point-to-multipoint downlink channel for transmitting data traffic from the network to the UE;

- A control channel, the Multicast Control CHannel (MCCH): a point-to-multipoint downlink channel used for transmitting MBMS control information from the network to the UE, for one or several MTCHs.

Both the MCCH and the MTCH are mapped to the MCH transport channel in MBSFN mode or to the normal Downlink Shared Channel (DL-SCH) transport channel in single-cell mode. This mapping applies both for a dedicated MBMS carrier and a mixed MBMS/unicast carrier.

The eNodeB performs MAC-level multiplexing for different MTCHs to be transmitted on an MCH. Multiple MBMS services can therefore be transmitted using a single MTCH, provided that they use the same MBSFN area. As noted above, one MCH can only contain data belonging to one MBSFN area; multiplexing between different MBSFN areas is not supported since MBSFN areas comprised of different sets of cells would have different composite channel responses, and would therefore need different channel estimates to be derived from different sets of reference signals.

However, MCHs transmitted over different MBSFN areas can be multiplexed in different subframes. The allocation of MTCH resources is signalled by the MCCH via MCH Subframe Allocation Patterns (MSAPs) which indicate the subframes in which MTCHs are transmitted; the mapping of subframes is known as an 'MSAP occasion'. Within each MSAP occasion, additional signalling indicates the order in which the MTCHs are multiplexed.

The MCCH design is outlined in the following section.

14.3.3.3   MCCH Design

In the same way that MBSFN transmission is beneficial for MBMS data transmission, it can also improve the reception of the MBMS control signalling transmitted on the MCCH if the signalling is MBSFN-area-specific rather than cell-specific.

---

[5] System Architecture Evolution – see Chapter 2.

Hence the number of MCCH channels may vary depending on the mode of data transmission in a cell:

- For MBSFN data transmission, there may be overlapping MBSFN areas, with a different set of MBMS services and hence a different MCCH corresponding to each MBSFN area;

- For single-cell point-to-multipoint MBMS transmission, there would be one MCCH channel for the MBMS services transmitted in single-cell mode.

On a mixed MBMS/unicast carrier, there can be a combination of single-cell point-to-multipoint transmissions and one or more multicell MBSFN transmissions of different MBMS services, where each separate configuration requires a different set of MCCH control information. This can create a complex structure for MBMS control information.

When MBMS is finalized in a later release of LTE, a hierarchical structure is therefore likely to be adopted for the MCCH, to address the variety of possible configurations for single-cell and MBSFN transmissions. This can be summarized as follows:

- **Broadcast Control CHannel (BCCH).** The BCCH indicates the scheduling of one or two Primary MCCHs: one for single-cell transmission on DL-SCH, and one for multicell transmission on MCH. The BCCH only points to the transmission resources where the primary MCCH(s) can be found, and does not indicate the availability of the services.

- **Primary MCCH(s) (P-MCCH).** A P-MCCH is transmitted on the DL-SCH for single-cell transmission and on an MCH for MBSFN transmission. The P-MCCH on the MCH could also point to optional additional S-MCCH(s) mapped to an MCH.

- **Secondary MCCH(s) (S-MCCH).** S-MCCH(s) signal the MBMS control information for each separate MBSFN area.

Figure 14.9 illustrates this possible MCCH hierarchical structure and the relationship between the logical channels.

Note that the MBSFN area used for transmission of an MCCH is not necessarily the same as the MBSFN area used for transmission of the MTCH carrying the services advertised by the MCCH.

According to such an MCCH design, a UE would only need to receive the P-MCCH in order to obtain MBMS control information on service availability. When an MBMS service starts which the user is interested in receiving, the UE would read the corresponding S-MCCH in order to obtain the location of the MSAP occasions for the corresponding MTCHs.

As described in Section 14.3.2.1, the overlap of two or more MBSFN areas would require a complex configuration of logical channels in order to signal services and resource allocations across the overlapping groups of cells, and indeed the whole service area. Some possibilities for optimizing the number of MCCHs (Primary and/or Secondary) in such a scenario could be to signal:

- One MCCH containing control information for each individual cell; this would require complex scheduling schemes;

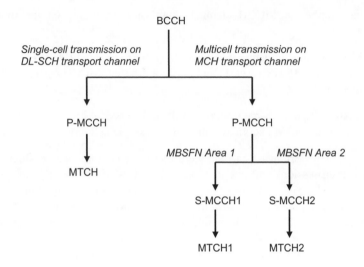

Figure 14.9 Possible hierarchical structure for MBMS signalling.

- One single MCCH containing combined control information for all MBMS services in an MBMS service area; this could create a rather large MCCH with unnecessary overhead.

An alternative to these extremes would be to signal a single MCCH covering a relatively static area that would need to be defined. Full details will be defined in a future release of the LTE specifications.

14.4   UE Capabilities for MBMS Reception

In order to support the introduction of MBMS into the LTE specifications in a forward-compatible way in a later release, all LTE UEs must be aware of some aspects of MBSFN operation even in the first release. In particular, all UEs must be aware of the possible existence of MBSFN subframes with their corresponding different Reference Signal (RS) pattern. UEs which are not 'MBSFN-capable' would therefore have the same knowledge as 'MBSFN-capable' UEs about which subframes in the serving cell contain MBSFN RS. UEs which are not 'MBSFN-capable' could operate on a mixed MBMS/unicast carrier, including receiving the PDCCH in the MBSFN subframes.

In the future, the actual reception of MBSFN data is likely to be an optional UE capability. Other optional capabilities are likely to include support for the 7.5 kHz subcarrier spacing for dedicated MBSFN carriers, and the simultaneous reception of unicast services when also receiving MBMS on a dedicated MBSFN carrier. The latter possibility is discussed in more detail in Section 14.4.1.

14.4.1 Dual Receiver Capability

On a mixed MBMS/unicast carrier, a UE can easily support simultaneous reception of unicast traffic and MBMS services with a single radio receiver, without interruption of the MBMS service. This is a major advantage of a mixed-carrier deployment. It also ensures the benefit of an uplink channel which can be used for interactive feedback.

On a dedicated MBSFN carrier, the ability to receive simultaneously unicast traffic (e.g. voice or video calls, or data services) and MBMS services would be optional, as simultaneous service reception involving a dedicated carrier is only possible with a dual radio receiver. A UE that does not support simultaneous reception may have to interrupt the reception of MBMS services in order to check for paging messages, which may result in degradation of the user experience.

14.4.2 Support of Emergency Services

One application for broadcast services over a cellular network is to support a public warning system for broadcasting emergency information in events such as natural disasters (e.g. earthquakes, tsunamis, floods).

It was initially intended that such emergency warnings would be carried over MBMS, with a minimum MBMS capability defined for the UEs. However, since MBMS will not be part of the first release of LTE, emergency services will instead be supported by a cell broadcast solution using the BCCH, independent of MBMS.

14.5 Comparison of Mobile Broadcast Modes

MBMS is not the only technology which is capable of efficient delivery of multimedia content to mobile consumers, and it is therefore instructive to consider briefly some of the relative strengths and weaknesses of the different general approaches.

14.5.1 Delivery by Cellular Networks

The key feature of using a cellular network for the content delivery is that the services can be deployed using a network operator's existing network infrastructure. Furthermore, it does not necessarily require the allocation of additional spectrum beyond that to which the operator already has access. However, the advantage of reusing the cellular network and spectrum for multimedia broadcasting is also its main drawback: it reduces the bandwidth available for other mobile services.

It can be noted that mobile operators have been providing mobile television over UTRAN networks for a number of years already; this has however generally been based on point-to-point unicast connections; this is feasible in lightly-loaded networks, but, as explained in Section 14.2.1, it does not make the most efficient use of the radio resources. MBMS transmission from a single source entity to multiple recipients on a single delivery channel allows radio resources to be shared, thus solving the capacity issues associated with the use of unicast transmissions for streaming of bandwidth-hungry services such as mobile television.

The use of a cellular network for MBMS also has the advantage of bringing the capability to send 'personalized' content to groups of a few users. In addition, a cellular network offers the possibility of a built-in return channel, which is a key feature for the provision of interactive services.

14.5.2   Delivery by Broadcast Networks

Mobile broadcast services may alternatively be provided by standalone broadcast systems such as Digital Video Broadcasting-Terrestrial (DVB-T) and Digital Audio Broadcasting (DAB). Originally developed for fixed receivers, specific mobile versions of these standards have been developed, namely Digital Video Broadcasting-Handheld (DVB-H) and Digital Mobile Broadcasting (DMB) respectively. Other proprietary broadcast solutions have also been developed, such as MediaFLO (Forward Link Only).

These systems typically assume the use of a relatively small number of higher-powered transmitters designed to cover a wide geographical area. For broadcast service provision, user capacity constraints do not require the use of small cells.

Such systems can therefore achieve relatively high data rates with excellent wide-area coverage. However, they clearly also require dedicated spectrum in which to operate, the construction and operation of a network of transmitters additional to that provided for the cellular network, and the presence of additional hardware in the mobile terminals.

Standalone broadcast systems are also by definition 'downlink-only': they do not provide a direct return channel for interactive services, although it may be possible to use the cellular network for this purpose.

14.5.3   Services and Applications

Both cellular delivery (MBMS) and standalone broadcast methods enable the delivery of a variety of multimedia services and applications, the most fashionable being mobile television. This gives network operators the opportunity to differentiate their service offerings.

In some ways, cellular delivery (MBMS) and standalone broadcast (e.g. DVB-H) systems can be seen as complementary. The mobile television industry and market are changing rapidly, and may be better described as part of the mobile entertainment industry, where new market rules prevail. An example of this is the concept of the 'long tail of content' [9] which describes the consumption of products or services across a large population, as illustrated in Figure 14.10: given a large availability of choice, a few very popular products or services will dominate the market for the majority of consumers; a large number of other products or services will only find a niche market.

This is a very appropriate representation of the usage of broadcast television, whether fixed or mobile: viewing figures for a few main channels (typically national broadcasters) are high, while usage is low for a profusion of specialist channels that gather a limited yet dedicated audience.

One obvious possibility then is to map this simple model to the earlier description of broadcast or multicast transmissions:

- Broadcast: for transmitting a few main channels to all users, e.g. via DVB-H or MBMS;

- Multicast: for transmitting numerous specialist channels on-demand to select groups of users, e.g. via MBMS.

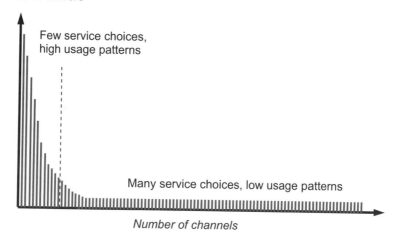

Figure 14.10 The 'Long Tail of Content'.

The deployment of mobile broadcast technology worldwide (whether by cellular MBMS or standalone broadcast) is very much dependent on the availability of spectrum and on the offer of services in each country, while infrastructure costs and content agreements between network operators and broadcasters are also significant factors.

Taking again the example of mobile television, what matters ultimately to the user is the quality of reception on their terminal, and mobility. Deployment of MBMS in LTE networks will bring an increase in capacity compared to what is achievable in a UMTS network, notably by exploiting the use of Single Frequency Network operation based on OFDM – which incidentally is the transmission method also used by DVB-H, DMB and MediaFLO.

References[6]

[1] 3GPP Technical Report 25.913, 'Requirements for Evolved UTRA (E-UTRA) and Evolved UTRAN (E-UTRAN) (Release 7)', www.3gpp.org.

[2] 3GPP Technical Specification 23.246, 'Multimedia Broadcast/Multicast Service (MBMS); Architecture and Functional Description (Release 6)', www.3gpp.org.

[3] 3GPP Technical Specification 23.228, 'IP Multimedia Subsystem (IMS); Stage 2 (Release 8)', www.3gpp.org.

[4] Internet Engineering Task Force, 'RFC3261 Session Initiation Protocol', www.ietf.org.

[5] Motorola, 'R1-070051: Performance of MBMS Transmission Configurations', www.3gpp.org, 3GPP TSG RAN WG1, meeting 47bis, Sorrento, Italy, January 2007.

[6] IEEE1588, 'IEEE 1588 Standard for a Precision Clock Synchronization Protocol for Networked Measurement and Control Systems', http://ieee1588.nist.gov.

---

[6]All web sites confirmed 18th December 2008.

[7] 3GPP Technical Specification 36.300, 'Evolved Universal Terrestrial Radio Access (E-UTRA) and Evolved Universal Terrestrial Radio Access (E-UTRAN); Overall description; Stage 2', www.3gpp.org.

[8] Internet Engineering Task Force, 'RFC4960 Stream Control Transmission Protocol', www.ietf.org.

[9] D. Anderson, *The Long Tail: Why the Future of Business is Selling Less of More*. New York, USA: Hyperion, 2006.

Part III

Physical Layer for Uplink

15

Uplink Physical Layer Design

Robert Love and Vijay Nangia

15.1 Introduction

While many of the requirements for the design of the LTE uplink physical layer and multiple-access scheme are similar to those of the downlink, the uplink also poses some unique challenges. Some of the desirable attributes for the LTE uplink include:

- Orthogonal uplink transmission by different User Equipment (UEs), to minimize intracell interference and maximize capacity.

- Flexibility to support a wide range of data rates, and to enable data rate to be adapted to the SINR (Signal-to-Interference plus Noise Ratio).

- Sufficiently low Peak-to-Average Power Ratio (PAPR) (or Cubic Metric (CM) – see Section 22.3.3) of the transmitted waveform, to avoid excessive cost, size and power consumption of the UE Power Amplifier (PA).

- Ability to exploit the frequency diversity afforded by the wideband channel (up to 20 MHz), even when transmitting at low data rates.

- Support for frequency-selective scheduling.

- Support for advanced multiple-antenna techniques, to exploit spatial diversity and enhance uplink capacity.

The multiple-access scheme selected for the LTE uplink so as to fulfil these principle characteristics is Single-Carrier Frequency Division Multiple Access (SC-FDMA).

A major advantage of SC-FDMA over the Direct-Sequence Code Division Multiple Access (DS-CDMA) scheme used in UMTS is that it achieves intra-cell orthogonality even in frequency-selective channels. SC-FDMA avoids the high level of intra-cell interference associated with DS-CDMA which significantly reduces system capacity and limits the use of adaptive modulation. A code-multiplexed uplink also suffers the drawback of an increased CM/PAPR if multi-code transmission is used from a single UE.

The use of OFDMA (Orthogonal Frequency Division Multiple Access) for the LTE uplink would have been attractive due to the possibility for full uplink-downlink commonality. In principle, an OFDMA scheme similar to the LTE downlink could satisfy all the uplink design criteria listed above, except for low CM/PAPR. As discussed in Section 5.2.2.1, much research has been conducted in recent years on methods to reduce the CM/PAPR of OFDM; however, in general the effectiveness of these methods requires careful practical evaluation against their associated complexity and/or overhead (for example, in terms of additional signalling or additional transmission bandwidth used to achieve the CM/PAPR reduction).

SC-FDMA combines the desirable characteristics of OFDM outlined in Section 5.2 with the low CM/PAPR of single-carrier transmission schemes.

Like OFDM, SC-FDMA divides the transmission bandwidth into multiple parallel subcarriers, with the orthogonality between the subcarriers being maintained in frequency-selective channels by the use of a Cyclic Prefix (CP) or guard period. The use of a CP prevents Inter-Symbol Interference (ISI) between SC-FDMA information blocks. It transforms the linear convolution of the multipath channel into a circular convolution, enabling the receiver to equalize the channel simply by scaling each subcarrier by a complex gain factor as explained in Chapter 5.

However, unlike OFDM, where the data symbols directly modulate each subcarrier independently (such that the amplitude of each subcarrier at a given time instant is set by the constellation points of the digital modulation scheme), in SC-FDMA the signal modulated onto a given subcarrier is a linear combination of all the data symbols transmitted at the same time instant. Thus in each symbol period, all the transmitted subcarriers of an SC-FDMA signal carry a component of each modulated data symbol. This gives SC-FDMA its crucial single-carrier property, which results in the CM/PAPR being significantly lower than pure multicarrier transmission schemes such as OFDM.

15.2   SC-FDMA Principles

15.2.1   SC-FDMA Transmission Structure

An SC-FDMA signal can, in theory, be generated in either the time-domain or the frequency-domain [1]. Although the two techniques are duals and 'functionally' equivalent, in practice, the time-domain generation is less bandwidth-efficient due to time-domain filtering and associated requirements for filter ramp-up and ramp-down times [2, 3]. Nevertheless, we describe both approaches here to facilitate understanding of the principles of SC-FDMA in both domains.

15.2.2   Time-Domain Signal Generation

Time-domain generation of an SC-FDMA signal is shown in Figure 15.1. It can be seen to be similar to conventional single-carrier transmission.

The input bit stream is mapped into a single-carrier stream of QPSK or QAM symbols, which are grouped into symbol-blocks of length M. This may be followed by an optional

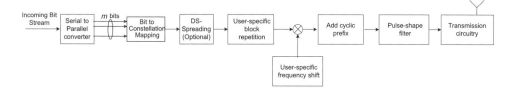

Figure 15.1 SC-FDMA time-domain transmit processing.

repetition stage, in which each block is repeated L times, and a user-specific frequency shift, by which each user's transmission may be translated to a particular part of the available bandwidth. A CP is then inserted. After filtering (e.g. with a root-raised cosine pulse-shaping filter), the resulting signal is transmitted.

The repetition of the symbol blocks results in the spectrum of the transmitted signal only being non-zero at certain subcarrier frequencies (namely every L^{th} subcarrier in this example) as shown in Figure 15.2.

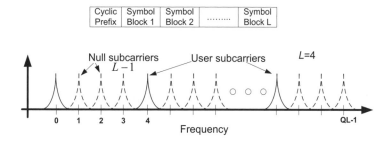

Figure 15.2 Distributed transmission with equal-spacing between occupied subcarriers.

Thus, the transmitted signal spectrum in this case is similar to what would be obtained if data symbols were only modulated on every Lth subcarrier of an OFDM signal.[1] Since such a signal occupies only one in every L subcarriers, the transmission is said to be 'distributed' and is one way of providing a frequency-diversity gain.

By varying the block length M and the repetition factor L, under the constraint that the total number of possible occupied subcarriers in the bandwidth is constant ($ML = $ constant), a wide range of data rates can be supported.

When no symbol-block repetition is performed ($L = 1$), the signal occupies consecutive subcarriers[2] and the transmission is said to be 'localized'. Localized transmissions are beneficial for supporting frequency-selective scheduling, for example when the eNodeB has knowledge of the uplink channel conditions (e.g. as a result of channel sounding as explained in Section 16.6), or for inter-cell interference coordination (as explained in Section 12.5). Localized transmission may also provide frequency diversity if the set of

---

[1] This type of OFDM transmission is sometimes called 'comb OFDM', owing to its comb-shaped spectrum.
[2] The occupied bandwidth then depends on the symbol rate.

consecutive subcarriers is hopped in the frequency domain, especially if the time interval between hops is shorter than the duration of a block of channel-coded data.

Different users' transmissions, using different repetition factors or bandwidths, remain orthogonal on the uplink when the following conditions are met:

- The users occupy different sets of subcarriers. This may in general be accomplished either by introducing a user-specific frequency shift (typically for the case of localized transmissions) or alternatively by arranging for different users to occupy interleaved sets of subcarriers (typically for the case of distributed transmissions). The latter method is known in the literature as Interleaved Frequency Division Multiple Access (IFDMA) [4].

- The received signals are properly synchronized in time and frequency.

- The CP is longer than the sum of the delay spread of the channel and any residual timing synchronization error between the users.

The SC-FDMA time-domain generated signal has a similar level of CM/PAPR as pulse-shaped single-carrier modulation. ISI in multipath channels is prevented by the CP, which enables efficient equalization at the receiver by means of a Frequency Domain Equalizer (FDE) [5, 6].

15.2.3   Frequency-Domain Signal Generation (DFT-S-OFDM)

Generation of an SC-FDMA signal in the frequency domain uses a Discrete Fourier Transform-Spread OFDM (DFT-S-OFDM) structure [7–9] as shown in Figure 15.3.

The first step of DFT-S-OFDM SC-FDMA signal generation is to perform an M-point DFT operation on each block of M QAM data symbols. Zeros are then inserted among the outputs of the DFT in order to match the DFT size to an N-subcarrier OFDM modulator (typically an Inverse Fast Fourier Transform (IFFT)). The zero-padded DFT output is mapped to the N subcarriers, with the positions of the zeros determining to which subcarriers the DFT-precoded data is mapped.

Usually N is larger than the maximum number of occupied subcarriers, thus providing for efficient oversampling and 'sinc' $(\sin(x)/x)$ pulse-shaping. The equivalence of DFT-S-OFDM and a time-domain-generated SC-FDMA transmission can readily be seen by considering the case of $M = N$, where the DFT operation cancels the IFFT of the OFDM modulator resulting in the data symbols being transmitted serially in the time domain. However, this simplistic construction would not provide any oversampling or pulse-shape filtering.

As with the time-domain approach, DFT-S-OFDM is capable of generating both localized and distributed transmissions:

- **Localized transmission.** The subcarrier mapping allocates a group of M adjacent subcarriers to a user. $M < N$ results in zero being appended to the output of the DFT spreader resulting in an upsampled/interpolated version of the original M QAM data symbols at the IFFT output of the OFDM modulator. The transmitted signal is thus similar to a narrowband single carrier with a CP (equivalent to time-domain generation with repetition factor $L = 1$) and 'sinc' pulse-shaping filtering (circular filtering).

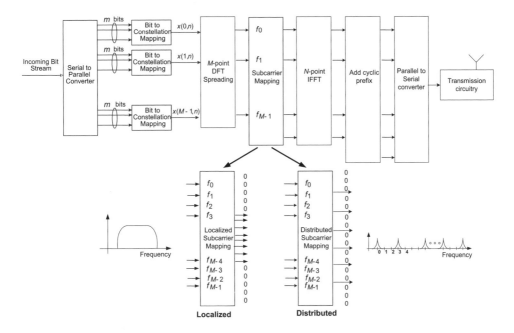

Figure 15.3 SC-FDMA frequency-domain transmit processing (DFT-S-OFDM) showing localized and distributed subcarrier mappings.

- **Distributed transmission.** The subcarrier mapping allocates M equally-spaced subcarriers (e.g. every L^{th} subcarrier). $(L-1)$ zeros are inserted between the M DFT outputs, and additional zeros are appended to either side of the DFT output prior to the IFFT ($ML < N$). As with the localized case, the zeros appended on either side of the DFT output provide upsampling or sinc interpolation, while the zeros inserted between the DFT outputs produce waveform repetition in the time domain. This results in a transmitted signal similar to time-domain IFDMA with repetition factor L and 'sinc' pulse-shaping filtering.

As for the time-domain SC-FDMA signal generation (in Section 15.2.2), orthogonality between different users with different data rate requirements can be achieved by assigning each user a unique set of subcarriers. The CP structure is the same as for the time-domain signal generation, and therefore the same efficient FDE techniques can be employed at the receiver [5, 6].

It is worth noting that, in principle, any unitary matrix can be used in the place of the DFT for the spreading operation with similar performance [10]. However, the use of non-DFT spreading would result in increased CM/PAPR since the transmitted signal would no longer have the single carrier characteristic.

15.3   SC-FDMA Design in LTE

Having outlined the key principles of SC-FDMA transmission, we now explain the application of SC-FDMA to the LTE uplink.

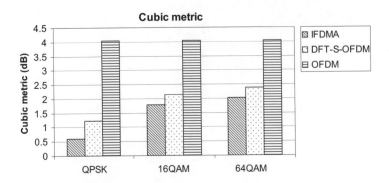

Figure 15.4 Cubic metric comparison of time-domain SC-FDMA generation (IFDMA), frequency-domain SC-FDMA generation (DFT-S-OFDM) and OFDM.

15.3.1 Transmit Processing for LTE

Although the frequency-domain generation of SC-FDMA (DFT-S-OFDM) is functionally equivalent to the time-domain SC-FDMA signal generation, each technique requires a slightly different parameterization for efficient signal generation [3]. The pulse-shaping filter used in the time domain SC-FDMA generation approach in practice has a non-zero excess bandwidth, resulting in bandwidth efficiency which is smaller than that achievable with the frequency domain method with its inherent 'sinc' (zero excess bandwidth) pulse-shaping filter which arises from the zero padding and IFFT operation.

For example, for a 5 MHz operating bandwidth, physical layer parameters optimized for time-domain implementation might have a sampling rate of 4.096 Mps (256 subcarriers with 16 kHz subcarrier spacing) resulting in bandwidth efficiency of 82% [2]. An equivalent set of parameters optimized for the frequency-domain generation can support a bandwidth efficiency of 90% (with 300 occupied subcarriers and 15 kHz subcarrier spacing). Thus, with frequency-domain processing, a 10% increase in bandwidth efficiency can be achieved, allowing higher data rates.

The non-zero excess bandwidth pulse-shaping filter in the time-domain generation also requires ramp-up and ramp-down times of 3–4 samples duration, while for DFT-S-OFDM there is no explicit pulse-shaping filter, resulting in a much shorter ramp time similar to OFDM. However, the pulse-shaping filter in the time-domain generation does provide the benefit of reduced CM by approximately 0.25–0.5 dB compared to DFT-S-OFDM, as shown in Figure 15.4. Thus there is a trade-off between bandwidth efficiency and CM/PAPR reduction between the time- and frequency-domain SC-FDMA generation methods.

Frequency-domain signal generation for the LTE uplink has a further benefit in that it allows a very similar parameterization to be adopted as for the OFDM downlink, including the same subcarrier spacing, number of occupied subcarriers in a given bandwidth, and CP lengths. This provides maximal commonality between uplink and downlink, including for example the same clock frequency.

For these reasons, the SC-FDMA parameters chosen for the LTE uplink have been optimized under the assumption of frequency-domain DFT-S-OFDM signal generation.

Table 15.1 LTE uplink SC-FDMA physical layer parameters.

Parameter	Value	Description
Subframe duration	1 ms	
Slot duration	0.5 ms	
Subcarrier spacing	15 kHz	
SC-FDMA symbol duration	66.67 µs	
CP duration	Normal CP:	5.2 µs first symbol in each slot, 4.69 µs all other symbols
	Extended CP:	16.67 µs all symbols
Number of symbols per slot	7 (Normal CP) 6 (Extended CP)	
Number of subcarriers per RB	12	

An important feature of the LTE SC-FDMA parameterization is that the numbers of subcarriers which can be allocated to a UE for transmission are restricted such that the DFT size in LTE can be constructed from multiples of 2, 3 and/or 5. This enables efficient, low-complexity mixed-radix FFT implementations.

15.3.2   SC-FDMA Parameters for LTE

The same basic transmission resource structure is used for the uplink as for the downlink: a 10 ms radio frame is divided into ten 1 ms subframes each consisting of two 0.5 ms slots. As LTE SC-FDMA is based on the same fundamental processing as OFDM, it uses the same 15 kHz subcarrier spacing as the downlink. The uplink transmission resources are also defined in the frequency domain (i.e. before the IFFT), with the smallest unit of resource being a Resource Element (RE), consisting of one SC-FDMA data block length on one subcarrier. As in the downlink, a Resource Block (RB) comprises 12 REs in the frequency domain for a duration of 1 slot, as detailed in Section 6.2. The LTE uplink SC-FDMA physical layer parameters for Frequency Division Duplex (FDD) and Time Division Duplex (TDD) deployments are detailed in Table 15.1.

Two CP durations are supported – a normal CP of duration 4.69 µs and an extended CP of 16.67 µs, as in the downlink (see Section 5.3.2). The extended CP is beneficial for deployments with large channel delay-spread characteristics, and for large cells.

The 1 ms subframe allows a 1 ms scheduling interval (or Transmission Time Interval (TTI)), as for the downlink, to enable low latency. However, one difference from the downlink is that the uplink coverage is more likely to be limited by the maximum transmission power of the UE. In some situations, this may mean that a single Voice-over-IP (VoIP) packet, for example, cannot be transmitted in a 1 ms subframe with an acceptable error rate. One solution to this is to segment the VoIP packet at higher layers to allow it to be transmitted over several subframes. However, such segmentation results in additional signalling overhead for each segment (including resource allocation signalling and Hybrid ARQ acknowledgement signalling). A more efficient technique for improving uplink VoIP coverage at the cell edge is to use so-called *TTI bundling*, where a single transport block from the MAC layer is

Table 15.2 LTE Uplink SC-FDMA parametrization for selected carrier bandwidths.

	Carrier bandwidth (MHz)					
	1.4	3	5	10	15	20
FFT size	128	256	512	1024	1536	2048
Sampling rate: M/N × 3.84 MHz	1/2	1/1	2/1	4/1	6/1	8/1
Number of subcarriers	72	180	300	600	900	1200
Number of RBs	6	15	25	50	75	100
Bandwidth efficiency (%)	77.1	90	90	90	90	90

transmitted repeatedly in multiple consecutive subframes, with only one set of signalling messages for the whole transmission. The LTE uplink allows groups of 4 TTIs to be 'bundled' in this way, in addition to the normal 1 ms TTI.

In practice in LTE, all the uplink data transmissions are localized, using contiguous blocks of subcarriers. This simplifies the transmission scheme, and enables the same RB structure to be used as in the downlink. Frequency-diversity can still be exploited by means of frequency hopping, which can occur both within one subframe (at the boundary between the two slots) and between subframes. In the case of frequency hopping within a subframe, the channel coding spans the two transmission frequencies, and therefore the frequency-diversity gain is maximized through the channel decoding process. The only instance of distributed transmission in the LTE uplink (using an IFDMA-like structure) is for the 'Sounding Reference Signals' (SRSs) which are transmitted to enable the eNodeB to perform uplink frequency-selective scheduling; these are discussed in Section 16.6.

Like the downlink, the LTE uplink supports scalable system bandwidths from approximately 1.4 MHz up to 20 MHz with the same subcarrier spacing and symbol duration for all bandwidths. The uplink scaling for the bandwidths supported in the first release of LTE is shown in Table 15.2. Note that the sampling rates resulting from the indicated FFT sizes are designed to be small rational multiples of the UMTS 3.84 MHz chip rate, for ease of implementation in a multimode UE.

Note that in the OFDM downlink parameter specification, the d.c. subcarrier is unused in order to support direct conversion (zero IF[3]) architectures. In contrast, no unused d.c. subcarrier is possible for SC-FDMA (as shown in Table 15.2) as it can affect the low CM/PAPR property of the transmit signal. These issues are discussed in Chapter 22.

15.3.3 d.c. Subcarrier in SC-FDMA

Direct conversion transmitters and receivers can introduce distortion at the carrier frequency (zero frequency or d.c. in baseband), for example arising from local oscillator leakage.

In this section we explore three possible configurations of the d.c. subcarrier which were considered in the design of the LTE uplink in order to minimize d.c. distortion effects on the packet error rate and the CM/PAPR [11].

- **Option 1.** The d.c. subcarrier distortion region falls in the middle of a RB, such that one of the RBs includes information modulated at d.c. (e.g. 600 subcarriers for 10 MHz

---

[3]Intermediate Frequency.

operation bandwidth with the d.c. subcarrier being one of the subcarriers for RB 26). The performance of the RB containing the d.c. subcarrier would be reduced at the receiver; this effect would be most noticeable with a narrow bandwidth transmission consisting of a single RB.

- **Option 2.** One more subcarrier is configured than is required for the number of RBs (e.g. 601 subcarriers for the 10 MHz bandwidth case). This option would be beneficial for the case of a system bandwidth with an even number of RBs where the additional subcarrier would be unused and correspond to the d.c. subcarrier located between RBs allocated to different UEs.

- **Option 3.** The subcarriers are frequency-shifted by half a subcarrier spacing (±7.5 kHz), resulting in an offset of 7.5 kHz for subcarriers relative to d.c. Thus two subcarriers straddle the d.c. location. This is the option used in LTE, and is illustrated in Figure 15.5 for deployments with even and odd numbers of RBs across the system bandwidth.

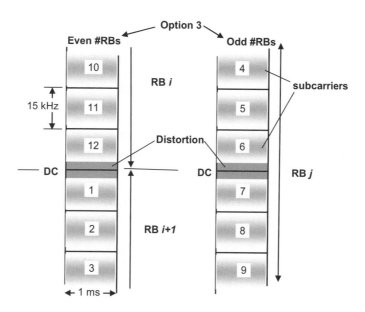

Figure 15.5 Distortion around d.c. with 7.5 kHz subcarrier shift for LTE system bandwidth with even or odd number of RBs.

15.3.4 Pulse Shaping

As explained in Sections 15.2.3 and 15.3.1, one of the benefits of frequency-domain processing for SC-FDMA is that, in principle, there is no need for explicit pulse-shaping thanks to the implicit 'sinc' pulse-shaping. Nevertheless, an additional explicit pulse-shaping

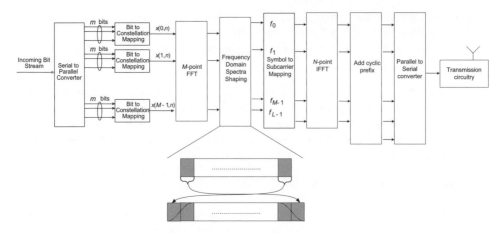

Figure 15.6 SC-FDMA (DFT-S-OFDM) with Pulse Shaping by Frequency Domain Spectral Shaping (FDSS).

filter can further reduce the CM/PAPR, but at the expense of spectral efficiency (due to the resulting non-zero excess filter bandwidth similar to that of the time domain SC-FDMA signal generation described in Section 15.2.2). As a result of this trade-off, additional pulse-shaping is not specified in LTE, but we review here some theoretical pulse-shaping techniques in order to elucidate the reasoning behind this decision.

Frequency-domain SC-FDMA transmit processing allows any pulse shaping to be imple-mented efficiently in the frequency-domain, after the DFT and prior to subcarrier mapping, by element-wise multiplication with the spectrum of the desired pulse-shaping filter. This is called Frequency Domain Spectral Shaping (FDSS) and is shown in Figure 15.6.

The use of pulse-shaping filters or FDSS window functions, such as Root Raised Cosine (RRC) [12] or Kaiser window [13], was considered for LTE in order to reduce CM/PAPR, especially for lower order modulations such as QPSK to enhance uplink coverage. The bandwidth expansion caused by FDSS requires the code rate to be increased for a given data rate. In order for FDSS to provide a net performance benefit (lower SINR requirement than without pulse shaping, for the same data rate), power boosting is required, with the size of the boost being approximately equal to the CM reduction achieved by FDSS.

We first consider the Kaiser window, which is defined as

$$w(n) = \begin{cases} K_1 \dfrac{I_0(\gamma)}{I_0(\beta)} & 0 \le n \le M \\ 0 & \text{otherwise} \end{cases} \qquad (15.1)$$

where $\gamma = \beta \sqrt{1 - ((n-p)/p)^2}$, $p = M/2$, $I_0(\cdot)$ is the modified Bessel function of the first kind and K_1 is the scaling factor needed to obtain the same power as a rectangular window of length $(M + 1)$ with unit amplitude. The parameter β is a shape factor which controls the smoothness of the window in a similar way to the roll-off factor α of a RRC filter.

The CM benefit obtained from the Kaiser window with different values of the shape factor β for QPSK modulation is shown in Figure 15.7 [14]. As can be seen, the benefit

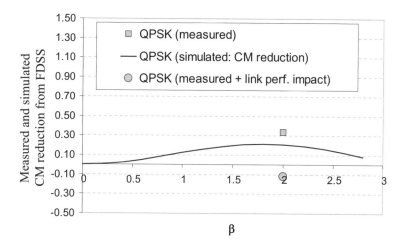

Figure 15.7 CM reduction from Kaiser windowing for QPSK modulation.

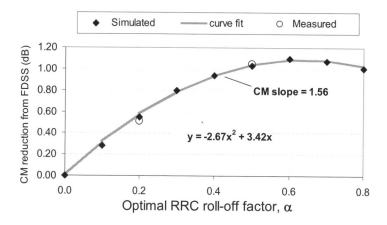

Figure 15.8 QPSK CM reduction versus RRC filter roll-off factor, α.

is small. The CM reduction obtained through simulation is only 0.20 dB for $\beta = 2$; measurements have shown about 0.35 dB gain, although with measurement error in the range of 0.1 to 0.2 dB. With a link performance loss ($\sim$0.45 dB) from Kaiser windowing at 10% BLER, virtually no net performance benefit can be achieved after power boosting by the amount of the CM reduction. If power boosting were not possible for any reason, then windowing for QPSK would result in a performance degradation of up to 0.45 dB. With an RRC filter, the CM benefit for a roll-off factor α ($0 \le \alpha \le 1$) for QPSK is shown in Figure 15.8 to be in the range 0.1 to 1.1 dB. The effective SINR gain with power boosting is shown in Figure 15.9 with optimal α selected for each code rate to maximize spectral efficiency. It can be seen from Figure 15.9 that, even with power boosting, the largest SINR

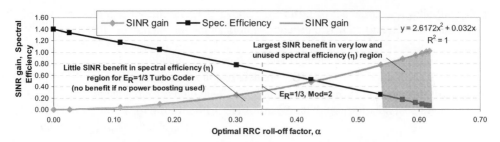

Figure 15.9 SINR gain and spectral efficiency η versus optimal RRC filter roll-off factor α for QPSK.

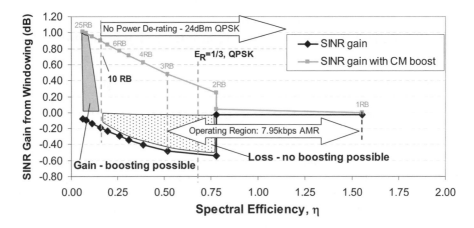

Figure 15.10 Maximum SINR gain from FDSS using optimal RRC filter roll-off factor (α) for QPSK.

benefit is in the very low spectral efficiency (η) region which is generally unused. Little SINR benefit is observed in the η region for code rates $> 1/3$. Without power boosting, there is no performance benefit or performance loss for RRC spectral shaping. Moreover, FDSS cannot be applied to reference symbols,in order to prevent degradation of the channel estimation.

As discussed in Chapter 22, CM is not a problem for UEs operating at a maximum power level for resource allocations of up to 10 RBs. This means that there is no benefit possible from FDSS given that UEs at the cell edge would often have less than 10 RBs allocated. With no power boosting, FDSS results in a throughput loss since the resulting spectral efficiency loss cannot be compensated as shown in Figure 15.10 [15].

Therefore, in LTE no pulse-shaping is used – in other words, the FDSS window is rectangular.

15.4 Summary

The important properties of the SC-FDMA transmission scheme used for the LTE uplink are derived from its multicarrier OFDM-like structure with single-carrier characteristic. The multicarrier-based structure gives the LTE uplink the same robustness against ISI as the LTE downlink, with low-complexity frequency-domain equalization being facilitated by the CP. At the same time, the DFT-based precoding ensures that the LTE uplink possesses the low CM required for efficient UE design. Crucially, the LTE uplink is designed to be orthogonal in the frequency domain between different UEs, thus virtually eliminating the intra-cell interference associated with CDMA.

The parameters of the LTE uplink are designed to ensure maximum commonality with the downlink, and to facilitate frequency-domain DFT-S-OFDM signal generation.

The localized resource allocation scheme of the LTE uplink allows both frequency-selective scheduling and the exploitation of frequency diversity, the latter being achieved by means of frequency hopping.

The following chapters explain the application of SC-FDMA in LTE in more detail, showing how reference signals for channel estimation, data transmissions and control signalling are multiplexed together.

References[4]

[1] Motorola, 'R1-050397: EUTRA Uplink Numerology and Frame Structure', www.3gpp.org, 3GPP TSG RAN WG1, meeting 41, Athens, Greece, May 2005.

[2] Motorola, 'R1-050584: EUTRA Uplink Numerology and Design', www.3gpp.org, 3GPP TSG RAN WG1, meeting 41bis, Sophia Antipolis, France, June 2005.

[3] Motorola, 'R1-050971: Single Carrier Uplink Options for E-UTRA', www.3gpp.org, 3GPP TSG RAN WG1, meeting 42, London, UK, August 2005.

[4] U. Sorger, I. De Broeck and M. Schnell, 'Interleaved FDMA – A New Spread-Spectrum Multiple-Access Scheme', in *Proc. IEEE International Conference on Communications*, pp. 1013–1017, 1998.

[5] H. Sari, G. Karam and I. Jeanclaude, 'Frequency-Domain Equalization of Mobile Radio and Terrestrial Broadcast Channels', in *Proc. IEEE Global Telecommunications Conference*, November 1994.

[6] D. Falconer, S. L. Ariyavisitakul, A. Benyamin-Seeyar and B. Eidson, 'Frequency Domain Equalization for Single-carrier Broadband Wireless Systems'. *IEEE Communications Magazine*, Vol. 40, pp. 58–66, April 2002.

[7] K. Bruninghaus and H. Rohling, 'Multi-carrier Spread Spectrum and its Relationship to Single-carrier Transmission', in *Proc. IEEE Vehicular Technology Conference*, May 1998.

[8] D. Galda and H. Rohling, 'A Low Complexity Transmitter Structure for OFDM-FDMA Uplink Systems', in *Proc. IEEE Vehicular Technology Conference*, May 2002.

[9] R. Dinis, D. Falconer, C. T. Lam and M. Sabbaghian, 'A Multiple Access Scheme for the Uplink of Broadband Wireless Systems', in *Proc. IEEE Global Telecommunications Conference*, November 2004.

---

[4]All web sites confirmed 18th December 2008.

[10] V. Nangia and K. L. Baum, 'Experimental Broadband OFDM System – Field Results for OFDM and OFDM with Frequency Domain Spreading', in *Proc. IEEE International Conference on Vehicular Technology*, Vancouver, BC, Canada, September 2002.

[11] Motorola, 'R1-062061: Uplink DC Subcarrier Distortion Considerations in LTE', www.3gpp.org, 3GPP TSG RAN WG1, meeting 46, Tallinn, Estonia, August 2006.

[12] NTT DoCoMo, NEC, and SHARP, 'R1-050702: DFT-spread OFDM with Pulse Shaping Filter in Frequency Domain in Evolved UTRA Uplink', www.3gpp.org, 3GPP TSG RAN WG1, meeting 42, London, UK, August 2005.

[13] Huawei, 'R1-051434: Optimum family of spectrum-shaping functions for PAPR reduction in SC-FDMA', www.3gpp.org, 3GPP TSG RAN WG1, meeting 43, Seoul, Korea, November 2005.

[14] Motorola, 'R1-070564: Frequency Domain Spectral Shaping for LTE', www.3gpp.org, 3GPP TSG RAN WG1, meeting 47bis, Sorrento, Italy, January 2007.

[15] Motorola, 'R1-070047: UE Power De-rating Reduction Techniques in LTE', www.3gpp.org, 3GPP TSG RAN WG1, meeting 47bis, Sorrento, Italy, January 2007.

16

Uplink Reference Signals

Robert Love and Vijay Nangia

16.1 Introduction

As in the downlink, the LTE Single-Carrier Frequency Division Multiple Access (SC-FDMA) uplink incorporates Reference Signals (RSs) for data demodulation and channel sounding. In this chapter the design principles behind these RSs are explained, including in particular features related to interference randomization and coordination, and the flexible configuration of channel sounding.

The roles of the uplink RSs include enabling channel estimation to aid coherent demodulation, channel quality estimation for uplink scheduling, power control, timing estimation and direction-of-arrival estimation to support downlink beamforming. Two types of RS are supported on the uplink:

- **DeModulation RS** (DM RS), associated with transmissions of uplink data on the Physical Uplink Shared CHannel (PUSCH) and/or control signalling on the Physical Uplink Control CHannel (PUCCH).[1] These RSs are primarily used for channel estimation for coherent demodulation.

- **Sounding RS** (SRS), not associated with uplink data and/or control transmissions, and primarily used for channel quality determination to enable frequency-selective scheduling on the uplink.

The uplink RSs are time-multiplexed with the data symbols. The DM RSs of a given UE (User Equipment) occupy the same bandwidth (i.e. the same Resource Blocks (RBs)) as its PUSCH/PUCCH data transmission. Thus, the allocation of orthogonal (in frequency) sets of RBs to different UEs for data transmission automatically ensures that their DM RSs

---

[1] See Chapter 17 for details of the uplink physical channels.

LTE – The UMTS Long Term Evolution: from Theory to Practice   Stefania Sesia, Issam Toufik and Matthew Baker
© 2009 John Wiley & Sons, Ltd

are also orthogonal to each other. The SRSs, if configured by higher layer signalling, are transmitted on the last SC-FDMA symbol in a subframe; SRS can occupy a bandwidth different from that used for data transmission. UEs transmitting SRS in the same subframe can be multiplexed via either Frequency or Code Division Multiplexing (FDM or CDM respectively), as explained in Section 16.6.

Desirable characteristics for the uplink RSs include:

- Constant amplitude in the frequency domain for equal excitation of all the allocated subcarriers for unbiased channel estimates.

- Low Cubic Metric (CM) in the time domain (at worst no higher than that of the data transmissions, while a lower CM for RSs than the data can be beneficial in enabling the transmission power of the RSs to be boosted at the cell-edge).

- Good autocorrelation properties for accurate channel estimation.

- Good cross-correlation properties between different RSs to reduce interference from RSs transmitted on the same resources in other (or, in some cases, the same) cells.

The following sections explain how these characteristics are achieved in LTE.

16.2   RS Signal Sequence Generation

The uplink reference signals in LTE are mostly based on Zadoff–Chu (ZC) sequences[2] [1,2]. The fundamental structure and properties of these sequences are described in Section 7.2.1.

These sequences satisfy the desirable properties for RS mentioned above, exhibiting 0 dB CM, ideal cyclic autocorrelation, and optimal cross-correlation. The cross-correlation property results in the impact of an interfering signal being spread evenly in the time domain after time-domain correlation of the received signal with the desired sequence; this results in more reliable detection of the significant channel taps. However, in practice the CM of a ZC sequence is degraded from the theoretical 0 dB value at the Nyquist sampling rate. This arises from the presence of unused guard subcarriers at each end of the sequence (due to the number of occupied RS subcarriers being less than the IFFT (Inverse Foot Fourier Transform) size of the OFDM (Orthogonal Frequency Division Multiplexing) modulator), and results in the ZC sequence effectively being oversampled in the time domain.

The RS sequence length, N_p, is equal to the number of assigned subcarriers, which is a multiple of the number of subcarriers per RB, $N_{sc}^{RB} = 12$, i.e.

$$N_p = M_{sc}^{RS} = m \cdot N_{sc}^{RB}, \quad 1 \le m \le N_{RB}^{UL} \tag{16.1}$$

where N_{RB}^{UL} is the uplink system bandwidth in terms of RBs.

The length-N_p RS sequence is directly applied (without Discrete Fourier Transform (DFT) spreading) to N_p reference signal subcarriers at the input of the IFFT as shown in Figure 16.1 [3]. Recall that a ZC sequence of odd-length N_{ZC} is given by

$$a_q(n) = \exp\left[-j2\pi q \frac{n(n+1)/2 + ln}{N_{ZC}}\right] \tag{16.2}$$

---

[2]Note that ZC sequences are a special case of the Generalized Chirp-Like (GCL) sequences.

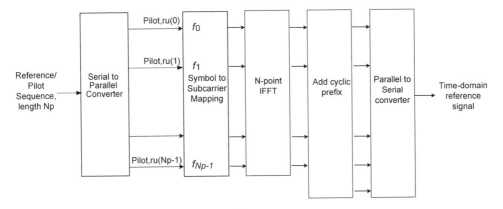

Figure 16.1 Transmitter structure for SC-FDMA reference signals. Note that no DFT spreading is applied to the RS sequence.

where $q = 1, \ldots, N_{ZC} - 1$ is the ZC sequence index (also known as the root index), $n = 0, 1, \ldots, N_{ZC} - 1$, and $l = 0$ in LTE.

In LTE, N_{ZC} is selected to be the largest prime number smaller than or equal to N_p. The ZC sequence of length N_{ZC} is then cyclically extended to the target length N_p as follows:

$$\bar{r}_q(n) = a_q(n \bmod N_{ZC}), \quad n = 0, 1, \ldots, N_p - 1 \tag{16.3}$$

The cyclic extension in the frequency domain preserves the constant amplitude properties (in the frequency domain) and also the zero autocorrelation cyclic shift orthogonality. Cyclic extension of the ZC sequences is used rather than truncation, as in general it provides better CM characteristics. For sequence lengths of three or more RBs, this provides at least 30 sequences with CM smaller than or close to that of QPSK.

However, for the shortest sequence lengths, suitable for resource allocations of just one or two RBs, only a small number of low-CM extended ZC sequences is available (six and 12 sequences respectively with CM less than that of QPSK). Therefore, in order to provide at least as large a number of sequences as for the 3 RB case, 30 special RS sequences are defined in LTE for resource allocations of one or two RBs. These special sequences are QPSK rather than ZC-based sequences, and were obtained from computer searches so as to have constant modulus in the frequency-domain, low CM, low memory/complexity requirements, and good cross-correlation properties.

The QPSK RS sequences are given by [4], Section 5.5.1.2,

$$\bar{r}(n) = e^{j\varphi(n)\pi/4}, \quad n = 0, 1, \ldots, M_{sc}^{RS} - 1, \tag{16.4}$$

where M_{sc}^{RS} is the number of subcarriers to which the sequence is mapped, and the values of $\varphi(n)$ can be found in Tables 5.5.1.2-1 and 5.5.1.2-2 in [4].

16.2.1 Base RS Sequences and Sequence Grouping

In order for a cell to support uplink transmissions of different bandwidths, it is necessary to assign a cell at least one *base RS sequence* for each possible RB allocation size. Multiple

RS sequences for each allocation size are then derived from each base sequence by means of different cyclic time shifts, as explained in Section 16.2.2.

The smallest number of available base sequences is for resource allocations of three RBs, where, as noted above, only 30 extended ZC sequences exist. As a result, the complete set of available base sequences across all RB allocation sizes is divided into 30 non-overlapping *sequence-groups*. A cell is then assigned one of the sequence-groups for uplink transmissions from UEs served by the cell.

For each resource allocation size up to and including five RBs, each of the 30 sequence-groups contains only one base sequence, since for five RBs (i.e. sequences of length 60) only 58 extended ZC-sequences are available. For sequence lengths greater than five RBs, more extended ZC-sequences are available, and therefore each of the 30 sequence-groups contains two base sequences per resource allocation size; this is exploited in LTE to support *sequence hopping* (within the sequence-group) between the two slots of a subframe.

The base sequences for resource allocations larger than three RBs are selected such that they are the sequences with high cross-correlation to the single 3 RB base sequence in the sequence-group [5]. Since the cross-correlation between the 3 RB base sequences of different sequence-groups is low due to the inherent properties of the ZC sequences, such a method for assigning the longer base sequences to sequence-groups helps to ensure that the cross-correlation between sequence-groups is kept low, thus reducing inter-cell interference.

The v base RS sequences of length 3 RBs or larger (i.e. $M_{sc}^{RS} \geq 36$) assigned to a sequence-group u are given by [4], Section 5.5.1,

$$\bar{r}_{u,v}(n) = a_q(n \bmod N_{ZC}), \quad n = 0, 1, \dots, M_{sc}^{RS} - 1 \tag{16.5}$$

where $u \in \{0, 1, \dots, 29\}$ is the sequence-group number, v is the index of the base sequence of length M_{sc}^{RS} within the sequence-group u, and is given by

$$v = \begin{cases} 0, 1 & \text{for } M_{sc}^{RS} \geq 72 \\ 0 & \text{otherwise} \end{cases} \tag{16.6}$$

N_{ZC} is the largest prime number smaller than M_{sc}^{RS}, and q is the root ZC sequence index (defined in [4], Section 5.5.1.1).

16.2.2   Orthogonal RS via Cyclic Time-Shifts of a Base Sequence

UEs which are assigned to different sets of subcarriers or RBs, transmit RS signals on these subcarriers and hence achieve separability of the RSs via FDM. However, in certain cases, UEs can be assigned to transmit on the same set of subcarriers, for example in the case of uplink multi-user MIMO[3] (sometimes also referred to as Spatial Division Multiple Access (SDMA) or 'Virtual MIMO'). In these cases the RSs can interfere with each other, and some means of separating the RSs from the different transmitters is required. Using different base sequences for different UEs transmitting in the same RBs is not ideal due to the non-zero cross-correlation between the base sequences which can degrade the channel estimation at the eNodeB. It is preferable that the RS signals from the different UEs are fully orthogonal. In theory, this could be achieved by FDM of the RSs within the same set of subcarriers,

---

[3]Multiple-Input Multiple-Output.

although this would reduce the RS sequence length and the number of different RS sequences available; this would be particularly undesirable for low-bandwidth transmissions.

Therefore in LTE, orthogonality between RSs occupying the same set of subcarriers is instead provided by exploiting the fact that the correlation of a ZC sequence with any Cyclic Shift (CS) of the same sequence is zero (see Section 7.2.1). As the channel impulse response is of finite duration, different transmitters can use different cyclic time shifts of the same base RS sequence, with the RSs remaining orthogonal provided that the cyclic shifts are longer than the channel impulse response.

If the RS SC-FDMA symbol duration is T_p and the channel impulse response duration is less than T_{cs}, then up to T_p/T_{cs} different transmitters can transmit in the same symbol, with different cyclic shift values, with separable channel estimates at the receiver. For example, Figure 16.2 shows that if $T_p/T_{cs} = 4$ and there are four transmitters, then each transmitter $t \in \{1, \ldots, 4\}$ can use a cyclic time shift of $(t - 1)T_p/4$ of the same base sequence. At the eNodeB receiver, by correlating the received composite signal from the different transmitters occupying the same set of subcarriers with the base sequence, the channel estimates from the different transmitters are separable in the time domain [6].

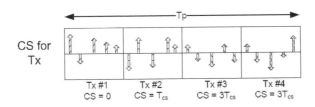

Figure 16.2 Illustration of cyclic time shift orthogonality of RS signals.

Since a cyclic time shift is equivalent to applying a phase ramp in the frequency domain, the frequency-domain representation of a base sequence with cyclic shift, α, is given by

$$r_{u,v}^{(\alpha)}(n) = e^{j\alpha n}\bar{r}_{u,v}(n) \tag{16.7}$$

where $\bar{r}_{u,v}(n)$ is the base (or unshifted) sequence of sequence-group u, with base sequence index v within the sequence-group, $\alpha = 2\pi n_t/P$ with n_t the cyclic time shift index for transmitter t, and P is the number of equally spaced cyclic time shifts supported. In the example in Figure 16.2, $P = 4$ and $n_1 = 0$, $n_2 = 1$, $n_3 = 2$, and $n_4 = 3$ for the four transmitters respectively.

In LTE, 12 equally spaced cyclic time shifts are defined for the DM RS on the PUSCH and PUCCH. This allows for delay spreads up to 5.55 μs.

The degree of channel estimate separability at the receiver between different cyclic time shifts depends in practice on the (circular) distance between the shifts (as well as the received power differences between the transmitters). Cyclic time shifts spaced the furthest apart (e.g. pairs (n_1, n_3) and (n_2, n_4) in the example in Figure 16.2) experience the least cross-talk between the channel estimates (for example, arising from practical issues such as channel estimation filtering and finite sampling granularity). Thus, when the number of UEs using different cyclic time shifts is less than the number of cyclic time shifts supported (P), it is beneficial to assign cyclic time shifts with the largest possible (circular) separation, and this is supported in LTE. Alternatively, interference randomization can be employed by cyclic time shift hopping, whereby the cyclic time shift used by a UE varies in time. In LTE, hopping of the cyclic time shifts between the two slots in a subframe is always enabled, for inter-cell interference randomization (see Section 16.4).

16.3 Sequence-Group Hopping and Planning

LTE supports both *RS sequence-group hopping* and *RS sequence-group planning* modes of system deployment [7, 8]; the mode is configurable by Radio Resource Control (RRC) signalling.

The sequence-group assigned to a cell is a function of its physical-layer cell identity (cell-ID) and can be different for PUSCH and PUCCH transmissions. A UE acquires knowledge of the physical layer cell-ID from its downlink synchronization signals, as described in Section 7.2.

16.3.1 Sequence-Group Hopping

The sequence-group hopping mode of deployment can be enabled in a cell by a 1-bit broadcast signalling parameter called 'groupHoppingEnabled'. This mode actually consists of a combination of hopping and shifting of the sequence-group according to one of 504 sequence-group hopping/shifting patterns corresponding to the 504 unique cell-IDs [9]. Since there are 30 base sequence-groups, $17(= \lceil 504/30 \rceil)$ unique sequence-group hopping patterns of length 20 are defined (corresponding to the duration of a radio frame with 20 slots), each of which can be offset by one of 30 sequence-group shift offsets. The sequence-group number u depends on the sequence-group hopping pattern f_{gh} and the sequence-group shift offset f_{ss} as defined in [4], Section 5.5.1.3. The sequence-group hopping pattern changes u from slot to slot in a pseudo-random manner, while the shift offset is fixed in all slots. Both f_{gh} and f_{ss} depend on the cell-ID.

The sequence-group hopping pattern f_{gh} is obtained from a length-31 Gold sequence generator (see Section 6.3) [10, 11], of which the second constituent M-sequence is initialized at the beginning of each radio frame by the sequence-group hopping pattern index of the cell. Up to 30 cell-IDs can have the same sequence-group hopping pattern (e.g. part of a planned coordinated cell cluster), with different sequence-group shift offsets being used to minimize RS collisions and inter-cell interference. The same sequence-group hopping pattern is used for PUSCH DM RS, SRS and PUCCH DM RS transmissions. The sequence-group shift offset can be different for PUSCH and PUCCH.

For PUSCH, it should be possible to assign cell-IDs such that the same sequence-group hopping pattern and sequence-group shift offset, and hence the same base sequences, are used in adjacent cells. This can enable the RSs from UEs in adjacent cells (for example, the cells of the same eNodeB) to be orthogonal to each other by using different cyclic time shifts of the same base sequence. Therefore in LTE the sequence-group shift offset for PUSCH is explicitly configured by a cell-specific 5-bit broadcast signalling parameter, 'groupAssignmentPUSCH'. As the sequence-group shift offset is a function of the cell-ID, the overhead for signalling the one of 504 sequence-group hopping patterns for PUSCH is reduced from nine to five bits, such that $f_{ss} = (\text{cell-ID mod } 30 + \Delta_{ss}) \text{ mod } 30$, where $\Delta_{ss} \in \{0, \ldots, 29\}$ is indicated by 'groupAssignmentPUSCH'.

For PUCCH transmissions, as described in Section 17.3 the same RBs at the edge of the system bandwidth are normally used by all cells. Thus, in order to randomize interference on the PUCCH between neighbouring cells which are using the same sequence-group hopping pattern, the sequence-shift offset for PUCCH is simply given by cell-ID mod 30. Similarly, for interference randomization on the SRS transmissions which occur in the same SC-FDMA symbol (see Section 16.6), the same sequence-group shift offset as PUCCH is used.

In Section 16.2.1, it was explained that there are two base sequences per sequence-group for each RS sequence length greater than 60 (5 RBs), with the possibility of interference randomization by sequence-hopping between the two base sequences at the slot boundary in the middle of each subframe. If sequence-group hopping is used, the base sequence automatically changes between each slot, and therefore additional sequence hopping within the sequence group is not needed; therefore only the first base sequence in the sequence group is used if sequence-group hopping is enabled.

16.3.2   Sequence-Group Planning

If sequence-group hopping is disabled, the same sequence-group number u, is used in all slots of a radio frame and is simply obtained from the sequence group shift offset, $u = f_{ss}$.

Since 30 sequence-groups are defined (see Section 16.2.1), planned sequence-group assignment is possible for up to 30 cells in LTE. This enables neighbouring cells to be assigned sequence groups with low cross-correlation to reduce RS interference, especially for small RB allocations.[4]

An example of sequence-group planning with a conventional six-sequence-group reuse plan is shown in Figure 16.3 [12]. The same sequence-group number (and hence base sequences) are used in the three cells of each eNodeB, with different cyclic time shifts assigned to each cell (only three cyclic time shifts D1, D2 and D3 are shown in this example).

With sequence-group planning, sequence hopping within the sequence group between the two slots of a subframe for interference randomization can be enabled by a 1-bit cell-specific parameter, 'sequenceHoppingEnabled'. The base sequence index for $M_{sc}^{RS} \geq 72$ used in slot n_s is then obtained from the length-31 Gold sequence generator. In order to enable the use of the same base RS sequence (and hopping pattern) in adjacent cells for PUSCH, the pseudo-random sequence generator is initialized at the beginning of each radio frame by the sequence-group hopping pattern index (based on part of the cell-ID), offset by the PUSCH sequence-group shift index of the cell (see [4], Section 5.5.1.4).

The same hopping pattern within the sequence-group is used for PUSCH DM RS, SRS, and PUCCH DM RS.

Further interference randomization in conjunction with sequence-group planning is provided by cyclic time shift hopping, which is always enabled in LTE as discussed in the following section.

---

[4]Power-limited cell-edge UEs are likely to have small RB allocations.

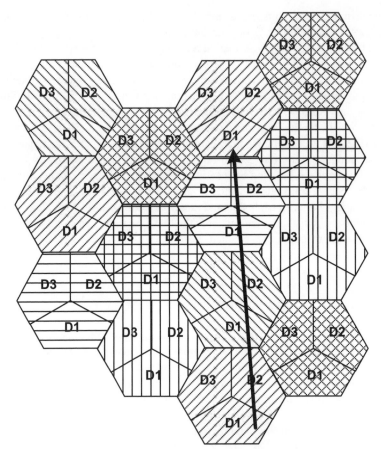

Pattern: Base Sequence-group Index
D1-D3: Different cyclic time shift values

Figure 16.3 Example of RS sequence-group planning – six-sequence-group reuse plan.

16.4   Cyclic Shift Hopping

Cyclic time shift hopping is always enabled for inter-cell interference randomization for PUSCH and PUCCH transmissions. For PUSCH with $P = 12$ evenly-spaced cyclic time shifts, the hopping is between the two slots in a subframe, with the cyclic shift (α in Equation (16.7)) for a UE being derived in each slot from a combination of a 3-bit cell-specific broadcast cyclic time shift offset parameter, a 3-bit cyclic time shift offset indicated in each uplink scheduling grant and a pseudo-random cyclic shift offset obtained from the output of the length-31 Gold sequence generator (see reference [4], Section 5.5.2.1.1).

As mentioned in Section 16.3, it should be possible to use different cyclic time shifts in adjacent cells with the same sequence-group (e.g. the cells of the same eNodeB), in order to support orthogonal RS transmissions from UEs in different cells. This requirement is similar to the case for initialization of the pseudo-random sequence generator for sequence hopping within a sequence-group, where the same hopping patterns are needed in neighbouring cells. Thus, the initialization of the PUSCH DM RS cyclic shift pseudo-random sequence generator is the same as that for the sequence hopping pattern generator in Section 16.3.2, initialized every radio frame (see [4], Section 5.5.1.4).

In the case of PUCCH transmission, cyclic time shift hopping (among the $P = 12$ evenly-spaced cyclic time shifts) is performed per SC-FDMA symbol, with the cyclic shift α for a given SC-FDMA symbol in a given slot being derived (as specified in reference [4], Section 5.4) from a combination of the assigned PUCCH resource index (see Section 17.3) and the output of the length-31 Gold sequence generator. In order to randomize interference on the PUCCH arising from the fact that the same band-edge RBs are used for PUCCH transmissions in all cells (see Section 17.3), the pseudo-random sequence generator is initialized at the beginning of each radio frame by the cell-ID.

In addition, to achieve intra-cell interference randomization for the PUCCH DM RS, the cyclic time shift used in the second slot is hopped such that UEs which are assigned adjacent cyclic time shifts in the first slot use non-adjacent cyclic time shifts (with large separation) in the second slot [13]. A further benefit of using a different cyclic time shift in each slot is that the non-ideal cross-correlation between different base RS sequences is averaged (as the cross-correlation is not constant for all time lags).

16.5 Demodulation Reference Signals (DM RS)

The DM RSs associated with uplink PUSCH data or PUCCH control transmissions are primarily provided for channel estimation for coherent demodulation, and are present in every transmitted uplink slot.

16.5.1 RS Symbol Duration

For a given RS overhead, RSs could in theory be concentrated in one position in each slot, or divided up and positioned in multiple locations in each slot. Two alternatives were considered in the design of RSs in LTE, as shown in Figure 16.4:

- One RS symbol per slot, having the same duration as a data SC-FDMA symbol (sometimes referred to as a 'Long Block' (LB)), with the RS symbol having the same subcarrier spacing as the data SC-FDMA symbols.

- Two RS symbols per slot, each of half the duration of a data SC-FDMA symbol (sometimes referred to as a 'Short Block' (SB)), with the subcarrier spacing in the RS symbols being double that of the data SC-FDMA symbols (i.e. resulting in only six subcarriers per RB in the RS symbols).

As can be seen from Figure 16.4, twice as many SB RSs can be supported in a slot in the time domain compared to the LB RS structure with the same number of data SC-FDMA symbols. However, the frequency resolution of the SB RS is half that of the LB RS due to the subcarrier spacing being doubled.

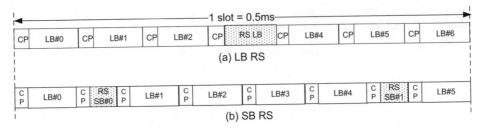

Figure 16.4 Example of slot formats considered with (a) one Long Block RS per slot and (b) two Short Block RSs per slot.

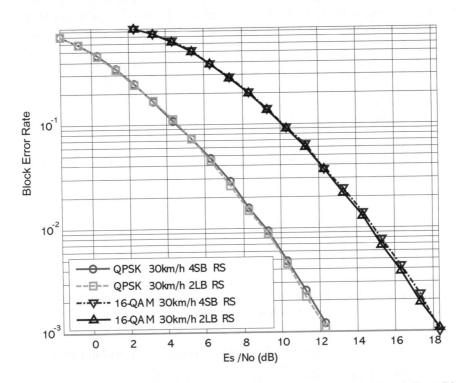

Figure 16.5 Demodulation performance comparison for Long Block RS and Short Block RS structure. Code rate $r = 1/2$, GSM Typical Urban (TU) channel model, 30 km/h, 2 GHz carrier frequency.

Figures 16.5 and 16.6 show a performance comparison of LB and SB RS for a resource allocation bandwidth of 1 RB for a duration of one subframe (1 ms), for UE speeds of 30 km/h and 250 km/h respectively. As can be seen from Figures 16.5 and 16.6, the performance of LB RS is similar to that for SB RS for medium speeds, with degradation at high speeds. However, with LB RS the signal parameterization remains the same as that of the downlink OFDM.

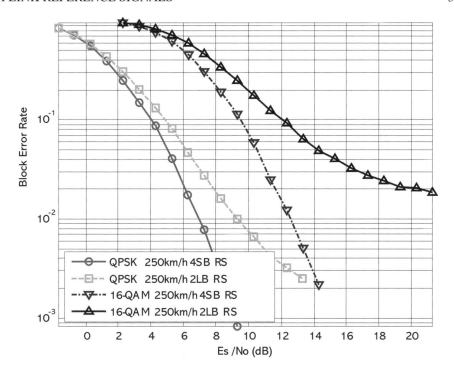

Figure 16.6 Demodulation performance comparison for Long Block RS and Short Block RS structure. Code rate $r = 1/2$, GSM Typical Urban (TU) channel model, 250 km/h, 2 GHz carrier frequency.

LB RSs also have the advantage of providing longer sequences for a given bandwidth allocation (due to there being twice as many RS subcarriers), and thus a larger number of RS sequences with desirable characteristics. Therefore the LB RS structure of Figure 16.4 was adopted in LTE for the PUSCH DM RS.

The exact position of the single PUSCH DM RS symbol in each uplink slot depends on whether the normal or extended CP is used, as shown in Figure 16.7. For the case of the normal CP with seven SC-FDMA symbols per slot, the PUSCH DM RS occupies the centre (i.e. fourth) SC-FDMA symbol. With six SC-FDMA symbols per slot in the case of the extended CP, the third SC-FDMA symbol is used. For PUCCH transmission, the position and number of DM RS depends on the type of uplink control information being transmitted, as discussed in Section 17.3.

The DM RS occupies the same RBs as the RB allocation for the uplink PUSCH data or PUCCH control transmission. Thus, the RS sequence length, $M_{\text{sc}}^{\text{RS}}$, is equal to the number of subcarriers allocated to the UE for PUSCH or PUCCH transmissions. Further, since the PUSCH RB allocation size is limited to multiples of two, three and/or five RBs (as explained in Section 15.3.1), the DM RS sequence lengths are also restricted to the same multiples.

As discussed in Section 16.2.2, the DM RS SC-FDMA symbol supports 12 cyclic time shifts with a spacing of 5.55 μs.

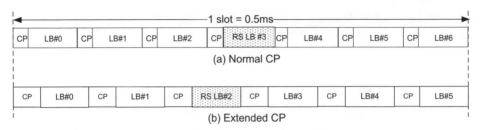

Figure 16.7 LTE uplink subframe configuration for PUSCH DM RS: (a) normal CP; (b) extended CP.

To support inter-cell interference randomization, cyclic time shift hopping is always enabled for DM RS as detailed in Section 16.4.

16.6   Uplink Sounding Reference Signals (SRS)

The SRS, which are not associated with uplink data and/or control transmission, are primarily used for channel quality estimation to enable frequency-selective scheduling on the uplink. However, they can be used for other purposes such as to enhance power control or to support various start-up functions for UEs not recently scheduled. Some examples include initial Modulation and Coding Scheme (MCS) selection, initial power control for data transmissions, timing advance, and so-called frequency semi-selective scheduling in which the frequency resource is assigned selectively for the first slot of a subframe and hops pseudo-randomly to a different frequency in the second slot [14].

16.6.1   SRS Subframe Configuration and Position

The subframes in which SRS are transmitted by any UE within the cell are indicated by cell-specific broadcast signalling. A 4-bit cell-specific 'srsSubframeConfiguration' parameter indicates 15 possible sets of subframes in which SRS may be transmitted within each radio frame (see reference [4], Section 5.5.3.3). This configurability provides flexibility in adjusting the SRS overhead depending on the deployment scenario. A 16th configuration switches the SRS off completely in the cell, which may for example be appropriate for a cell serving primarily high-speed UEs.

The SRS transmissions are always in the last SC-FDMA symbol in the configured subframes, as shown in Figure 16.8. Thus the SRS and DM RS are located in different SC-FDMA symbols. PUSCH data transmission is not permitted on the SC-FDMA symbol

Figure 16.8 Uplink subframe configuration with SRS symbol.

designated for SRS, resulting in a worst-case sounding overhead (with an SRS symbol in every subframe) of ~7%.

16.6.2 Duration and Periodicity of SRS Transmissions

The eNodeB in LTE may either request an individual SRS transmission from a UE or configure a UE to transmit SRS periodically until terminated; a 1-bit UE-specific signalling parameter, 'duration', indicates whether the requested SRS transmission is single or periodic. If periodic SRS transmissions are configured for a UE, the periodicity may be any of 2, 5, 10, 20, 40, 80, 160 or 320 ms; the SRS periodicity and SRS subframe offset within the period in which the UE should transmit its SRS are configured by a 10-bit UE-specific dedicated signalling parameter called 'srsConfigurationIndex'.

16.6.3 SRS Symbol Structure

In order to support frequency-selective scheduling between multiple UEs, it is necessary that SRS from different UEs with different sounding bandwidths can overlap. In order to support this, Interleaved FDMA (IFDMA, introduced in Section 15.2) is used in the SRS SC-FDMA symbol, with a RePetition Factor (RPF) of 2. The (time-domain) RPF is equivalent to a frequency-domain decimation factor, giving the spacing between occupied subcarriers of an SRS signal with a comb-like spectrum. Thus, RPF = 2 implies that the signal occupies every 2nd subcarrier within the allocated sounding bandwidth as shown by way of example in Figure 16.9. Using a larger RPF could in theory have provided more flexibility in how the bandwidth could be allocated between UEs, but it would have reduced the sounding sequence length (for a given sounding bandwidth) and the number of available SRS sequences (similar to the case for DM RS), and therefore the RPF was limited to 2 in LTE.

Due to the IFDMA structure of the SRS symbol, a UE is assigned, as part of its configurable SRS parameters, the 'transmissionComb' index (0 or 1) on which to transmit the SRS. The RS sequences used for the SRS are the same as for the DM RS, resulting in the SRS sequence length being restricted to multiples of two, three and/or five times the resource block size. In addition, the SRS bandwidth (in RBs) must be an even number, due to the RPF of 2 and the minimum SRS sequence length being 12. Therefore, the possible SRS bandwidths, N_{RB}^{SRS} (in number of RBs), and the SRS sequence length, M_{sc}^{SRS}, are respectively given by,

$$N_{RB}^{SRS} = 2^{(1+\alpha_2)} \cdot 3^{\alpha_3} \cdot 5^{\alpha_5}$$

$$M_{sc}^{SRS} = \tfrac{1}{2} \cdot N_{RB}^{SRS} \cdot 12 \tag{16.8}$$

where α_2, α_3, α_5 is a set of positive integers. Similarly to the DM RS, simultaneous SRS can be transmitted from multiple UEs using the same RBs and the same offset of the comb, using different cyclic time shifts of the same base sequence to achieve orthogonal separation (see Section 16.2.2). For the SRS, eight (evenly-spaced) cyclic time shifts per SRS-comb are supported (see [4], Section 5.5.3.1), with the cyclic shift being configured individually for each UE.

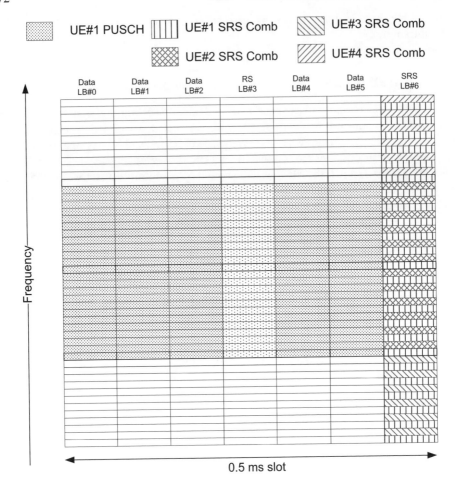

Figure 16.9 Sounding RS symbol structure with RPF = 2.

16.6.3.1   SRS Bandwidths

Some of the factors which affect the SRS bandwidth are the maximum power of the UE, the number of supportable sounding UEs, and the sounding bandwidth needed to benefit from uplink channel-dependent scheduling. Full bandwidth sounding provides the most complete channel information when the UE is sufficiently close to the eNodeB, but degrades as the path-loss increases when the UE cannot further increase its transmit power to maintain the transmission across the full bandwidth. Full bandwidth transmission of SRS also limits the number of simultaneous UEs whose channels can be sounded, due to the limited number of cyclic time shifts (eight cyclic time shifts per SRS-comb as explained in Section 16.6.3).

To improve the SNR and support a larger number of SRS, up to four SRS bandwidths can be simultaneously supported in LTE depending on the system bandwidth. To provide flexibility with the values for the SRS bandwidths, eight sets of four SRS bandwidths are

Table 16.1 Sounding RS BandWidth (BW) configurations for system bandwidths 40–60 RBs. (See [4], Table 5.5.3.2-2.) Reproduced by permission of © 3GPP.

SRS BW configuration	SRS-BW 0	SRS-BW 1	SRS-BW 2	SRS-BW 3
0	48	24	12	4
1	48	16	8	4
2	40	20	4	4
3	36	12	4	4
4	32	16	8	4
5	24	4	4	4
6	20	4	4	4
7	16	4	4	4

defined for each possible system bandwidth. RRC signalling indicates which of the eight sets is applicable in the cell by means of a 3-bit cell-specific parameter 'srsBandwidthConfiguration'. This allows some variability in the maximum SRS bandwidths, which is important as the SRS region does not include the PUCCH region near the edges of the system bandwidth (see Section 17.3), which is itself variable in bandwidth. An example of the eight sets of four SRS bandwidths applicable to uplink system bandwidths in the range 40–60 RBs is shown in Table 16.1 (see [4], Table 5.5.3.2-2).

The specific SRS bandwidth to be used by a given UE is configured by a further 2-bit UE-specific parameter, 'srsBandwidth'.

As can be seen from Table 16.1, the smallest sounding bandwidth supported is 4 RBs. A small sounding bandwidth of 4 RBs provides for higher-quality channel information from a power-limited UE. The sounding bandwidths are constrained to be multiples of each other, i.e. following a tree-like structure, to support frequency hopping of the different narrowband SRS bandwidths (see Section 5.5.3.2 of reference [4]). Frequency hopping can be enabled or disabled for an individual UE based on the value of the parameter 'frequencyDomainPosition'. The tree structure of the SRS bandwidths limits the possible starting positions for the different SRS bandwidths, reducing the overhead for signalling the starting position to 5 bits (signalled to each UE by the parameter 'Frequency-domain position'.

Table 16.2 summarizes the various SRS configurable parameters which are signalled to a UE [15].

16.7 Summary

The uplink reference signals provided in LTE fulfil an important function in facilitating channel estimation and channel sounding. The ZC-based sequence design can be seen to be a good match to this role, with constant amplitude in the frequency domain and the ability to provide a large number of sequences with zero or low correlation. This enables both interference randomization and interference coordination techniques to be employed in LTE system deployments, as appropriate to the scenario. A high degree of flexibility is provided for configuring the reference signals, especially for the sounding reference signals, where

Table 16.2 UL sounding RS configurable parameters.

Sounding RS parameter name	Significance	Signalling type
srsBandwidthConfiguration	Maximum SRS bandwidth in the cell	Cell-specific
srsSubframeConfiguration	Sets of subframes in which SRS may be transmitted in the cell	Cell-specific
srsBandwidth	SRS transmission bandwidth for a UE	UE-specific
frequencyDomainPosition	Frequency-domain position	UE-specific
srsHoppingBandwidth	Frequency hop size	UE-specific
duration	Single SRS or periodic	UE-specific
srsConfigurationIndex	Periodicity and subframe offset	UE-specific
transmissionComb	Transmission comb offset	UE-specific
$n_{\text{SRS}}^{\text{cs}}$	Cyclic shift	UE-specific

the overhead arising from their transmission can be traded off against the improvements in system efficiency which may be achievable from frequency-selective uplink scheduling.

References[5]

[1] D. C. Chu, 'Polyphase Codes With Good Periodic Correlation Properties'. *IEEE Trans. on Information Theory*, pp. 531–532, July 1972.

[2] B. M Popovic, 'Generalized Chirp-like Polyphase Sequences with Optimal Correlation Properties'. *IEEE Trans. on Information Theory*, Vol. 38, pp. 1406–1409, July 1992.

[3] Motorola, 'R1-060878: EUTRA SC-FDMA Uplink Pilot/Reference Signal Design & TP', www.3gpp.org, 3GPP TSG RAN WG1, meeting 44bis, Athens, Greece, March 2006.

[4] 3GPP Technical Specification 36.211, 'Evolved Universal Terrestrial Radio Access (E-UTRA); Physical channels and modulation (Release 8)', www.3gpp.org.

[5] Huawei, LG Electronics, NTT DoCoMo, and Panasonic, 'R1-080576: Way Forward on the Sequence Grouping for UL DM RS', www.3gpp.org, 3GPP TSG RAN WG1, meeting 51bis, Sevilla, Spain, January 2008.

[6] K. Fazel and G. P. Fettweis, *Multi-Carrier Spread-Spectrum*. Kluwer Academic Publishers, Dordrecht, Holland, 1997.

---

[5] All web sites confirmed 18th December 2008.

[7] Alcatel-Lucent, Ericsson, Freescale, Huawei, LGE, Motorola, Nokia, Nokia-Siemens Networks, NTT DoCoMo, Nortel, Panasonic, Qualcomm, and TI, 'R1-072584: Way Forward for PUSCH RS', www.3gpp.org, 3GPP TSG RAN WG1, meeting 49, Kobe, Japan, May 2007.

[8] Alcatel-Lucent, Ericsson, Freescale, Huawei, LGE, Motorola, Nokia, Nokia-Siemens Networks, NTT DoCoMo, Nortel, Panasonic, Qualcomm, and TI, 'R1-072585: Way forward for PUCCH RS', www.3gpp.org, 3GPP TSG RAN WG1, meeting 49, Kobe, Japan, May 2007.

[9] NTT DoCoMo, 'R1-074278: Hopping and Planning of Sequence Groups for Uplink RS', www.3gpp.org, 3GPP TSG RAN WG1, meeting 50bis, Shanghai, China, October 2007.

[10] Motorola, 'R1-080719: Hopping Patterns for UL RS', www.3gpp.org, 3GPP TSG RAN WG1, meeting 52, Sorrento, Italy, February 2008.

[11] Qualcomm Europe, Ericsson, Motorola, Samsung, Panasonic, and NTT DoCoMo, 'R1-081133: WF on UL DM-RS Hopping Pattern Generation', www.3gpp.org, 3GPP TSG RAN WG1, meeting 52, Sorrento, Italy, February 2008.

[12] Motorola, 'R1-071341: Uplink Reference Signal Planning Aspects', www.3gpp.org, 3GPP TSG RAN WG1, meeting 48bis, St. Julian's, Malta, March 2007.

[13] Panasonic, Samsung, and ETRI, 'R1-080983: Way forward on the Cyclic Shift Hopping for PUCCH', www.3gpp.org, 3GPP TSG RAN WG1, meeting 52, Sorrento, Italy, February 2008.

[14] Motorola, 'R1-073756: Benefit of Non-Persistent UL Sounding for Frequency Hopping PUSCH', www.3gpp.org, 3GPP TSG RAN WG1, meeting 50, Athens, Greece, August 2007.

[15] Ericsson, 'R1-082199: Physical-layer parameters to be configured by RRC', www.3gpp.org, 3GPP TSG RAN WG1, meeting 53, Kansas City, USA, May 2008.

17

Uplink Physical Channel Structure

Robert Love and Vijay Nangia

17.1 Introduction

The LTE Single-Carrier Frequency Division Multiple Access (SC-FDMA) uplink provides separate physical channels for the transmission of data and control signalling, the latter being predominantly to support the downlink data transmissions. The detailed structure of these channels, as explained in this chapter, is designed to make efficient use of the available frequency-domain resources, and to support effective multiplexing between data and control signalling.

LTE also introduces multiple antenna techniques to the uplink, including closed-loop antenna selection and Spatial Division Multiple Access (SDMA) or Multi-User Multiple-Input Multiple-Output (MU-MIMO). The physical layer transmissions of the LTE uplink are comprised of three physical channels and two signals:

- PRACH – Physical Random Access CHannel (see Chapter 19);

- PUSCH – Physical Uplink Shared CHannel (see Section 17.2);

- PUCCH – Physical Uplink Control CHannel (see Section 17.3);

- DM RS – DeModulation Reference Signal (see Section 16.5);

- SRS – Sounding Reference Signal (see Section 16.6).

The uplink physical channels, and their relationship to the higher-layer channels, are summarized in Figure 17.1.

LTE – The UMTS Long Term Evolution: from Theory to Practice   Stefania Sesia, Issam Toufik and Matthew Baker
© 2009 John Wiley & Sons, Ltd

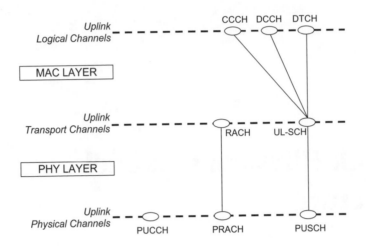

Figure 17.1 Summary of uplink physical channels and mapping to higher layers.

17.2   Uplink Shared Data Channel Structure

The Physical Uplink Shared CHannel (PUSCH), which carries data from the Uplink Shared Channel (UL-SCH) transport channel, uses DFT-Spread OFDM (DFT-S-OFDM), as described in Chapter 15. The transmit processing chain is shown in Figure 17.2. As explained in Chapter 10, the information bits are first channel-coded with a turbo code of mother code rate $r = 1/3$, which is adapted to a suitable final code rate by a rate-matching process. This is followed by symbol-level channel interleaving which follows a simple 'time-first' mapping [1] – in other words, adjacent data symbols end up being mapped first to adjacent SC-FDMA symbols in the time domain, and then across the subcarriers (see [2], Section 5.2.2.8). The coded and interleaved bits are then scrambled by a length-31 Gold code (as in Section 6.3) prior to modulation mapping, DFT-spreading, subcarrier mapping[1] and OFDM modulation. The signal is frequency-shifted by half a subcarrier prior to transmission, to avoid the distortion caused by the d.c. subcarrier being concentrated in one RB, as described in Section 15.3.3. The modulations supported are QPSK, 16QAM and 64QAM (the latter being only for the highest category of User Equipment (UE)).

The baseband SC-FDMA transmit signal for SC-FDMA symbol ℓ is thus of the form (see [3], Section 5.6),

$$s_\ell(t) = \sum_{k=-\lfloor N_{\mathrm{RB}}^{\mathrm{UL}} N_{\mathrm{sc}}^{\mathrm{RB}}/2 \rfloor}^{k=-\lceil N_{\mathrm{RB}}^{\mathrm{UL}} N_{\mathrm{sc}}^{\mathrm{RB}}/2 \rceil - 1} a_{k-,\ell} \exp[j 2\pi (k + 1/2) \Delta f (t - N_{\mathrm{CP},\ell} T_s)] \qquad (17.1)$$

for $0 \leq t < (N_{\mathrm{CP},\ell} + N) T_s$, where $N_{\mathrm{CP},\ell}$ is the number of samples of the Cyclic Prefix (CP) in SC-FDMA symbol ℓ (see Section 15.3), $N = 2048$ is the Inverse Fast Fourier

---

[1]Only localized mapping (i.e. to contiguous blocks of subcarriers) is supported for PUSCH and PUCCH transmissions in LTE.

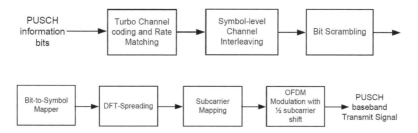

Figure 17.2 Uplink physical data channel processing.

Transform (IFFT) size, $\Delta f = 15$ kHz is the subcarrier spacing, $T_s = 1/(N \cdot \Delta f)$ is the sampling interval, N_{RB}^{UL} is the uplink system bandwidth in RBs, $N_{sc}^{RB} = 12$ is the number of subcarriers per resource block, $k^{(-)} = k + \lfloor N_{RB}^{UL} N_{sc}^{RB}/2 \rfloor$ and $a_{k,\ell}$ is the content of subcarrier k on symbol ℓ. For PUSCH data SC-FDMA symbols, $a_{k,\ell}$ is obtained by DFT-spreading the data QAM symbols, $[d_{0,\ell}, d_{1,\ell}, \ldots, d_{M_{sc}^{PUSCH}-1,\ell}]$ to be transmitted on data SC-FDMA symbol ℓ (see [3], Section 5.3.3),

$$a_{k,\ell} = \frac{1}{\sqrt{M_{sc}^{PUSCH}}} \sum_{i=0}^{M_{sc}^{PUSCH}-1} d_{i,\ell} e^{-j2\pi ik/M_{sc}^{PUSCH}} \qquad (17.2)$$

for $k = 0, 1, 2, \ldots, M_{sc}^{PUSCH} - 1$, where $M_{sc}^{PUSCH} = M_{RB}^{PUSCH} \cdot N_{sc}^{RB}$ and M_{RB}^{PUSCH} is the allocated PUSCH bandwidth in RBs.

As explained in Section 4.4.1, a Hybrid Automatic Repeat reQuest (HARQ) scheme is used, which in the uplink is synchronous, using N-channel stop and wait. This means that retransmissions occur in specific periodically-occurring subframes (HARQ channels). Further details of the HARQ operation are given in Section 10.3.2.5.

17.2.1 Scheduling Supported in LTE SC-FDMA Uplink

In the LTE uplink, both frequency-selective scheduling and non-frequency-selective scheduling are supported. The former is based on the eNodeB exploiting available channel knowledge to schedule a UE to transmit using specific Resource Blocks (RBs) in the frequency domain where the channel response is good. The latter does not make use of frequency-specific channel knowledge, but rather aims to benefit from frequency diversity during the transmission of each transport block. The possible techniques supported in LTE are discussed in more detail below. Intermediate approaches are also possible.

17.2.1.1 Frequency-Selective Scheduling

With frequency-selective scheduling, the same localized[2] allocation of transmission resources is typically used in both slots of a subframe – there is no frequency hopping during a subframe. The frequency-domain RB allocation and the Modulation and Coding Scheme (MCS) are chosen based on the location and quality of an above-average gain in the uplink

---

[2]Localized means that allocated RBs are consecutive in the frequency domain.

channel response [4]. In order to enable frequency-selective scheduling, timely channel quality information is needed at the eNodeB. One method for obtaining such information in LTE is by uplink channel sounding using the SRS described in Section 16.6. The performance of frequency-selective scheduling using the SRS depends on the sounding bandwidth and the quality of the channel estimate, the latter being a function of the transmission power spectral density used for the SRS. With a large sounding bandwidth, link quality can be evaluated on a larger number of RBs. However, this is likely to lead to the SRS being transmitted at a lower power density, due to the limited total UE transmit power, and this reduces the accuracy of the estimate for each RB within the sounding bandwidth. Conversely, sounding a smaller bandwidth can improve channel estimation on the sounded RBs but results in missing channel information for certain parts of the channel bandwidth, thus risking exclusion of the best quality RBs. As an example, experiments performed in reference [5] show that at least for a bandwidth of 5 MHz, frequency-selective scheduling based on full-band sounding outperforms narrower bandwidth sounding.

17.2.1.2   Frequency-Diverse or Non-Selective Scheduling

There are cases when no, or limited, frequency-specific channel quality information is available, for example because of SRS overhead constraints or high Doppler conditions. In such cases, it is preferable to exploit the frequency diversity of LTE's wideband channel.

In LTE, frequency hopping of a localized transmission is used to provide frequency-diversity. Two hopping modes are supported – hopping only between subframes (inter-subframe hopping), or hopping both between and within subframes (inter- and intra-subframe hopping). These modes are illustrated in Figure 17.3. Cell-specific broadcast signalling is used to configure the hopping mode (see [6], Section 8.4).

In case of intra-subframe hopping, a frequency hop occurs at the slot boundary in the middle of a subframe; this provides frequency diversity within a codeword (i.e. within a single transmission of transport block). On the other hand, inter-subframe hopping provides frequency diversity between HARQ retransmissions of a transport block, as the frequency allocation hops every allocated subframe.

Two methods are defined for the frequency hopping allocation (see [6], Section 8.4): either a pre-determined pseudo-random frequency hopping pattern (see reference [3], Section 5.3.4), or an explicit hopping offset signalled in the UL resource grant on the PDCCH. For uplink system bandwidths less than 50 RBs, the size of the hopping offset (modulo the system bandwidth) is approximately half the number of RBs available for PUSCH transmissions (i.e. $\lfloor N_{RB}^{PUSCH}/2 \rfloor$), while for uplink system bandwidths of 50 RBs or more, the possible hopping offsets are $\lfloor N_{RB}^{PUSCH}/2 \rfloor$, and $\pm \lfloor N_{RB}^{PUSCH}/4 \rfloor$ (see [6], Section 8.4).

Signalling the frequency hop via the uplink resource grant can be used for frequency semi-selective scheduling [7], in which the frequency resource is assigned selectively for the first slot of a subframe and frequency diversity is also achieved by hopping to a different frequency in the second slot. In some scenarios this may yield intermediate performance between that of fully frequency-selective and fully non-selective scheduling; this may be seen as one way to reduce the sounding overhead typically needed for fully frequency-selective scheduling.

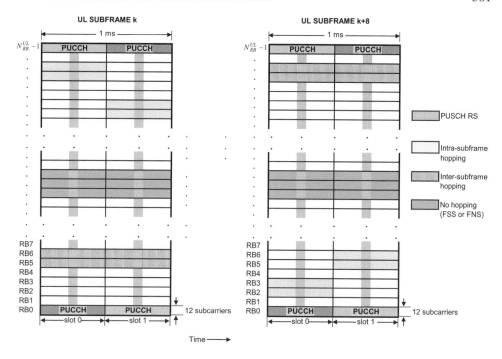

Figure 17.3 Uplink physical data channel processing.

17.3 Uplink Control Channel Design

In general, uplink control signalling in mobile communications systems can be divided into two categories:

- **Data-associated control signalling.** Control signalling which is always transmitted together with uplink data, and is used in the processing of that data. Examples include transport format indications, 'new data' indicators, and MIMO parameters.

- **Data non-associated control signalling.** Control signalling not associated with uplink data, transmitted independently of any uplink data packet. Examples include HARQ Acknowledgments (ACK/NACK) for downlink data packets, Channel Quality Indicators (CQIs), and MIMO feedback (such as Rank Indicator (RI) or Precoding Matrix Indicator (PMI)) for downlink transmissions. Scheduling Requests (SRs) for uplink transmissions also fall into this category.

In LTE, the low signalling latency afforded by the short subframe duration of 1 ms, together with the orthogonal nature of the uplink multiple access scheme which necessitates centralized resource allocation, make it appropriate for the eNodeB to be in full control of the uplink transmission parameters. Consequently uplink data-associated control signalling is not necessary in LTE, as the relevant information is already known to the eNodeB. Therefore only data non-associated control signalling is supported in the LTE uplink.

When simultaneous uplink PUSCH data and control signalling is scheduled for a UE, the control signalling is multiplexed together with the data prior to the DFT spreading, in order to preserve the single-carrier low-Cubic Metric (CM) property of the uplink transmission. The uplink control channel, PUCCH, is used by a UE to transmit any necessary control signalling only in subframes in which the UE has not been allocated any RBs for PUSCH transmission.

In the design of the PUCCH, special consideration was given to maintaining a low CM [8].

17.3.1 Physical Uplink Control Channel (PUCCH) Structure

The control signalling on the PUCCH is transmitted in a frequency region on the edges of the system bandwidth.

In order to minimize the resources needed for transmission of control signalling, the PUCCH in LTE is designed to exploit frequency diversity: each PUCCH transmission in one subframe is comprised of a single (0.5 ms) RB at or near one edge of the system bandwidth, followed (in the second slot of the subframe) by a second RB at or near the opposite edge of the system bandwidth, as shown in Figure 17.5; together, the two RBs are referred to as a *PUCCH region*. This design can achieve a frequency diversity benefit of approximately 2 dB compared to transmission in the same RB throughout the subframe.

At the same time, the narrow bandwidth of the PUCCH in each slot (only a single resource block) maximizes the power per subcarrier for a given total transmission power (see Figure 17.4), and therefore helps to fulfil stringent coverage requirements.

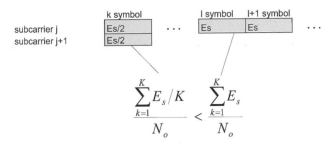

Figure 17.4 The link budget of a two-slot narrowband transmission exceeds that of a one-slot wider-band transmission, given equal coding gain.

Positioning the control regions at the edges of the system bandwidth has a number of advantages, including the following:

- The frequency diversity achieved through frequency hopping is maximized by allowing hopping from one edge of the band to the other.

- Out-Of-Band (OOB) emissions are smaller if a UE is only transmitting on a single RB per slot compared to multiple RBs. The PUCCH regions can therefore serve as a kind of guard band between the wider-bandwidth PUSCH transmissions of adjacent carriers, and therefore can improve coexistence [9].

Table 17.1 Typical numbers of PUCCH regions.

Bandwidth (MHz)	Number of 0.5 ms RBs subframe	Number of PUCCH regions
1.4	2	1
3	4	2
5	8	4
10	16	8
20	32	16

- Using control regions on the band edges maximizes the achievable PUSCH data rate, as the entire central portion of the band can be allocated to a single UE. If the control regions were in the central portion of a carrier, a UE bandwidth allocation would be limited to one side of the control region in order to maintain the single-carrier nature of the signal, thus limiting the maximum data achievable rate.

- Control regions on the band edges impose fewer constraints on the uplink data scheduling, both with and without inter-/intra-subframe frequency hopping.

The number of resource blocks (in each slot) that can be used for PUCCH transmission within the cell is N_{RB}^{PUCCH} (parameter 'pusch-HoppingOffset'). This is indicated to the UEs in the cell through broadcast signalling. Note that the number of PUCCH RBs per slot is the same as the number of PUCCH regions per subframe. Some typical expected numbers of PUCCH regions for different LTE bandwidths are shown in Table 17.1.

Figures 17.5 and 17.6 show respectively examples of even and odd numbers of PUCCH regions being configured in a cell. In the case of an even number of PUCCH regions (Figure 17.5), both RBs of each RB-pair (e.g. RB-pair 2 and RB-pair $N_{RB}^{UL} - 3$) are used for PUCCH transmission. However, for the case of an odd number of PUCCH regions (Figure 17.6), one RB of an RB-pair in each slot is not used for PUCCH (e.g. one RB of RB-pair 2 and RB-pair $N_{RB}^{UL} - 3$ is unused). In order to exploit the unused RBs in each slot in the case of an odd number of PUCCH regions, the eNodeB may schedule a UE with an intra-subframe frequency hopping (i.e. mirror hopping) PUSCH allocation in the unused RBs.

Alternatively, a UE can be assigned a localized allocation which includes the unused RB-pair, (e.g. RB-pair 2 or RB-pair $N_{RB}^{UL} - 3$). In this case, the UE will transmit PUSCH data on both RBs of the RB-pair, assuming that neither of the RBs are used for PUCCH by any UE in the subframe. Thus, the eNodeB scheduler can appropriately schedule PUSCH transmission (mirror hopping or localized) on the PUCCH RBs when they are under-utilized. The eNodeB may also choose to schedule low-power PUSCH transmission (e.g. from UEs close to the eNodeB) in the outer RBs of the configured PUCCH region, while the inner PUCCH region is used for PUCCH signalling. This can provide further reduction in OOB emissions which is necessary in some frequency bands, by moving higher-power PUCCH transmission (e.g. those from cell-edge UEs) slightly away from the edge of the band.

17.3.1.1 Multiplexing of UEs within a PUCCH Region

Control signalling from multiple UEs can be multiplexed into a single PUCCH region using orthogonal Code Division Multiplexing (CDM). In some scenarios this can have benefits

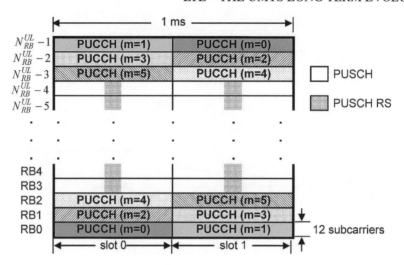

Figure 17.5   PUCCH uplink control structure with an even number of 'PUCCH Control Regions' ($N_{\mathrm{RB}}^{\mathrm{PUCCH}} = 6$). Reproduced by permission of © 3GPP.

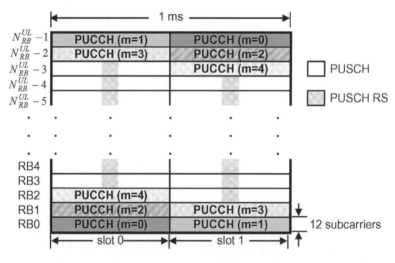

Figure 17.6 Example of odd number of PUCCH control RBs or regions ($N_{\mathrm{RB}}^{\mathrm{PUCCH}} = 5$).

over a pure FDM approach, as it reduces the need to limit the power differentials between the PUCCH transmissions of different UEs. One technique to provide orthogonality between UEs is by using cyclic time shifts of a sequence with suitable properties, as explained in Section 16.2.2. In a given SC-FDMA symbol, different cyclic time shifts of a waveform (e.g. a Zadoff–Chu (ZC) sequence as explained in Section 7.2.1) are modulated with a UE-specific QAM symbol carrying the necessary control signalling information, with the supported number of cyclic time shifts determining the number of UEs which can be multiplexed per

SC-FDMA symbol. As the PUCCH RB spans 12 subcarriers, and assuming the channel is approximately constant over the RB (i.e. a single-tap channel), the LTE PUCCH supports up to 12 cyclic shifts per PUCCH RB.

For control information transmissions with a small number of control signalling bits, such as 1- or 2-bit positive/negative acknowledgments (ACK/NACK), orthogonality is achieved between UEs by a combination of cyclic time shifts within an SC-FDMA symbol and SC-FDMA symbol time-domain spreading with orthogonal spreading codes, i.e. modulating the SC-FDMA symbols by elements of an orthogonal spreading code [10]. CDM of multiple UEs is used rather than Time Domain Multiplexing (TDM) because CDM enables the time duration of the transmission to be longer, which increases the total transmitted energy per signalling message in the case of a power-limited UE.

Thus, the LTE PUCCH control structure uses frequency-domain code multiplexing (different cyclic time shifts of a base sequence) and/or time-domain code multiplexing (different orthogonal block spreading codes), thereby providing an efficient, orthogonal control channel which supports small payloads (up to 22 coded bits) from multiple UEs simultaneously, together with and good operational capability at low SNR.

17.3.1.2 Control Signalling Information Carried on PUCCH

The control signalling information carried on the PUCCH can consist of:

- Scheduling Requests (SRs) (see Section 4.4.2.2).

- HARQ ACK/NACK in response to downlink data packets on (PDSCH). One ACK/NACK bit is transmitted in case of single codeword downlink transmission while two ACK/NACK bits are used in case of two codeword downlink transmission.

- CQI, which for the purposes of control signalling categorization, is taken to include the MIMO-related feedback consisting of RIs and PMI. 20 bits per subframe are used for the CQI.

The amount of control information which a UE can transmit in a subframe depends on the number of SC-FDMA symbols available for transmission of control signalling data (i.e. excluding SC-FDMA symbols used for reference signal transmission for coherent detection of the PUCCH). The PUCCH supports seven different formats depending on the information to be signalled. The mapping between the PUCCH format and the Uplink Control Information (UCI) supported in LTE is shown in Table 17.2 (see [6] Section 10.1, [3] Table 5.4-1).

The physical mapping of the PUCCH formats to the PUCCH regions is shown in Figure 17.7.

It can be seen that the PUCCH CQI formats 2/2a/2b are mapped and transmitted on the band-edge RBs (e.g. PUCCH region $m = 0, 1$) followed by a mixed PUCCH RB (if present, e.g. region $m = 2$) of CQI format 2/2a/2b and SR/HARQ ACK/NACK format 1/1a/1b, and then by PUCCH SR/HARQ ACK/NACK format 1/1a/1b (e.g. region $m = 4, 5$). The number of PUCCH RBs available for use by CQI format 2/2a/2b, N_{RB}^2, is indicated to the UEs in the cell by broadcast signalling.

Table 17.2 Supported uplink control information formats on PUCCH.

PUCCH Format	Uplink Control Information (UCI)
Format 1	Scheduling request (SR) (unmodulated waveform)
Format 1a	1-bit HARQ ACK/NACK with/without SR
Format 1b	2-bit HARQ ACK/NACK with/without SR
Format 2	CQI (20 coded bits)
Format 2	CQI and 1- or 2-bit HARQ ACK/NACK (20 bits) for extended CP only
Format 2a	CQI and 1-bit HARQ ACK/NACK (20 + 1 coded bits)
Format 2b	CQI and 2-bit HARQ ACK/NACK (20 + 2 coded bits)

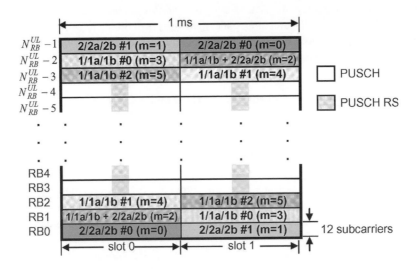

Figure 17.7 Physical mapping of PUCCH formats to PUCCH RBs or regions.

17.3.2 Channel Quality Indicator Transmission on PUCCH (Format 2)

The PUCCH CQI channel structure (Format 2) for one slot with normal CP is shown in Figure 17.8. SC-FDMA symbols 1 and 5 are used for DM RS transmissions in the case of the normal CP (while in the case of the extended CP only one RS is transmitted, on SC-FDMA symbol 3).

The number of RS symbols per slot results from a trade-off between channel estimation accuracy and the supportable code rate for the control signalling bits. For a small number of control information bits with a low SNR operating point (for a typical 1% target error rate), improving the channel estimation accuracy by using more RS symbols (e.g. 3) is more beneficial than being able to use a lower channel code rate. However, with larger numbers of control information bits the required SNR operating point increases, and the higher code rate resulting from a larger overhead of RS symbols becomes more critical, thus favouring fewer RS symbols. In view of these factors, two RS symbols per slot (in case of normal CP) was

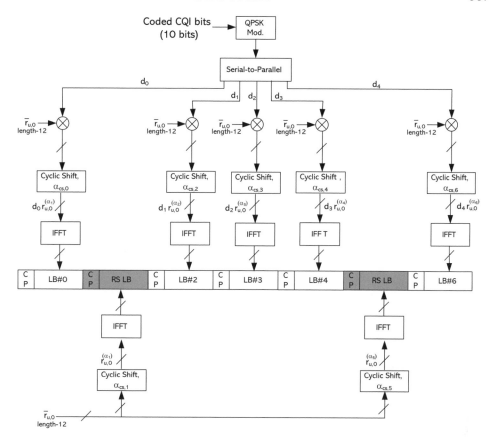

Figure 17.8 CQI channel structure for PUCCH format 2/2a/2b with normal CP for one slot.

considered to provide the best trade-off in terms of performance and RS overhead, given the payload sizes required.

10 CQI information bits are channel coded with a rate $1/2$ punctured $(20, k)$ Reed–Muller code (see reference [2], Section 5.2.3.3) to give 20 coded bits, which are then scrambled (in a similar way to PUSCH data with a length-31 Gold sequence) prior to QPSK constellation mapping. One QPSK modulated symbol is transmitted on each of the 10 SC-FDMA symbols in the subframe by modulating a cyclic time shift of the base RS sequence of length-12 prior to OFDM modulation. The 12 equally-spaced cyclic time shifts allow 12 different UEs to be orthogonally multiplexed on the same CQI PUCCH RB.

The DM RS signal sequence (on the 2nd and 6th SC-FDMA symbols for the normal CP, or the 4th symbol for the extended CP) is similar to the frequency domain CQI signal sequence but without the CQI data modulation.

In order to provide inter-cell interference randomization, cell-specific symbol-level cyclic time shift hopping is used, as described in Section 16.4. For example, the PUCCH cyclic time shift index on SC-FDMA symbol l in even slots n_s is obtained by adding (modulo-12) a pseudo-random cell-specific PUCCH cyclic shift offset to the assigned cyclic time shift n_{RS}^{PUCCH}. Intra-cell interference randomization is achieved by cyclic time shift remapping in the second slot as explained in Section 16.4.

A UE is semi-statically configured by higher layer signalling to report periodically different CQI, PMI, and RI types (see Section 10.2.1) on CQI PUCCH using a PUCCH *resource index* $n_{PUCCH}^{(2)}$, which indicates both the PUCCH region and the cyclic time shift to be used. The PUCCH region m used for the PUCCH format 2/2a/2b transmission (see Figure 17.7), is given by (see [3], Section 5.4.3)

$$m = \left\lfloor \frac{n_{PUCCH}^{(2)}}{12} \right\rfloor \tag{17.3}$$

and the assigned cyclic time shift, n_{RS}^{PUCCH}, is given by

$$n_{RS}^{PUCCH} = n_{PUCCH}^{(2)} \mod 12 \tag{17.4}$$

17.3.3 Multiplexing of CQI and HARQ ACK/NACK from a UE on PUCCH

In LTE, the simultaneous transmission of HARQ ACK/NACK and CQI by a UE can be enabled by UE-specific higher layer signalling. In case simultaneous transmission is not enabled, and the UE needs to transmit HARQ ACK/NACK on the PUCCH in the same subframe in which a CQI report has been configured, the CQI is dropped and only HARQ ACK/NACK is transmitted using the transmission structure detailed in Section 17.3.4.

In subframes where the eNodeB scheduler allows for simultaneous transmission of CQI and HARQ ACK/NACK from a UE, the CQI and the 1- or 2-bit HARQ ACK/NACK information needs to be multiplexed in the same PUCCH RB, while maintaining the low CM single carrier property of the signal. The method used to achieve this is different for the case of normal CP and extended CP as described in the following sections.

17.3.3.1   Multiplexing of CQI and HARQ ACK/NACK – Normal CP (Format 2a/2b)

The transmission structure for CQI data is the same as described in Section 17.3.2. In order to transmit a 1- or 2-bit HARQ ACK/NACK together with CQI (Format 2a/2b), the HARQ ACK/NACK bits (which are not scrambled) are BPSK/QPSK modulated as shown in Figure 17.9, resulting in a single HARQ ACK/NACK modulation symbol, d_{HARQ}. A positive acknowledgement (ACK) is encoded as a binary '1' and a negative acknowledgement (NACK) is encoded as a binary '0' (see [2], Section 5.2.3.4).

The single HARQ ACK/NACK modulation symbol, d_{HARQ}, is then used to modulate the second RS symbol (SC-FDMA symbol 5) in each CQI slot – i.e. ACK/NACK is signalled using the RS. It can be seen from Figure 17.9 that the modulation mapping is such that a NACK (or NACK, NACK in the case of two downlink MIMO codewords) is mapped to +1, resulting in a default NACK in case neither ACK nor NACK is transmitted (so-called Discontinuous Transmission (DTX)), as happens if the UE fails to detect the downlink grant on the Physical Downlink Control CHannel (PDCCH). In other words, a DTX (no RS modulation) is interpreted as a NACK by the eNodeB, triggering a downlink retransmission.

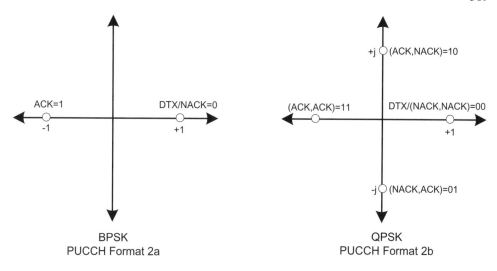

Figure 17.9 Constellation mapping for HARQ ACK/NACK.

As the one of the RS in a CQI slot is modulated by the HARQ ACK/NACK modulation symbol, some different ACK/NACK and CQI detection schemes are possible. In low-Doppler environments with little channel variation over the 0.5 ms slot, coherent detection of ACK/NACK and CQI can be achieved by using only the first RS symbol in the slot as the phase reference. Alternatively, to improve the channel estimation quality for CQI detection, an estimate of the HARQ ACK/NACK symbol can be used to undo the modulation on the second RS in the slot so that both RS symbols can be used for channel estimation and demodulation of CQI. In high-Doppler environments in which significant channel variations occur over a slot, relying on a single RS symbol for coherent detection degrades performance of ACK/NACK and CQI. In such cases blind decoding or multiple hypothesis testing of the different ACK/NACK combinations can be used to decode the ACK/NACK and CQI, selecting the hypothesis that maximizes the correlation between the received signal and the estimated CQI information [11] (i.e. a Maximum Likelihood detection).

17.3.3.2 Multiplexing of CQI and HARQ ACK/NACK – Extended CP (Format 2)

In the case of the extended CP (with one RS symbol per slot), the 1- or 2-bit HARQ ACK/NACK is jointly encoded with the CQI resulting in a (20, $k_{CQI} + k_{ACK/NACK}$) Reed–Muller based block code. A 20-bit codeword is transmitted on the PUCCH using the CQI channel structure in Section 17.3.2. The joint coding of the ACK/NACK and CQI is performed as shown in Figure 17.10. The largest number of information bits supported by the block code is 13, corresponding to $k_{CQI} = 11$ CQI bits and $k_{ACK/NACK} = 2$ bits (for two-codeword transmission in the downlink).

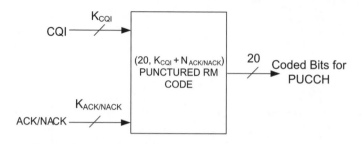

Figure 17.10 Joint coding of HARQ ACK/NACK and CQI for extended CP.

17.3.4 HARQ ACK/NACK Transmission on PUCCH (Format 1a/1b)

The PUCCH channel structure for HARQ ACK/NACK transmission with no CQI is shown in Figure 17.11 for one slot with normal CP. Three (two in case of extended CP) SC-FDMA symbols are used in the middle of the slot for RS transmission, with the remaining four SC-FDMA symbols being used for ACK/NACK transmission. Due to the small number of ACK/NACK bits, three RS symbols are used to improve the channel estimation accuracy for a lower SNR operating point than for the CQI structure in Section 17.3.2.

Both 1- and 2-bit acknowledgements are supported using BPSK and QPSK modulation respectively. The HARQ ACK/NACK bits (which are not scrambled) are BPSK/QPSK modulated according to the modulation mapping shown in Figure 17.9 (see [3], Table 5.4.1-1) resulting in a single HARQ ACK/NACK modulation symbol. A positive ACK is encoded as a binary '1' and a negative ACK (NACK) as a binary '0' (see [2], Section 5.2.3.4). The modulation mapping is the same as the mapping for 1- or 2-bit HARQ ACK/NACK when multiplexed with CQI for PUCCH formats 2a/2b.

As in the case of CQI transmission, the one BPSK/QPSK modulated symbol (which is phase-rotated by 90 degrees in the second slot) is transmitted on each SC-FDMA data symbol by modulating a cyclic time shift of the base RS sequence of length-12 (i.e. frequency-domain CDM) prior to OFDM modulation. In addition, as mentioned in Section 17.3.1.1, time-domain spreading with orthogonal (Walsh–Hadamard or DFT) spreading codes is used to code-division-multiplex UEs. Thus, a large number of UEs (data and RSs) can be multiplexed on the same PUCCH RB using frequency-domain and time-domain code multiplexing. The RSs from the different UEs are multiplexed in the same way as the data SC-FDMA symbols.

For the cyclic time shift multiplexing, the number of cyclic time shifts supported in an SC-FDMA symbol for PUCCH HARQ ACK/NACK RBs is configurable by a cell-specific higher-layer signalling parameter $\Delta_{\text{shift}}^{\text{PUCCH}} \in \{1, 2, 3\}$, indicating 12, 6, or 4 shifts respectively (see [3], Section 5.4.1). The value selected by the eNodeB for $\Delta_{\text{shift}}^{\text{PUCCH}}$ can be based on the expected delay spread in the cell.

For the time-domain spreading CDM, the number of supported spreading codes for ACK/NACK data is limited by the number of RS symbols, as the multiplexing capacity of RS is smaller than that of the data symbols due to smaller number of RS symbols. For example, in the case of six supportable cyclic time shifts and three (or two) orthogonal time spreading codes in case of normal (or extended) CP with three (or two) RS symbols, acknowledgments from 18 (or 12) different UEs can be multiplexed within one PUCCH RB. The length-2

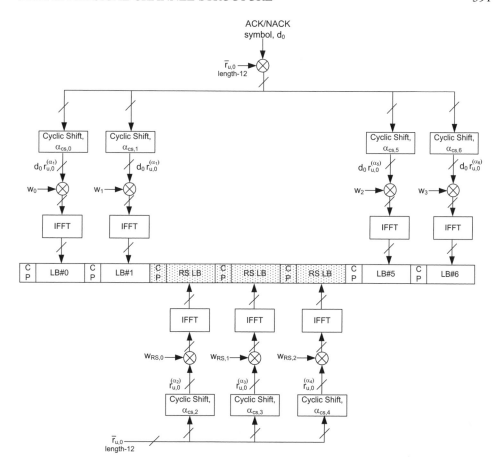

Figure 17.11 ACK/NACK structure – users are multiplexed using different cyclic shifts and time-domain spreading.

and length-4 orthogonal block spreading codes are based on Walsh–Hadamard codes, and the length-3 spreading codes are based on DFT codes as shown in Table 17.3. A subset of size-s orthogonal spreading codes of a particular length L ($s \leq L$) is used depending on the number of RS SC-FDMA symbols. For the normal CP with four data SC-FDMA symbols and three supportable orthogonal time spreading codes (due to there being three RS symbols), the indices 0, 1, 2 of the length-4 orthogonal spreading codes are used for the data time-domain block spreading.

Similarly, for the extended CP case with four data SC-FDMA symbols but only two RS symbols, orthogonal spreading code indices 0 and 2 of length-4 are used for the data block spreading codes. For the length-4 orthogonal codes, the code sequences used are such that subsets of the code sequences result in the minimum inter-code interference in high Doppler conditions where generally the orthogonality between the code sequences breaks down [12]. Table 17.4 summarizes the time-domain orthogonal spreading code lengths (i.e. spreading

Table 17.3 Time-domain orthogonal spreading code sequences. Reproduced by permission of © 3GPP.

Orthogonal code sequence index	Length-2 Walsh–Hadamard	Length-3 DFT	Length-4 Walsh–Hadamard
0	$\begin{bmatrix} +1 & +1 \end{bmatrix}$	$\begin{bmatrix} +1 & +1 & +1 \end{bmatrix}$	$\begin{bmatrix} +1 & +1 & +1 & +1 \end{bmatrix}$
1	$\begin{bmatrix} +1 & -1 \end{bmatrix}$	$\begin{bmatrix} 1 & e^{j2\pi/3} & e^{j4\pi/3} \end{bmatrix}$	$\begin{bmatrix} +1 & -1 & +1 & -1 \end{bmatrix}$
2	N/A	$\begin{bmatrix} 1 & e^{j4\pi/3} & e^{j2\pi/3} \end{bmatrix}$	$\begin{bmatrix} +1 & -1 & -1 & +1 \end{bmatrix}$
3	N/A	N/A	$\begin{bmatrix} +1 & +1 & -1 & -1 \end{bmatrix}$

Table 17.4 Spreading factors for time-domain orthogonal spreading codes for data and RS for PUCCH formats 1/1a/1b for normal and extended CP.

	Normal CP		Extended CP	
	Data, N_{SF}^{PUCCH}	RS, N_{RS}^{PUCCH}	Data, N_{SF}^{PUCCH}	RS, N_{RS}^{PUCCH}
Spreading factor	4	3	4	2

factors) for data and RS. The number of supportable orthogonal spreading codes is equal to the number of RS SC-FDMA symbols, N_{RS}^{PUCCH}.

It should be noted that it is possible for the transmission of HARQ ACK/NACK and SRS to be configured in the same subframe. If this occurs, the eNodeB can also configure (by cell-specific broadcast signalling) the way in which these transmissions are to be handled by the UE. One option is for the ACK/NACK to take precedence over the SRS, such that the SRS is not transmitted and only HARQ ACK/NACK is transmitted in the relevant subframe, according to the PUCCH ACK/NACK structure in Figure 17.11. The alternative is for the eNodeB to configure the UEs to use a shortened PUCCH transmission in such subframes, whereby the last SC-FDMA symbol of the ACK/NACK (i.e. the last SC-FDMA symbol in the second slot of the subframe is not transmitted; this is shown in Figure 17.12).

This maintains the low CM single-carrier property of the transmitted signal, by ensuring that a UE never needs to transmit both HARQ ACK/NACK and SRS symbols simultaneously, even if both signals are configured in the same subframe. If the last symbol of the ACK/NACK is not transmitted in the second slot of the subframe, this is known as a *shortened* PUCCH format, as shown in Figure 17.13.[3] For the shortened PUCCH, the length of the time-domain orthogonal block spreading code is reduced by one (compared to the first slot shown in Figure 17.11). Hence, it uses the length-3 DFT basis spreading codes in Table 17.3 in place of the length-4 Walsh–Hadamard codes.

The frequency-domain HARQ ACK/NACK signal sequence on data SC-FDMA symbol n is defined in [3], Section 5.4.1.

---

[3] Note that configuration of SRS in the same subframe as CQI or SR is not valid. Therefore the shortened PUCCH formats are only applicable for PUCCH formats 1a and 1b.

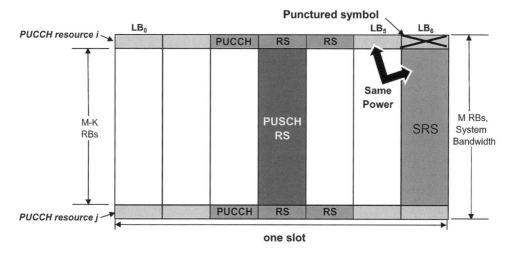

Figure 17.12 A UE may not simultaneously transmit on SRS and PUCCH or PUSCH, in order to avoid violating the single-carrier nature of the signal. Therefore, a PUCCH or PUSCH symbol may be punctured if SRS is transmitted.

The number of HARQ ACK/NACK resource indices $N_{\mathrm{PUCCH, RB}}^{(1)}$ corresponding to cyclic-time-shift/orthogonal-code combinations that can be supported in a PUCCH RB is given by

$$N_{\mathrm{PUCCH, RB}}^{(1)} = c \cdot P, \quad c = \begin{cases} 3 & \text{normal cyclic prefix} \\ 2 & \text{extended cyclic prefix} \end{cases} \tag{17.5}$$

where $P = 12/\Delta_{\mathrm{shift}}^{\mathrm{PUCCH}}$, and $\Delta_{\mathrm{shift}}^{\mathrm{PUCCH}} \in \{1, 2, 3\}$ is the number of equally spaced cyclic time shifts supported.

As in the case of CQI (see Section 17.3.2), cyclic time shift hopping (described in Section 16.4) is used to provide inter-cell interference randomization.

In the case of semi-persistently scheduled downlink data transmissions on the PDSCH (see Section 4.4.2.1) without a corresponding downlink grant on the control channel (PDCCH), the PUCCH ACK/NACK resource index $n_{\mathrm{PUCCH}}^{(1)}$ to be used by a UE is semi-statically configured by higher layer signalling. This PUCCH ACK/NACK resource is used for ACK/NACK transmission corresponding to initial HARQ transmission. For dynamically-scheduled downlink data transmissions (including HARQ retransmissions for semi-persistent data) on PDSCH (indicated by downlink assignment signalling on the PDCCH), the PUCCH HARQ ACK/NACK resource index $n_{\mathrm{PUCCH}}^{(1)}$ is implicitly determined based on the index of the first Control Channel Element (CCE, see Section 9.3) of the downlink control assignment.

The PUCCH region m used for the HARQ ACK/NACK with format 1/1a/1b transmission for the case with no mixed PUCCH region (shown in Figure 17.7), is given by [3], Section 5.4.3

$$m = \left\lfloor \frac{n_{\mathrm{PUCCH}}^{(1)}}{N_{\mathrm{PUCCH, RB}}^{(1)}} \right\rfloor + N_{\mathrm{RB}}^{(2)} \tag{17.6}$$

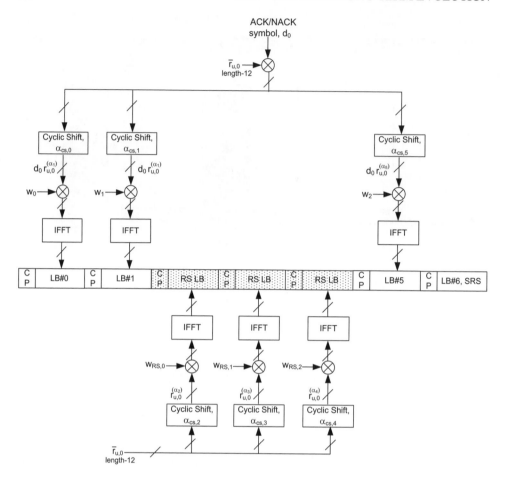

Figure 17.13 Shortened PUCCH ACK/NACK Structure when simultaneous SRS and ACK/NACK is enabled in the cell.

where $N_{\mathrm{RB}}^{(2)}$ is the number of RBs that are available for PUCCH formats 2/2a/2b and is a cell-specific broadcast parameter (see [3], Section 5.4).

The PUCCH resource index $n^{(1)}(n_s)$, corresponding to a combination of a cyclic time shift and orthogonal code ($n_{\mathrm{RS}}^{\mathrm{PUCCH}}$ and n_{oc}), within the PUCCH region m in even slots is given by

$$n^{(1)}(n_s) = n_{\mathrm{PUCCH}}^{(1)} \bmod N_{\mathrm{PUCCH, \, RB}}^{(1)} \qquad \text{for } n_s \bmod 2 = 0 \qquad (17.7)$$

The PUCCH resource index ($n_{\mathrm{RS}}^{\mathrm{PUCCH}}$, n_{oc}) allocation within a PUCCH RB format 1/1a/1b, is shown in Tables 17.5, 17.6, and 17.7, for $\Delta_{\mathrm{shift}}^{\mathrm{PUCCH}} \in \{1, 2, 3\}$ with 36, 18, and 12 resource indices respectively for the normal CP case [13]. For the extended CP, with two time-domain orthogonal spreading code sequences, only the first two columns of the orthogonal code

Table 17.5 PUCCH RB format 1/1a/1b resource index allocation, $\Delta_{\text{shift}}^{\text{PUCCH}} = 1$, 36 resource indices, normal CP.

Cyclic shift index, $n_{\text{RS}}^{\text{PUCCH}}$	Orthogonal code sequence index, n_{oc}		
	$n_{\text{oc}} = 0$	$n_{\text{oc}} = 1$	$n_{\text{oc}} = 2$
0	0	12	24
1	1	13	25
2	2	14	26
3	3	15	27
4	4	16	28
5	5	17	29
6	6	18	30
7	7	19	31
8	8	20	32
9	9	21	33
10	10	22	34
11	11	23	35

Table 17.6 PUCCH RB format 1/1a/1b resource index allocation, $\Delta_{\text{shift}}^{\text{PUCCH}} = 2$, 18 resource indices, normal CP.

Cyclic shift index, $n_{\text{RS}}^{\text{PUCCH}}$	Orthogonal code sequence index, n_{oc}		
	$n_{\text{oc}} = 0$	$n_{\text{oc}} = 1$	$n_{\text{oc}} = 2$
0	0		12
1		6	
2	1		13
3		7	
4	2		14
5		8	
6	3		15
7		9	
8	4		16
9		10	
10	5		17
11		11	

sequence index, $n_{\text{oc}} = 1, 2$ are used, resulting in 24, 12 and 8 resource indices for $\Delta_{\text{shift}}^{\text{PUCCH}} \in \{1, 2, 3\}$ respectively.

The PUCCH resources are first indexed in the cyclic time shift domain, followed by the orthogonal time spreading code domain.

The cyclic time shifts used on *adjacent* orthogonal codes can also be staggered, providing the opportunity to separate the channel estimates prior to de-spreading. As high Doppler breaks down the orthogonality between the spread blocks, offsetting the cyclic time shift

Table 17.7 PUCCH RB format 1/1a/1b resource index allocation, $\Delta_{\text{shift}}^{\text{PUCCH}} = 3$, 12 resource indices, normal CP.

Cyclic shift index, $n_{\text{RS}}^{\text{PUCCH}}$	Orthogonal code sequence index, n_{oc}		
	$n_{\text{oc}} = 0$	$n_{\text{oc}} = 1$	$n_{\text{oc}} = 2$
0	0		
1		4	
2			7
3	1		
4		5	
5			8
6	2		
7			
8			
9	3		
10		6	
11			9

values within each SC-FDMA symbol can restore orthogonality at moderate delay spreads. This can enhance the tracking of high Doppler channels [14].

In order to randomize intra-cell interference, PUCCH resource index remapping is used in the second slot [15]. Index remapping includes both cyclic shift remapping and orthogonal block spreading code remapping (similar to the case of CQI – see Section 17.3.2).

The PUCCH resource index remapping function in an odd slot is based on the PUCCH resource index in the even slot of the subframe, as defined in [3, Section 5.4.1].

17.3.5 Multiplexing of CQI and HARQ ACK/NACK in the Same PUCCH RB (Mixed PUCCH RB)

The multiplexing of CQI and HARQ ACK/NACK in different PUCCH RBs can in general simplify the system. However, in the case of small system bandwidths such as 1.4 MHz the control signalling overhead can become undesirably high with separate CQI and ACK/NACK RB allocations (two out of a total of six RBs for control signalling in 1.4 MHz system bandwidths). Therefore, multiplexing of CQI and ACK/NACK from different UEs in the same mixed PUCCH RB is supported in LTE to reduce the total control signalling overhead.

The ZC cyclic time shift structure facilitates the orthogonal multiplexing of CQI and ACK/NACK signals with different numbers of RS symbols. This is achieved by assigning different sets of adjacent cyclic time shifts to CQI and ACK/NACK signals [16] as shown in Table 17.8. As can be seen from this table, $N_{\text{cs}}^{(1)} \in \{0, 1, \ldots, 7\}$ cyclic time shifts are used for PUCCH ACK/NACK formats 1/1a/1b in the mixed PUCCH RB case, where $N_{\text{cs}}^{(1)}$ is a cell-specific broadcast parameter (see [3], Section 5.4) restricted to integer multiples of $\Delta_{\text{shift}}^{\text{PUCCH}}$. A guard cyclic time shift is used between the ACK/NACK and CQI cyclic shift resources to improve orthogonality and channel separation between UEs transmitting CQI and those transmitting ACK/NACK. To avoid mixing of the cyclic time shifts for ACK/NACK and

Table 17.8 Multiplexing of ACK/NACK (format 1/1a/1b) and CQI (format 2/2a/2b) from different UEs in the same (mixed) PUCCH RB by using different sets of cyclic time shifts.

Cyclic shift index	Cyclic shift index allocation
0	Format 1/1a/1b (HARQ ACK/NACK, SR) cyclic shifts
1	
2	
$\vdots$	
$N_{cs}^{(1)}$	
$N_{cs}^{(1)} + 1$	Guard cyclic shift
$N_{cs}^{(1)} + 2$	Format 2/2a/2b (CQI) cyclic shifts
$\vdots$	
10	
11	Guard cyclic shift

CQI, the cyclic time shift (i.e. the PUCCH resource index remapping function) for the odd slot of the subframe is not used; the same cyclic time shift as in the first slot of the subframe is used.

17.3.6 Scheduling Request (SR) Transmission on PUCCH (Format 1)

The structure of the SR PUCCH format 1 is the same as that of the ACK/NACK PUCCH format 1a/1b explained in Section 17.3.4, where a cyclic time shift of the base RS sequence is modulated with time-domain orthogonal block spreading. The SR uses simple On–Off keying, with the UE transmitting a SR with modulation symbol $d(0) = 1$ to request a PUSCH resource (positive SR transmission), and transmitting nothing when it does not request to be scheduled (negative SR).

Since the HARQ ACK/NACK structure is reused for the SR, different PUCCH resource indices (i.e. different cyclic time shift/orthogonal code combinations) in the same PUCCH region can be assigned for SR (Format 1) or HARQ ACK/NACK (Format 1a/1b) from different UEs. This results in orthogonal multiplexing of SR and HARQ ACK/NACK in the same PUCCH region. The PUCCH resource index to be used by a UE for SR transmission, $m_{\text{PUCCH,SRI}}^{(1)}$, is configured by UE-specific higher-layer signalling.

In case a UE needs to transmit a positive SR in the same subframe as a scheduled CQI transmission, the CQI is dropped and only the SR is transmitted, in order to maintain the low CM of the transmit signal. Similarly, in the case of simultaneous SR and SRS configuration, the UE does not transmit SRS and transmits only SR (see [6], Section 8.2).

If an SR and ACK/NACK happen to coincide in the same subframe, the UE transmits the ACK/NACK on the assigned SR PUCCH resource for a positive SR and transmits ACK/NACK on its assigned ACK/NACK PUCCH resource in case of a negative SR (see [6], Section 8.3). The constellation mapping for simultaneous HARQ ACK/NACK and SR is shown in Figure 17.14.

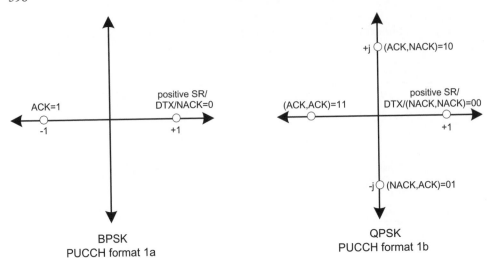

Figure 17.14 Constellation Mapping for ACK/NACK and SR for PUCCH format 1/1a/1b.

The modulation mapping is such that a NACK (or NACK, NACK in the case of two downlink MIMO codewords) is mapped to +1 resulting in a default NACK in case of DTX. This is similar to the case of multiplexing CQI and HARQ ACK/NACK for the normal CP case as described in Section 17.3.3.1.

17.4 Multiplexing of Control Signalling and UL-SCH Data on PUSCH

When control signalling is to be transmitted in a subframe in which the UE has been allocated transmission resources for the PUSCH, the control signalling is multiplexed together with the UL-SCH data prior to DFT-spreading, in order to preserve the low CM single-carrier property; the PUCCH is never transmitted in the same subframe as the PUSCH. The multiplexing of CQI/PMI, HARQ ACK/NACK, and RI with the PUSCH data symbols onto uplink resource elements is shown in Figure 17.15.

The number of resource elements used for each of CQI/PMI, ACK/NACK and RI is based on the MCS assigned for PUSCH and an offset parameter, Δ_{offset}^{CQI}, $\Delta_{offset}^{HARQ-ACK}$, or Δ_{offset}^{RI}, which is semi-statically configured by higher-layer signalling (see [2], Section 5.2.2.6). This allows different code rates to be used for the control signalling. PUSCH data and control information are never mapped to the same resource element. Control information is mapped in such a way that control is present in both slots of the subframe. Since the eNodeB has prior knowledge of uplink control signaling transmission, it can easily de-multiplex control and data packets.

As shown in Figure 17.15, CQI/PMI resources are placed at the beginning of the UL-SCH data resources and mapped sequentially to all SC-FDMA symbols on one subcarrier before continuing on the next subcarrier. The UL-SCH data is rate-matched (see Section 10.3.2.4)

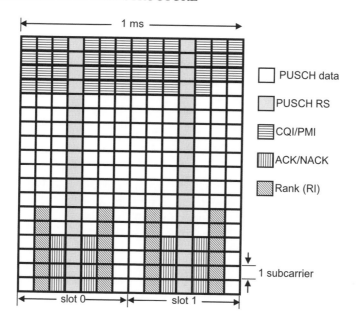

Figure 17.15 Multiplexing of control signalling with UL-SCH data.

around the CQI/PMI data. The same modulation order as UL-SCH data on PUSCH is used for CQI/PMI. For small CQI and/or PMI report sizes up to 11 bits, a $(32, k)$ block code, similar to the one used for PUCCH, is used, with optional circular repetition of encoded data (see [2], Section 5.2.2.6.4); no CRC is applied. For large CQI reporting modes (> 11 bits), an 8-bit CRC is attached and channel coding and rate matching is performed using the tail-biting convolutional code as described in Chapter 10.

The HARQ ACK/NACK resources are mapped to SC-FDMA symbols, by puncturing the UL-SCH PUSCH data. Positions next to the RS are used, so as to benefit from the best possible channel estimation. The maximum amount of resource for HARQ ACK/NACK is 4 SC-FDMA symbols.

The coded RI symbols are placed next to the HARQ ACK/NACK symbol positions irrespective of whether ACK/NACK is actually present in a given subframe. The modulation of the 1- or 2-bit ACK/NACK or RI is such that the Euclidean distance of the modulation symbols carrying ACK/NACK and RI is maximized (see [2], Section 5.2.2.6). The outermost constellation points of the higher-order 16/64-QAM PUSCH modulations are used, resulting in increased transmit power for ACK/NACK/RI relative to the average PUSCH data power.

The coding of the RI and CQI/PMI are separate, with the UL-SCH data being rate-matched around the RI resource elements similarly to the case of CQI/PMI.

In the case of 1-bit ACK/NACK or RI, repetition coding is used. For the case of 2-bit ACK/NACK/RI, a $(3, 2)$ simplex code is used with optional circular repetition of the encoded data (see [2], Section 5.2.2.6). The resulting code achieves the theoretical maximum values of the minimum Hamming distance of the output codewords in an efficient way. The $(3, 2)$ simplex codeword mapping is shown in Table 17.9.

Table 17.9 (3, 2) Simplex code for 2-bit ACK/NACK and RI.

2-bit Information Bit Sequence	3-bit Output Codeword
00	000
01	011
10	101
11	110

In LTE, control signalling (using QPSK modulation) can also be scheduled to be transmitted on PUSCH without UL-SCH data. The control signalling (CQI/PMI, RI, and/or HARQ ACK/NACK) are multiplexed prior to DFT-spreading, in order to preserve the low CM single-carrier property. The multiplexing of HARQ ACK/NACK and RI with the CQI/PMI QPSK symbols onto uplink resource elements is similar to that shown in Figure 17.15. HARQ ACK/NACK is mapped to SC-FDMA symbols next to the RS, by puncturing the CQI data and RI symbols, irrespective of whether ACK/NACK is actually present in a given subframe. The number of resource elements used for each of ACK/NACK and RI is based on a reference MCS for CQI/PMI and offset parameters, $\Delta_{\text{offset}}^{\text{CQI}}$, $\Delta_{\text{offset}}^{\text{HARQ-ACK}}$, or $\Delta_{\text{offset}}^{\text{RI}}$. The reference CQI/PMI MCS is computed from the CQI payload size and resource allocation. The channel coding and rate matching of the control signalling without UL-SCH data is the same as that of multiplexing control with UL-SCH data as described above.

17.5 Multiple-Antenna Techniques

In the first version of LTE, simultaneous transmissions from multiple-transmit antennas of a single UE are not supported. Only a single power-amplifier is assumed to be available at the UE. However, LTE does support closed-loop antenna selection transmit diversity in the uplink from UEs which have multiple transmit antennas.

LTE is also designed to support uplink SDMA, or Virtual Multi-User MIMO (MU-MIMO), and this is discussed in more detail in Section 17.5.2.

17.5.1 Closed-Loop Switched Antenna Diversity

Uplink closed-loop antenna selection (for up to two transmit antennas) is supported as an optional UE capability in LTE (see [17], Section 4.3.4.1).

If a UE signals that it supports uplink antenna selection, the eNodeB may take this capability into consideration when configuring and scheduling the UE.[4]

If the eNodeB enables a UE's closed-loop antenna selection capability, the SRS transmissions then alternate between the transmit antennas in successive configured SRS transmission subframes, irrespective of whether frequency hopping is enabled or disabled, except when the UE is configured for a single one-shot SRS transmission (see [6], Section 8.2).

---

[4]Alternatively, the eNodeB may permit the UE to use open-loop antenna selection, in which case the UE is free to determine which antenna to transmit from. This may be based on uplink-downlink channel reciprocity, for example in the case of TDD operation (see Section 23.5.2.5).

Table 17.10 UE transmit antenna selection CRC mask. Reproduced by permission of © 3GPP.

UE transmit antenna selection	Antenna selection mask $\langle x_{AS,0}, x_{AS,1}, \ldots\ldots\ldots, x_{AS,15} \rangle$
UE transmit antenna 0	$\langle 0, 0, 0, 0, 0, 0, 0, 0, 0, 0, 0, 0, 0, 0, 0, 0 \rangle$
UE transmit antenna 1	$\langle 0, 0, 0, 0, 0, 0, 0, 0, 0, 0, 0, 0, 0, 0, 0, 1 \rangle$

17.5.1.1 PUSCH UE Antenna Selection Indication

When closed-loop antenna selection is enabled, the eNodeB indicates which antenna should be used for the PUSCH by implicit coding in the uplink scheduling grant (Downlink Control Information (DCI) Format 0 – see Section 9.3): the 16 CRC parity bits are scrambled (modulo-2 addition) by an antenna selection mask [18], shown in Table 17.10 [19]. The antenna selection mask is applied in addition to the UE-ID masking which indicates for which UE the scheduling grant is intended. This implicit encoding avoids the use of an explicit antenna selection bit which would result in an increased overhead for UEs not supporting (or not configured) for transmit antenna selection.

It can be seen from Table 17.10 that the minimum Hamming distance between the antenna selection masks is only 1 rather than the maximum possible Hamming distance of 16. Since the CRC is masked by both the antenna selection indicator and the 16-bit UE-ID, the minimum Hamming distance between the correct UE-ID/antenna selection mask and the nearest erroneous UE-ID/antenna selection mask is 1 for any antenna selection mask. Out of the possible $2^{16} - 1$ incorrect masks, a vast majority ($2^{16} - 2$) result in the misidentification of the UE-ID, such that the performance is similar regardless of the Hamming distance between antenna selection masks. The primary advantage of using the mask $\langle 0, 0, 0, 0, 0, 0, 0, 0, 0, 0, 0, 0, 0, 0, 0, 1 \rangle$ is the ease of implementation due to simpler half-space identification, as the eNodeB can allocate UE-IDs with a fixed Most Significant Bit (e.g. MSB set to '0', or equivalently UE IDs from 0 to $2^{15} - 1$). The UE-ID can be detected directly from the 15 least significant bits of the decoded mask without needing to use the transmitted antenna selection mask (bit 16).

The UE behaviour for adaptive/non-adaptive HARQ retransmissions when configured for antenna selection is as follows [18]:

- **Adaptive HARQ.** The antenna indicator (via CRC masking) is always sent in the UL grant to indicate which antenna to use. For example, for a high Doppler UE with adaptive HARQ, the eNodeB might instruct the UE to alternate between the transmit antennas, or alternatively select the primary antenna. In typical UE implementations, a transmit antenna gain imbalance of 3 to 6 dB between the secondary and primary antenna is not uncommon.

- **Non-adaptive HARQ.** The UE behaviour is unspecified as to which antenna to use. Thus, for low Doppler conditions, the UE could use the same antenna as that signalled in the UL grant, while at high Doppler the UE could hop between antennas or just select the primary antenna. For large numbers of retransmissions with non-adaptive HARQ, the antenna indicated on the UL grant may not be the best and it is better to let the UE select the antenna to use. If the eNodeB wishes to instruct the UE to use a specific antenna for the retransmissions, it can use adaptive HARQ.

17.5.2   Multi-User 'Virtual' MIMO or SDMA

Uplink MU-MIMO consists of multiple UEs transmitting on the same set of RBs, each using a single transmit antenna. From the point of view of an individual UE, such a mode of operation is hardly visible, being predominantly a matter for the eNodeB to handle in terms of scheduling and uplink reception.

However, in order to support uplink MU-MIMO, LTE specifically provides orthogonal DM RS using different cyclic time shifts (see Section 16.2.2) to enable the eNodeB to derive independent channel estimates for the uplink from each UE.

A cell can assign up to eight different cyclic time shifts using the 3-bit PUSCH cyclic time shift offset on the uplink scheduling grant. As a maximum of eight cyclic time shifts can be assigned, SDMA of up to eight UEs can be supported in a cell. SDMA between cells (i.e. uplink inter-cell cooperation) is supported in LTE by assigning the same base sequence-groups and/or RS hopping patterns to the different cells as explained in Section 16.3.

17.6   Summary

The main uplink physical channels are the PUSCH for data transmission and the PUCCH for control signalling.

The PUSCH supports resource allocation for both frequency-selective scheduling and frequency-diverse transmissions, the latter being by means of intra- and/or inter-subframe frequency hopping.

Control signalling (consisting of ACK/NACK, CQI/PMI and RI) is carried by the PUCCH when no PUSCH resources have been allocated. The PUCCH is deliberately mapped to resource blocks near the edge of the system bandwidth, in order to reduce out-of-band emissions caused by data transmissions on the inner RBs, as well as maximizing flexibility for PUSCH scheduling in the central part of the band. In all cases of multiplexing different kinds of control signalling, the single-carrier property of the uplink signal is preserved. The control signalling from multiple UEs is multiplexed via orthogonal coding by using cyclic time shift orthogonality and/or time-domain block spreading.

LTE also introduces multiple antenna techniques in the uplink, in particular through the support of closed-loop switched antenna diversity and SDMA. These techniques are also cost-effective for a UE implementation, as they neither assume simultaneous transmissions from multiple UE antennas.

References[5]

[1] Motorola, 'R1-072671: Uplink Channel Interleaving', www.3gpp.org, 3GPP TSG RAN WG1, meeting 49bis, Orlando, USA, June 2007.

---

[5] All web sites confirmed 18th December 2008.

[2] 3GPP Technical Specification 36.212, 'Evolved Universal Terrestrial Radio Access (E-UTRA); Multiplexing and Channel Coding (Release 8)', www.3gpp.org.

[3] 3GPP Technical Specification 36.211, 'Evolved Universal Terrestrial Radio Access (E-UTRA); Physical Channels and Modulation (Release 8)', www.3gpp.org.

[4] B. Classon, P. Sartori, V. Nangia, X. Zhuang and K. Baum, 'Multi-dimensional Adaptation and Multi-User Scheduling Techniques for Wireless OFDM Systems' in *Proc. IEEE International Conference on Communications*, Anchorage, Alaska, May 2003.

[5] Motorola, 'R1-071340: Considerations and Recommendations for UL Sounding RS', www.3gpp.org, 3GPP TSG RAN WG1, meeting 48bis, St Julian's, Malta, March 2007.

[6] 3GPP Technical Specification 36.213, 'Evolved Universal Terrestrial Radio Access (E-UTRA); Physical Layer Procedures (Release 8)', www.3gpp.org.

[7] Motorola, 'R1-073756: Benefit of Non-Persistent UL Sounding for Frequency Hopping PUSCH', www.3gpp.org, 3GPP TSG RAN WG1, meeting 50, Athens, Greece, August 2007.

[8] A. Ghosh, R. Ratasuk, W. Xiao, B. Classon, V. Nangia, R. Love, D. Schwent and D. Wilson, 'Uplink Control Channel Design for 3GPP LTE' in *Proc. IEEE International Symposium on Personal, Indoor and Mobile Radio Communications*, Athens, Greece, September 2007.

[9] Motorola, 'R1-070084: Coexistance Simulation Results for 5MHz E-UTRA → UTRA FDD Uplink with Revised Simulation Assumptions', www.3gpp.org, 3GPP TSG RAN WG4, meeting 42, St Louis MO, USA, February 2007.

[10] S. Zhou, G. B. Giannakis and Martret C. L., 'Chip-Interleaved Block-Spread Code Division Multiple Access'. *IEEE Trans. on Communications*, Vol. 50, pp. 235–248, February 2002.

[11] Texas Instruments, 'R1-080190: Embedding ACK/NAK in CQI Reference Signals and Receiver Structures', www.3gpp.org, 3GPP TSG RAN WG1, meeting 51bis, Sevilla, Spain, January 2008.

[12] Samsung, 'R1-073564: Selection of Orthogonal Cover and Cyclic Shift for High Speed UL ACK Channels', www.3gpp.org, 3GPP TSG RAN WG1, meeting 50, Athens, Greece, August 2007.

[13] Samsung, Panasonic, Nokia, Nokia Siemens Networks and Texas Instruments, 'R1-080035: Joint Proposal on Uplink ACK/NACK Channelization', www.3gpp.org, 3GPP TSG RAN WG1, meeting 51bis, Sevilla, Spain, January 2008.

[14] Motorola, 'R1-062072: Uplink Reference Signal Multiplexing Structures for E-UTRA', www.3gpp.org, 3GPP TSG RAN WG1, meeting 46, Tallinn, Estonia, August 2006.

[15] Panasonic, Samsung and ETRI, 'R1-080983: Way forward on the Cyclic Shift Hopping for PUCCH', www.3gpp.org, 3GPP TSG RAN WG1, meeting 52, Sorrento, Italy, February 2008.

[16] Nokia Siemens Networks, Nokia and Texas Instruments, 'R1-080931: ACK/NACK channelization for PRBs containing both ACK/NACK and CQI', www.3gpp.org, 3GPP TSG RAN WG1, meeting 52, Sorrento, Italy, February 2008.

[17] 3GPP Technical Specification 36.306, 'Evolved Universal Terrestrial Radio Access (E-UTRA); User Equipment (UE) Radio Access Capabilities (Release 8)', www.3gpp.org.

[18] Motorola, Mitsubishi Electric and Nortel, 'R1-081928: Way Forward on Indication of UE Antenna Selection for PUSCH', www.3gpp.org, 3GPP TSG RAN WG1, meeting 53, Kansas City, USA, May 2008.

[19] Motorola, 'R1-082109: UE Transmit Antenna Selection', www.3gpp.org, 3GPP TSG RAN WG1, meeting 53, Kansas City, USA, May 2008.

18

Uplink Capacity and Coverage

Robert Love and Vijay Nangia

18.1 Introduction

This chapter considers the LTE uplink system performance in light of its different technology enablers and their respective practical constraints.

As introduced in Chapter 15, the LTE uplink multiple access employs a Single-Carrier Frequency Division Multiple Access (SC-FDMA) waveform, also known as DFT[1]-Spread OFDM (DFT-S-OFDM). This technology enables intra-cell orthogonality between the transmissions of different User Equipments (UEs) by means of a per-symbol Cyclic Prefix (CP), provided that each UE's transmissions are adequately time-aligned with its serving cell (see Figure 18.1). Intra-cell orthogonality is the main reason why a spectral efficiency two to three times higher than that of a WCDMA uplink can be achieved. The CP of the SC-FDMA waveform also facilitates the use of a simplified receiver structure with frequency-domain equalization in the eNodeB, which further improves uplink spectral efficiency.

The main drivers for the LTE uplink performance (many of which are generic to other FDM systems) are summarized in Table 18.1.

The discussion in this chapter includes a number of evaluations for typical LTE deployment scenarios, using a 10 MHz bandwidth, 20 dB building penetration loss and 3 km/h UE speed. Following the terminology used in 3GPP [1], we refer to the scenarios as *Case 1* and *Case 3* for inter-eNodeB distances of 500 m and 1732 m respectively.

18.2 Uplink Capacity

A wide range of factors affect the achievable uplink capacity. In this section we briefly discuss such aspects and explain their impact in the context of LTE. We then show some evaluations of the typical capacity of the LTE uplink.

[1]Discrete Fourier Transform.

LTE – The UMTS Long Term Evolution: from Theory to Practice   Stefania Sesia, Issam Toufik and Matthew Baker
© 2009 John Wiley & Sons, Ltd

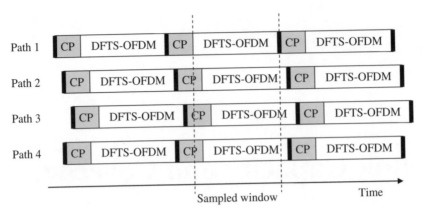

Figure 18.1 Timing of the sampled window if perfect timing estimation is assumed given multipath.

Table 18.1 LTE uplink capacity drivers.

LTE uplink capacity drivers	Remarks
Short subframe duration (1 ms) Low HARQ Round-Trip Time, RTT (8 ms)	Reduces latency and increases number of transmission opportunities
Low demodulation reference signal overhead	Only 2/14 of subframe resources
Low control channel overhead	Typically 16/100 of subframe resources
MMSE equalization	Attribute of basic DFT receiver
Intra-cell orthogonality (low inter-user interference)	Due to SC-FDMA + CP + time alignment
Power control and interference avoidance	Enable frequency reuse of 1
Semi-persistent scheduling	Avoids control channel limitation for services with periodic and frequent transmission of small packets (e.g. VoIP)
FDM resource allocation with fine frequency granularity (180 kHz)	Narrow bandwidth allocations improve cell edge performance when noise-limited

18.2.1 Factors Affecting Uplink Capacity

18.2.1.1 Uplink Frequency Reuse and Interference Mitigation

A frequency reuse factor of one is necessary to maximize spectral efficiency. However, it means that data and control channels in one cell will experience interference from other cells, especially the closest neighbour cells. In order to avoid low cell-edge throughput performance, it is important to employ interference mitigation techniques which allow an efficient trade-off between cell-edge performance and average spectral efficiency across the whole cell. Such interference mitigation techniques include:

- Coordination/avoidance;

- Inter-cell interference randomization;

- Frequency-domain spreading;

- Slow power control.

Uplink cell capacity in an SC-FDMA or OFDMA based network is constrained by interference levels from other active UEs. The ratio between the total received power spectral density I_o, including signal and interference, and the thermal noise level N_0 at an eNodeB receiver j in the time-frequency region k composed of N_{symb}^{UL} symbols and N_{sc}^{UL} subcarriers, is denoted $I_oT(j, k)$ and defined as

$$I_oT(j, k) = (I_o(j, k) + N_0)/N_0 \qquad (18.1)$$

where

$$I_o(j, k) = \sum_{i=1}^{N_{cells}} P_{i,k} T_{i,j}$$

$P_{i,k}$ is the average total power transmitted by UE i in region k, $T_{i,j}$ is the channel gain between UE i and cell j, and N_{cells} is the total number of cells.

Cell $I_oT(j, k)$ levels must be managed to maintain cell-edge coverage for uplink control channels including random access channels (which are vital for enabling the scheduling of the uplink data channels and supporting handovers between cells), as well as maintaining the minimum cell edge data rates for crucial services such as VoIP. Figure 18.2 shows the $I_oT(j, k)$ distribution for a three-sectored 19 cell-site LTE network based on deployment scenario Case 3. Without loss of generality the indices j and k in Equation (18.1) can be dropped, and I_oT represents the level for an arbitrary time-frequency region.

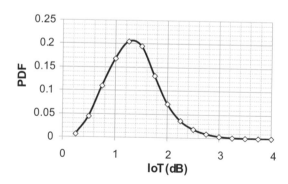

Figure 18.2 I_oT of arbitrary time plus frequency region for deployment scenario Case 3 with six UEs/cell using full buffer traffic.

Figure 18.3 shows how the average I_oT varies versus VoIP loading for deployment scenarios Case 1 and Case 3 with two different scheduling methods, A and B. Scheduling

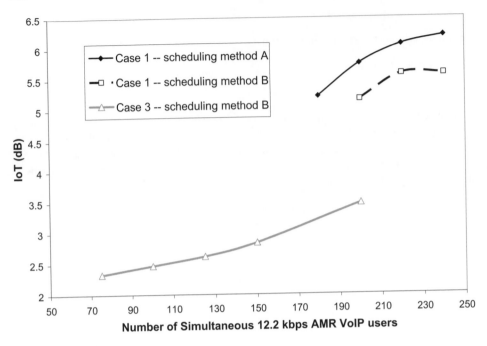

Figure 18.3 I_oT versus VoIP traffic loading for deployment scenarios Cases 1 and 3.

method A uses semi-persistent scheduling and method B uses persistent scheduling with a signalled resource assignment bitmap. The VoIP users were modelled using a 12.2 kbps Adaptive Multirate (AMR) codec. Fractional power control (see Section 20.3) was used and parameterized to minimize I_oT. As the load increases, it becomes more likely that the same time and frequency resource regions will be occupied in neighbouring cells, such that average I_oT increases.

18.2.1.2 Interference Management with Separate Control and Data Regions

Section 17.3.1 outlined the benefits of the fact that the main uplink control channel (the PUCCH[2]) is mapped to frequency resources which are separate from those of the data transmissions (on the PUSCH[3]) and located at the outer edges of the system bandwidth. Additionally, the use of separate frequency resources for control and data enables the I_oT to be separately managed and interference mitigation techniques to be optimized independently for control and data transmissions (e.g. by optimizing different power control algorithms for PUSCH and PUCCH as explained in Section 20.3). In addition, the specific control channel resources and codes used for transmission of Channel Quality Indicators (CQIs) and

---

[2]Physical Uplink Control CHannel.
[3]Physical Uplink Shared CHannel.

ACK/NACK[4] are designed to be assigned in such a manner as to reduce inter-cell and intra-cell control channel interference under light to moderately loaded conditions, as explained in Section 17.3.

18.2.1.3 Uplink Power Control and Interference Management

Uplink power control plays an important role in optimizing uplink system capacity. Each UE uses path-loss measurements based on the serving cell's Reference Symbols (RSs) to determine the transmission power needed to compensate for a fraction of the path-loss; such 'fractional' power control can be parameterized by the eNodeB to effect a tradeoff between overall spectral efficiency and cell edge performance, in conjunction with explicit closed-loop power control commands. These mechanisms are explained in detail in Section 20.3.

Power control can be combined with frequency-domain resource allocation strategies which allow interference coordination to further enhance cell edge performance and allow higher overall spectral efficiency. Uplink resource allocation techniques for interference coordination/avoidance are implementation-dependent and are not specified in LTE. One possible interference coordination technique for the uplink is to schedule UEs with comparable path-loss in adjacent cells to transmit in the same time-frequency resources [2]. On average, such a grouping of UEs with similar channel quality in adjacent cells results in the best cell edge performance, since it avoids strong interference from UEs close to the eNodeB in adjacent cells (especially if the front-to-back ratio of the eNodeBs antennas is low, resulting in significant interference between the cells).

Conversely, aligning UEs with different channel quality between cells will benefit the UEs with good channel quality, and hence improve peak data rates and average cell throughput.

Residual uplink interference after such interference coordination techniques may be mitigated by receiver techniques at the eNodeB, for example employing multiple receive antennas for beamforming, or using an Interference Rejection Combining (IRC) receiver [3] as discussed in Section 23.3.1.2.

18.2.1.4 Uplink Control Channel Overhead

In any wireless system there is a spectral efficiency trade-off between the fraction of transmission resources available for data and the fraction used for control signalling. The amount of transmission resources needed for control signalling depends on the error rate requirements (typically at or below 1% for control signalling), the size of the data packets (small packets generally result in a higher percentage overhead from signalling), and the time considered acceptable to switch from idle to active states. Data transmissions can make use of whatever transmission resources are left over after resources have been allocated for the control signalling. Hence minimizing control signalling is key to maximizing data spectral efficiency.

---

[4]ACKnowledgement/Negative ACKnowledgement.

18.2.1.5   Modulation and Number of HARQ Transmissions

Uplink data capacity is limited by the maximum available modulation order (16QAM) in the LTE uplink, except for the highest UE category, which supports 64QAM), the number of receive antennas at the eNodeB and the scheduling algorithm. The latter can trade off system capacity against fairness and cell edge throughput (typically considered as the 5-percentile user throughput) by appropriate configuration of the power control and resource allocation. Any Quality of Service (QoS) constraints (usually a delay limit) also affect the achievable uplink capacity.

Figure 18.4 shows the Shannon bound for the uplink rate, plotted versus the total transmission gain (i.e. the UE's maximum transmit power plus the UE and eNodeB antenna gains minus the path-loss). By enabling the use of 64QAM, the peak rate is increased to 14.3 Mbps for (from 9.6 Mbps for 16QAM) UEs closest to their serving cell. The figure also shows the number of HARQ transmissions necessary per packet for successful reception for each of three data rates given by 1.92 Mbps, 576 kbps, 12.2 kbps AMR RTP[5] Codec (320 bits/20 ms). The VoIP data rate assumes Transmission Time Interval (TTI) bundling (see Section 15.3) with a 244 bit full-rate AMR codec and a 10-byte RTP/UDP/RLC header.[6]

18.2.1.6   Delay Constraints and VoIP

The HARQ Round Trip Time (RTT) (i.e. the time between retransmissions) and the dropped packet delay bound (i.e. the maximum time for which a packet will be maintained in the transmitter queue before it is dropped) also affect capacity for delay-sensitive services like VoIP or other quasi-real-time services, especially in coverage-limited deployments. For a delay bound of 50 ms (typical for VoIP), the LTE uplink HARQ RTT of 8 ms (see Section 10.3.2.5) means that up to six transmissions per voice packet are possible.

Packet segmentation and TTI bundling (use of multiple HARQ processes per packet) may be used to improve coverage, as discussed in Section 15.3.

Best-effort data services (such as HTTP and FTP traffic) are not subject to tight delay constraints like VoIP, and can therefore trade off delay for higher throughput, since their transmissions can more effectively take advantage of changes in instantaneous channel conditions through appropriate scheduling.

18.2.1.7   Number of eNodeB Receive Antennas

Increasing the number of receive antennas from two to four in each cell of an LTE system can result in a 50% throughput improvement given full buffer traffic, even with single transmit antenna UEs. This improvement arises from the two-fold increase in energy obtained with four-antenna diversity compared to two antennas, as well as a diversity gain in fading channels. This is illustrated in Section 18.2.2.

---

[5] Adaptive Multi-Rate Real-time Transport Protocol.

[6] The 10-byte header is composed of 10 bits plus padding (RTP pre-header) plus 4 bytes (RTP/UDP/IP) plus 2 bytes (RLC/security) plus 16 bits (CRC).

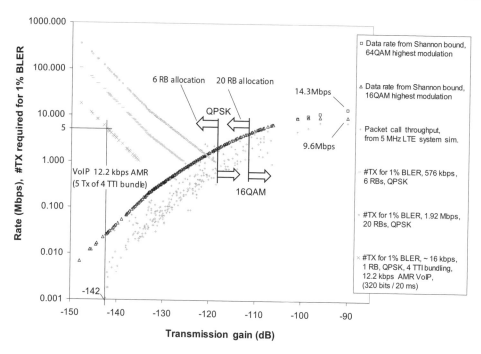

Figure 18.4 Maximum rate derived from Shannon bound conditioned on whether 16QAM or 64QAM is the highest-order modulation allowed. The total number of HARQ transmissions required to achieve 1% BLER for a given transmission gain is also shown. An Error Vector Magnitude (EVM) of 6.3% was assumed for the transmitted signal.

18.2.1.8 Minimum Size of Resource Allocation

In high Signal-to-Interference plus Noise Ratio (SINR) conditions, the maximum achievable capacity can be limited by the minimum amount of transmission resource which can be allocated to a single UE. This is defined by the TTI length in the time domain and the number of subcarriers per Resource Block (RB) in the frequency domain.

Figure 18.5 shows an example of the nominal VoIP capacity for a 1.25 MHz bandwidth for 0.5 ms and 1 ms TTI lengths, for a range of deployment scenarios with different SINR levels (see [4] for more details on the deployment scenarios considered). For the high SINR scenarios, it can clearly be seen that the supportable number of VoIP users in the cell reaches a *plateau* in the case of a 1 ms TTI, as the minimum resource allocation is larger than necessary to transmit a single VoIP packet when the SINR is very high.[7] This is the opposite effect of the benefits of 'TTI bundling' which we observed earlier, whereby a long TTI has the potential to increase the received energy per packet, and therefore to improve coverage at the cell edge when the UE is power-limited. The choice of a 1 ms TTI length in LTE is

---

[7]Note that this evaluation used an RB size of 15 subcarriers in the frequency domain, which gives a slightly larger minimum resource allocation size than is provided by the 12-subcarrier RBs in LTE.

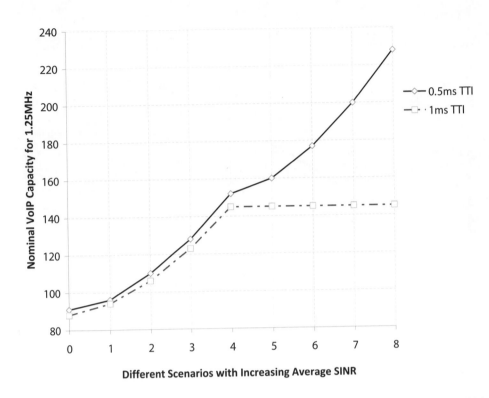

Figure 18.5 Impact of RB granularity on capacity [4]. Reproduced by permission of © 2006 Motorola.

therefore a compromise between high capacity in high-SINR conditions and good coverage at the cell edge.

In the frequency domain, the choice of a small RB size (only 12 subcarriers) helps to ensure that the minimum resource allocation size does not unduly limit capacity.

18.2.1.9 Transmitter and Receiver Impairments

Practical impairments in the transmitter and receiver are another important constraint in a real system. Here we introduce a method for modelling such impairments for the LTE uplink.

In the absence of receiver impairments such as non-ideal channel estimation, the DFT-S-OFDM symbol SINR, γ_{mmse}, based on an Minimum Mean-Squared Error (MMSE) Frequency Domain Equalizer (FDE) (see for example [5, 6]) is given by

$$\gamma_{mmse} = \frac{|G|^2}{M_{sc}|G| - |G|^2} \tag{18.2}$$

where $G = \sum_{k=0}^{M_{sc}-1} w_k |H_k|^2$, $w_k = (H_k + \gamma_o)^{-1}$ is the FDE tap for symbol k, and $\gamma_o = P/\sigma_n^2$ with P being the transmit power, σ_n^2 the noise variance, M_{sc} the number of subcarriers and H_k the DFT of the channel response.

The variable γ_{mmse} in Equation (18.2) can also be written as

$$\gamma_{mmse} = \frac{1}{M_{sc}/\gamma_{sum} - 1} \tag{18.3}$$

where

$$\gamma_{sum} = \sum_{k=0}^{M_{sc}-1} \gamma_k^{ideal}/(\gamma_k^{ideal} + 1)$$

and $\gamma_k^{ideal} = \gamma_o|H_k|^2$. H_k is the DFT of the channel response and w_k is the FDE tap for symbol k.

The Error Vector Magnitude (EVM, see Section 22.3.1.1) introduces a limitation in effective SINR which cannot exceed γ_{max}. More precisely the EVM-limited effective SINR γ_k^{lim} is given by

$$\frac{1}{\gamma_k^{lim}} = \frac{1}{\gamma_k^{ideal}} + \frac{1}{\gamma_{max}} \tag{18.4}$$

The symbol SINR accounting for non-ideal channel estimation and the SINR limit are then computed by

$$\frac{1}{\gamma_k^{chan}} = \frac{1}{\gamma_k^{lim}} + \frac{1}{\gamma_k^{pilot}} + \frac{1}{\gamma_k^{lim}\gamma_k^{pilot}} \tag{18.5}$$

where $\gamma_k^{pilot} = \gamma_k^{ideal}/b$, and b is the mean noise gain of the channel estimation filter [8]. Substituting γ_k^{chan} from Equation (18.5) into (18.3) results in a more realistic MMSE SINR, $\gamma_{mmse,n}$, for an uplink DFT-S-OFDM symbol.

The effect on the Block Error Rate (BLER) of the link can then be predicted by applying a technique such as Exponential Effective SINR Mapping (EESM) to the M_{symb} symbol SINRs of a packet (see Equation (18.6)) and then determining the corresponding BLER from AWGN reference performance curves for the resulting effective SINR, i.e. SINR$_{eff}$, given by

$$SINR_{eff} = -\beta \ln\left(\frac{1}{M_{symb}} \sum_{n=0}^{M_{symb}-1} \exp\left\{-\frac{\gamma_{mmse,n}}{\beta}\right\}\right) \tag{18.6}$$

where β is a parameter that must be optimized from link-level simulation results for every combination of modulation and coding rate.

18.2.2 LTE Uplink Capacity Evaluation

In this section a number of evaluations of the LTE uplink capacity are shown for different scenarios.

Tables 18.2 and 18.3 show spectral efficiency and cell-edge user throughput for deployment scenarios Case 1 and Case 3 (see Section 18.1) respectively, based on best-effort full buffer traffic. The tables also show the increase in performance relative to the 'Release-6' UMTS uplink (HSUPA). This improvement is attributable to the uplink capacity drivers summarized in Table 18.1.

Table 18.2 Summary of LTE performance evaluation results for Case 1 [7].

Tx-Rx configuration	Spectrum efficiency		Cell-edge user throughput		Avg. I_oT (dB)
	bps/Hz/cell	×HSUPA	bps/Hz/user	×HSUPA	
HSUPA baseline	0.332	×1.0	0.009	×1.0	5.1
LTE 1×2	0.735	×2.2	0.024	×2.5	5.2
LTE 1×2 MU-MIMO	0.675	×2.0	0.023	×2.4	5.2
LTE 1×4	1.103	×3.3	0.052	×5.5	5.1
LTE 2×2 SU-MIMO	0.776	×2.3	0.010	×1.1	5.5

Table 18.3 Summary of LTE performance evaluation results for Case 3 [7].

Tx-Rx configuration	Spectrum efficiency		Cell-edge user throughput		Avg. I_oT (dB)
	bps/Hz/cell	×HSUPA	bps/Hz/user	×HSUPA	
HSUPA baseline	0.316	×1.0	0.0023	×1.0	5.1
LTE 1 × 2	0.681	×2.2	0.0044	×2.0	4.5
LTE 1 × 2 MU-MIMO	0.622	×2.0	0.0023	×1.0	5.0
LTE 1 × 4	1.38	×3.3	0.0094	×4.2	2.7

Table 18.4 Time to upload 10 MB for different radio access technologies, based on system simulations.

Network	Normalized bandwidth (MHz)	Maximum throughput (Mbps)	Avg. user throughput[a] (kbps)	10 MB upload time[b] (s)
UMTS (Rel-99)	5.0	0.384	32–90	~890
CDMA2000 EV-DO Rev B	5.0	5.4	50–200	~385
HSUPA (Rel-6)	5.0	5.74	60–200	~385
WiMAX	5.0 (TDD)	1.7	48–135	~590
LTE	5.0	10.0	147–460	~174

[a] 25 to 8 FTP users/sector. [b] 8 FTP users/sector.

Table 18.4 gives a comparison of LTE against some other radio access technologies, in terms of the time needed to upload 10 MB of data via an FTP connection, given eight FTP users in each cell with a 5 MHz carrier. LTE shows the lowest delay of the systems considered.

The VoIP capacity is typically quoted in terms of a number of users per cell, as shown in Table 18.5. VoIP services on the LTE uplink benefit from HARQ, which compensates for the lack of uplink soft handover in LTE (i.e. no macro-diversity from reception at multiple cell sites). For coverage-limited UEs, the use of narrow resource allocations in the frequency domain (only 180 kHz bandwidth, corresponding to a single 12 subcarrier RB) helps to improve coverage by increasing the power per subcarrier.

Table 18.5 Summary of uplink VoIP capacity [9]. Reproduced by permission of © 2007
Motorola.

Deployment scenario	Average uplink VoIP capacity (UEs/cell)
Case 1	241
Case 3	123

18.3 LTE Uplink Coverage and Link Budget

Effective operation of a cellular communication system like LTE requires not only good
cell-edge performance, but also that the different channels are well balanced in terms of the
coverage they each provide.

Figure 18.6 shows the coverage provided by the CQI and ACK/NACK information
transported on the PUCCH, compared to the Physical Random Access Channel (PRACH),[8]
and a 12.2 kbps AMR VoIP service transported on the PUSCH. The maximum supportable
PUSCH cell-edge data rate and maximum data rate for a 5 MHz bandwidth FDD LTE carrier
are also shown. More details on the link budget are given in Table 18.8. Table 18.6 provides

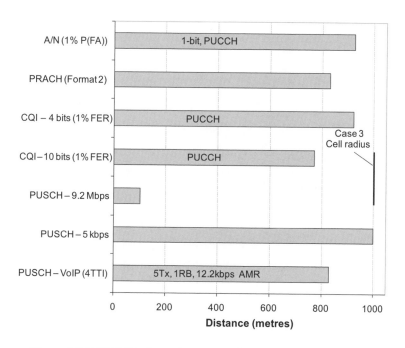

Figure 18.6 LTE UL channel coverage for Case 3 (5 MHz FDD).

---

[8]PRACH using format 2 in this example – see Section 19.4.2.2.

references for the SINR targets used in the uplink link budget and Table 18.7 provides the detection and false detection requirements for the uplink control channels. For comparison, Tables 18.9 and 18.10 provide similar information for the downlink link budget given in Table 18.11.

Table 18.6 Required UL E_s/N_0 for target error rate, 5 MHz LTE carrier.

Uplink control channel	Required E_s/N_0 for Target ER	Reference
ACK/NACK – PUCCH	$E_s/N_0 = -7.5$ dB $P(\text{NACK} \rightarrow \text{ACK}) = 10^{-4}$ $P(\text{ACK} \rightarrow \text{NACK}) = 10^{-2}$ $P(\text{DTX} \rightarrow \text{ACK}) = 10^{-2}$	[15–18]
CQI – PUCCH	$E_s/N_0 = -7.5$ dB (5-bit) for 1% BLER $E_s/N_0 = -4.5$ dB (10-bit) for 1% BLER	[17, 19, 20]
PRACH	$E_s/N_0 = -13.5$ dB for 1% PER	[10, 15, 17]

Table 18.7 BER targets for UL ACK/NACK signalling for LTE.

Event	Target quality
ACK missed detection	10^{-2}
DTX to ACK error	10^{-2}
NACK to ACK error	10^{-4}
CQI block error rate	Likely to be around 10^{-2}–10^{-1}

The link budget results are for deployment scenario Case 3, and assume a log-normal shadowing margin of 12.1 dB corresponding to 98-percentile single-cell area coverage reliability and a propagation model given by

$$\text{Propagation loss} = 128.1 + 37.6\log(\text{distance}) \qquad (18.7)$$

with distance given in metres. Given the other uplink link budget assumptions from Table 18.8, then only an average throughput of 5 kbps can be supported based on a single (1 ms) RB PUSCH transmission as shown in Figure 18.6. The downlink SINR at 1000 m given the corresponding downlink conditions is about -8.3 dB, which is also the SINR required for a 1% BLER on the Physical Broadcast CHannel (PBCH) SINR [10] as shown in Figure 18.7 and the link budget in Table 18.11. At 1000 m the transmission loss (i.e. the propagation loss minus the antenna gains plus the log-normal shadowing margin plus the penetration/body loss) is 146.2 dB, which must be supported by the uplink and downlink control channels in order to achieve the 98-percentile area coverage reliability.

Under these assumptions coverage gaps can be discussed:

Table 18.8 LTE uplink link budget for deployment scenario Case 3 for eNodeB with 2 receive antennas. Antenna gain + cable loss = 14 dB, penetration + body loss = 20 dB, interference margin = 3 dB, log-normal shadowing margin = 12.1 dB (98% area coverage reliability for propagation model given by Equation (18.7)); here VoIP uses 4 TTI bundling.

Uplink channel type	N. retx	N. RBs	Mod.	Dist.	Tx Pwr	Tx Loss	Per subcarrier Rx Pwr	$(I_o + N_0)W$	SINR
Unit				metres	dBm	dB	dBm	dBm	dB
VoIP, 12.2 kbps	4	1	2	829	24.0	143.1	-122.9	-124.2	1.3
VoIP, 12.2 kbps	2	1	2	724	24.0	140.9	-122.9	-124.2	1.3
PUSCH, 5 kbps	0	1	2	1000	24.0	146.2	-133.0	-124.2	-8.8
PUSCH, 9 Mbps	0	20	4	101	24.0	108.8	-108.6	-124.2	15.6
CQI, 10-bits (1% FER)	0	1	2	770	24.0	141.9	-128.7	-124.2	-4.5
CQI, 4-bits (1% FER)	0	1	2	922	24.0	144.9	-131.7	-124.2	-7.5
PRACH, Format 2	0	6	2	830	24.0	143.2	-137.7	-124.2	-13.5
A/N 1 bit (1% P(FA))	0	1	1	925	24.0	144.9	-131.7	-124.2	-7.5

Table 18.9 Required DL E_s/N_0 for target error rate, 5 MHz LTE carrier.

Downlink control channel	Required E_s/N_0 for target ER	Reference
PCFICH	$E_s/N_0 = -2.0$ dB for 0.1% BER	[11]
PBCH	$E_s/N_0 = -7.3$ dB for 1% BLER	(48-bit payload) [10]
PHICH, SF = 4	$E_s/N_0 = -2.8$ dB for 0.1% BER	[12]
	$E_s/N_0 = -4.0$ dB (48-bit) for 1% BLER	[13]
PDCCH, 8 CCEs	$E_s/N_0 = -5.3$ dB (36-bit)	[14]
	$E_s/N_0 = -6.8$ dB (24-bit)	
PCH Message	$E_s/N_0 = -8.3$ dB for 1% BLER	Same as PDSCH

Table 18.10 BER targets for downlink ACK/NACK signalling for LTE.

Event	Target quality
DL scheduling info. missed detection	10^{-2}
UL scheduling grant missed detection	10^{-2}
NACK to ACK error	10^{-4}
ACK to NACK error	10^{-4}

- **1-bit ACK/NACK and 4-bit CQI Coverage via PUCCH.** The 1-bit ACK/NACK and the 4-bit CQI, shown in Figure 18.6, each transported on a PUCCH, achieve almost the same coverage as the PBCH as shown in Figure 18.7, but fall short by approximately 1.3 dB as determined by the difference in transmission loss (146.2–144.9 dB) given in the uplink link budget in Table 18.8. A repetition factor of 2 may be used to close the coverage gap.

Table 18.11 LTE downlink link budget for deployment scenario Case 3 for 1 eNodeB transmit antenna. Antenna gain + cable loss = 14 dB, penetration + body loss = 20 dB, $I_o/N_0 = 1$ dB, log-normal shadowing margin = 12.1 dB (98% area coverage reliability for propagation model given by Equation (18.7)), control region average interference reduction due to power sharing and randomization = 2 dB.

Downlink channel type	N. retx	N. RBs	N. Sym	Mod	Dist.	Tx Pwr	Tx Loss	Per subcarrier Rx Pwr	$(I_o + N_0)W$	SINR
Unit					metres	dBm	dB	dBm	dBm	dB
PCH 96-bits	0	10	11	2	1000	43.0	146.2	−128.0	−119.7	−8.3
PDCCH 48-bits	0	8	3	2	868	43.0	143.9	−125.7	−119.7	−4.0
PDCCH 36-bits	0	8	3	2	940	43.0	145.2	−127.0	−119.7	−5.3
PDCCH 24-bits	0	8	3	2	1000	43.0	146.2	−128.0	−119.7	−6.3
PBCH (1.0% BLER)	3	6	4	2	1000	43.0	146.2	−122.0	−119.7	−2.3
PHICH (0.1% BER)	0	1	1	1	760	43.0	141.7	−123.5	−119.7	−1.8
PCFICH (0.1% BER)	0	1.33	1	2	770	43.0	141.9	−123.7	−119.7	−2.0

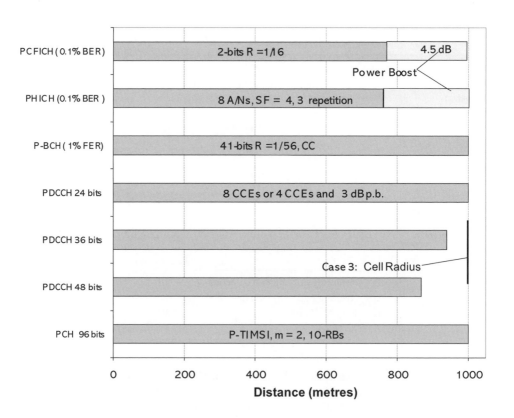

Figure 18.7 LTE DL channel coverage for Case 3 (5 MHz FDD).

- **12.2 kbps AMR VoIP Coverage via PUSCH.** Figure 18.6 shows that even TTI bundling over four subframes (i.e. allocating four HARQ processes so that each VoIP packet can occupy four consecutive TTIs) combined with five HARQ transmissions per packet is not enough to close the coverage gap for a VoIP 12.2 kbps AMR service. Some relaxation in error rate or reduction of the AMR codec rate (e.g. 7.95 kbps) may be enough to close the coverage gap.

- **PRACH coverage.** A repeated RACH preamble burst (2×800 μs) is needed for the PRACH to achieve 98-percentile or better area coverage reliability, since PRACH format 2 (see Section 19.4.2.2), as shown in Figure 18.7, only supports a cell radius of about 0.8 km which is close to the range supported by one Zadoff–Chu root sequence (0.78 km). With a single sequence a total received preamble energy per sequence of approximately 18 dB ($E_s/N_0 \sim -11.5$ dB) is required to meet missed detection and false alarm probabilities of less than 1% [21]. PRACH format 2 with repetition is slightly better with a required $E_s/N_0 = -13.5$ dB for the same error probabilities.

18.4 Summary

This chapter highlights the main factors affecting uplink capacity and coverage for the SC-FDMA-based LTE uplink. It is possible to observe how uplink coverage issues have led to particular design choices in LTE, such as the size of a resource block, the length of the scheduling interval (TTI), and the design of the control channels.

Evaluations can be carried out to examine the effect of each relevant factor, leading to the LTE performance being able to be characterized by a variety of metrics. Such metrics include average throughput, cell-edge throughput, number of VoIP users, and FTP download time. This enables LTE to be compared against other radio access technologies.

Depending on the requirements of particular deployments, the eNodeB has the freedom to balance average throughput against cell-edge coverage and fairness.

References[9]

[1] 3GPP Technical Report 25.814, 'Physical Layer Aspects for Evolved UTRA, (Release 7)', www.3gpp.org.

[2] W. Xiao, R. Ratasuk, A. Ghosh, R. Love, S. Yakun and R. Nory, 'Uplink Power Control, Interference Coordination and Resource Allocation for 3GPP E-UTRA' in *Proc. IEEE Vehicular Technology Conference*, September 2006.

[3] J. Winters, 'Optimum Combining in Digital Mobile Radio with Cochannel Interference'. *IEEE Journal on Selected Areas in Communications*, Vol. 2, July 1984.

[4] Motorola, 'R1-062058: E-UTRA TTI Size and Number of TTIs', www.3gpp.org, 3GPP TSG RAN WG1, meeting 46, Tallinn, Estonia, August 2006.

[5] Motorola, 'R1-050719: Simulation Methodology for E-UTRA UL: IFDMA and DFT-SOFDM', www.3gpp.org, 3GPP TSG RAN WG1, meeting 42, London, UK, August 2005.

[6] Samsung, 'R1-051352: Simulation Methodology for E-UTRA UL: IFDMA and DFT-SOFDM', www.3gpp.org, 3GPP TSG RAN WG1, meeting 43, Seoul, Korea, November 2005.

---

[9]All web sites confirmed 18th December 2008.

[7] Nokia, 'R1-072261: Summary of UL LTE Performance Evaluation', www.3gpp.org, 3GPP TSG RAN WG1, meeting 49, Kobe, Japan, May 2007.

[8] Motorola, 'R1-061551: LTE Uplink System Performance of VoIP', www.3gpp.org, 3GPP TSG RAN WG1, meeting 45, Shanghai, China, May 2006.

[9] Motorola, 'R1-072188: Performance Evaluation Checkpoint: VoIP Summary', www.3gpp.org, 3GPP TSG RAN WG1, meeting 49, Kobe, Japan, May 2007.

[10] Motorola, 'R1-072665: PBCH Design', www.3gpp.org, 3GPP TSG RAN WG1, meeting 49bis, Orlando FL, USA, June 2007.

[11] Motorola, 'R1-072692: Coding and Transmission of Control Channel Format Indicator', www.3gpp.org, 3GPP TSG RAN WG1, meeting 49bis, Orlando FL, USA, June 2007.

[12] Motorola, 'R1-072689: Downlink Acknowledgement Mapping to RE's', www.3gpp.org, 3GPP TSG RAN WG1, meeting 49bis, Orlando FL, USA, June 2007.

[13] Motorola, 'R1-072690: Support of Precoding for E-UTRA DL L1/L2 Control Channel', www.3gpp.org, 3GPP TSG RAN WG1, meeting 49bis, Orlando FL, USA, June 2007.

[14] Motorola, 'R1-073374: PDCCH Channel Estimation Impact (1 versus 2 DRS) on System Performance', www.3gpp.org, 3GPP TSG RAN WG1, meeting 50, Athens, Greece, August 2007.

[15] Ericsson, 'R1-081529: On Ack/Nack Repetition', www.3gpp.org, 3GPP TSG RAN WG1, meeting 52bis, Shenzhen, China, August 2008.

[16] Motorola, 'R1-073371: E-UTRA Coverage', www.3gpp.org, 3GPP TSG RAN WG1, meeting 50, Athens, Greece, August 2007.

[17] Nokia, 'R1-081461: Coverage Comparison between PUSCH, PUCCH, and RACH', www.3gpp.org, 3GPP TSG RAN WG1, meeting 52bis, Shenzhen, China, April 2008.

[18] Motorola, 'R1-081291: Repetition of UL ACK/NACK on PUCCH', www.3gpp.org, 3GPP TSG RAN WG1, meeting 52bis, Shenzhen, China, April 2008.

[19] Motorola, 'R1-073385: CQI Coding Schemes', www.3gpp.org, 3GPP TSG RAN WG1, meeting 50, Athens, Greece, August 2007.

[20] Motorola, 'R1-072705: Uplink CQI Channel Structure', www.3gpp.org, 3GPP TSG RAN WG1, meeting 49bis, Orlando FL, USA, June 2007.

[21] Panasonic and NTT DoCoMo, 'R1-062175: Random Access Burst Design for E-UTRA', www.3gpp.org, 3GPP TSG RAN WG1, meeting 46, Tallinn, Estonia, August 2006.

19

Random Access

Pierre Bertrand and Jing Jiang

19.1 Introduction

An LTE User Equipment (UE) can only be scheduled for uplink transmission if its uplink transmission timing is synchronized. The LTE Random Access CHannel (RACH) therefore plays a key role as an interface between non-synchronized UEs and the orthogonal transmission scheme of the LTE uplink radio access. In this chapter, the main roles of the LTE RACH are elaborated, together with its differences from the Wideband Code Division Multiple Access (WCDMA) RACH. The rationale for the design of the LTE Physical Random Access CHannel (PRACH) is explained, and some possible implementation options are discussed for both the UE and the eNodeB.

19.2 Random Access Usage and Requirements in LTE

In WCDMA, the RACH is primarily used for initial network access and short message transmission. LTE likewise uses the RACH for initial network access, but in LTE the RACH cannot carry any user data, which is exclusively sent on the Physical Uplink Shared CHannel (PUSCH). Instead, the LTE RACH is used to achieve uplink time synchronization for a UE which either has not yet acquired, or has lost, its uplink synchronization. Once uplink synchronization is achieved for a UE, the eNodeB can schedule orthogonal uplink transmission resources for it. Relevant scenarios in which the RACH is used are therefore:

(1) A UE in RRC_CONNECTED state, but not uplink-synchronized, needing to send new uplink data or control information (e.g. an event-triggered measurement report);

(2) A UE in RRC_CONNECTED state, but not uplink-synchronized, needing to receive new downlink data, and therefore to transmit corresponding ACK/NACK in the uplink;

(3) A UE in RRC_CONNECTED state, handing over from its current serving cell to a target cell;

(4) A transition from RRC_IDLE state to RRC_CONNECTED, for example for initial access or tracking area updates;

(5) Recovering from radio link failure.

One additional exceptional case is that an uplink-synchronized UE is allowed to use the RACH to send a Scheduling Request (SR) if it does not have any other uplink resource allocated in which to send the SR. These roles require the LTE RACH to be designed for low latency, as well as good detection probability at low Signal-to-Noise (SNR) (for cell edge UEs undergoing handover) in order to guarantee similar coverage to that of the PUSCH and Physical Uplink Control CHannel (PUCCH).[1]

A successful RACH attempt should allow subsequent UE transmissions to be inserted among the scheduled synchronized transmissions of other UEs. This sets the required timing estimation accuracy which must be achievable from the RACH, and hence the required RACH transmission bandwidth: due to the Cyclic Prefix (CP) of the uplink transmissions, the LTE RACH only needs to allow for round-trip delay estimation (instead of the timing of individual channel taps), and this therefore reduces the required RACH bandwidth compared to WCDMA.

This is beneficial in minimizing the overhead of the RACH, which is another key consideration. Unlike in WCDMA, the RACH should be able to be fitted into the orthogonal time-frequency structure of the uplink, so that an eNodeB which wants to avoid interference between the RACH and scheduled PUSCH/PUCCH transmissions can do so. It is also important that the RACH is designed so as to minimize interference generated to adjacent scheduled PUSCH/PUCCH transmissions.

19.3 Random Access Procedure

The LTE random access procedure comes in two forms, allowing access to be either *contention-based* (implying an inherent risk of collision) or *contention-free*.

A UE initiates a contention-based random access procedure for all use-cases listed in Section 19.2. In this procedure, a random access preamble signature is randomly chosen by the UE, with the result that it is possible for more than one UE simultaneously to transmit the same signature, leading to a need for a subsequent contention resolution process.

For the use-cases (2) (new downlink data) and (3) (handover) the eNodeB has the option of preventing contention occurring by allocating a dedicated signature to a UE, resulting in contention-free access. This is faster than contention-based access – a factor which is particularly important for the case of handover, which is time-critical.

Unlike in WCDMA, a fixed number (64) of preamble signatures is available in each LTE cell, and the operation of the two types of RACH procedure depends on a partitioning of these signatures between those for contention-based access and those reserved for allocation to specific UEs on a contention-free basis.

The two procedures are outlined in the following sections.

---

[1] See Chapter 18 for a comparison of the coverage of the LTE uplink channels.

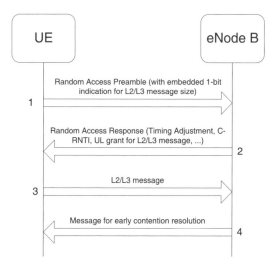

Figure 19.1 Contention-based Random Access Procedure. Reproduced by permission of © 3GPP.

19.3.1 Contention-Based Random Access Procedure

The contention-based procedure consists of four-steps as shown in Figure 19.1:

- Step 1: Preamble transmission;

- Step 2: Random access response;

- Step 3: Layer 2 / Layer 3 (L2/L3) message;

- Step 4: Contention resolution message.

19.3.1.1 Step 1: Preamble Transmission

The UE selects one of the $64 - N_{cf}$ available PRACH contention-based signatures, where N_{cf} is the number of signatures reserved by the eNodeB for contention-free RACH. The set of contention-based signatures is further subdivided into two subgroups, so that the choice of signature can carry one bit of information to indicate information relating to the amount of transmission resource needed to transmit the message at Step 3. The broadcast system information indicates which signatures are in each of the two subgroups (each subgroup corresponding to one value of the one bit of information), as well as the meaning of each subgroup. The UE selects a signature from the subgroup corresponding to the size of transmission resource needed for the appropriate RACH use case (some use cases require only a few bits to be transmitted at Step 3, so choosing the small message size avoids allocating unnecessary uplink resources), which may also take into account the observed downlink radio channel conditions. The eNodeB can control the number of signatures in each subgroup according to the observed loads in each group.

The initial preamble transmission power setting is based on an open-loop estimation with full compensation for the path-loss. This is designed to ensure that the received power of the preambles is independent of the path-loss; this is designed to help the eNodeB to detect several simultaneous preamble transmissions in the same time-frequency PRACH resource. The UE estimates the path-loss by averaging measurements of the downlink Reference Signal Received Power (RSRP). The eNodeB may also configure an additional power offset, depending for example on the desired received Signal to Interference plus Noise Ratio (SINR), the measured uplink interference and noise level in the time-frequency slots allocated to RACH preambles, and possibly also on the preamble format (see Section 19.4.2.2).

19.3.1.2   Step 2: Random Access Response

The Random Access Response (RAR) is sent by the eNodeB on the Physical Downlink Shared CHannel (PDSCH), and addressed with an ID, the Random Access Radio Network Temporary Identifier (RA-RNTI), identifying the time-frequency slot in which the preamble was detected. If multiple UEs had collided by selecting the same signature in the same preamble time-frequency resource, they would each receive the RAR.

The RAR conveys the identity of the detected preamble, a timing alignment instruction to synchronize subsequent uplink transmissions from the UE, an initial uplink resource grant for transmission of the Step 3 message, and an assignment of a Temporary Cell Radio Network Temporary Identifier (C-RNTI) (which may or may not be made permanent as a result of the next step – contention resolution). The RAR message can also include a 'backoff indicator' which the eNodeB can set to instruct the UE to back off for a period of time before retrying a random access attempt.

The UE expects to receive the RAR within a time window, of which the start and end are configured by the eNodeB and broadcast as part of the cell-specific system information. The earliest subframe allowed by the specifications occurs 2 ms after the end of the preamble subframe, as illustrated in Figure 19.2. However, a typical delay (measured from the end of the preamble subframe to the beginning of the first subframe of RAR window) is more likely to be 4 ms. Figure 19.2 shows the RAR consisting of the step 2 message (on PDSCH) together with its downlink transmission resource allocation message 'G' (on the Physical Downlink Control CHannel (PDCCH) – see Section 9.3.2.2).

Figure 19.2 Timing of the Random Access Response (RAR) window.

If the UE does not receive a RAR within the configured time window, it retransmits the preamble. The minimum delay for preamble retransmission after the end of the RAR window is 3 ms. (If the UE receives the PDCCH signalling the downlink resource used for the RAR but cannot satisfactorily decode the RAR message itself, the minimum delay before preamble re-transmission is increased to 4 ms, to allow for the time taken by the UE in attempting to decode the RAR.)

The eNodeB may configure *preamble power ramping* so that the transmission power for each retransmitted preamble is increased by a fixed step. However, since the random access preambles in LTE are normally orthogonal to other uplink transmissions, it is less critical than it was in WCDMA to ensure that the initial preamble power is kept low to control interference. Therefore, the proportion of random access attempts which succeed at the first preamble transmission is likely to be higher than in WCDMA, and the need for power ramping is likely to be reduced.

19.3.1.3 Step 3: Layer 2/Layer 3 (L2/L3) Message

This message is the first scheduled uplink transmission on the PUSCH and makes use of Hybrid Automatic Repeat reQuest (HARQ). It conveys the actual random access procedure message, such as an RRC connection request, tracking area update, or scheduling request. It includes the Temporary C-RNTI allocated in the RAR at Step 2 and either the C-RNTI if the UE already has one (RRC_CONNECTED UEs) or the (unique) 48-bit UE identity. In case of a preamble collision having occurred at Step 1, the colliding UEs will receive the same Temporary C-RNTI through the RAR and will also collide in the same uplink time-frequency resources when transmitting their L2/L3 message. This may result in such interference that no colliding UE can be decoded, and the UEs restart the random access procedure after reaching the maximum number of HARQ retransmissions. However, if one UE is successfully decoded, the contention remains unresolved for the other UEs. The following downlink message (in Step 4) allows a quick resolution of this contention.

If the UE successfully receives the RAR, the UE minimum processing delay before message 3 transmission is 5 ms minus the round-trip propagation time. This is shown in Figure 19.3 for the case of the largest supported cell size of 100 km.

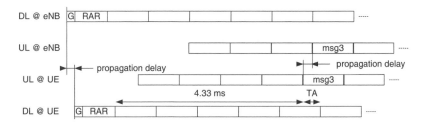

Figure 19.3 Timing of the message 3 transmission.

19.3.1.4 Step 4: Contention Resolution Message

The contention resolution message is addressed to the C-RNTI or Temporary C-RNTI, and, in the latter case, echoes the UE identity contained in the L2/L3 message. It supports HARQ. In case of a collision followed by successful decoding of the L2/L3 message, the HARQ feedback is transmitted only by the UE which detects its own UE identity (or C-RNTI); other UEs understand there was a collision, transmit no HARQ feedback, and can quickly exit the current random access procedure and start another one. The UE's behaviour upon reception of the contention resolution message therefore has three possibilities:

- The UE correctly decodes the message and detects its own identity: it sends back a positive ACKnowledgement, 'ACK'.

- The UE correctly decodes the message and discovers that it contains another UE's identity (contention resolution): it sends nothing back (Discontinuous Transmission, 'DTX').

- The UE fails to decode the message or misses the DL grant: it sends nothing back ('DTX').

19.3.2   Contention-Free Random Access Procedure

The slightly unpredictable latency of the random access procedure can be circumvented for some use cases where low latency is required, such as handover and resumption of downlink traffic for a UE, by allocating a dedicated signature to the UE on a per-need basis. In this case the procedure is simplified as shown in Figure 19.4. The procedure terminates with the RAR.

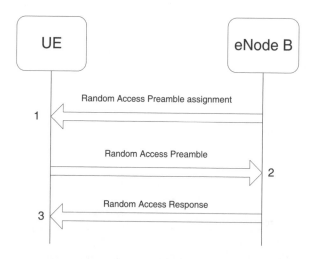

Figure 19.4   Contention-free Random Access Procedure. Reproduced by permission of © 3GPP.

19.4   Physical Random Access Channel Design

The random access preamble part of the random access procedure is mapped at the physical layer onto the PRACH. The design of the preamble is crucial to the success of the RACH procedure, and therefore the next section focuses on the details of the PRACH design.

19.4.1 Multiplexing of PRACH with PUSCH and PUCCH

The PRACH is time- and frequency-multiplexed with PUSCH and PUCCH as illustrated in Figure 19.5. PRACH time-frequency resources are semi-statically allocated within the PUSCH region, and repeat periodically. The possibility of scheduling PUSCH transmissions within PRACH slots is left to the eNodeB's discretion.

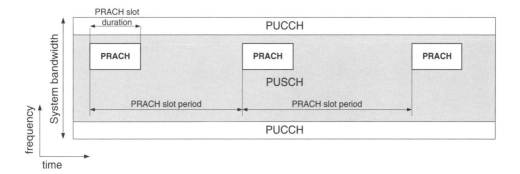

Figure 19.5 PRACH multiplexing with PUSCH and PUCCH.

19.4.2 The PRACH Structure

19.4.2.1 DFT-S-OFDM PRACH Preamble Symbol

Similarly to WCDMA, the LTE PRACH preamble consists of a complex sequence. However, it differs from the WCDMA preamble in that it is also an OFDM symbol, built with a CP, thus allowing for an efficient frequency-domain receiver at the eNodeB. As shown in Figure 19.6, the end of the sequence is appended at the start of the preamble, thus allowing a periodic correlation at the PRACH receiver (as opposed to a less efficient aperiodic correlation) [1].

The UE aligns the start of the random access preamble with the start of the corresponding uplink subframe at the UE assuming a timing advance of zero (see Section 20.2), and the preamble length is shorter than the PRACH slot in order to provide room for a Guard Time (GT) to absorb the propagation delay. Figure 19.6 shows two preambles at the eNodeB received with different timings depending on the propagation delay: as for a conventional OFDM symbol, a single observation interval can be used regardless of the UE's delay, within which periodic correlation is possible.

As further elaborated in Section 19.4.3, the LTE PRACH preamble sequence is optimized with respect to its periodic autocorrelation property. The dimensioning of the CP and GT is addressed in Section 19.4.2.4.

19.4.2.2 PRACH Formats

Four Random Access (RA) preamble formats are defined for Frequency Division Duplex (FDD) operation [2]. Each format is defined by the durations of the sequence and its CP, as listed in Table 19.1.

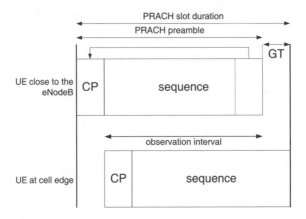

Figure 19.6 PRACH preamble received at the eNodeB.

Table 19.1 Random access preamble formats. Reproduced by permission of © 3GPP.

Preamble format	T_{CP} (μs)	T_{SEQ} (μs)	Typical usage
0	103.13	800	Normal 1 ms RA burst with 800 μs preamble sequence, for small–medium cells (up to ∼14 km)
1	684.38	800	2 ms RA burst with 800 μs preamble sequence, for large cells (up to ∼77 km) without a link budget problem
2	203.13	1600	2 ms RA burst with 1600 μs preamble sequence, for medium cells (up to ∼29 km) supporting low data rates
3	684.38	1600	3 ms RA burst with 1600 μs preamble sequence, for very large cells (up to ∼100 km)

The rationale behind these choices of sequence duration, CP and GT, are discussed below.

19.4.2.3 Sequence Duration

The sequence duration, T_{SEQ}, is driven by the following factors:

- Trade-off between sequence length and overhead: a single sequence must be as long as possible to maximize the number of orthogonal preambles (see Section 7.2.1), while still fitting within a single subframe in order to keep the PRACH overhead small in most deployments;

- Compatibility with the maximum expected round-trip delay;

- Compatibility between PRACH and PUSCH subcarrier spacings;

- Coverage performance.

Maximum round-trip time. The lower bound for T_{SEQ} must allow for unambiguous round-trip time estimation for a UE located at the edge of the largest expected cell (i.e. 100 km radius), including the maximum delay spread expected in such large cells, namely 16.67 μs. Hence

$$T_{SEQ} \geq \frac{200 \cdot 10^3}{3 \cdot 10^8} + 16.67 \cdot 10^{-6} = 683.33 \text{ μs} \quad (19.1)$$

Subcarrier spacing compatibility. Further constraints on T_{SEQ} are given by the Single-Carrier FDMA (SC-FDMA – see Chapter 15) signal generation principle (see Section 19.5), such that the size of the DFT and IDFT, N_{DFT}, must be an integer number:

$$N_{DFT} = f_s T_{SEQ} = k, \quad k \in \mathbb{N} \quad (19.2)$$

where f_s is the system sampling rate (e.g. 30.72 MHz). Additionally, it is desirable to minimize the orthogonality loss in the frequency domain between the preamble subcarriers and the subcarriers of the surrounding uplink data transmissions. This is achieved if the PUSCH data symbol subcarrier spacing Δf is an integer multiple of the PRACH subcarrier spacing Δf_{RA}:

$$\Delta f_{RA} = \frac{f_s}{N_{DFT}} = \frac{1}{T_{SEQ}} = \frac{1}{kT_{SYM}} = \frac{1}{k}\Delta f, \quad k \in \mathbb{N} \quad (19.3)$$

where $T_{SYM} = 66.67$ μs is the uplink subframe symbol duration. In other words, the preamble duration must be an integer multiple of the uplink subframe symbol duration:

$$T_{SEQ} = kT_{SYM} = \frac{k}{\Delta f}, \quad k \in \mathbb{N} \quad (19.4)$$

An additional benefit of this property is the possibility to reuse the FFT/IFFT[2] components from the SC-FDMA signal processing for the scheduled data. Moreover, it should be possible to implement the large DFT/IDFT[3] blocks involved in the PRACH transmitter and receiver (see Section 19.5) using a combination of the elementary FFT/IFFT blocks. For example, an $n \cdot 2^m$ DFT can be implemented with an FFT of 2^m samples combined with a DFT of n samples [3], since from Equation (19.4) it follows that

$$N_{DFT} = kf_s T_{SYM} = kN_{FFT}, \quad k \in \mathbb{N} \quad (19.5)$$

where N_{FFT} is the FFT size for a PUSCH symbol.

Coverage performance. In general a longer sequence gives better coverage, but better coverage requires a longer CP and GT in order to absorb the corresponding round-trip delay (Figure 19.6). The required CP and GT lengths for PRACH format 0, for example, can therefore be estimated from the maximum round-trip delay achievable by a preamble sequence which can fit into a 1 ms subframe.

Under a noise-limited scenario, as is typical of a low density, medium to large suburban or rural cell, coverage performance can be estimated from a link budget calculation. Under the assumption of the Okumura-Hata empirical model of distance-dependent path-loss $L(r)$ [4,5]

---

[2]Fast Fourier Transform / Inverse Fast Fourier Transform.
[3]Discrete Fourier Transform / Inverse Discrete Fourier Transform.

(where r is the cell radius in km), the PRACH signal power P_{RA} received at the eNodeB baseband input can be computed as follows:

$$P_{RA}(r) = P_{max} + G_a - L(r) - LF - P_L \quad \text{(dB)} \tag{19.6}$$

where the parameters are listed in Table 19.2 (mainly from [6]).

The required PRACH preamble sequence duration T_{SEQ} is then derived from the required preamble sequence energy to thermal noise ratio E_p/N_0 to meet a target missed detection and false alarm probability, as follows:

$$T_{SEQ} = \frac{N_0 N_f}{P_{RA}(r)} \frac{E_p}{N_0} \tag{19.7}$$

where N_0 is the thermal noise power density (in mW/Hz) and N_f is the receiver noise figure (in linear scale).

Assuming that $E_p/N_0 = 18$ dB is required to meet missed detection and false alarm probabilities of 10^{-2} and 10^{-3} respectively (see Section 19.4.3.3), Figure 19.7 plots the coverage performance of the PRACH sequence as a function of the sequence length T_{SEQ}.

It can be observed from Figure 19.7 that the potential coverage performance of a 1 ms PRACH preamble is in the region of 14 km. As a consequence, the required CP and GT lengths are approximately $(2 \cdot 14000)/(3 \cdot 10^8) = 93.3$ μs, so that the upper bound for T_{SEQ} is given by

$$T_{SEQ} \leq 1000 - 2 \cdot 93.33 = 813 \text{ μs} \tag{19.8}$$

Therefore, the longest sequence simultaneously satisfying Equations (19.1), (19.4) and (19.8) is $T_{SEQ} = 800$ μs, as used for preamble formats 0 and 1. The resulting PRACH sub-carrier spacing is $\Delta f_{RA} = 1/T_{SEQ} = 1.25$ kHz.

The 1600 μs preamble sequence of formats 2 and 3 is implemented by repeating the baseline 800 μs preamble sequence. These formats can provide up to 3 dB link budget improvement, which is useful in large cells and/or to balance PUSCH/PUCCH and PRACH coverage at low data rates.

Table 19.2 Link budget parameters for analysis of PRACH preamble coverage.

Parameter	Value
Carrier frequency (f)	2000 MHz
Antenna height (h_b)	30 m / 60 m
UE antenna height (h_m)	1.5 m
UE transmitter EIRP[a] (P_{max})	24 dBm (250 mW)
eNodeB Receiver Antenna Gain (including cable loss) (G_a)	14 dBi
Receiver noise figure (N_f)	5.0 dB
Thermal noise density (N_0)	−174 dBm/Hz
Percentage of the area covered by buildings (α)	10%
Required E_p/N_0 (eNodeB with 2 Rx antenna diversity)	18 dB (six-path Typical Urban channel model)
Penetration loss (P_L)	0 dB
Log-normal fade margin (LF)	0 dB

[a]Equivalent Isotropic Radiated Power.

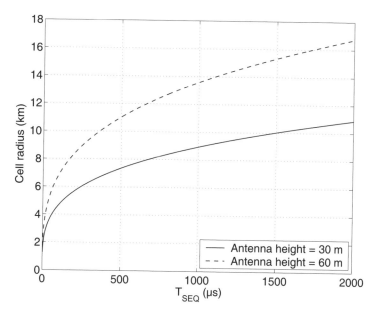

Figure 19.7 PRACH coverage performance versus sequence duration.

19.4.2.4 CP and GT Duration

Having chosen T_{SEQ}, the CP and GT dimensioning can be specified more precisely.

For formats 0 and 2, the CP is dimensioned to maximize the coverage, given a maximum delay spread d: $T_{CP} = (1000 - 800)/2 + d/2$ µs, with $d \approx 5.2$ µs (corresponding to the longest normal CP of a PUSCH SC-FDMA symbol). The maximum delay spread is used as a guard period at the end of CP, thus providing protection against multipath interference even for the cell-edge UEs.

In addition, for a cell-edge UE, the delay spread energy at the end of the preamble is replicated at the end of the CP (see Figure 19.8) and is therefore within the observation interval. Consequently, there is no need to include the maximum delay spread in the GT dimensioning. Hence, instead of locating the sequence in the centre of the PRACH slot, it is shifted later by half the maximum delay spread, allowing the maximum Round-Trip Delay (RTD) to be increased by the same amount. Note that, as for a regular OFDM symbol, the residual delay spread at the end of the preamble from a cell-edge UE spills over into the next subframe, but this is taken care of by the CP at the start of the next subframe to avoid any inter-symbol interference.

For formats 1 and 3, the CP is dimensioned to address the maximum cell range in LTE, 100 km, with a maximum delay spread of $d \approx 16.67$ µs. In practice, format 1 is expected to be used with a 3-subframe PRACH slot; the available GT in 2 subframes can only address a 77 km cell range. It was chosen to use the same CP length for both format 1 and format 3 for implementation simplicity. Of course, handling larger cell sizes than 100 km with suboptimal CP dimensioning is still possible and is left to implementation.

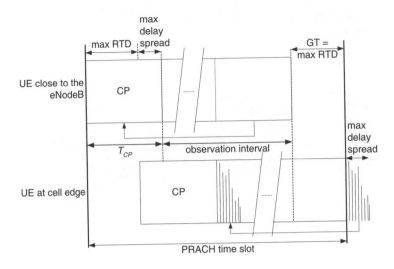

Figure 19.8 PRACH CP/GT dimensioning for formats 0 and 2.

Table 19.3 Field durations and achievable cell radius of the PRACH preamble formats.

Preamble format	Number of allocated subframes	CP duration in μs	CP duration as multiple of T_S	GT duration in μs	GT duration as multiple of T_S	Max. delay spread (μs)	Max. cell radius (km)
0	1	103.13	3168	96.88	2976	6.25	14.53
1	2	684.38	21024	515.63	15840	16.67	77.34
2	2	203.13	6240	196.88	6048	6.25	29.53
3	3	684.38	21024	715.63	21984	16.67	100.16

Table 19.3 shows the resulting cell radius and delay spread ranges associated with the four PRACH preamble formats. The CP lengths are designed to be an integer multiple of the assumed system sampling period for LTE, $T_S = 1/30.72$ μs.

These are also illustrated in Figure 19.9.

19.4.2.5 PRACH Resource Configurations

The PRACH slots shown in Figure 19.5 can be configured to occur in up to 16 different layouts, or *resource configurations*. Depending on the RACH load, one or more PRACH resources may need to be allocated per PRACH slot period. The eNodeB has to process the PRACH very quickly so that message 2 of the RACH procedure can be sent within the required window, as shown in Figure 19.2. In case of more than one PRACH resource per PRACH period it is generally preferable to multiplex the PRACH resources in time (Figure 19.10 – right) rather than in frequency (Figure 19.10 – left). This helps to avoid processing peaks at the eNodeB.

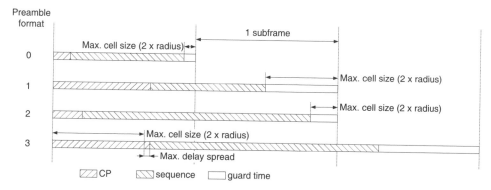

Figure 19.9 PRACH preamble formats and cell size dimensioning.

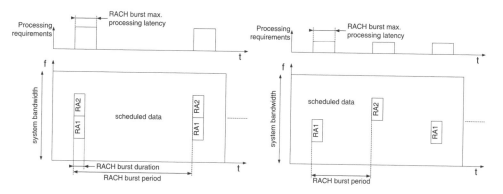

Figure 19.10 Processing peaks of the random access preamble receiver.

Extending this principle, the available slot configurations are designed to facilitate a PRACH receiver which may be used for multiple cells of an eNodeB, assuming a periodic pattern with period 10 ms or 20 ms.

Assuming an operating collision probability per UE, $p_{\text{coll}}^{\text{UE}} = 1\%$, one PRACH time-frequency resource (with 64 signatures) per 10 ms per 5 MHz can handle an offered load $G = -64 \ln(1 - p_{\text{coll}}^{\text{UE}}) = 0.6432$ average PRACH attempts, which translates into 128 attempts per second in 10 MHz. This is expected to be a typical PRACH load in LTE.

Assuming a typical PRACH load, example usages and cell allocations of the 16 available resource configurations are shown in Figure 19.11 for different system bandwidths and different numbers of cells per eNodeB. Resource configurations 0–2 and 15 use a 20 ms PRACH period, which can be desirable for small bandwidths (e.g. 1.4 MHz) in order to reduce the PRACH overhead at the price of higher waiting times.

As can be observed from Figure 19.11, in a three-cell scenario, time collision of PRACH resources can always be avoided except for the 20 MHz case, where collisions can be minimized to two PRACH resources occurring in the same subframe. It should also be noted

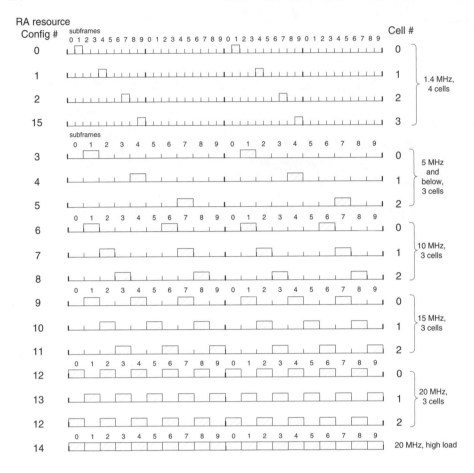

Figure 19.11 Random Access resource configurations.

that in a scenario with six cells per eNodeB, at most two PRACH resources will occur in the same subframe for bandwidths below 20 MHz.

The variety of configurations provided therefore enables efficient dimensioning of a multicell PRACH receiver.

19.4.3 Preamble Sequence Theory and Design

As noted above, 64 PRACH signatures are available in LTE, compared to only 16 in WCDMA. This can not only reduce the collision probability, but also allow for 1 bit of information to be carried by the preamble and some signatures to be reserved for contention-free access (see Section 19.3.2). Therefore, the LTE PRACH preamble called for an improved sequence design with respect to WCDMA. While Pseudo-Noise (PN) based sequences were used in WCDMA, in LTE prime-length Zadoff–Chu (ZC) [7,8] sequences have been chosen (see Chapter 7 for an overview of the properties of ZC sequences). These sequences enable improved PRACH preamble detection performance. In particular:

- The power delay profile is built from periodic instead of aperiodic correlation;

- The intra-cell interference between different preambles received in the same PRACH resource is reduced;

- Intra-cell interference is optimized with respect to cell size: the smaller the cell size, the larger the number of orthogonal signatures and the better the detection performance;

- The eNodeB complexity is reduced;

- The support for high-speed UEs is improved.

The 800 μs LTE PRACH sequence is built from cyclicly-shifting a ZC sequence of prime-length N_{ZC}, defined as

$$x_u(n) = \exp\left[-j\frac{\pi un(n+1)}{N_{ZC}}\right], \quad 0 \le n \le N_{ZC} - 1 \qquad (19.9)$$

where u is the ZC sequence index and the sequence length $N_{ZC} = 839$ for FDD.

The reasons that led to this design choice are elaborated in the next sections.

19.4.3.1   Preamble Bandwidth

In order to ease the frequency multiplexing of the PRACH and the PUSCH resource allocations, a PRACH slot must be allocated a bandwidth BW_{PRACH} equal to an integer multiple of Resource Blocks (RBs), i.e. an integer multiple of 180 kHz.

For simplicity, BW_{PRACH} in LTE is constant for all system bandwidths; it is chosen to optimize both the detection performance and the timing estimation accuracy. The latter drives the lower bound of the PRACH bandwidth. Indeed, a minimum bandwidth of ~1 MHz is necessary to provide a one-shot accuracy of about ±0.5 μs, which is an acceptable timing accuracy for PUCCH/PUSCH transmissions.

Regarding the detection performance, one would intuitively expect that the higher the bandwidth, the better the detection performance, due to the diversity gain. However, it is important to make the comparison using a constant signal energy to noise ratio, E_p/N_0, resulting from accumulation (or despreading) over the same preamble duration, and the same false alarm probability, p_{fa}, for all bandwidths. The latter requires the detection threshold to be adjusted with respect to the search window size, which increases with the bandwidth. Indeed, it will become clear from the discussion in Section 19.5.2.3 that the larger the search window, the higher p_{fa}. In other words, the larger the bandwidth the higher the threshold relative to the noise floor, given a false alarm target p_{fa_target} and cell size L; equivalently, the larger the cell size, the higher the threshold relative to the noise floor, given a target p_{fa_target} and bandwidth. As a result, under the above conditions, a smaller bandwidth will perform better than a large bandwidth in a single-path static AWGN channel, given that no diversity improvement is to be expected from such a channel.

Figure 19.12 shows simulation results for the TU-6 fading channel,[4] comparing detection performance of preamble bandwidths BW_{PRACH} of 6, 12, 25 and 50 RBs. For each bandwidth, the sequence length is set to the largest prime number smaller than $1/BW_{PRACH}$, the false alarm rate is set to $p_{fa\_target} = 0.1\%$, the cell radius is 0.7 km and the receiver searches for 64 signatures constructed from 64 cyclic shifts of one root ZC sequence.

---

[4]The six-path Typical Urban channel model [9].

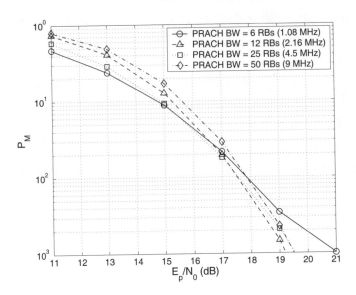

Figure 19.12 PRACH missed detection performance comparison for different bandwidths.

We can observe that the best detection performance is achieved by preambles of 6 RBs and 12 RBs for low and high SNRs respectively. The 25-RB preamble has the overall best performance considering the whole SNR range. Thus the diversity gain of large bandwidths only compensates the increased detection threshold in the high SNR region corresponding to misdetection performances in the range of 10^{-3} and below. At a typical 10^{-2} detection probability target, the 6-RB allocation only has 0.5 dB degradation with respect to the best case.

Therefore, a PRACH allocation of 6 RBs provides a good trade-off between PRACH overhead, detection performance and timing estimation accuracy. Note that for the smallest system bandwidth (1.4 MHz, 6 RBs) the PRACH overlaps with the PUCCH; it is left to the eNodeB implementation whether to implement scheduling restrictions during PRACH slots to avoid collisions, or to let PRACH collide with PUCCH and handle the resulting interference.

Finally, the exact preamble transmission bandwidth is adjusted to isolate PRACH slots from surrounding PUSCH/PUCCH allocations through guard bands, as elaborated in the following section.

19.4.3.2   Sequence Length

The sequence length design should address the following requirements:

- Maximize the number of ZC sequences with optimal cross-correlation properties;

- Minimize the interference to/from the surrounding scheduled data on the PUSCH.

The former requirement is guaranteed by choosing a prime-length sequence. For the latter, since data and preamble OFDM symbols are neither aligned nor have the same durations, strict orthogonality cannot be achieved. At least, fixing the preamble duration to an integer multiple of the PUSCH symbol provides some compatibility between preamble and PUSCH subcarriers. However, with the 800 μs duration, the corresponding sequence length would be 864, which does not meet the prime number requirement. Therefore, shortening the preamble to a prime length slightly increases the interference between PUSCH and PRACH by slightly decreasing the preamble sampling rate.

The interference from PUSCH to PRACH is further amplified by the fact that the operating E_s/N_0 of PUSCH (where E_s is the PUSCH symbol energy) is much greater than that of the PRACH (typically as much as 24 dB greater if we assume 13 dB E_s/N_0 for 16QAM PUSCH, while the equivalent ratio for the PRACH would be -11 dB assuming $E_p/N_0 = 18$ dB and adjusting by $-10 \log_{10}(864)$ to account for the sequence length). This is illustrated in Figure 19.13 showing the missed detection rate (P_m) with and without data interference adjacent to the PRACH. The simulations assume a TU-6 channel, two receive antennas at the eNodeB and 15 km/h UE speed. The PRACH shows about 1 dB performance loss at $P_m = 1\%$.

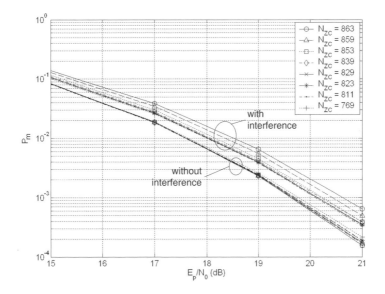

Figure 19.13   Missed detection rates of PRACH preamble with and without 16QAM interferer for different sequence lengths (cell radius of 0.68 km).

The PRACH uses *guard bands* to avoid the data interference at preamble edges. A cautious design of preamble sequence length not only retains a high inherent processing gain, but also allows avoidance of strong data interference. In addition, the loss of spectral efficiency (by reservation of guard subcarriers) can also be well controlled at a fine granularity ($\Delta f_{RA} = 1.25$ kHz). Figure 19.13 shows the missed detection rate for a cell radius of 0.68 km, for various preamble sequence lengths with and without 16QAM data interference.

In the absence of interference, there is no significant performance difference between sequences of similar prime length. In the presence of interference, it can be seen that reducing

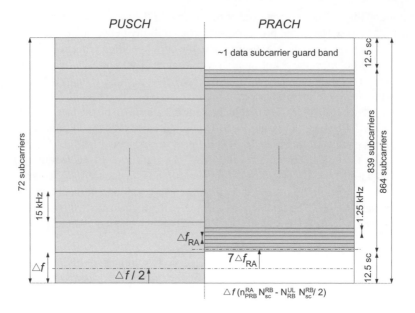

Figure 19.14 PRACH preamble mapping onto allocated subcarriers.

the sequence length below 839 gives no further improvement in detection rate. No effect is observed on the false alarm rate.

Therefore the sequence length of 839 is selected for LTE PRACH, corresponding to 69.91 PUSCH subcarriers in each SC-FDMA symbol, and offers $72 - 69.91 = 2.09$ PUSCH subcarriers protection, which is very close to one PUSCH subcarrier protection on each side of the preamble. This is illustrated in Figure 19.14; note that the preamble is positioned centrally in the block of 864 available PRACH subcarriers, with 12.5 null subcarriers on each side.

Finally, the PRACH preamble signal $s(t)$ can therefore be defined as follows [2]:[5]

$$s(t) = \beta_{\text{PRACH}} \sum_{k=0}^{N_{\text{ZC}}-1} \sum_{n=0}^{N_{\text{ZC}}-1} x_{u,v}(n) \cdot \exp\left[-j\frac{2\pi nk}{N_{\text{ZC}}}\right]$$

$$\times \exp[j2\pi[k + \varphi + K(k_0 + \tfrac{1}{2})]\Delta f_{\text{RA}}(t - T_{\text{CP}})] \qquad (19.10)$$

where $0 \leq t < T_{\text{SEQ}} + T_{\text{CP}}$, β_{PRACH} is an amplitude scaling factor and $k_0 = n_{\text{PRB}}^{\text{RA}} N_{\text{SC}}^{\text{RB}} - N_{\text{RB}}^{\text{UL}} N_{\text{SC}}^{\text{RB}}/2$. The location in the frequency domain is controlled by the parameter $n_{\text{PRB}}^{\text{RA}}$, expressed as a resource block number configured by higher layers and fulfilling $0 \leq n_{\text{PRB}}^{\text{RA}} \leq N_{\text{RB}}^{\text{UL}} - 6$. The factor $K = \Delta f/\Delta f_{\text{RA}}$ accounts for the ratio of subcarrier spacings between the PUSCH and PRACH. The variable φ (equal to 7 for LTE FDD) defines a fixed offset determining the frequency-domain location of the random access preamble within the resource blocks. $N_{\text{RB}}^{\text{UL}}$ is the uplink system bandwidth (in RBs) and $N_{\text{SC}}^{\text{RB}}$ is the number of subcarriers per RB, i.e. 12.

---

[5]Equation (19.10) reproduced by permission of © 3GPP.

19.4.3.3 Cyclic Shift Dimensioning (N_{CS}) for Normal Cells

Sequences obtained from cyclic shifts of *different* ZC sequences are not orthogonal (see Section 7.2.1). Therefore, orthogonal sequences obtained by cyclically shifting a single root sequence should be favoured over non-orthogonal sequences; additional ZC root sequences should be used only when the required number of sequences (64) cannot be generated by cyclic shifts of a single root sequence. The cyclic shift dimensioning is therefore very important in the RACH design.

The cyclic shift offset N_{CS} is dimensioned so that the Zero Correlation Zone (ZCZ) of the sequences guarantees the orthogonality of the PRACH sequences regardless of the delay spread and time uncertainty of the UEs. The minimum value of N_{CS} should therefore be the smallest integer number of sequence sample periods that is greater than the maximum delay spread and time uncertainty of an uplink non-synchronized UE, plus some additional guard samples provisioned for the spill-over of the pulse shaping filter envelope present in the PRACH receiver (Figure 19.15).

The resulting lower bound for cyclic shift N_{CS} can be written as

$$N_{CS} \geq \left\lceil \left(\frac{20}{3}r - \tau_{ds} \right) \frac{N_{ZC}}{T_{SEQ}} \right\rceil + n_g \tag{19.11}$$

where r is the cell size (km), τ_{ds} is the maximum delay spread, $N_{ZC} = 839$ and T_{SEQ} are the PRACH sequence length and duration (measured in μs) respectively, and n_g is the number of additional guard samples due to the receiver pulse shaping filter.

The delay spread can generally be assumed to be constant for a given environment. However, the larger the cell, the larger the cyclic shift required to generate orthogonal sequences, and consequently, the larger the number of ZC root sequences necessary to provide the 64 required preambles.

The relationship between cell size and the required number of ZC root sequences allows for some system optimization. In general, the eNodeB should configure N_{CS} independently in each cell, because the expected inter-cell interference and load (user density) increases as cell size decreases; therefore smaller cells need more protection from co-preamble interference than larger cells.

Some practical examples of this optimization are given in Table 19.4, showing four cell scenarios resulting from different N_{CS} values configured by the eNodeB. For each scenario,

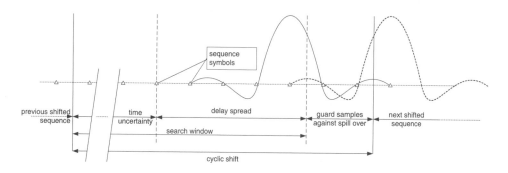

Figure 19.15 Cyclic shift dimensioning.

Table 19.4 Cell scenarios with different cyclic shift increments.

Cell scenario	Number of cyclic shifts per ZC sequence	Number of ZC root sequences	Cyclic shift size N_{CS} (samples)	Cell radius (km)
1	64	1	13	0.7
2	32	2	26	2.5
3	18	4	46	5
4	9	8	93	12

the total number of sequences is 64, but resulting from different combinations of the number of root sequences and cyclic shifts.

As an example, we evaluate in Figure 19.16 the effect of UEs selecting orthogonal and non-orthogonal sequences in scenario 3 by comparing the missed-detection rate in each of the following cases:

- Case 1: Only one UE transmits a preamble;

- Case 2: Two UEs transmit a preamble, and the two preamble sequences are generated from the same root ZC sequence;

- Case 3: Two UEs transmit a preamble, and the two preamble sequences are generated from different root ZC sequences.

In all cases, the false-alarm rate is set at around 0.1%. The channel model used is TU-6 with a UE speed of 3 km/h.

It can be observed from Figure 19.16 that when two preambles are transmitted which are cyclic shifts of the same root sequence (Case 2), the performance does not degrade compared to the case of only one preamble being transmitted, confirming the ZCZ property of the shifted sequences. By contrast, when the two preambles are generated from different root sequences (Case 3), a degradation of 0.25–0.4 dB is observed at 1% – 0.1% missed detection rates.

N_{CS} set design.   Given the sequence length of 839, allowing full flexibility in signalling N_{CS} would lead to broadcasting a 10-bit parameter, which is over-dimensioning. As a result, in LTE the allowed values of N_{CS} are quantized to a predefined set of just 16 configurations. The 16 allowed values of N_{CS} were chosen so that the number of orthogonal preambles is as close as possible to what could be obtained if there were no restrictions on the value of N_{CS} [10]. This is illustrated in Figure 19.17 (left), where the cell radii are derived assuming a delay spread of 5.2 μs and 2 guard samples n_g for the pulse shaping filter.

The effect of the quantization is shown in Figure 19.17 (right), which plots the probability p_2 that two UEs randomly select two preambles on the same root sequence, as a function of the cell radius, for both the quantized N_{CS} set and an ideal unquantized set. The larger p_2, the better the detection performance. The figure also shows an ideal unquantized set. It can be seen that the performance loss due to the quantization is negligible.

Figure 19.18 illustrates the range of N_{CS} values and their usage with the various preamble formats. Note that this set of N_{CS} values is designed for use in low-speed cells. LTE also

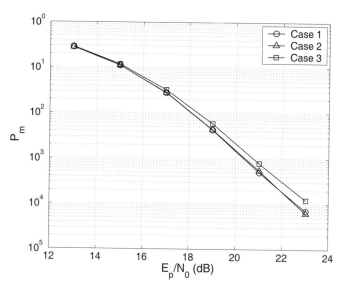

Figure 19.16 Detection performance loss from using ZCZ and non-ZCZ sequences – Cell scenario 3: 5 km cell radius.

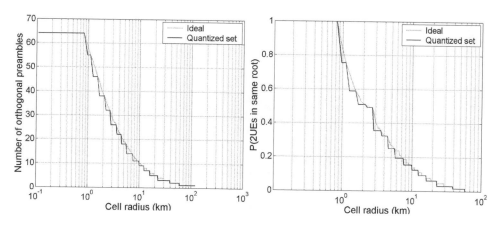

Figure 19.17 Number of orthogonal preambles (left) and probability that two UEs select two orthogonal preambles (right).

provides a second N_{CS} set specially designed for high-speed cells, as elaborated in the following sections.

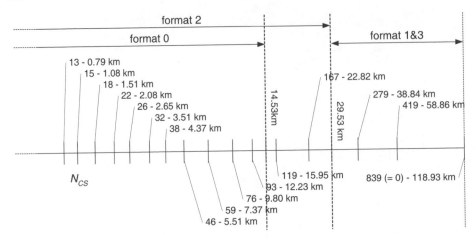

Figure 19.18 N_{CS} values and usage with the various preamble formats.

19.4.3.4 Cyclic Shift (N_{CS}) Restriction for High-Speed Cells

The support of 64 RACH preambles as described above assumes little or no frequency shifting due to Doppler spread, in the presence of which ZC sequences lose their zero auto-correlation property. In the presence of a frequency offset δf, it can be shown that the PRACH ZC sequence in Equation (19.9) is distorted as follows:

$$x_u(n, \delta f) = \exp\left[-j\pi \frac{u(n-1/u)(n-1/u+1)}{N_{ZC}}\right]$$

$$\times \exp\left[j2\pi \frac{n}{N_{ZC}}(\delta f\, T_{SEQ} - 1)\right] \exp\left[-\frac{j\pi}{N_{ZC}} \frac{u-1}{u}\right]$$

$$= x_u(n-1/u) \exp\left[j2\pi \frac{n}{N_{ZC}}(\delta f\, T_{SEQ} - 1)\right] e^{j\Phi_u} \qquad (19.12)$$

where T_s is the PRACH preamble sampling period.

A similar expression can be written for the opposite frequency offset.

As can be observed, frequency offsets as large as one PRACH subcarrier ($\delta f = \pm\Delta f_{RA} = \pm 1/T_{SEQ} = \pm 1.25$ kHz) result in cyclic shifts $d_u = (\pm 1/u) \bmod N_{ZC}$ on the ZC sequence $x_u(n)$. (Note that $u \cdot d_u \bmod N_{ZC} = \pm 1$.) This frequency offset δf can be due to the accumulated frequency uncertainties at both UE transmitter and eNodeB receiver, δf_{LO}, and the Doppler shift resulting from the UE motion in a Line of Sight (LOS) radio propagation condition, δf_D:

$$\delta f = \delta f_{LO} + \delta f_D \qquad (19.13)$$

where δf_D corresponds to a UE speed $s(\delta f_D)$ given by

$$s(\delta f_D) = \delta f_D \frac{c}{2 f_c} \qquad (19.14)$$

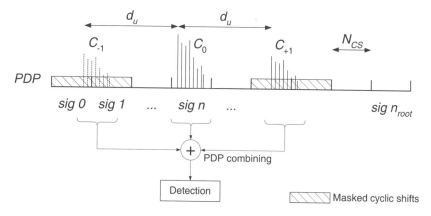

Figure 19.19 Side peaks in PDP due to frequency offset.

where c is the speed of light and f_c the carrier frequency. The factor 2 in the denominator results from the Doppler shift being accounted twice at the eNodeB: first at the UE when locking its local oscillator on the downlink received signal, then on the uplink transmission.

Figure 19.19 illustrates the impact of the cyclic shift distortion on the received Power Delay Profile (PDP): it creates false alarm peaks whose relative amplitude to the correct peak depends on the frequency offset. The solution adopted in LTE to address this issue is referred to as 'cyclic shift restriction' and consists of 'masking' some cyclic shift positions in the ZC root sequence. This makes it possible to retain an acceptable false alarm rate, while also combining the PDPs of the three uncertainty windows, thus also maintaining a high detection performance even for very high-speed UEs.

It should be noted that at $|\delta f| = \Delta f_{RA}$, the preamble peak completely disappears at the desired location (as per Equation (19.12)). However, the false image peak begins to appear even with $|\delta f| < \Delta f_{RA}$. Another impact of the side peaks is that they restrict the possible cyclic shift range so as to prevent from side peaks from falling into the cyclic shift region (see Figure 19.20). This restriction on N_{CS} is captured by Equations (19.15) and (19.16) and is important for the design of the high-speed N_{CS} set (explained in Section 19.4.3.5) and the order in which the ZC sequences are used (explained in Section 19.4.3.6):

$$N_{CS} \leq d < (N_{ZC} - N_{CS})/2 \qquad (19.15)$$

where

$$d = \begin{cases} d_u, & 0 \leq d_u < N_{ZC}/2 \\ N_{ZC} - d_u, & d_u \geq N_{ZC}/2 \end{cases} \qquad (19.16)$$

We use C_{-1} and C_{+1} to denote the two wrong cyclic shift windows arising from the frequency offset, while C_0 denotes the correct cyclic shift window (Figure 19.19). The cyclic restriction rule must be such that the two wrong cyclic shift windows C_{-1} and C_{+1} of a cyclicly-shifted ZC sequence overlap none of C_0, C_{-1} or C_{+1} of other cyclicly-shifted ZC sequences, nor the correct cyclic shift window C_0 of the same cyclicly-shifted ZC sequence, nor each other [11]. Finally, the restricted set of cyclic shifts is obtained such

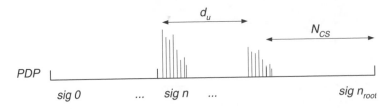

Figure 19.20 Side peaks within the signature search window.

that the minimum difference between two cyclic shifts is still N_{CS} but the cyclic shifts are not necessarily multiples of N_{CS}.

It is interesting to check the speed limit beyond which it is worth considering a cell to be a high-speed cell. This is done by assessing the performance degradation of the PRACH at the system-level as a function of the UE speed when no cyclic shift restriction is applied.

For this analysis, we model the RACH access attempts of multiple concurrent UEs with a Poisson arrival rate. A preamble detection is considered to be correct if the timing estimation is within 2 µs. A target E_p/N_0 of 18 dB is used for the first preamble transmission, with a power ramping step of 1 dB for subsequent retransmissions. The cell radius is random between 0.5 and 12 km, with either AWGN or a six-path TU channel, and a 2 GHz carrier frequency. eNodeB and UE frequency errors are modelled randomly within ±0.05 ppm. The access failure rate is the measure of the number of times a UE unsuccessfully re-tries access attempts (up to a maximum of three retransmissions), weighted by the total number of new access attempts.

Figure 19.21 shows the access failure rate performance for both channel types as a function of the UE speed, for various offered loads G. It can be observed that under fading conditions, the RACH failure rates experience some degradation with the UE speed (which translates into Doppler spread), but remains within acceptable performance even at 350 km/h. For the AWGN channel (where the UE speed translates into Doppler shift) the RACH failure rate stays below 10^{-2} up to UE speeds in the range 150 to 200 km/h. However, at 250 km/h and above, the throughput collapses. Without the cyclic shift restrictions the upper bound for useful performance is around 150–200 km/h.

19.4.3.5   Cyclic Shift Configuration for High-Speed Cells

The cyclic shift dimensioning for high-speed cells in general follows the same principle as for normal cells, namely maximizing the sequence reuse when group quantization is applied to cyclic shift values. However, for high-speed cells, the cyclic shift restriction needs to be considered when deriving the sequence reuse factor with a specific cyclic shift value. Note that there is no extra signalling cost to support an additional set of cyclic shift configurations for high-speed cells since the one signalling bit which indicates a 'high-speed cell configuration' serves this purpose.

The N_{CS} values for high-speed cells are shown in Figure 19.22 for the number of available and used preambles, with both consecutive and non-consecutive (quantized) cyclic shift values. The number of available preambles assumes no cyclic shift restriction at all, as in low-speed cells. It should be noted that with the cyclic shift restriction above, the largest

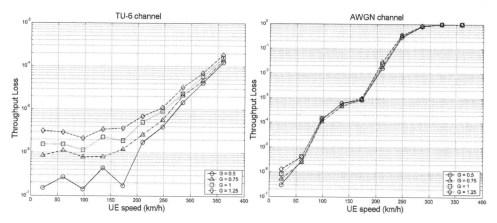

Figure 19.21 Random access failure rate as a function of UE speed.

usable high-speed cyclic shift value among all root sequences is 279 (from Equation (19.15)). As is further elaborated in the next section, only the preambles with Cubic Metric (CM) (see Section 22.3.3) below 1.2 dB are considered in Figure 19.22.

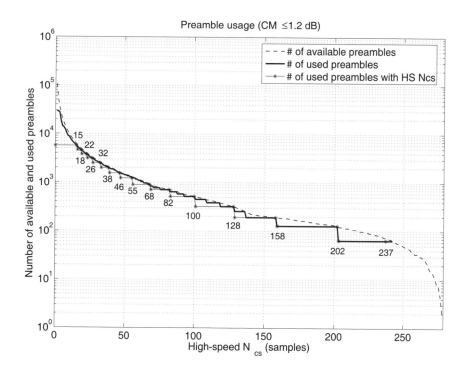

Figure 19.22 Number of available and used preambles in low CM group.

Since for small N_{CS} values the sequence usage is not so tight with a generally high sequence reuse factor, a way to simplify design, while still achieving a high reuse factor, is to reuse the small N_{CS} values for normal cells. In Figure 19.22, N_{CS} values up to 46 are from the normal cyclic shift values, corresponding to a cell radius up to 5.8 km. At the high end, the value of 237 rather than 242 is chosen to support a minimum of two high-speed cells when all the 838 sequences are used. The maximum supportable high-speed cell radius is approximately 33 km, providing sufficient coverage for preamble formats 0 and 2.

19.4.3.6 Sequence Ordering

A UE using the contention-based random access procedure described in Section 19.3.1 needs to know which sequences are available to select from. As explained in Section 19.4.3.3, the full set of 64 sequences may require the use of several ZC root sequences, the identity of each of which must be broadcast in the cell. Given the existence of 838 root sequences, signalling each individual sequence index requires 10 bits per root sequence, which can lead to a large signalling overhead. Therefore in LTE the signalling is streamlined by broadcasting only the index of the *first* root sequence in a cell, and the UE derives the other preamble signatures from it given a predefined ordering of all the sequences.

Two factors are taken into account for the root sequence ordering, namely the CM [12,13] of the sequence, and the maximum supportable cell size for high-speed cells (or equivalently the maximum supported cyclic shift). Since CM has a direct impact on cell coverage, the first step in ordering the root sequences is to divide the 838 sequences into a low CM group and a high CM group, using the CM of Quadrature Phase Shift Keying (QPSK) (1.2 dB) as a threshold. The low CM group would be used first in sequence planning (and also for high-speed cells) since it is more favourable for coverage.

Then, within each CM group, the root sequences are classified into subgroups based on their maximum supportable cell radius, to facilitate sequence planning including high-speed cells. Specifically, a sequence subgroup g is the set of all root sequences with their maximum allowed cyclic shifts (S_{max}) derived from Equation (19.15) lying between two consecutive high-speed N_{CS} values according to

$$N_{CS}(g) \le S_{max} < N_{CS}(g+1), \text{ for } g = 0, 1, \ldots, G-2, \text{ and}$$

$$S_{max} \ge N_{CS}(G-1) \tag{19.17}$$

for G cyclic shift values, with the set of N_{CS} values being those for high-speed cells. Sequences in each subgroup are ordered according to their CM values.

Figure 19.23 shows the CM and maximum allowed cyclic shift values at high speed for the resulting root sequence index ordering used in LTE. This ordering arrangement reflects a continuous CM transition across subgroups and groups, which ensures that consecutive sequences always have close CM values when allocated to a cell. Thus, consistent cell coverage and preamble detection can be achieved in one cell.

The LTE specifications define the mapping from root sequence index u to a reordered index in a table [2], an enhanced extract of which is given in Figure 19.24. The first 16 subgroups are from the low CM group, and the last 16 from the high CM group. Figure 19.24 shows the corresponding high-speed N_{CS} value for each subgroup. Note that sequences with S_{max} less than 15 cannot be used by any high-speed cells, but they can be used by any normal cells which require no more than 24 root sequences from this group for

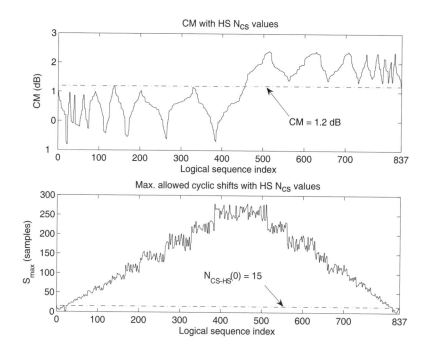

Figure 19.23 Cubic metric and maximum allowed cyclic shift of reordered group sequences.

a total of 64 preambles. Note also that the ordering of the physical ZC root sequence indices is pairwise, since root sequence indices u and $N_{ZC} - u$ have the same CM and S_{max} values. This helps to simplify the PRACH receiver, as elaborated in Section 19.5.2.2.

19.5 PRACH Implementation

This section provides some general principles for practical implementation of the PRACH function.

19.5.1 UE Transmitter

The PRACH preamble can be generated at the system sampling rate, by means of a large IDFT as illustrated in Figure 19.25.

Note that the DFT block in Figure 19.25 is optional as the sequence can be mapped directly in the frequency domain at the IDFT input, as explained in Section 7.2.1. The cyclic shift can be implemented either in the time domain after the IDFT, or in the frequency domain before the IDFT through a phase shift. For all possible system sampling rates, both CP and sequence durations correspond to an integer number of samples.

CM group	Sub-group no.	N_{CS} (High-Speed)	Logical index (i.e. re-ordered)	Physical root sequence index u (in increasing order of the corresponding logical index number)
Low	0	-	0~23	129, 710, 140, 699, 120, 719, 210, 629, 168, 671, 84, 755, 105, 734, 93, 746, 70, 769, 60, 779, 2, 837, 1, 838
	1	15	24~29	56, 783, 112, 727, 148, 691
	2	18	30~35	80, 759, 42, 797, 40, 799
	3	22	36~41	35, 804, 73, 766, 146, 693
	4	26	42~51	31, 808, 28, 811, 30, 809, 27, 812, 29, 810
	...	...	...	...
	15	23 7	384~455	3, 836, 19, 820, 22, 817, 41, 798, 38, 801, 44, 795, 52, 787, 45, 794, 63, 776, 67, 772, 72 767, 76, 763, 94, 745, 102, 737, 90, 749, 109, 730, 165, 674, 111, 728, 209, 630, 204, 635, 117, 722, 188, 651, 159, 680, 198, 641, 113, 726, 183, 656, 180, 659, 177, 662, 196, 643, 155, 684, 214, 625, 126, 713, 131, 708, 219, 620, 222, 617, 226, 613
High	16	23 7	456~513	230, 609, 232, 607, 262, 577, 252, 587, 418, 421, 416, 423, 413, 426, 411, 428, 376, 463, 395, 444, 283, 556, 285, 554, 379, 460, 390, 449, 363, 476, 384, 455, 388, 451, 386, 453, 361, 478, 387, 452, 360, 479, 310, 529, 354, 485, 328, 511, 315, 524, 337, 502, 349, 490, 335, 504, 324, 515
	...	...	...	...
	29	18	810~815	309, 530, 265, 574, 233, 606
	30	15	816~819	367, 472, 296, 543
	31	-	820~837	336, 503, 305, 534, 373, 466, 280, 559, 279, 560, 419, 420, 240, 599, 258, 581, 229, 610

Figure 19.24 Example of mapping from logical index to physical root sequence index.

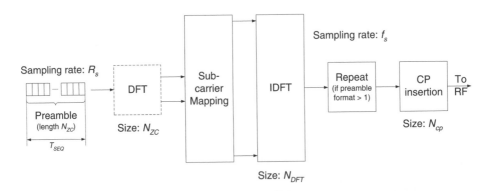

Figure 19.25 Functional structure of PRACH preamble transmitter.

The method of Figure 19.25 does not require any time-domain filtering at baseband, but leads to large IDFT sizes (up to 24 576 for a 20 MHz spectrum allocation), which are cumbersome to implement in practice.

Therefore, another option for generating the preamble consists of using a smaller IDFT, actually an IFFT, and shifting the preamble to the required frequency location through time-domain upsampling and filtering (hybrid frequency/time-domain generation, shown in Figure 19.26). Given that the preamble sequence length is 839, the smallest IFFT size that can be used is 1024, resulting in a sampling frequency $f_{IFFT} = 1.28$ Msps. Both the CP and sequence durations have been designed to provide an integer number of samples at this sampling rate. The CP can be inserted before the upsampling and time-domain frequency shift, so as to minimize the intermediate storage requirements.

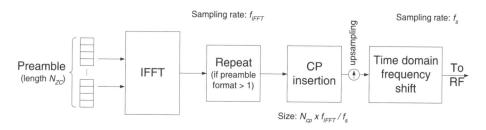

Figure 19.26 Hybrid frequency/time domain PRACH generation.

19.5.2   eNodeB PRACH Receiver

19.5.2.1   Front-End

In the same way as for the preamble transmitter, a choice can be made for the PRACH receiver at the eNodeB between full frequency-domain and hybrid time/frequency domain approaches. As illustrated in Figure 19.27, the common parts to both approaches are the CP removal, which always occurs at the front-end at the system sampling rate f_s, the PDP computation and signature detection. The approaches differ only in the computation of the frequency tones carrying the PRACH signal(s).

The full frequency-domain method computes, from the 800 μs worth of received input samples during the observation interval (Figure 19.6), the full range of frequency tones used for UL transmission given the system bandwidth. As a result, the PRACH tones are directly extracted from the set of UL tones, which does not require any frequency shift or time-domain filtering but involves a large DFT computation. Note that even though $N_{DFT} = n \cdot 2^m$, thus allowing fast and efficient DFT computation algorithms inherited from the building-block construction approach [3], the DFT computation cannot start until the complete sequence is stored in memory, which increases delay.

On the other hand, the hybrid time-frequency domain method first extracts the relevant PRACH signal through a time-domain frequency shift with down-sampling and anti-aliasing filtering. There follows a small-size DFT (preferably an FFT), computing the set of frequency tones centered on the PRACH tones, which can then be extracted. The down-sampling ratio and corresponding anti-aliasing filter are chosen to deliver a number of PRACH time samples suitable for an FFT or simple DFT computation at a sampling rate which is an integer fraction of the system sampling rate. Unlike the full frequency-domain approach, the hybrid time/frequency-domain computation can start as soon as the first samples have been received, which helps to reduce latency.

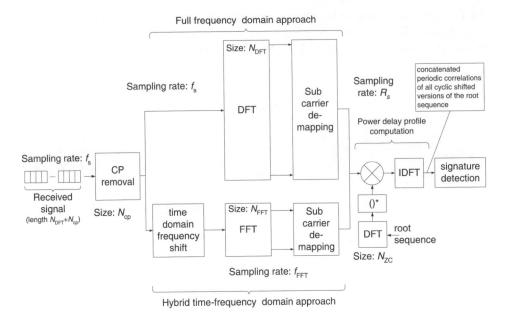

Figure 19.27 PRACH receiver options.

19.5.2.2 Power Delay Profile Computation

The LTE PRACH receiver can benefit from the PRACH format and Constant Amplitude Zero AutoCorrelation (CAZAC) properties as described in the earlier sections by computing the PRACH power delay profile through a frequency-domain periodic correlation. The PDP of the received sequence is given by

$$\text{PDP}(l) = |z_u(l)|^2 = \left| \sum_{n=0}^{N_{ZC}-1} y(n) x_u^*[(n+l)_{N_{ZC}}] \right|^2 \tag{19.18}$$

where $z_u(l)$ is the discrete periodic correlation function at lag l of the received sequence $y(n)$ and the reference searched ZC sequence $x_u(n)$ of length N_{ZC}, and where $(\cdot)^*$ denotes the complex conjugate. Given the periodic convolution of the complex sequences $y(n)$ and $x_u(n)$ defined as

$$[y(n) * x_u(n)](l) = \sum_{n=0}^{N_{ZC}-1} y(n) x_u[(l-n)_{N_{ZC}}], \tag{19.19}$$

$z_u(l)$ can be expressed as follows:

$$z_u(l) = [y(n) * x_u^*(-n)](l) \tag{19.20}$$

Let $X_u(k) = R_{X_u}(k) + jI_{X_u}(k)$, $Y_u(k) = R_Y(k) + jI_Y(k)$ and $Z_u(k)$ be the DFT coefficients of the time-domain ZC sequence $x_u(n)$, the received baseband samples $y(n)$, and the discrete periodic correlation function $z_u(n)$ respectively. Using the properties of the DFT, $z_u(n)$ can

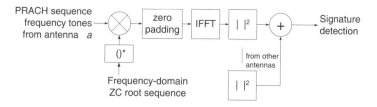

Figure 19.28 PDP computation per root sequence.

be efficiently computed in the frequency domain as

$$\begin{cases} Z_u(k) = Y(k)X_u^*(k) & \text{for } k = 0, \ldots, N_{ZC} - 1 \\ z_u(n) = \text{IDFT}\{Z_u(k)\}_n & \text{for } n = 0, \ldots, N_{ZC} - 1 \end{cases} \qquad (19.21)$$

The PDP computation is illustrated in Figure 19.28.

The zero padding aims at providing the desired oversampling factor and/or adjusting the resulting number of samples to a convenient IFFT size. Note that for high-speed cells, additional non-coherent combining over three timing uncertainty windows can be performed for each receive antenna, as shown in Figure 19.19.

In addition, the pairwise sequence indexing (Section 19.4.3.6) allows further efficient 'paired' matched filtering.

Let $W_u(k)$ be defined as

$$W_u(k) = Y(k)X_u(k) \quad \text{for } k = 0, \ldots, N_{ZC} - 1 \qquad (19.22)$$

The element-wise multiplications $Z_u(k)$ and $W_u(k)$ can be computed jointly [14], where the partial products $R_Y(k)R_{X_u}(k)$, $I_Y(k)I_{X_u}(k)$, $I_Y(k)R_{X_u}(k)$ and $I_{X_u}(k)R_Y(k)$ only need to be computed once for both $Z_u(k)$ and $W_u(k)$ frequency-domain matched filters.

Now let $w_u(l)$ be defined as

$$w_u(l) = \text{IDFT}\{W_u(k)\}_l \quad \text{for } l = 0, \ldots, N_{ZC} - 1 \qquad (19.23)$$

By substituting Equation (19.22) into Equation (19.23) and using the property $\text{DFT}\{x_{N_{ZC}-u}(n-1)\}_k = X_u^*(k)$, we can obtain [15]

$$w_u(l) = \sum_{n=0}^{N_{ZC}-1} y(n)x_{N_{ZC}-u}(l + n - 1) = [y(n) * x_{N_{ZC}-u}(-n)](l - 1) \qquad (19.24)$$

As a result, the IDFTs of the joint computation of the frequency domain matched filters $Z_u(k)$ and $W_u(k)$ provide the periodic correlations of time domain ZC sequences $x_u(n)$ and $x_{N_{ZC}-u}(n)$, the latter being shifted by one sequence sample.

19.5.2.3 Signature Detection

The fact that different PRACH signatures are generated from cyclic shifts of a common root sequence means that the frequency-domain computation of the PDP of a root sequence

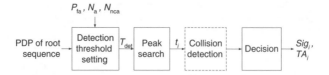

Figure 19.29 Signature detection per root sequence.

provides in one shot the concatenated PDPs of all signatures derived from the same root sequence.

Therefore, the signature detection process consists of searching, within each ZCZ defined by each cyclic shift, the PDP peaks above a detection threshold over a search window corresponding to the cell size. Figure 19.29 shows the basic functions of the signature detector.

Detection threshold setting. The target false alarm probability $p_{fa}(T_{det})$ drives the setting of the detection threshold T_{det}.

Under the assumption that the L samples in the uncertainty window are uncorrelated[6] Gaussian noise with variance σ_n^2 in the absence of preamble transmission, the complex sample sequence $z_a^m(\tau)$ received from antenna a (delayed to reflect a targeted time offset τ of the search window, and despread over a coherent accumulation length (in samples) N_{ca} against the reference code sequence) is a complex Gaussian random variable with variance $\sigma_{n,ca}^2 = N_{ca}\sigma_n^2$. In practice, N_{ca} is the size of the IFFT in Figure 19.28. The non-coherent accumulation $z_{nca}(\tau)$ is modelled as follows:

$$z_{nca}(\tau) = \sum_{a=1}^{N_a} \sum_{m=0}^{N_{nca}-1} |z_a^m(\tau)|^2 \tag{19.25}$$

where N_a is the number of antennas and N_{nca} is the number of additional non-coherent accumulations (e.g. in case of sequence repetition).

$z_{nca}(\tau)$ follows a central chi-square distribution with $2N = 2N_a \cdot N_{nca}$ degrees of freedom, with mean (defining the noise floor) $\gamma_n = N\sigma_{n,ca}^2$ and Cumulative Density Function (CDF) $F(T_{det}) = 1 - p_{fa}(T_{det})^L$. It is worth noticing that instead of the absolute threshold we can consider the threshold T_r relative to the noise floor γ_n as follows:

$$T_r = \frac{T_{det}}{\gamma_n} = \frac{T_{det}}{N_a \cdot N_{nca}N_{ca}\sigma_n^2} \tag{19.26}$$

This removes the dependency of $F(T_r)$ on the noise variance [16]:

$$F(T_r) = 1 - e^{-N_a \cdot N_{nca}T_r} \sum_{k=0}^{N_a \cdot N_{nca}-1} \frac{1}{k!}(N_a \cdot N_{nca}T_r)^k \tag{19.27}$$

As a result, the relative detection threshold can be precomputed and stored.

---

[6]The assumption of no correlation between samples holds true in practice up to an oversampling factor of 2.

Noise floor estimation. For the PDP arising from the transmissions of each root sequence, the noise floor can be estimated as follows:

$$\gamma_n = \frac{1}{N_s} \sum_{\substack{i=0, z_{nca}(\tau_i) < T_{det\_ini}}}^{L-1} z_{nca}(\tau_i) \qquad (19.28)$$

where the summation is over all samples less than the absolute noise floor threshold T_{det_ini} and N_s is the number of such samples. In a real system implementation, the number of additions can be made a power of two by repeating some additions if needed. The initial absolute threshold T_{det_ini} is computed using an initial noise floor estimated by averaging across all search window samples.

Collision detection. In any cell, the eNodeB can be made aware of the maximum expected delay spread. As a result, whenever the cell size is more than twice the distance corresponding to the maximum delay spread, the eNodeB may in some circumstances be able to differentiate the PRACH transmissions of two UEs if they appear distinctly apart in the PDP. This is illustrated in Figure 19.30, where the upper PDP reflects a small cell, where collision detection is never possible, while the lower PDP represents a larger cell where it may sometimes be possible to detect two distinct preambles within the same ZCZ. If an eNodeB detects a collision, it would not send any random access response, and the colliding UEs would each randomly reselect their signatures and retransmit.

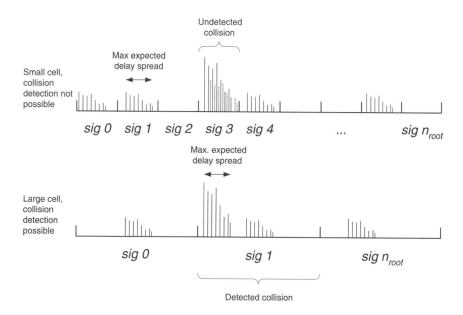

Figure 19.30 Collision detection.

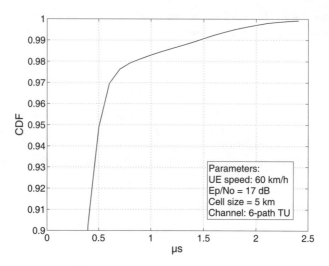

Figure 19.31 CDF of timing estimation error from the PRACH preamble.

19.5.2.4 Timing Estimation

The primary role of the PRACH preamble is to enable the eNodeB to estimate a UE's transmission timing. Figure 19.31 shows the CDF of the typical performance achievable for the timing estimation. From Figure 19.31 one can observe that the timing of 95% of UEs can be estimated to within 0.5 μs, and more than 98% within 1 μs. These results are obtained assuming a very simple timing advance estimation algorithm, using only the earliest detected peak of a detected signature. No collision detection algorithm is implemented here. The IFFT size is 2048 and the system sampling rate 7.68 MHz, giving an oversampling rate of 2.44.

19.5.2.5 Channel Quality Estimation

For each detected signature, the relative frequency-domain channel quality of the transmitting UE can be estimated from the received preamble. This allows the eNodeB to schedule the L2/L3 message (message 3) in a frequency-selective manner within the PRACH bandwidth.

Figure 19.32 shows the BLER performance of the L2/L3 message of the RACH procedure when frequency-selectively scheduled or randomly scheduled, assuming a typical 10 ms delay between the PRACH preamble and the L2/L3 message. A Least Squares (LS) filter is used for the frequency-domain interpolation, and a single RB is assumed for the size of the L2/L3 message.

It can be seen that the performance of a frequency-selectively scheduled L2/L3 message at 10% BLER can be more than 2 dB better than 'blind scheduling' at 3 km/h, and 0.5 dB at 10 km/h.

19.6 Time Division Duplex (TDD) PRACH

As discussed in Chapter 23, one design principle of LTE is to maximize the commonality of FDD and TDD transmission modes. With this in mind, the random access preamble formats

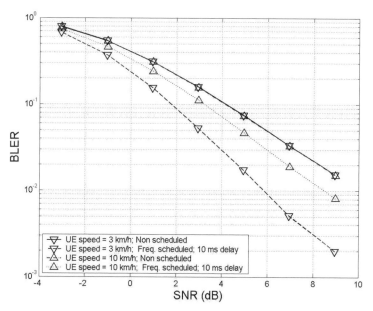

Figure 19.32 BLER performance of post-preamble scheduled data.

Table 19.5 Random Access preamble format 4.

Preamble format	T_{CP}	T_{SEQ}
4	14.6 μs	133 μs

0 to 3 are supported in both FDD and TDD operation. In addition, a short preamble format, 'format 4', is supported for TDD operation. Format 4 is designed to fit into the short uplink special field known as UpPTS for small cells (see Section 23.4).

19.6.1 Preamble Format 4

Table 19.5 lists the parameters for preamble format 4, for which a ZC sequence of length 139 is used. The preamble starts 157 μs before the end of the UpPTS field at the UE.

Unlike preamble formats 0 to 3, a restricted preamble set for high-speed cells is not necessary for preamble format 4, which uses a 7.5 kHz subcarrier spacing. With a random access duration of two OFDM symbols (157 μs), the preamble format 4 is mainly used for small cells with a cell radius less than 1.5 km, and where cyclic shift restrictions for high UE velocities (see Section 19.4.3.4) are not needed. Therefore, considering that Layer 2 always sees 64 preambles and a sequence length of 139, a smaller set of cyclic shift configurations can be used as shown in Table 19.6.

Unlike for preamble formats 0 to 3, the root ZC sequence index for preamble format 4 follows the natural pairwise ordering of the physical ZC sequences, with no special restrictions related to the CM or high-speed scenarios. The sequence mapping for preamble

Table 19.6 Cyclic shift configuration for preamble format 4. Reproduced by permission of © 3GPP.

N_{CS} configuration	N_{CS} value	Required number of ZC root sequences per cell
0	2	1
1	4	2
2	6	3
3	8	4
4	10	5
5	12	6
6	15	8

format 4 can be formulated as

$$u = \left((-1)^v \left\lfloor \frac{v}{2} + 1 \right\rfloor \right) \bmod N_{ZC} \tag{19.29}$$

from logical sequence index v to physical sequence index u, therefore avoiding the need to prestore a mapping table.

19.7 Concluding Remarks

In this chapter the detailed design choices of the LTE PRACH have been explained, based on theoretical derivations and performance evaluations. In particular, it can be seen how the PRACH preamble addresses the high performance targets of LTE, such as high user density, very large cells, very high speed, low latency and a plurality of use cases, while fitting with minimum overhead within the uplink SC-FDMA transmission scheme.

Many of these aspects benefit from the choice of ZC sequences for the PRACH preamble sequences in place of the pseudo-noise sequences used in earlier systems. The properties of these sequences enable substantial numbers of orthogonal preambles to be transmitted simultaneously.

Considerable flexibility exists in the selection of the PRACH slot formats and cyclic shifts of the ZC sequences to enable the LTE PRACH to be dimensioned appropriately for different cell radii and loadings.

Some options for the implementation are available, by which the complexity of the PRACH transmitter and receiver can be minimized without sacrificing the performance.

References[6]

[1] A. V. Oppenheim and R. W. Schafer, *Discrete-Time Signal Processing*. Englewood Cliffs, NJ: Prentice Hall, 1999.
[2] 3GPP Technical Specification 36.211, 'Physical Channels and Modulation (Release 8)', www.3gpp.org.

[6] All web sites confirmed 18th December 2008.

[3] W. W. Smith and J. M. Smith, *Handbook of Real-time Fast Fourier Transforms*. New York: Wiley Inter-Science, 1995.

[4] M. Shafi, S. Ogose and T. Hattori, *Wireless Communications in the 21st Century*. New York: Wiley Inter-Science, 2002.

[5] Panasonic and NTT DoCoMo, 'R1-062175: Random Access Burst Design for E-UTRA', www.3gpp.org, 3GPP TSG RAN WG1, meeting 46, Tallinn, Estonia, August 2006.

[6] 3GPP Technical Report 25.814 'Physical layer aspects for E-UTRA (Release 7)', www.3gpp.org.

[7] R. L. Frank, S. A. Zadoff and R. Heimiller, 'Phase Shift Pulse Codes with Good Periodic Correlation Properties'. *IRE IEEE Trans. on Information Theory*, Vol. 7, pp. 254–257, October 1961.

[8] D. C. Chu, 'Polyphase Codes with Good Periodic Correlation Properties'. *IEEE Trans. on Information Theory*, Vol. 18, pp. 531–532, July 1972.

[9] ETSI EN 300 910, 'Radio Transmission and Reception (Release 1999),' www.etsi.org.

[10] Huawei, 'R1-072325: Multiple Values of Cyclic Shift Increment NCS', www.3gpp.org, 3GPP TSG RAN WG1, meeting 49, Kobe, Japan, May 2007.

[11] Panasonic and NTT DoCoMo, 'R1-073624: Limitation of RACH Sequence Allocation for High Mobility Cell', www.3gpp.org, 3GPP TSG RAN WG1, meeting 50, Athens, Greece, August 2007.

[12] Motorola, 'R1-040642: Comparison of PAR and Cubic Metric for Power De-rating', www.3gpp.org, 3GPP TSG RAN WG1, meeting 37, Montreal, Canada, May 2004.

[13] Motorola, 'R1-060023: Cubic Metric in 3GPP-LTE', www.3gpp.org, 3GPP TSG RAN WG1, LTE adhoc meeting, Helsinki, Finland, January 2006.

[14] Panasonic, 'R1-071517: RACH Sequence Allocation for Efficient Matched Filter Implementation', www.3gpp.org, 3GPP TSG RAN WG1, meeting 48bis, St Julians, Malta, March 2007.

[15] Huawei, 'R1-071409: Efficient Matched Filters for Paired Root Zadoff–Chu Sequences,' www.3gpp.org, 3GPP TSG RAN WG1, meeting 48bis, St Julians, Malta, March 2007.

[16] J. G. Proakis, *Digital Communications*. New York: McGraw–Hill, 1995.

20

Uplink Transmission Procedures

Matthew Baker

20.1 Introduction

In this chapter two procedures are explained which are fundamental to the efficient operation of the LTE uplink.

Timing control is of paramount significance for the orthogonal uplink intra-cell multiple access scheme, while *power control* is important for maintaining Quality-of-Service (QoS), ensuring an acceptable User Equipment (UE) battery life and controlling inter-cell interference.

20.2 Uplink Timing Control

20.2.1 Overview

As explained in Chapters 15 and 17, a key feature of the uplink transmission scheme in LTE is that it is designed for orthogonal multiple-access in time and frequency between the different UEs.

This is fundamentally different from WCDMA, in which the uplink is non-orthogonal; in WCDMA, from the point of view of the multiple access there is therefore no need to arrange for the uplink signals from different UEs to be received with any particular timing at the NodeB receiver. The dominant consideration for the uplink transmission timing in WCDMA is the operation of the power control loop, which was designed (in most cases) for a loop delay of just one timeslot (0.666 ms). This is achieved by setting the uplink transmission timing as close as possible to a fixed offset relative to the received downlink timing, without taking into account any propagation delays. Propagation delays in uplink and downlink are absorbed at

LTE – The UMTS Long Term Evolution: from Theory to Practice   Stefania Sesia, Issam Toufik and Matthew Baker
© 2009 John Wiley & Sons, Ltd

the NodeB, by means of reducing the time spent measuring the Signal-to-Interference Ratio (SIR) to derive the next power control command.

For LTE, uplink orthogonality is maintained by ensuring that the transmissions from different UEs in a cell are time-aligned at the receiver of the eNodeB. This avoids intra-cell interference occurring, both between UEs assigned to transmit in consecutive subframes and between UEs transmitting on adjacent subcarriers.

Time alignment of the uplink transmissions is achieved by applying a *timing advance* at the UE transmitter, relative to the received downlink timing. The main role of this is to counteract differing propagation delays between different UEs, as shown in Figure 20.1. A similar approach is used in GSM.

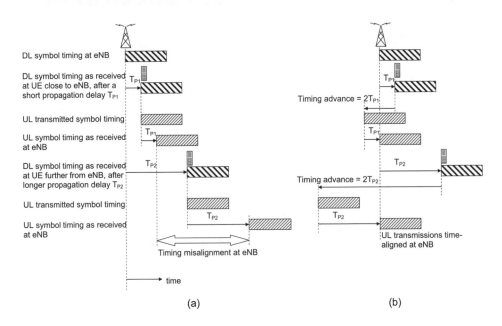

Figure 20.1 Time alignment of uplink transmissions by means of timing advance: (a) without timing advance; (b) with timing advance.

20.2.2   Timing Advance Procedure

20.2.2.1   Initial Timing Advance

After a UE has first synchronized its receiver to the downlink transmissions received from the eNodeB (see Section 7.2), the initial timing advance is set by means of the random access procedure as described in detail in Section 19.3. This involves the UE transmitting a random access preamble from which the eNodeB can estimate the uplink timing and respond with an 11-bit initial timing advance command contained within the Random Access Response (RAR) message. This allows the timing advance to be configured by the eNodeB with a granularity of 0.52 µs from 0 up to a maximum of 0.67 ms, corresponding to a cell radius of 100 km.[1] Larger cell radii would in theory be possible, but in the first release of LTE the timing advance was limited to this range in order to avoid further restricting the processing time available at the UE between receiving the downlink signal and having to make a corresponding uplink transmission. Nevertheless, a cell range of 100 km is sufficient for practical scenarios, and is far beyond what could be achieved with the early versions of GSM in which the range of the timing advance restricted the cell range to about 35 km.

The granularity of 0.52 μs enables the uplink transmission timing to be set with an accuracy well within the length of the uplink Cyclic Prefix (CP) (the smallest value of which is 4.7 μs). This granularity is also significantly finer than the length of a cyclic shift of the uplink reference signals (see Chapter 16). Simulations have shown [1, 2] that timing misalignment of up to at least 1 μs does not cause significant degradation in system performance due to increased interference. Thus the granularity of 0.52 μs is sufficiently fine to allow for additional timing errors arising from the uplink timing estimation in the eNodeB and the accuracy with which the UE sets its transmission timing – the latter being required to be better than 0.39 μs in LTE [3].

20.2.2.2 Timing Advance Updates

After the timing advance has first been set for each UE, it will then need to be updated from time to time to counteract changes in the arrival time of the uplink signals at the eNodeB. Such changes may arise from:

- The movement of a UE, causing the propagation delay to change at a rate dependent on the velocity of the UE relative to the eNodeB; at 500 km/h (the highest speed considered for LTE), the round-trip propagation delay would change by a maximum of 0.93 μs/s.

- Abrupt changes in propagation delay due to existing propagation paths disappearing and new ones coming into play; such changes typically occur most frequently in dense urban environments as the UEs move around the corners of buildings.

- Oscillator drift in the UE, where small frequency errors accumulated over time result in timing errors; the frequency accuracy of the oscillator in an LTE UE is required to be better than 0.1 ppm, which would result in a maximum accumulated timing error of 0.1 μs/s.

- Doppler shift arising from the movement of the UE, especially in Line-Of-Sight (LOS) propagation conditions (in non-LOS conditions, this becomes a Doppler spread, where the error is typically a zero-mean random variable); this results in an additional frequency offset of the uplink signals when they are received at the eNodeB.

The updates of the timing advance to counteract these effects are performed by a closed-loop mechanism whereby the eNodeB measures the received uplink timing and issues timing

---

[1]In theory it would also be possible to support small negative timing advances (i.e. a timing delay) up to the duration of the Cyclic Prefix (CP), for UEs very close to the eNodeB, without causing loss of uplink time-domain orthogonality. However, this is not supported in the first version of LTE.

advance update commands to instruct the UE to adjust its transmission timing accordingly, relative to its previous transmission timing.[2]

In deriving the timing advance update commands, the eNodeB may measure any uplink signal which is useful. This may include the Sounding Reference Signals (SRSs), Channel Quality Indicator (CQI), ACKnowledgements/Negative ACKnowledgements (ACK/NACKs) sent in response to downlink data, or the uplink data transmissions themselves. In general, wider-bandwidth uplink signals enable a more accurate timing estimate to be made, although this is not, in itself, likely to be a sufficient reason to configure all UEs to transmit wideband SRS very frequently. The benefit of highly accurate timing estimation has to be traded off against the uplink overhead from such signals. In addition, cell-edge UEs are power-limited and therefore also bandwidth-limited for a given uplink SINR; in such cases, the timing estimation accuracy of narrower-bandwidth uplink signals can be increased through averaging multiple measurements over time and interpolating the resulting power delay profile. The details of the uplink timing measurements at the eNodeB are not specified, but rather left to the implementation of the eNodeB.

A timing advance command received at the UE is applied at the beginning of the uplink subframe which begins 4–5 ms later (depending on the propagation delay), as shown in Figure 20.2. For a TDD or half-duplex FDD system configuration, the new uplink transmission timing would take effect at the start of the first uplink transmission after this point. In the case of an increase in the timing advance relative to the previous transmission, the first part of the subframe in which the new timing is applied is skipped.

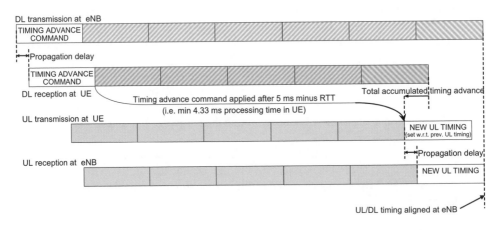

Figure 20.2 Application of timing advance commands.

The timing advance update commands are generated at the Medium Access Control (MAC) layer in the eNodeB and transmitted to the UE as MAC control packets which may be multiplexed together with data on the Physical Downlink Shared Channel (PDSCH). Like the initial timing advance command in the response to the Random Access Channel

---

[2]Only unsynchronized UEs set their transmission timing relative to the received downlink timing; all subsequent adjustments are made relative to the latest uplink timing.

(RACH) preamble, the update commands have a granularity of 0.52 μs. The range of the update commands is ±16 μs, allowing a step change in uplink timing equivalent to the length of the extended CP. They would typically not be sent more frequently than about 2 Hz. In practice, fast updates are unlikely to be necessary, as even for a UE moving at 500 km/h the change in round-trip path length is not more than 278 m/s, corresponding to a change in round-trip time of 0.93 μs/s.

The eNodeB must balance the overhead of sending regular timing update commands to all the UEs in the cell against a UE's ability to transmit quickly when data arrives in its transmit buffer. The eNodeB therefore configures a timer for each UE, which the UE restarts each time a timing advance update is received; if the UE does not receive another timing advance update before the timer expires, it must then consider its uplink to have lost synchronization. In such a case, in order to avoid the risk of generating interference to uplink transmissions from other UEs, the UE is not permitted to make another uplink transmission of any sort without first transmitting a random access preamble to reinitialize the uplink timing.

One further use of timing advance is to create a switching time between uplink reception at the eNodeB and downlink transmission for TDD and half-duplex FDD operation. This switching time can be generated by applying an additional timing advance offset to the uplink transmissions, to increase the amount of timing advance beyond what is required to compensate for the round-trip propagation delay. Typically a switching time of up to 20 μs may be needed. This is discussed in more detail in Section 23.4.1.

20.3 Power Control

20.3.1 Overview

Uplink transmitter power control in a mobile communication system serves an important purpose: it balances the need for sufficient transmitted energy per bit to achieve the required Quality-of-Service (QoS), against the needs to minimize interference to other users of the system and to maximize the battery life of the mobile terminal.

In achieving this purpose, uplink power control has to adapt to the characteristics of the radio propagation channel, including path loss, shadowing and fast fading, as well as overcoming interference from other users – both within the same cell and in neighbouring cells.

The requirements for uplink interference management in LTE are quite different from those for WCDMA. In WCDMA, the uplink is basically non-orthogonal,[3] and the primary source of interference which has to be managed is intra-cell interference between different users in the same cell. Uplink users in WCDMA share the same time-frequency resources, and they generate an interference rise above thermal noise at the Node B receiver; this is known as 'Rise over Thermal' (RoT), and it has to be carefully controlled and shared between users. The primary mechanism for increasing the uplink data rate for a given user in WCDMA is to reduce the spreading factor and increase the transmission power accordingly.

By contrast, in LTE the uplink is basically orthogonal by design, and intra-cell interference management is consequently less critical than in WCDMA. The primary mechanisms for

---

[3]The later releases of WCDMA do, however, introduce a greater element of orthogonality into the uplink transmissions, by means of lower spreading factors and greater use of time-division multiplexing of different users in HSDPA and HSUPA.

varying the uplink data rate in LTE are varying the transmitted bandwidth and varying the Modulation and Coding Scheme (MCS), while the transmitted power per unit bandwidth (i.e. the Power Spectral Density (PSD)) could typically remain approximately constant for a given MCS.

Moreover, in WCDMA the power control [4] was primarily designed with continuous transmission in mind for circuit-switched services, while in LTE fast scheduling of different UEs is applied at 1 ms intervals. This is reflected in the fact that power control in WCDMA is periodic with a loop delay of 0.67 ms and a normal power step of ± 1 dB, while LTE allows for larger power steps (which do not have to be periodic), with a minimum loop delay of about 5 ms.

With these considerations in mind, the power control scheme provided in LTE employs a combination of open-loop and closed-loop control. This in theory requires less feedback than a purely closed-loop scheme, as the closed-loop feedback is only needed to compensate for cases when the UE's own estimate of the required power setting is not satisfactory.

A typical mode of operation for power control in LTE involves setting a coarse operating point for the transmission PSD[4] by open-loop means, based on path-loss estimation. This would give a suitable PSD for an average MCS in the prevailing path-loss and shadowing conditions.

Faster adaptation can then be applied around the open-loop operating point by closed-loop power control. This can control interference and fine-tune the power setting to suit the channel conditions (including fast fading). However, due to the orthogonal nature of the LTE uplink, the LTE closed loop power control does not need to be as fast as in WCDMA – in LTE it would typically be expected to operate at no more than a few hundred Hertz.

Meanwhile, the fastest and most frequent adaptation of the uplink transmissions is by means of the uplink scheduling grants, which vary the transmitted bandwidth (and accordingly the total transmitted power), together with setting the MCS, in order to reach the desired transmitted data rate.

With this combination of mechanisms, the power control scheme in LTE in practice provides support for more than one mode of operation. It can be seen as a 'toolkit' from which different power control strategies can be selected and used depending on the deployment scenario, system loading and operator preference.

20.3.2   Detailed Power Control Behaviour

Detailed power control formulae are specified in LTE for the Physical Uplink Shared Channel (PUSCH), Physical Uplink Control Channel (PUCCH) and the Sounding Reference Signals (SRSs) [5]. The formula for each of these uplink signals follows the same basic principles; though they appear complex, in all cases they can be considered as a summation of two main terms: a basic open-loop operating point derived from static or semi-static parameters signalled by the eNodeB, and a dynamic offset updated from subframe to subframe:

Power per resource block = basic open-loop operating point + dynamic offset

---

[4]In LTE the PSD is set as a power per Resource Block (RB), although if multiple RBs are transmitted by a UE in a subframe the power per RB is the same for all RBs.

20.3.2.1 Basic Open-Loop Operating Point

The basic open-loop operating point for the transmit power per resource block depends on a number of factors including the inter-cell interference and cell load. It can be further broken down into two components:

- a semi-static base level, P_0, further comprised of a common power level for all UEs in the cell (measured in dBm) and a UE-specific offset;

- an open-loop path-loss compensation component.

Different base levels can be configured for PUSCH data transmissions which are dynamically scheduled (i.e. using Physical Downlink Control Channel (PDCCH) signalling) and those which are persistently scheduled (i.e. using Radio Resource Control (RRC) signalling). This in principle allows different BLER (BLock Error Rate) operating points to be used for dynamically-scheduled and persistently-scheduled transmissions. One possible use for different BLER operating points is to achieve a lower probability of retransmission for persistently-scheduled transmissions, hence avoiding the PDCCH signalling overhead associated with dynamically-scheduled retransmissions; this is consistent with using persistent scheduling for delivery of services such as VoIP with minimal signalling overhead.

The UE-specific offset component of the base level P_0 enables the eNodeB to correct for systematic offsets in a UE's transmission power setting, for example arising from errors in path-loss estimation or in absolute output power setting.

The path-loss compensation component is based on the UE's estimate of the downlink path-loss, which can be derived from the UE's measurement of Reference Signal Received Power (RSRP) (see Section 13.4.1.1) and the known transmission power of the downlink Reference Signals (RSs), which is broadcast by the eNodeB. In order to obtain a reasonable indication of the uplink path-loss, the UE should filter the downlink path-loss estimate with a suitable time-window to remove the effect of fast fading but not shadowing. Typical filter lengths are between 100 and 500 ms for effective operation.

For the PUSCH and SRS, the degree to which the uplink PSD is adapted to compensate for the path-loss can be set by the eNodeB, on a scale from 'no compensation' to 'full compensation'. This is achieved by means of a path-loss compensation factor, referred to as α.

In principle, the combination of the base level P_0 and the path-loss compensation component together allow the eNodeB to configure the degree to which the UE measures and responds to the path-loss. At one extreme, the eNodeB could configure the base level to the lowest level (-126 dBm) and rely entirely on the UE's path-loss measurement to raise the power towards the cell edge, while alternatively the eNodeB can set the base level to a higher value, possibly in conjunction with only partial path-loss compensation.

The range of the base level P_0 for the PUSCH (-126 dBm to $+23$ dBm per RB) is designed to cover the full range of target SINR values for different degrees of path-loss compensation, transmission bandwidths and interference levels. For example, the highest value of P_0, $+23$ dBm, corresponds to the maximum transmission power of an LTE UE, and would typically only be used if the path-loss compensation was not being used at all. The lowest value of P_0 for the PUSCH, -126 dBm, is relevant to a case when full path-loss compensation is used and the uplink transmission and reception conditions are optimal: for example, taking a single RB transmission, with a target SINR at the eNodeB of -5 dB (around the lowest useful SINR), interference-free reception and a 0 dB noise figure for the eNodeB receiver, then the required value of P_0 is the thermal noise level in one RB (180 kHz) minus 5 dB, which gives $P_0 = -126$ dBm.

In general, the maximum path-loss that can be compensated (either by P_0 or by the path-loss compensation component) depends on the required SINR and the transmission bandwidth. Some examples are shown in Figure 20.3, for typical ranges of SINR from -5 dB to $+30$ dB, interference rise above thermal noise from 0 dB to $+30$ dB, and transmission bandwidth from one RB (180 kHz) to the maximum LTE system bandwidth of 110 RBs (19.8 MHz). Note that this assumes full path-loss compensation and ignores the dynamic offset.

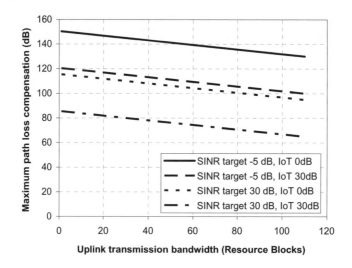

Figure 20.3 Maximum path-loss compensation in typical scenarios for 23 dBm UE.

The fractional path-loss compensation factor α can be seen as a tool to trade off the fairness of the uplink scheduling against the total cell capacity. Full path-loss compensation maximizes fairness for cell-edge UEs. However, when considering multiple cells together as a system, the use of only partial path-loss compensation can increase the total system capacity in the uplink, as less resources are spent ensuring the success of transmissions from cell-edge UEs and less inter-cell interference is caused to neighbouring cells. Path-loss compensation factors around 0.7–0.8 typically give a close-to-maximal uplink system capacity without causing significant degradation to the cell-edge data rate that can be achieved.

Inter-cell interference is of particular concern for UEs located near the edge of a cell, as they may disrupt the uplink transmissions in neighbouring cells. LTE consequently provides an interference coordination mechanism whereby a frequency-dependent 'overload indicator' may be signalled directly between eNodeBs to warn a neighbouring eNodeB of high uplink interference levels in specific RBs. In response to this, the neighbouring eNodeB may reduce the permitted energy per RB of the UEs which are scheduled in the corresponding RBs in its cell(s). It is also possible for eNodeBs to cooperate to avoid scheduling cell-edge UEs in neighbouring cells to transmit in the same resource blocks. This is discussed further in Section 12.5.

In summary, the basic operating point for the transmit power per RB can be defined as follows:

$$\text{Basic operating point} = P_0 + \alpha \cdot \text{PL} \tag{20.1}$$

where α is the path-loss compensation factor which allows the trade-off between total uplink capacity and cell-edge data rate, as discussed above, and PL is the path-loss estimate.

For the low-rate PUCCH (carrying ACK/NACK and CQI signalling), the path-loss compensation is handled separately from that for the PUSCH, as the PUCCH transmissions from different users are code-division-multiplexed. In order to provide good control of the interference between the different users, and hence to maximize the number of users which can be accommodated simultaneously on the PUCCH, full path-loss compensation is always used. A different base level P_0 is also provided for the PUCCH compared to those used for the PUSCH.

20.3.2.2 Dynamic Offset

The dynamic offset part of the power per resource block can also be further broken down into two components:

- A component dependent on the MCS;

- Explicit Transmitter Power Control (TPC) commands.

MCS-dependent component. The MCS-dependent component (referred to in the LTE specifications as Δ_{TF}, where TF stands for 'Transport Format') allows the transmitted power per RB to be adapted according to the transmitted information data rate. Ideally, the transmission power required for a given information data rate should follow Shannon's theorem, such that

$$R_N = \log_2(1 + \text{SNR}) \tag{20.2}$$

where R_N is the normalized information data rate per unit bandwidth and can be calculated as the number of information bits per Resource Element (RE) in the RB, denoted here BPRE (Bits Per RE) and SNR is the Signal-to-Noise Ratio.

Practical limitations of the system and receiver can be modelled with a scaling factor k:

$$\text{BPRE} = \frac{1}{k} \log_2(1 + \text{SNR}) \tag{20.3}$$

It follows that

$$\text{Transmit power required per RB} \propto 2^{k \cdot \text{BPRE}} - 1 \tag{20.4}$$

A suitable value for k is taken as 1.25 for the MCS-dependent power offset [6].

The MCS-dependent component of the transmit power setting can act like a power control command, as the MCS is under the direct control of the eNodeB scheduler: by changing the MCS which the UE has permission to transmit, the eNodeB can quickly apply an indirect adjustment to the UE's transmit power spectral density via the MCS-dependent component of the transmit power setting. This may be done to take into account the instantaneous buffer status and QoS requirements of the UE.

The MCS-dependent component can also be used to allow an element of frequency-dependent power control, for example in cases where explicit power control commands (discussed in more detail below) are not transmitted frequently and are therefore following only the wideband fading characteristics; for example, by scheduling a low-rate MCS when the UE is granted permission to transmit in a particular part of the band, the eNodeB can dictate a low transmission power in those RBs.

Another use for the MCS-dependent component is in cases where the number of uplink RBs allocated to a UE in a subframe is not matched to the desired data rate and SIR. One example is to enable the transmit power to be reduced if the amount of data to be transmitted is less than the rate supported by the radio channel in a single RB.

The MCS-dependent component for the PUSCH can be set to zero if it is not needed, for example if fast Adaptive Modulation and Coding (AMC) is used instead.

MCS-dependent power control is also particularly relevant for the PUCCH, as the PUCCH bandwidth for a UE does not vary depending on the amount of information to be transmitted in a given subframe (ranging from a single bit for a scheduling request or ACK/NACK, to 22 bits for combined dual-codeword ACK/NACK and CQI together – see Section 17.3). Moreover, for the PUCCH the magnitude of the power offset for each combination of control information can be adjusted semi-statically by the eNodeB, in order to set a suitable error-rate operating point for each type of control signalling. This is analogous to the different power offsets which may be set in High Speed Packet Access (HSPA) for ACK/NACK and CQI signalling according to the error rate desired by the network.

UE-specific power control commands.   The other component of the dynamic offset is the UE-specific TPC commands. These can operate in two different modes: *accumulative* TPC commands (available for PUSCH, PUCCH and SRS) and *absolute* TPC commands (available for PUSCH only). For the PUSCH, the switch between these two modes is configured semi-statically for each UE by RRC signalling – i.e. the mode cannot be changed dynamically.

With the accumulative TPC commands, each TPC command signals a power step relative to the previous level. This is the default mode and is particularly well-suited to situations where a UE receives power control commands in groups of successive subframes. This mode is similar to the closed-loop power control operation in WCDMA, except that the exact values of the power steps are different: in LTE, two sets of power step values are provided: $\{-1, +1\}$ dB and $\{-1, 0, +1, +3\}$ dB (compared to the sets $\{-1, +1\}$ dB and $\{-2, +2\}$ dB in WCDMA). Which of these two sets of power steps is used is determined by the format of the TPC commands and RRC configuration. The maximum size of power step that can be made using accumulative TPC commands is therefore $+3/-1$ dB, but the range over which the power can be adjusted relative to the semi-static operating point is unlimited (except for the maximum and minimum power limits according to the UE power class). Larger power steps can be achieved by combining an accumulative TPC command with an MCS-dependent power step, by changing the MCS. The provision of one set of power step values containing a 0 dB step size enables the transmit power to be kept constant if needed (i.e. without necessarily having to change the transmission power every time a scheduling grant is sent). This is useful, for example, in scenarios where the interference is not expected to vary significantly over time.

By contrast, the transmit power setting that results from an absolute TPC command is independent of the sequence of TPC commands that may have been received previously; the transmit power setting depends only on the most recently-received absolute TPC command, which independently signals a power offset relative to the semi-static operating point.[5] The set of offsets which can be signalled by absolute TPC commands is $\{-4, -1, +1, +4\}$ dB. Thus the absolute power control mode can only control the power within a range of ± 4 dB from the semi-static operating point, but a relatively large power step can be triggered by a single command (up to ± 8 dB). This mode is therefore well suited to scenarios where the scheduling of the UE's uplink transmissions may be intermittent; an absolute TPC command enables the UE's transmission power to be adjusted to a suitable level in a single step after each transmission gap.

20.3.2.3 Total Transmit Power Setting

Finally, for the PUSCH and SRS, the total transmit power of the UE in each subframe is scaled up linearly from the power level derived from the semi-static operating point and dynamic offset, according to the number of RBs actually scheduled for transmission from the UE in the subframe.

Thus the overall power control equation is as follows:

$$\text{UE transmit power} = \underbrace{P_0 + \alpha \cdot \text{PL}}_{\substack{\text{basic open-loop} \\ \text{operating point}}} + \underbrace{\Delta_{\text{TF}} + f(\Delta_{\text{TPC}})}_{\text{dynamic offset}} + \underbrace{10 \log_{10} M}_{\text{bandwidth factor}}$$

where Δ_{TPC} denotes a TPC command, $f(\cdot)$ represents accumulation in the case of accumulative TPC commands, and M is the number of allocated RBs.

This overall power control formula allows the UE's transmit power to be controlled with an accuracy of 1 dB within a range set by the applicable performance requirements for the UE, typically -50 dBm to $+23$ dBm (corresponding to a maximum transmission power of 0.2 W). The maximum transmission power of the UEs in a cell may be restricted to a lower level by RRC signalling.

For the SRS, an additional semi-static offset may be configured by RRC signalling.

20.3.2.4 Transmission of TPC Commands

TPC commands for the dynamic offset part of the power control are sent to the UE in messages on the PDCCH. The UE is required to check for a TPC command in every subframe unless it is specifically configured in Discontinuous Reception (DRX – see Section 4.4.1.1). However, unlike in WCDMA, the TPC commands in LTE are not necessarily periodic.

One method by which TPC commands are transmitted to the UEs is in the uplink resource scheduling assignment messages for each specific UE. This is logical as it results in all the applicable information for an uplink transmission (set of resource blocks, transport format, and power setting) being included in a single message.

Additionally, individual accumulative TPC commands for multiple UEs can be jointly coded into a special PDCCH message dedicated to power control (PDCCH Formats 3 and 3A – see Section 9.3.2.3); furthermore, for the PUCCH only, TPC commands can be sent in downlink resource assignment messages on the PDCCH. The latter two methods for TPC

---

[5]The absolute TPC mode can be seen as a low-overhead way to adjust the UE-specific offset in the base level component of the semi-static operating point.

command transmission enable the power control loop to track changes in channel conditions even when the UE is not scheduled for uplink data transmission, and can therefore be seen as an alternative to the use of absolute TPC commands. The LTE specifications do not allow the jointly-coded TPC commands on PDCCH to be used if the UE is configured in the absolute power control mode.

Due to the structure of the PDCCH signalling (see Section 9.3.2.2), in all cases the TPC commands are protected by a CRC (Cyclic Redundancy Check); this means that they should be considerably more reliable than in WCDMA. The only likely source of error in LTE would be the UE's failure to detect a PDCCH message, which should typically have a probability around 10^{-2} (compared to a typical power control error rate of 4–10% in WCDMA).

The eNodeB can use a number of techniques to determine how to command each UE to adjust its transmit power. One method will be the received SIR, based, for example, on measurements of the SRS and uplink demodulation RSs; in addition the BLER experienced on the decoding of uplink data packets may be used.

The eNodeB may also take into account interference coordination with neighbouring cells, for example if it has received an 'overload indicator' indicating that interference from a UE is causing a problem in a neighbouring cell. Note, however, that although eNodeBs may signal overload indicators to each other, an eNodeB receiving an overload indicator cannot know for certain whether the overload situation is caused by a UE in its cell or not; it can only infer that, if the received overload indicator relates to a group of RBs where it has scheduled a cell-edge UE, then it is possible that the interference arises from its cell and it should therefore react. Further details of interference coordination are explained in Section 12.5.

20.3.3   UE Power Headroom Reporting

In order to assist the eNodeB to schedule the uplink transmission resources to different UEs in an appropriate way, it is important that the UE can report its available power headroom to the eNodeB.

The eNodeB can use the power headroom reports to determine how much more uplink bandwidth per subframe a UE is capable of using. This can help to avoid allocating uplink transmission resources to UEs which are unable to use them; as the uplink is basically orthogonal in LTE, no other UE would be able to use such resources, so system capacity would be wasted.

The range of the power headroom report is from $+40$ to -23 dB. The negative part of the range enables the UE to signal to the eNodeB the extent to which it has received an uplink resource grant which would require more transmission power than the UE has available. This would enable the eNodeB to reduce the size (i.e. the number of RBs in the frequency domain) of a subsequent grant, thus freeing up transmission resources to allocate to other UEs.

A power headroom report can only be sent in subframes in which a UE has an uplink transmission grant; the report relates to the subframe in which it is sent. The headroom report is therefore a prediction rather than a direct measurement; the UE cannot directly measure its actual transmission power headroom for the subframe in which the report is transmitted. It therefore relies on reasonably accurate calibration of the UE's power amplifier output, especially at high output powers when reliable knowledge of the headroom is more critical to system performance.

A number of criteria are defined to trigger a power headroom report. These include:

- A significant change in estimated path loss since the last power headroom report;

- More than a configured time has elapsed since the previous power headroom report;

- More than a configured number of closed-loop TPC commands have been implemented by the UE.

The eNodeB can configure parameters to control each of these triggers depending on, for example, the system loading and the requirements of its scheduling algorithm.

20.3.4 Summary of Uplink Power Control Strategies

In summary, a variety of degrees of freedom are available for power control in the LTE uplink. It is likely that not every parameter will be actively used in every network deployment, but each deployment will select a mode of power control appropriate to the scenario or scheduling strategy.

One typical mode of operation would be to set the semi-static operating point (via P_0 and the path-loss compensation factor α) to achieve at least the required SINR at the eNodeB for the required QoS for each UE, compensating for path-loss and wideband shadowing. Further control for interference management and rate adaptation can be exercised by means of frequency-domain scheduling and bandwidth adaptation – these being degrees of freedom for power management which were not available in WCDMA. Bandwidth adaptation may also be used in conjunction with changing the MCS to set different BLER operating points for different HARQ processes.

Finally, dynamic transmission power offsets can be used to give a finer degree of control, by means of the MCS-dependent offsets and the closed-loop corrections using the explicit TPC commands.

In practice, implementation issues must also be taken into account when considering the accuracy with which a UE can set its transmission power, and the speed at which the transmission power can be changed.

References[6]

[1] Nokia, 'R1-063377: UL Timing Control Accuracy and Update Rate', www.3gpp.org, 3GPP TSG RAN WG1, meeting 47, Riga, Latvia, November 2006.
[2] Texas Instruments, 'R1-072841: Simulation of Uplink Timing Error Impact on PUSCH', www.3gpp.org, 3GPP TSG RAN WGI, meeting 49bis, Orlando, USA, June 2007.
[3] 3GPP Technical Specification 36.133, 'Requirements for Support of Radio Resource Management (Release 8)', www.3gpp.org.
[4] M. P. J. Baker and T. J. Moulsley, 'Power Control in UMTS Release 99' in *Proc. First IEE Int. Conf. on 3G Communications*, March 2000.
[5] 3GPP Technical Specification 36.213, 'Evolved Universal Terrestrial Radio Access (E-UTRA); Physical Layer Procedures (Release 8)', www.3gpp.org.
[6] Ericsson, 'R1-080881: Range and Representation of Delta_MCS', www.3gpp.org, 3GPP TSG RAN WG1, meeting 52, Sorrento, Italy, February 2008.

[6]All web sites confirmed 18th December 2008.

Part IV

Practical Deployment Aspects

21

The Radio Propagation Environment

Juha Ylitalo and Tommi Jämsä

21.1 Introduction

Realistic modelling of propagation characteristics is essential for two main reasons when developing the LTE system. Firstly, the link- and system-level performance of LTE can be evaluated accurately only when the radio channel models are realistic. In particular, the spatial characteristics of the channel models have a significant effect on the performance of multi-antenna systems. Secondly, the model used for radio propagation plays an important role in the network planning phase of LTE deployment.

Different environments such as rural, suburban, urban, and indoor set different requirements for network planning, antenna configuration, and the preferred spatial transmission mode. Moreover, the actual deployment scenarios for LTE will cover numerous special cases ranging from mountainous rural surroundings to dense urban and outdoor-to-indoor situations. As an example of the range of different propagation environments likely to be relevant for high-bandwidth Multiple-Input Multiple-Output (MIMO) systems such as LTE, the IST-WINNER[1] project [1] developed wideband radio channel models for 13 propagation environments.

The propagation characteristics are in a large part affected by the carrier frequency. In November 2007 the International Telecommunication Union (ITU) World Radiocommunication Conference (WRC-2007) allocated new frequency bands for International Mobile Telecommunications (IMT) radio access systems between 450 MHz and 3.6 GHz. Thus depending on the deployed carrier frequency, the LTE radio channel characteristics will vary

---

[1]Information Society Technologies – Wireless world INitiative NEw Radio.

LTE – The UMTS Long Term Evolution: from Theory to Practice   Stefania Sesia, Issam Toufik and Matthew Baker
© 2009 John Wiley & Sons, Ltd

significantly even within a particular type of propagation environment. The carrier frequency used may also have an impact on the deployment of MIMO for LTE, due to the fact that the size of the antenna arrangement depends strongly on the signal wavelength.

As discussed in Chapter 11, the ability to use a variety of MIMO techniques is an important feature of LTE. While conventional wireless systems are designed to counteract multipath fading, MIMO approaches such as those used in LTE are able to take advantage of multipath scattering to increase capacity [2, 3]. However, the theoretical MIMO performance is obtainable only with fully uncorrelated transmit and receive antennas, which is not the case in practice.

Therefore, it is important to create realistic standardized MIMO radio channel models for the evaluation of the performance of LTE and its future enhancements.

It is especially important to model accurately the correlation between the signals of different antenna branches, since this dictates, to a large extent, the preferred spatial transmission mode and its performance. The spatial correlation properties of the MIMO radio channel model define the ultimate limit of the theoretical channel capacity. The applied channel model has to reflect all the instantaneous space-time-frequency characteristics which affect the configuration of diversity, beamforming, and spatial multiplexing techniques.

In the following sections we discuss first the 3GPP Single-Input Single-Output (SISO) and Single-Input Multiple-Output (SIMO) channel models which include only a single antenna at the transmitter and one or two antennas at the receiver. Next, the characteristics and modelling principles of a MIMO radio channel are discussed; in this context we also address the WINNER model, which represents the state of the art in MIMO modelling and from which insights may be gained into relevant channel models for future versions of LTE such as LTE-Advanced (see Chapter 24). We also discuss the practical emulation of MIMO channels to enable efficient real-time conformance testing of eNodeBs and UEs.

21.2   SISO and SIMO Channel Models

In practice, realistic modelling of the radio channel propagation characteristics requires extensive measurement campaigns with appropriate carrier frequencies and bandwidths, in the radio environments which are identified to be the most relevant for deployment. The research projects COST207, COST231, COST259, CODIT (UMTS Code Division Testbed), and ATDMA (Advanced TDMA Mobile Access) created extensive wideband measurement datasets for SISO and SIMO channel modelling in the 1980s and 1990s [4–8]. The corresponding channel models form a basis for the ITU models which were largely applied in the development of the third generation wireless communication systems.

Figure 21.1 illustrates a multipath propagation scenario in which the UE has an approximately omnidirectional antenna. The transmitted signal traverses three paths with different delays. As explained in Chapter 5, in wideband communications the delay spread of the propagation paths is larger than the symbol period and the receiver observes the multipath components separately. In the frequency domain this corresponds to frequency-selective fading since the coherence bandwidth of the radio channel is smaller than the signal bandwidth. The coherence bandwidth is proportional to the inverse of the root mean square (r.m.s.) delay spread which can be calculated from the Power Delay Profile (PDP) of the radio channel (see Section 8.2.1). The PDP also defines the *maximum excess delay*, which

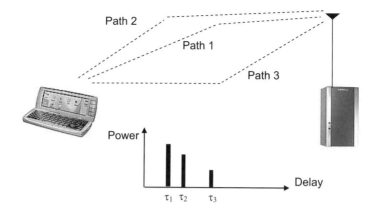

Figure 21.1 Multipath propagation and PDP.

is a measure of the largest delay difference between the propagation paths. This measure is important when considering the length of the Cyclic Prefix (CP) in LTE, as discussed in Section 5.3.1.

Three main phenomena can be identified which affect the received signal properties:

- Propagation path-loss;

- Shadow (slow, or 'large-scale') fading;

- Multipath (fast, or 'small-scale') fading.

The propagation loss determines the average signal level at the receiver. Typical radio access systems can cope with a path-loss up to around 150 dB. The received signal level varies relatively slowly due to shadowing effects from large obstructions such as buildings and hills. Finally, the multipath characteristics define the frequency, time, and space correlations. As several multipath signal components with approximately the same propagation delay add up incoherently at the receiver antenna, the composite signal level can vary by up to 30–40 dB as the User Equipment (UE) moves.

21.2.1 ITU Channel Model

ITU channel models were used in developing the Third Generation 'IMT-2000' family[2] of radio access systems [9]. The main user scenarios covered indoor office, outdoor-to-indoor, pedestrian and vehicular radio environments. The key parameters to describe each propagation model include time delay spread and its statistical variability, path-loss and shadow fading characteristics, multipath fading characteristics, and operating radio frequency. Each environment is defined for two cases which have different probabilities of occurrence: a smaller delay spread case and a larger delay spread case. The numbers of paths are as for the 3GPP Channel Models introduced in the following section.

---

[2]The initial 'IMT-2000' systems of the IMT family were proposed around 2000.

Table 21.1 Propagation conditions for multipath fading in the ITU models.

Tap number	ITU Pedestrian A		ITU Pedestrian B		ITU Vehicular A	
	Relative delay (ns)	Relative mean power (dB)	Relative delay (ns)	Relative mean power (dB)	Relative delay (ns)	Relative mean power (dB)
1	0	0	0	0	0	0
2	110	−9.7	200	−0.9	310	−1.0
3	190	−19.2	800	−4.9	710	−9.0
4	410	−22.8	1200	−8.0	1090	−10.0
5			2300	−7.8	1730	−15.0
6			3700	−23.9	2510	−20.0

21.2.2 3GPP Channel Model

The 3GPP specifications for UMTS make use of propagation conditions largely based on the ITU models. As an example, Table 21.1 shows the propagation conditions used for performance evaluation in different multipath fading environments [10]. All taps use the classical Doppler spectrum [11, 12] (see Section 8.3.1), which assumes that the propagation paths at a receiver antenna are uniformly distributed over 360° in azimuth. The fading of the signals at different antennas is assumed to be independent, which is only appropriate for SISO and some antenna diversity cases. MIMO was not introduced in UMTS until Release 7.

21.2.3 Extended ITU Models

The evaluation of LTE techniques demands channel models with increased bandwidth compared to UMTS models, to reflect the fact that the characteristics of the radio channel frequency response are connected to the delay resolution of the receiver. In 3GPP the 20 MHz LTE channel models were based on a synthesis of existing models such as the ITU and 3GPP models. Specifically the six ITU models covering an excess (maximum) delay spread from 35 ns to 4000 ns were chosen as a starting point, together with the Typical Urban (TU) model from GSM[3] which has a maximum excess delay of 1000 ns. In this way, extended wideband models with low, medium, and large delay spread values could be identified. The low delay spread gives an Extended Pedestrian A (EPA) model which is employed in an urban environment with fairly small cell sizes (or even up to about 2 km in suburban environments with low delay spread), while the medium and large delay spreads give an Extended Vehicular A (EVA) model and Extended TU (ETU) model respectively. The ETU model has a large maximum excess delay of 5000 ns, which in fact is not very typical in urban environments. Instead it applies to some extreme urban, suburban, and rural cases which occur seldom but which are important in evaluating LTE performance in the most challenging environments. Table 21.2 shows the r.m.s. delay spread values for the three extended models [13].

It was also decided that the extended channel models are applied with low, medium, and high Doppler shifts, namely 5 Hz, 70 Hz and 300 Hz, which at a 2.5 GHz carrier frequency correspond roughly to mobile velocities of 2, 30 and 130 km/h respectively. Combinations which are likely to be used are EPA 5 Hz, EVA 5 Hz, EVA 70 Hz and ETU 70 Hz.

---

[3]Global System for Mobile Communications.

Table 21.2 r.m.s. delay spread for the extended ITU models.

Category	Channel model	Acronym	r.m.s. delay spread (ns)
Low delay spread	Extended Pedestrian A	EPA	43
Medium delay spread	Extended Vehicular A model	EVA	357
High delay spread	Extended Typical Urban model	ETU	991

Table 21.3 Power delay profiles of extended ITU models.

Tap number	EPA model		EVA model		ETU model	
	Excess tap delay (ns)	Relative power (dB)	Excess tap delay (ns)	Relative power (dB)	Excess tap delay (ns)	Relative power (dB)
1	0	0.0	[0	0.0	0	−1.0
2	30	−1.0	30	−1.5	50	−1.0
3	70	−2.0	150	−1.4	120	−1.0
4	80	−3.0	310	−3.6	200	0.0
5	110	−8.0	370	−0.6	230	0.0
6	190	−17.2	710	−9.1	500	0.0
7	410	−20.8	1090	−7.0	1600	−3.0
8			1730	−12.0	2300	−5.0
9			2510	−16.9	5000	−7.0

The classical Doppler spectrum is again assumed. The tapped delay line propagation conditions for the LTE performance evaluation are hence summarized in Table 21.3.

LTE conformance tests also include propagation conditions for two high-speed train scenarios. These are based on 300 and 350 km/h, with different Doppler shift trajectories, in non-fading propagation channels [14].

21.3 MIMO Channel

As already mentioned, the performance of MIMO systems depends strongly on the underlying propagation conditions. First of all, the Carrier-to-Interference-and-Noise Ratio (CINR) at the MIMO receiver determines the ultimate gains of spatial multiplexing. The CINR is a direct function of the path-loss, so that a terminal far from the base station tends to have a relatively small CINR compared to a terminal close to the base station. A second important factor is the correlation between signals at different antennas. Fading decorrelation is facilitated by the presence of a large number of multipath components at both the transmitter and the receiver, as typically experienced in a non-Line-Of-Sight (NLOS) situation. The antenna separation also has a strong impact on the spatial correlation. The largest MIMO gains are obtained in scenarios with large CINR and low spatial correlation (where the channel matrix has a high rank), which in practice may occur rarely.

There are two commonly applied types of MIMO channel model:

- Correlation matrix based channel models;

- Geometry-based channel models.

Models of both types are used for LTE. The 3GPP Spatial Channel Model (SCM) and Spatial Channel Model – Extension (SCME) (see Sections 21.3.2 and 21.3.3 respectively) are geometry-based stochastic models, whereas the extended ITU models (EPA, EVA, ETU) are correlation matrix based models. In Section 21.3.4 we then discuss the geometry-based IST-WINNER channel model to give an insight into the state of the art in channel modelling. It should be noted that it is possible to create a correlation matrix based channel model from a geometry-based one, but not vice versa.

21.3.1   Effect of Spatial Correlation

In MIMO systems the spatial correlation has a strong impact on the capacity limit of the radio channel. If the transmitter has no knowledge about the radio channel $\mathbf{H}$, the capacity is calculated as [2]:

$$C = \log\left[\det\left(\mathbf{I}_{N_R} + \frac{\gamma}{N_T}\mathbf{H}\mathbf{H}^H\right)\right] \tag{21.1}$$

where N_R is the number of receive antennas, N_T is the number of transmit antennas, $\mathbf{I}_{N_R}$ is the $N_R \times N_R$ identity matrix, γ is the Signal-to-Interference plus Noise Ratio (SINR), $\mathbf{H}$ is the matrix of channel transfer functions between the N_T transmit antennas and the N_R receive antennas and $\{\cdot\}^H$ denotes the Hermitian transpose operation.

It is obvious that for a given SINR γ at the receiver and for a given number of transmit and receive antennas, it is solely the spatial correlation characteristics of the MIMO channel $\mathbf{H}$ which determine the theoretical capacity limit. Figure 21.2 shows the channel capacity in three cases with different Angular Spreads (ASs) at the eNodeB. In all cases the angular spread at the UE is large, consisting of three signal clusters in random directions, while the angular spread at the eNodeB is varied. Both the eNodeB and the UE have four antennas, the spacing of which is half a carrier wavelength. For eNodeB angular spreads of 2°, 5° and 15°, the ergodic capacities at an SNR of 25 dB are 11.7, 13.9 and 16.6 bps/Hz respectively.[4] Thus this simple example confirms that the spatial modelling of the radio channel plays an important role when evaluating performance.

Figure 21.3 shows another example of the impact of spatial correlation on MIMO performance at the radio link level. It illustrates the QPSK Bit Error Rate (BER) curves (with rate 1/2 convolutional coding) of 2×1 MISO and 2×2 MIMO scenarios in an ITU Pedestrian B radio channel. A Space-Time Transmit Diversity (STTD) Alamouti-type OFDM transmission scheme was used with a bandwidth of 7 MHz and three different spatial correlation values (0, 0.8 and 1.0), measured with a real hardware transmitter and receiver and a fading emulator. The results show that the SNR required to support a BER of 10^{-4} may vary by as much as 10 dB depending on the spatial correlation. The dependence of spatial correlation on the angular spread and antenna separation is shown in Figure 21.4.

---

[4]The ergodic capacity is defined as the capacity obtained when using a coding scheme whose codewords have a length much longer than the channel coherence time – i.e. the instantaneous capacity averaged over the fading statistic. The capacity values given here were obtained for a duration of 10 s with a mobile speed of 120 km/h.

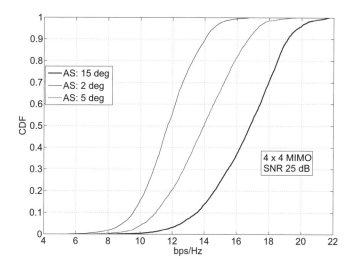

Figure 21.2 Channel capacity for base station angular spreads of 2°, 5° and 15°.

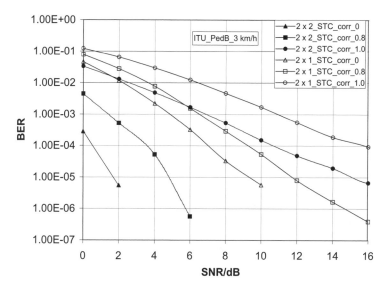

Figure 21.3 BER curves for 2×2 MIMO in an ITU Pedestrian B channel with different channel correlations.

21.3.2 SCM Channel Model

In order to model MIMO channel characteristics realistically, originally for the evaluation of different MIMO schemes for High Speed Downlink Packet Access (HSDPA), 3GPP developed jointly with 3GPP2 a geometry-based SCM [15]. The SCM includes simple

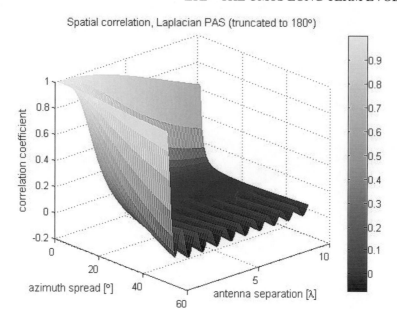

Figure 21.4 Spatial correlation as a function of angular spread and antenna separation.

tapped-delay line models for calibration purposes and a geometry-based stochastic model for system-level simulations. The time-delay properties of the SCM calibration model are the same as in the 3GPP SISO/SIMO models discussed in Section 21.2.2. The spatial characteristics of the MIMO channel are defined by the angular spread and directional distribution of the propagation paths at both the base station and the UE.

System simulations typically consist of multiple cells/sectors, multiple base stations and multiple UEs. Performance metrics such as throughput and delay are gathered over a large number of simulation runs, called 'drops', which may consist of a predefined number of radio frames. During a drop, the large-scale parameters are fixed but the channel undergoes fast fading according to the motion of the terminals. The UEs may feed back channel state information about the instantaneous radio channel conditions and the base station can schedule its transmissions accordingly. The overall procedure for generating the channel realizations consists of three basic steps, shown in Figure 21.5. Figure 21.6 defines the geometrical framework for the spatial parameters.

The SCM includes three main propagation environments: urban microcell, urban macrocell, and suburban macrocell. Additionally, it is possible to modify the basic scenarios by applying a LOS component in an urban micro case and a far-scattering cluster or an urban canyon option in an urban macro case. The SCM is a stochastic geometry-based model which enables realistic modelling of spatial correlation at both the base station and the UE. It is based on the ITU SISO models described in Section 21.2.1, so the number of propagation paths in each environment is six. Delay and angular spreads as well as directions of the scattering clusters are random variables with normal or log-normal distribution. The elevation (i.e. vertical) domain is not modelled. Table 21.4 shows the main SCM parameters.

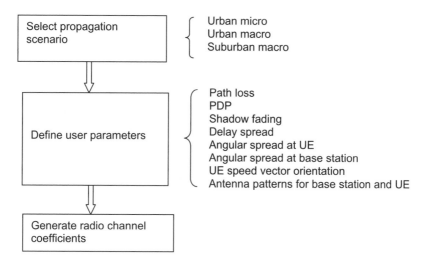

Figure 21.5 Method for generating SCM parameters.

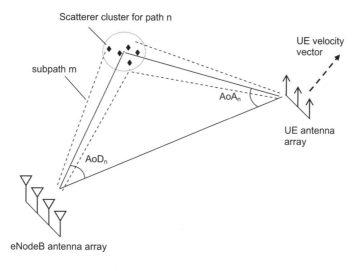

Figure 21.6 Geometry for generating spatial parameters for base station and UE.

The SCM model is designed so that different antenna constellations can be applied. For example, the base station antenna spacing and beam pattern can be varied, including beam patterns for 3-sector and 6-sector cells. Typical values for the antenna spacing at the base station vary from 0.5 to 10 carrier wavelengths. At the base station the per-path angular spread is rather small ($<5°$) in all three environments, which is typical of antennas well above rooftop level. The mean angular spread is also relatively small, the largest value being $19°$ in the urban micro scenario, which is small compared to the beamwidth of a 4-antenna array.

Table 21.4 Main parameters for SCM.

Parameter	Suburban macro	Urban macro	Urban micro
Number of paths	6	6	6
Number of subpaths per path	20	20	20
Mean AS at base station	5°	8°, 15°	19°
Per-path AS at base station (fixed)	2°	2°	5°
Mean AS at UE	68°	68°	68°
Per-path AS at UE (fixed)	35°	35°	35°
Mean total r.m.s. delay spread	0.17 µs	0.65 µs	0.25 µs
Standard deviation for log-normal shadowing	8 dB	8 dB	NLOS: 10 dB; LOS: 4 dB
Path-loss model (dB, d = distance in metres from base station to UE)	$31.5 + 35\log_{10}(d)$	$34.5 + 35\log_{10}(d)$	NLOS: $34.53 + 38\log_{10}(d)$; LOS: $30.18 + 26\log_{10}(d)$

At the UE, the per-path and mean angular spreads are large (35°), which reflects the fact that the UE antennas are embedded in the scattering environment.

In all environments the angular spread is composed of 6×20 subpaths. The per-path angular spread at both the base station and the UE follows a Laplacian distribution, which is generated by giving the 20 subpaths an equal power and fixed azimuth directions with respect to the nominal direction of the corresponding path. Figure 21.7 illustrates the per-path AS at the base station in the urban micro case. A Matlab® implementation of the SCM model can be found in [1].

21.3.3 SCM-Extension Channel Model

An extension to the 3GPP/3GPP2 SCM model was developed in the IST-WINNER project. Known as the SCME (SCM-Extension) model, it is described in [16]. The extension was designed to remain backward-compatible, simple and consistent with the initial SCM concept. The main development is a broadening of the channel model bandwidth from 5 MHz to 100 MHz. Even though the first release of LTE employs only 20 MHz bandwidth, the wider bandwidth extensions are likely to become relevant for later releases.

The bandwidth extension is carried out by introducing so-called midpaths which define the intra-cluster delay spread (i.e. the delay spread within a cluster of paths in a similar direction). The midpaths have fixed delay and power offsets in order to keep the SCME model backward-compatible with SCM. Therefore, the low-pass-filtered SCME impulse response corresponds closely to the respective SCM impulse response. As a result of the bandwidth extension the number of delay taps increases from six in SCM to 18 or 24, depending on

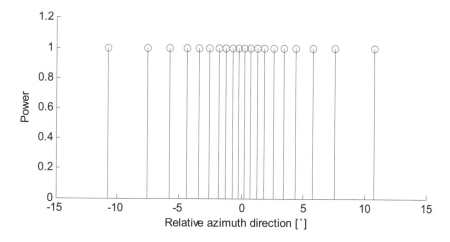

Figure 21.7 Relative azimuth directions of 20 subpaths for 5° angular spread at base station in the SCM model.

Table 21.5 SCME midpath powers and delays.

Scenario	Urban macro, suburban macro		Urban micro	
No. of midpaths per path	3		4	
	Midpath power	Relative delay (ns)	Midpath power	Relative delay (ns)
Midpath 1	10/20	0	6/20	0
Midpath 2	6/20	7	6/20	5.8
Midpath 3	4/20	26.5	4/20	13.5
Midpath 4	—	—	4/20	27.6

the scenario. Table 21.5 shows the midpath delays and powers for the SCME propagation scenarios. Frequency-dependent factors are also added to the path-loss formulae, to extend the frequency range from 2 GHz to 5 GHz.

Another important contribution of the SCME model is the introduction of fixed spatio-temporal Tapped Delay Line (TDL) models, referred to as 'Clustered Delay Line (CDL) models'. The model parameters include the arrival and departure angles in addition to the traditional power and delay, thus covering all the MIMO propagation channel parameters except the polarization information. The TDL models are intended for the calibration of simulators. The original SCM specification also includes calibration models, mainly for comparison with the 3GPP SISO models. The spatio-temporal TDL models of the SCM extension resemble closely the SCM system model; it can actually be viewed as a realization of the system model which is optimized for low correlation in the frequency domain.

The SCME model has a number of optional features which can be applied depending on simulation purposes. In addition to the SCM 'drop' concept, the SCME model introduces optional drifting of delays and arrival/departure angles of the propagation paths. In the original SCM all the propagation parameters remain fixed during a drop and the only variation is the fast fading caused by the Doppler effects. Time evolution of the delay taps may be used for example for evaluating receiver synchronization algorithms. Drifting of the path arrival/departure angles is targeted to the testing of beamforming algorithms. Time evolution of shadow fading is also an optional feature; since the correlation of the shadow fading at two locations is related to the distance between them, this is modelled by an exponentially-shaped spatial auto-correlation function applied to the shadow fading such that the correlation of the shadow fading drops exponentially with distance. Finally, the SCME model includes parameterization of the LOS condition for all the macro-cellular scenarios. A Matlab® implementation of the SCM model can be found in [1].

21.3.4   WINNER Model

The WINNER radio channel models represent the latest development of MIMO channel modelling. They were developed step-by-step during four years in the two phases of the IST-WINNER project. Since the initial SCME model was not adequate for more advanced simulations, new measurement-based models had to be developed.

The Phase I model, known as the WINNER generic model, was the first step forward from the SCME and was based on channel measurements performed at 2 and 5 GHz. The generic model makes it possible to create an arbitrary geometry-based radio channel model. The model is a ray-based double-directional multilink model which is antenna-independent, scalable and capable of modelling MIMO channels. Statistical distributions and channel parameters extracted from measurements in any propagation scenarios can be fitted to the generic model. The model covers a limited set of propagation scenarios.

The Phase II model increased the number of scenarios to 13, added new features to the model, and extended the frequency range to cover frequencies from 2 to 6 GHz. These models cover a wide range of propagation scenarios and environments, including indoor, outdoor-to-indoor (and vice versa), urban micro- and macrocell and corresponding difficult urban scenarios, suburban and rural macrocell, feeder links, and moving networks. Table 21.6 lists the most important scenarios which belong to Local Area (LA), Metropolitan Area (MA), or Wide Area (WA) environments. They are based on a generic channel modelling approach, which means that it is possible to vary the number of antennas, the antenna configurations, geometry and the antenna beam pattern without changing the basic propagation model. This method enables the use of the same channel data in different link- and system-level simulations. The details of the model are described in [17]. The models include CDL models for system calibration.

Path-losses have been specified for all the scenarios, divided into two subscenarios according to propagation conditions (LOS/NLOS) when applicable. The general structure of the path-loss is given by

$$PL = A \log_{10}(d) + B + C \log_{10}\left(\frac{f_c}{5}\right) + X \qquad (21.2)$$

Table 21.6 Selected WINNER Propagation scenarios.

Scenario	Definition	Environment	LOS/NLOS	Mobility (km/h)	Notes
A1	Indoor office	LA	LOS/NLOS	0–5	
B1	Urban microcell	LA, MA	LOS/NLOS	0–70	
B4	Outdoor to indoor microcell	MA	NLOS	0–5	
B5	LOS stationary feeder	MA	LOS	0	Below-rooftop to street-level
C1	Suburban	WA	LOS/NLOS	0–120	
C2	Typical urban macrocell	MA, WA	LOS/NLOS	0–120	
D1	Rural macrocell	WA	LOS/NLOS	0–200	
D2	(a) Moving networks: BS-MRS*, rural	WA	LOS	0–350	Very large Doppler variability
	(b) Moving networks: MRS*-UE, rural	LA	LOS/OLOS**/NLOS	0–5	Same as A1 NLOS

* MRS: Mobile Relay Station. ** OLOS: Obstructed Line-Of-Sight.

where d is the distance in metres and f_c is the carrier frequency in GHz. For each (sub)scenario, a set of constants A, B and C are defined in [17], as is the additional term X, for special cases. X is an environment-dependent factor, for example modelling attenuation due to walls or floors (e.g. if the transmitter and receiver are located on different floors). A Matlab® implementation of the WINNER model can be found in [1].

21.3.5 LTE Evaluation Model

Channel models are used in the design of LTE for many purposes, including evaluation of UE and eNodeB performance requirements, Radio Resource Management (RRM) requirements (see Chapter 13), system concepts and RF system scenarios (see Chapters 22 and 23). Conformance tests of the UE and base station are discussed in Section 21.5.2.

Fixed tapped delay line models of the SCME were proposed for both link- and system-level simulations in [18]. Two simplification steps from the SCM approach were taken. Firstly, the fixed TDL models, which were originally defined for calibration simulations only, were planned to be applied also in system-level simulations. The stochastic drop concept of SCM was discarded. As a slight modification to the SCME TDL models, the delays were quantized to the resolution which corresponds to fourfold over-sampling of the

LTE bandwidth. Secondly, the MIMO characteristics were described by correlation matrices instead of directional propagation parameters such as the angles of arrival/departure and the polarization characteristics. For this purpose, antenna array characteristics were defined for the eNodeB and UE for different scenarios. The LTE channel model scenarios and antenna configurations are listed in Table 21.7 where λ is the wavelength defined as $\lambda = c/f_c$, c is the speed of light ($3 \cdot 10^8$ m/s) and f_c is the carrier frequency.

Table 21.7 Scenarios of LTE performance evaluation model.

Name	Propagation scenario	eNodeB arrangement	UE arrangement
SCM-A	Suburban Macro	3-sector, 0.5λ spacing	Handset, talk position
SCM-B	Urban Macro (low spread)	6-sector, 0.5λ spacing	Handset, data position
SCM-C	Urban Macro (high spread)	3-sector, 4λ spacing	Laptop
SCM-D	Urban Micro	6-sector, 4λ spacing	Laptop

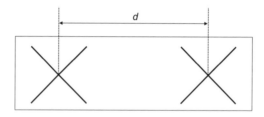

Figure 21.8 Base station antenna configuration.

The eNodeB and UE antenna configurations are illustrated in Figures 21.8–21.10. In the eNodeB, four antenna elements are used, exploiting both space and polarization diversity. Two dual-polarized slanted $+45°$ and $-45°$ elements are spatially separated by a distance d (in metres). The polarization of each element is for simplicity assumed to be unchanged over all departure angles. The antenna element patterns are identical to those proposed for the link calibration model of SCM and can be either 3-sector or 6-sector. The radiation pattern as a function of angle, $A(\theta)$, is as follows:

$$A(\theta) = -\min\left[12\left(\frac{\theta}{\theta_{3\ \text{dB}}}\right)^2, A_m \right] \quad \text{where } -180° \leq \theta \leq 180° \tag{21.3}$$

where, for a 3-sector arrangement, the 3 dB beamwidth $\theta_{3\ \text{dB}} = 70°$, $A_m = 20$ dB and the maximum gain is 14 dBi,[5] while for a 6-sector arrangement $\theta_{3\ \text{dB}} = 35°$, $A_m = 23$ dB and the maximum gain is 17 dBi. Two types of mobile antenna scenarios are assumed: a small handset with two orthogonally-polarized antennas (see Figure 21.9), and a laptop with two spatially

---

[5]This indicates the maximum forward gain of the antenna relative to the gain of a hypothetical omnidirectional antenna.

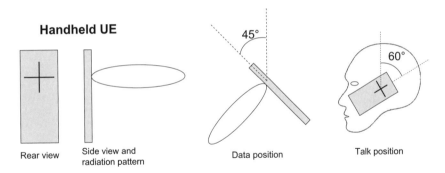

Figure 21.9 Handset antenna configuration.

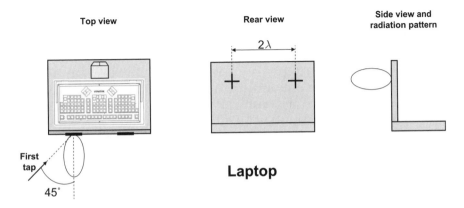

Figure 21.10 Laptop antenna configuration.

separated dual-polarized antennas (see Figure 21.10). It is assumed that the polarizations are purely horizontal and vertical in all directions when the antennas are in the nominal position. In the 'talk position' case, the lobe is in the horizontal direction and the handset is rotated 60° (with the polarizations also being rotated). In the data position, the mobile is tilted 45° such that the radiation lobe has its maximum slanted downwards. The azimuthal orientations of the mobile antennas are selected such that the angle of arrival of the first tap occurs at +45° in all scenarios. The antenna element patterns are given by the same function as the eNodeB patterns, with the following parameters:

- Handheld, talk position: $\theta_{3\,\mathrm{dB}} = 120°$, $A_m = 15$ dB, maximum gain = 3 dBi (vertical), 0 dBi (horizontal);

- Handheld, data position: $\theta_{3\,\mathrm{dB}} = 120°$, $A_m = 5$ dB, maximum gain = 3 dBi (vertical), 0 dBi (horizontal);

- Laptop: $\theta_{3\,\mathrm{dB}} = 90°$, $A_m = 10$ dB, maximum gain = 7 dBi, spatial separation 2λ.

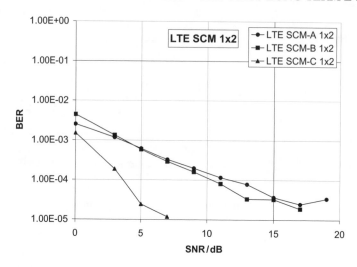

Figure 21.11 BER performance for LTE evaluation channel models in 1 × 2 SIMO case.

Spatial correlations of the LTE performance evaluation model are calculated based on the antenna characteristics and the path arrival/departure angles defined in the SCME model. Polarization covariance matrices are determined based on the antenna polarizations and a cross polarization discrimination value of 8 dB. The Kronecker assumption is applied – i.e. fully separable transmitter and receiver power azimuth spectra. Finally 8 × 8 correlation matrices for SCM-A and -B scenarios and 16 × 16 correlation matrices for SCM-C and -D scenarios can be determined, where the scenarios SCM-A, -B, -C, and -D are introduced in Table 21.7.

Figures 21.11 and 21.12 show an example of the measured BER performance of some of the LTE evaluation channel models for 1 × 2 SIMO (receive diversity), 2 × 1 MISO (transmit diversity) and 2 × 2 MIMO (transmit and receive diversity) OFDM configurations. A 7 MHz bandwidth was used at 2.4 GHz carrier frequency, with a rate $\frac{1}{2}$ convolutional code and QPSK modulation. A total of 500 000 channel impulse responses were generated per SNR value.

The results show the beneficial effect of the low spatial correlation of the LTE-SCM-C model compared to the LTE-SCM-A and LTE-SCM-B models. Moreover, the excellent performance achieved by means of simultaneous transmit and receive diversity is clearly demonstrated.

21.3.6   Comparison of MIMO Channel Models

The SCM, SCME, and WINNER channel models are compared in [19]. Table 21.8 compares the main features of the SCM, 3GPP LTE-evaluation, SCME, and WINNER channel models, and Table 21.9 shows the numerical values of some of the key parameters of the models.

The LTE evaluation channel models are simplified models which have fixed delay and angle spreads. In that sense they apply well to link-level calibration models and also to conformance testing. However, it could be argued that they are too simplified for the most thorough of system-level performance evaluations. The SCM, SCME and WINNER models provide random delay and angle spreads for different users and therefore apply well to system

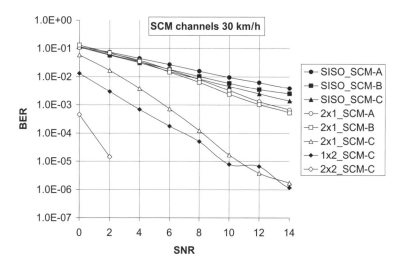

Figure 21.12 BER performance for LTE evaluation channel models with different antenna configurations.

Table 21.8 Feature comparison of SCM, LTE evaluation, SCME and WINNER models.

Feature	SCM	LTE eval	SCME	WINNER
Bandwidth >20 MHz	No	Yes	Yes	Yes
Indoor scenarios	No	No	No	Yes
Outdoor-to-indoor and indoor-to-outdoor scenarios	No	No	No	Yes
AoA/AoD elevation	No	No	No	Yes
Intra-cluster delay spread	No	Yes	Yes	Yes
Cross-correlation between large-scale parameters	No	No	No	Yes
Time evolution of model parameters	No	No	Yes*	Yes

*Continuous time evolution.

level (multicell, multi-user) simulations. Thus the delay spread can take quite different values during different drops for a single user or during parallel drops to different users – this increases the variability and dynamics of the radio channels in multi-user simulations. From the MIMO modelling point of view, the LTE and SCM/SCME/WINNER models differ significantly in the way the spatial correlation is defined. In SCM/SCME/WINNER the spatial correlation is defined by the nominal direction and the angular geometry of the sub-paths of each delay tap, while the LTE evaluation model generates deterministic correlation values for different clusters. Another significant difference is that the SCM/SCME/WINNER models specify the path-loss and shadow fading for each of the user environments, which are important in system-level performance evaluation.

Table 21.9 Numerical comparison of SCM, LTE, SCME and WINNER models.

Parameter	SCM	LTE eval	SCME	WINNER
Max. bandwidth (MHz)	5	20*	100*	100**
Frequency range (GHz)	2	n/a	2–6	2–6
No. of scenarios	3	4	3	12
No. of clusters	6	6	6	4–20
No. of midpaths per cluster	1	3	3–4	1–3
No. of subpaths per cluster	20	n/a	20	20
No. of taps	6	18	18–24	4–24
Base station angle spread (°)	5–19	4.7–18.2	4.7–18.2	2.5–53.7
UE angle spread (°)	68	62.2–67.8	62.2–67.8	11.7–52.5
Delay spread (ns)	170–650	231–841	231–841	16–630
Shadow fading standard deviation (dB)	4–10	n/a	4–10	2–8

* Artificial extension from 5 MHz bandwidth. ** Based on 100 MHz measurements.

21.3.7   Extended ITU Models with Spatial Correlation

The extended ITU models (EPA, EVA, ETU) were developed for MIMO conformance tests in a simple way by allocating the same correlation matrix to all the multipath components [20]. Three degrees of spatial correlation between different antenna signals are defined: high, medium and low. The low correlation case actually represents a case where the antennas are fully uncorrelated. The high and medium correlations are defined by specific correlation matrices based on practical antenna constellations. In the high correlation case the base station and UE antenna arrays are co-polarized and have small inter-antenna distances of 1.5 and 0.5 wavelengths respectively. Medium correlation is achieved by means of orthogonally polarized antennas. These cases are illustrated in Figure 21.13.

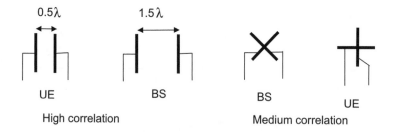

Figure 21.13 Antenna configurations for medium and high spatial correlation for extended ITU models.

In the 2×2 MIMO case the base station and UE correlation matrices are defined respectively as follows:

$$\mathbf{R}_{\text{BS}} = \begin{pmatrix} 1 & \alpha \\ \alpha^* & 1 \end{pmatrix}, \quad \mathbf{R}_{\text{UE}} = \begin{pmatrix} 1 & \beta \\ \beta^* & 1 \end{pmatrix} \tag{21.4}$$

and accordingly the spatial correlation is defined as

$$\mathbf{R}_{\text{spat}} = \mathbf{R}_{\text{BS}} \otimes \mathbf{R}_{\text{UE}} = \begin{pmatrix} 1 & \alpha \\ \alpha^* & 1 \end{pmatrix} \otimes \begin{pmatrix} 1 & \beta \\ \beta^* & 1 \end{pmatrix} \tag{21.5}$$

where $\otimes$ is the Kronecker product. The spatial correlation in the different cases can be represented as shown in Table 21.10.

Table 21.10 Spatial correlation with extended ITU channel models.

Low correlation		Medium correlation		High correlation	
α	β	α	β	α	β
0	0	0.3	0.9	0.9	0.9

Further, by applying the polarization matrices for the medium and high correlation cases, the following spatial correlation matrices are obtained for the 2×2 MIMO case:

- Low correlation:

$$\mathbf{R}_{\text{low}} = \begin{pmatrix} 1 & 0 & 0 & 0 \\ 0 & 1 & 0 & 0 \\ 0 & 0 & 1 & 0 \\ 0 & 0 & 0 & 1 \end{pmatrix} \tag{21.6}$$

- Medium correlation:

$$\mathbf{R}_{\text{medium}} = \begin{pmatrix} 1 & 0.9 & 0.3 & 0.27 \\ 0.9 & 1 & 0.27 & 0.3 \\ 0.3 & 0.27 & 1 & 0.9 \\ 0.27 & 0.3 & 0.9 & 1 \end{pmatrix} \tag{21.7}$$

- High correlation:

$$\mathbf{R}_{\text{high}} = \begin{pmatrix} 1 & 0.9 & 0.9 & 0.81 \\ 0.9 & 1 & 0.81 & 0.9 \\ 0.9 & 0.81 & 1 & 0.9 \\ 0.81 & 0.9 & 0.9 & 1 \end{pmatrix} \tag{21.8}$$

Similar spatial channel models for the 4×2 and 4×4 MIMO cases are also specified. The above correlation models are relatively simple. They apply reasonably well to basic MIMO concepts but may not be adequate for advanced beamforming/MIMO concepts due to the fact that all the delay taps have the same correlation properties. In real environments each delay tap will have different directions and angular spreads which define the per-path spatial correlation properties.

21.4   ITU Channel Models for IMT-Advanced

ITU-R[6] is defining channel models which will be needed in the evaluation of the candidate radio interface technologies for IMT-Advanced in relevant environments (see Chapter 24). The channel model needs to cover all required test environments and scenarios of the IMT-Advanced evaluations [21]. The following four environments are proposed:

- Base coverage urban: an urban macrocellular model for continuous coverage for pedestrian and vehicular users;

- Microcellular: an urban microcellular model with higher user density focusing on pedestrian and slow vehicular users;

- Indoor: an indoor model for isolated cells in offices and/or hotspots, based on stationary and pedestrian users;

- High speed: macrocell environment with high-speed vehicles and trains.

The IMT-Advanced modelling approach includes a primary module and an extension module. The primary module covers mandatory features including the parameters needed for the evaluations of the IMT-Advanced candidate radio interface technologies for all the four environments. The extension module extends the capabilities of the model enabling it to be applied in specific cases not covered by mandatory features.

The ITU-R IMT-Advanced channel model is a geometry-based stochastic model which is targeted towards multi-cell, multi-user system simulations. In a similar way to the 3GPP/3GPP2 SCM, it does not explicitly specify the positions of the scatterers, but instead models the directions of the propagation paths. It allows separate definition of the antenna constellation and radiation patterns. The channel parameters for individual snapshots of the radio channel are determined stochastically, based on statistical distributions extracted from channel measurements. Channel realizations are generated by summing contributions of individual propagation paths with specific small-scale parameters like delay, power, Angle-of-Arrival (AoA) and Angle-of-Departure (AoD). Superposition results in realistic spatial correlation between antenna elements and also in temporal fading which is defined by the geometry-dependent Doppler spectrum.

The IMT-Advanced channel model is based on the 'drop' concept as used in the SCM model. In a simulation the number of drops and the length of each drop have to be selected properly in order to obtain statistically representative results (i.e. to collect a sufficient statistic). Some of the main parameters related to system evaluation for IMT-Advanced are listed in Table 21.11.

21.5   MIMO Channel Emulation

Radio channel phenomena can be mimicked by software (simulation) or by hardware (emulation). Software simulation enables easy programmability and low cost, and is typically utilized for link- and system-level performance evaluations. On the other hand, hardware emulation may be used to test real hardware implementations, to verify and validate products, and to speed up testing. It typically offers more realistic testing.

---

[6]ITU Radio Communication Sector – see Section 1.1.2.

Table 21.11 Key parameters of IMT-Advanced system evaluation (draft).

	Base coverage urban	Microcellular	Indoor	High speed
Deployment scenario	Urban macrocell	Urban microcell	Hotspot	Rural macrocell
Network layout	Hexagonal grid	Hexagonal grid	Indoor floor	Hexagonal grid
Channel model	Urban macro (LOS/NLOS)	Urban micro (LOS/NLOS, outdoor-to-indoor)	Indoor hotspot (LOS/NLOS)	Rural macro (LOS/NLOS)
Site-to-site distance	500 m	200 m	60 m	1732 m
Carrier frequency	2 GHz	2.5 GHz	3.4 GHz	800 MHz
BS antenna height	25 m, above rooftop	10 m, below rooftop	6 m, ceiling-mounted	35 m, above rooftop
No of BS antennas	up to 8	up to 8	up to 8	up to 8
BS Tx power	46 dBm/10 MHz 49 dBm/20 MHz	41 dBm/10 MHz 44 dBm/20 MHz	21 dBm/10 MHz 24 dBm/20 MHz	46 dBm/10 MHz 49 dBm/20 MHz
No of UE antennas	up to 2	up to 2	up to 2	up to 2
UE transmit power	24 dBm	24 dBm	21 dBm	24 dBm
UE velocity	30 km/h	3 km/h	3 km/h	120 km/h

21.5.1 Performance and Conformance Testing

Performance testing is a general term for all testing where some performance figures are measured. The figures can be, for example, power, voltage, sensitivity, BER, BLock Error Rate (BLER), outage, adjacent channel power, system throughput or handover efficiency. To measure the performance accurately it is necessary to use calibrated test instruments such as signal generators and fading emulators. Standardized channel models are needed to compare different products and technologies.

Conformance testing is carried out to determine whether a product meets some specified standard, typically resulting in either a 'pass' or a 'fail'. It is often performed by an external organization, such as a certified conformance test laboratory. Thus, it does not define the performance accurately, but just indicates whether some specified performance goes above or below the specified threshold. Obviously, standardized radio channel models are needed. Traditionally, these models have been rather simple tapped-delay line models in order to allow their direct implementation in hardware based radio channel emulators. However, more recent hardware-based channel emulators implement much more realistic channel models.

21.5.2 LTE Channel Models for Conformance Testing

Extended ITU channel models EPA, EVA and ETU will be used in LTE conformance tests. In addition, the spatial correlation models described above will be employed. Since the

LTE MIMO capability can cover up to four transmit and receive antennas supporting beamforming, transmit diversity and spatial multiplexing, the current 2×2 channel models have to be extended to 4×4 models in a way that supports different spatial modes. Then the extended SCM model or a derivative of it could be employed. The development of the conformance test specifications for systems such as LTE typically lags behind the main system specifications.

21.5.3   Requirements for a Channel Emulator

A radio channel emulator is an item of test equipment which is employed in radio performance and conformance testing to generate the radio channel response between a transmitter and a receiver. The channel emulator has to support advanced channel modelling capability across wide bandwidths with adjustable carrier frequencies. For example, beamforming and precoding techniques require calibrated and accurate phase shifts between parallel antenna elements. Any phase angle and amplitude imperfections cause errors in the spatial correlation. Advanced MIMO schemes with feedback channels require a large number of parallel synchronized channels to be emulated. Moreover, in the case of OFDM, the high Peak-to-Average-Power Ratio (PAPR) necessitates a high dynamic range for the channel emulator (see Section 5.2.2). Linearity is a key requirement of a channel emulator.

21.5.4   MIMO Conformance Testing

3GPP LTE will base the conformance test requirements for the eNodeBs and UEs on the existing principles and test structures used for UMTS. In this section we discuss some aspects of conformance testing which are related to propagation characteristics.

Propagation conditions for LTE conformance tests are specified for the different MIMO cases. The test configurations and channel emulation requirements are likely to become more complicated due to the increasing number of transmit and receive antennas, and due to the fact that in advanced MIMO systems with feedback both the downlink and the uplink need to be modelled.

In future the most important aspect of LTE conformance testing from the viewpoint of the radio channel emulator will be how the MIMO functionality is implemented. The first issue is the number of required fading channels. A 4×4 MIMO link requires 16 parallel fading channels only in the downlink. A handover scenario with two cells requires two sets of 16 channels. Moreover, in some cases it is desirable that both the downlink and uplink are emulated which further increases the required number of parallel channels.

Figures 21.14 and 21.15 show two examples of test connection setups from the UMTS Release 7 conformance test specifications [22]. The first example illustrates a simple SISO case test setup which is used to verify the ability of the receiver to receive the transmitted signal with a BLER not exceeding a specified value. It includes several adjustable attenuators, a fading channel emulator, a noise generator, a combiner, and a circulator. The test setup can be implemented using properly-calibrated analogue components; alternatively, a digital channel emulator is often used to simplify the test procedure.

The second example, in Figure 21.15, shows a connection diagram for a uni-directional full duplex 2×2 MIMO connection for a MIMO receiver performance test. If realized using analogue components, the test requires four fading channels and numerous attenuators,

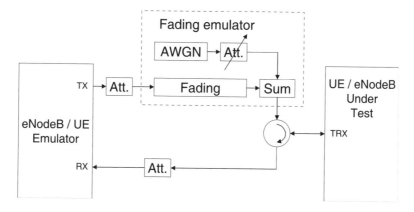

Figure 21.14 Conformance test set-up for SISO case.

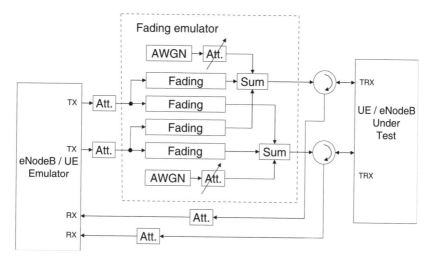

Figure 21.15 Conformance test set-up for 2 × 2 MIMO case.

combiners and splitters. It can be simplified if implemented in a single digital channel emulator.

As mentioned above, the standardized channel models typically assume specific antenna polarization and positions. These, however, are typically product-specific, and therefore a MIMO capable channel emulator can offer more accurate performance test results by providing channel models which reflect the actual antenna design implementations.

21.6 Concluding Remarks

It is important to specify realistic propagation conditions for the evaluation of LTE performance. Accurate modelling of the spatial characteristics of the MIMO radio channel is needed in order to determine the best-performing multi-antenna transmission schemes, and to evaluate the LTE system performance in different radio environments.

3GPP specifications define advanced geometry-based radio channel models (SCM, SCME) which fulfil most of the sophisticated demands for system performance testing of LTE. However, most realistic models tend to be rather complex, leading to long simulation runs. Therefore, simple extended ITU models with specific spatial correlation characteristics are also specified for performance evaluation and conformance testing. These simple models are unlikely to be adequate for the testing of advanced multiple antenna concepts in the future. For example, the use of simple correlation-based models in the testing of open-loop beamforming or closed-loop precoding techniques could be improved by using different spatial properties for the delay taps. In these cases, geometry-based models in which each of the delay taps are modelled separately are superior to correlation-based models.

SCME, WINNER and IMT-Advanced channel models provide additional features for further evaluations and simulations of LTE. The area of MIMO propagation modelling will continue to be important as the LTE conformance tests are developed, and as new versions of LTE with even larger numbers of antennas are standardized in the future.

References[7]

[1] Commission of the European Communities, 'IST-WINNER Project', www.ist-winner.org.

[2] G. J. Foschini, 'Layered Space-time Architecture for Wireless Communication in a Fading Environment when Using Multi-element Antennas'. *Bell Labs Tech. Journal*, Vol. 1, pp. 41–59, 1996.

[3] E. Telatar, 'Capacity of multi-antenna Gaussian channels'. *European Trans. on Telecommunications*, Vol. 10, pp. 585–595, November 1999.

[4] Commission of the European Communities, 'Digital Land Mobile Radio Communications-COST 207 Final Report', 1989.

[5] Commission of the European Communities, 'Digital Mobile Radio towards Future Generation Systems-COST 231 Final Report', 1993.

[6] Commission of the European Communities, 'Wireless Flexible Personalised Communications-COST 259 Final Report', 2001.

[7] W. R. Braun and U Dersch, 'A Physical Mobile Radio Channel Model'. *IEEE Trans. on Vehicular Technology*, Vol. 40, pp. 472–482, 1991.

[8] Commission of the European Communities, 'RACE-II advanced TDMA mobile access project: An Approach for UMTS'. Springer, Berlin/Heidelberg, 1994.

[9] ITU-R M.1225 International Telecommunication Union, 'Guidelines for evaluation of radio transmission technologies for IMT-2000', 1997.

[10] 3GPP Technical Specification 25.101, 'UE Radio Transmission and Reception (FDD) (Release 7)', www.3gpp.org.

[11] W. C. Jakes, *Microwave Mobile Communications*. New York: John Wiley & Sons Ltd/Inc, 1974.

[12] R. H. Clarke, 'A Statistical Theory of Mobile-radio Reception'. *Bell Syst. Tech. Journal.*, pp. 957–1000, July 1968.

[13] Ericsson, Nokia, Motorola, and Rohde & Schwarz, 'R4-070572: Proposal for LTE Channel Models', www.3gpp.org, 3GPP TSG RAN WG4, meeting 43, Kobe, Japan, May 2007.

---

[7] All web sites confirmed 18th December 2008.

[14] 3GPP Technical Specification 36.141, 'Base Station (BS) conformance testing (Release 8)', www.3gpp.org.

[15] 3GPP Technical Specification 25.996, 'Spatial Channel Model for Multiple Input Multiple Output (MIMO) Simulations (Release 6)', www.3gpp.org.

[16] D. S. Baum, J. Salo, G. Del Galdo, M. Milojevic, P. Kyösti and J. Hansen, 'An Interim Channel Model for Beyond-3G Systems' in *Proc. IEEE Vehicular Technology Conference* (Stockholm, Sweden), May 2005.

[17] Commission of the European Communities, IST-WINNER II project deliverable D1.1.2 'WINNER II Channel Models', www.ist-winner.org.

[18] Ericsson, Elektrobit, Nokia, Motorola, and Siemens, 'R4-060334: LTE Channel Models and simulations', www.3gpp.org, 3GPP TSG RAN WG4, meeting 38, Denver, USA, February 2006.

[19] M. Narandzic, C. Schneider, R. Thomä, T. Jämsä, P. Kyösti and X. Zhao, 'Comparison of SCM, SCME, and WINNER Channel Models', in *Proc. IEEE Vehicular Technology Conference* (Dublin, Ireland), April 2007.

[20] Agilent Technologies, 'R4-071318: LTE MIMO Correlation Matrices', August 2007.

[21] NTT DOCOMO, 'R1-082713: New Evaluation Models (Micro cell, indoor, Rural/High-speed)', www.3gpp.org, 3GPP TSG RAN WG1, meeting 43bis, Warsaw, Poland, June 2008.

[22] 3GPP Technical Specification 25.141, 'Base Station (BS) Conformance Testing (FDD) (Release 8)', www.3gpp.org.

22

Radio Frequency Aspects

Tony Sayers, Adrian Payne, Stefania Sesia, Robert Love, Vijay Nangia and Gunnar Nitsche

22.1 Introduction

Radio Frequency (RF) signal processing constitutes the final interface between the baseband techniques described in earlier chapters and the transmission medium – the air. The RF processing presents its own unique challenges for practical transceiver design, arising in particular from the fact that the air interface is a shared resource between multiple RF carriers, each with their own assigned portion of spectrum. The RF transmitter must be designed in such a way as not only to generate a clean signal within the assigned spectrum portion, but also to keep inter-carrier interference within acceptable levels. The receiver likewise must reliably demodulate the wanted signal, in order to avoid requiring excessive energy to be transmitted, whilst also rejecting interference from neighbouring carriers. Performance requirements for these RF aspects aim to ensure that equipment authorized to operate on an LTE carrier meets certain minimum standards.

In general, the performance requirements for LTE transceivers are intended not to be significantly more complex than for UMTS in terms of implementation and testing. Consequently, many of the RF requirements for LTE are derived from those already defined for UMTS.

There are, however, a number of key differences between LTE and UMTS which affect the RF complexity and performance.

One such difference is the use of a variable channel bandwidth in LTE, up to a maximum of 20 MHz. Even the lowest category of LTE User Equipment (UE) is required to support all the bandwidths specified, which in Release 8 means 1.4, 3.0, 5.0, 10.0, 15.0 and 20.0 MHz. This requires a set of values to be defined for each requirement, in contrast to the Frequency Division Duplex (FDD) mode of UMTS which only supports a single channel bandwidth of 5 MHz. For LTE, the variable bandwidth represents a new challenge for the RF design and testing. For example, a transceiver for a constant-bandwidth radio system can potentially employ fixed filters at a number of points in the signal processing. Such filters are designed to pass a signal with known characteristics and reject particular frequencies. If, however the bandwidth of the transceiver is variable over a wide range, it is clear that fixed filters cannot be used. Frequencies which must be passed in 20 MHz operation may need to be rejected in the narrower bandwidth modes. This implies that LTE transceivers must be more adaptable than those of previous systems, while also being cleaner in transmission and having better selectivity in reception.

A second difference between UMTS and LTE is that in LTE it is assumed that the UE has at least two receive antennas. This means that multiple RF signal paths are needed at all times.

Thirdly, LTE is even more adaptable than UMTS in terms of the range of data rates it supports in order to suit different SINR (Signal-to-Interference plus Noise Ratio) conditions (e.g. from 4×2 Multiple-Input Multiple-Output (MIMO) 64QAM in high SINR conditions to Single-Input Single-Output (SISO) QPSK in low SINR conditions (see Section 10.2)). Together with the ability to vary the instantaneous bandwidth to a given user, this implies a large number of modes of operation and flexibility in signal handling capabilities. The reception of the maximum data rate requires high SINR at the highest bandwidth, which is particularly challenging for the analogue-to-digital converter in the receiver.

Fourthly, the LTE signal structure itself alters which specific RF aspects are the most critical. As already discussed extensively in Chapters 5 and 15, the use of Orthogonal Frequency Division Multiplexing (OFDM) in the downlink and Single Carrier Frequency Division Multiple Access (SC-FDMA) in the uplink provides an inherent robustness against multipath propagation, which means that amplitude and phase distortions from receiver and transmitter filters are not as critical as for UMTS's single-carrier Wideband Code Division Multiple Access (WCDMA) modulation. On the other hand, OFDM requires better frequency synchronization and is more sensitive to phase noise.

It is also worth noting that for WCDMA the RF requirement specifications are separate for Frequency and Time Division Duplex (FDD and TDD) [1–4], while in LTE the commonality between FDD and TDD is such that they can be handled in the same specifications [5, 6]. A more detailed discussion of aspects specific to TDD operation is given in Chapter 23.

Some further useful background on the transmitter and receiver RF characteristics can be found in references [7] and [8] for the LTE UE and eNodeB respectively.

This chapter seeks to describe the impact of the LTE physical layer radio requirements on the implementation complexities of an LTE transceiver, compared to the well-known UMTS system. The relevant spectrum bands for LTE are introduced in Section 22.2. Radio requirements for the transmitter and receiver are discussed in Sections 22.3 and 22.4 respectively, and some of the implementation challenges are highlighted. Both the UE and eNodeB are addressed, although with the greater emphasis on the UE side. This is followed by a discussion in Section 22.5 of the RF impairments for both transmitter and receiver, including mathematical models of the most common impairments to help elucidate the impact of RF imperfections on the overall radio behaviour.

22.2 Frequency Bands and Arrangements

LTE and UMTS are both defined for a wide range of different frequency bands, in each of which one or more independent carriers may be operated. Tables 22.1 and 22.2 give details of the frequency bands for FDD and TDD operation respectively. For FDD, the duplex separation is not actually defined, but typically the uplink and downlink carriers in a pair are in a similar position in their respective bands so that the duplex separation is usually approximately $F_{DL\,low}-F_{UL\,low}$ as shown in Table 22.1.

Table 22.1 UMTS and LTE frequency bands for FDD.

Band Number	Uplink, (MHz) $F_{UL\,low}-F_{UL\,high}$	Downlink, (MHz) $F_{DL\,low}-F_{DL\,high}$	Band Gap (MHz)	Duplex Separation (MHz)	UMTS Usage	LTE Usage
1	1920–1980	2110–2170	130	190	Y	Y
2	1850–1910	1930–1990	20	80	Y	Y
3	1710–1785	1805–1880	20	95	Y	Y
4	1710–1755	2110–2155	355	400	Y	Y
5	824–849	869–894	20	45	Y	Y
6	830–840	875–885	35	45	Y	Y
7	2500–2570	2620–2690	50	120	Y	Y
8	880–915	925–960	10	45	Y	Y
9	1749.9–1784.9	1844.9–1879.9	60	95	Y	Y
10	1710–1770	2110–2170	340	400	Y	Y
11	1427.9–1452.9	1475.9–1500.9	23	48	Y	Y
12	698–716	728–746	12	30	Y	Y
13	777–787	746–756	21	31	Y	Y
14	788–798	758–768	20	30	Y	Y
17	704–716	734–746	18	30	N	Y

A typical UE would support a certain subset of these bands depending on the desired market, since supporting all would be challenging for the transceiver, in particular for the front-end components such as Power Amplifiers (PAs), filters and duplexers. The set of frequency bands chosen will then define the capability of the UE to switch bands, roam between national operators and roam internationally.

Each frequency band is regulated to allow operation in only a certain set of channel bandwidths. Some frequency bands do not allow operation in the narrow bandwidth modes below 5 MHz, while others do not allow operation in the wider bandwidths, generally 15 MHz or above. The status (at the time of going to press) is given in Table 22.3, which shows the number of supported non-overlapping channels for each frequency band and bandwidth. '—' denotes that the channel bandwidth is not supported in the specific band; 'X' indicates the channel bandwidth is too wide to be supported in the band; values in square brackets denote bandwidths for which UE receiver sensitivity requirement can be relaxed.

Table 22.2 UMTS and LTE frequency bands for TDD.

Band	F_{low}–F_{high} (MHz)	UMTS	LTE
33	1900–1920	Y	Y
34	2010–2025	Y	Y
35	1850–1910	Y	Y
36	1930–1990	Y	Y
37	1910–1930	Y	Y
38	2570–2620	Y	Y
39	1880–1920	N	Y
40	2300–2400	Y	Y

Table 22.3 Number of supported non-overlapping channels in each frequency band and bandwidth.

LTE band	Downlink bandwidth	Channel bandwidth (MHz)					
		1.4	3	5	10	15	20
1	60	—	—	12	6	4	3
2	60	42	20	12	6	[4]	[3]
3	75	53	23	15	7	[5]	[3]
4	45	32	15	9	4	3	2
5	25	17	8	5	[2]	—	—
6	10	—	—	2	[1]	X	X
7	70	—	—	14	7	4	[3]
8	35	25	11	7	[3]	—	—
9	35	—	—	7	3	[2]	[1]
10	60	—	—	12	6	4	3
11	25	—	—	5	[2]	[1]	[1]
12	18	12	6	[3]	[1]	—	X
13	10	7	3	[2]	[1]	X	X
14	10	7	3	[2]	[1]	X	X
...							
17	12	8	4	[2]	[1]	X	X
...							
33	20	—	—	4	2	1	1
34	15	—	—	3	1	1	X
35	60	42	20	12	6	4	3
36	60	42	20	12	6	4	3
37	20	—	—	4	2	1	1
38	50	—	—	10	5	—	—
39	40	—	—	8	4	3	2
40	100	—	—	—	10	6	5

22.3 Transmitter RF Requirements

This section explains the implications of the LTE RF performance specification for the design of its transmitters.

Each transmitter must satisfy two categories of requirements: those relating to the power level and quality of the intended transmissions, and those prescribing the level of unwanted emissions which can be tolerated. The first of these aspects is usually easier to deal with: the radio must transmit a precisely-defined radio signal in its designated channel. The second aspect, unwanted emissions, is usually more challenging for the radio designer.

22.3.1 Requirements for the Intended Transmissions

22.3.1.1 Signal Quality: Error Vector Magnitude

The quality of the transmitted radio signal has to fulfil certain requirements. The main parameter used to measure this quality is the Error Vector Magnitude (EVM). The EVM is a measure of the distortion introduced by the RF imperfections of practical implementations. It is defined as the magnitude of the difference between a reference signal (i.e. the signal defined by the physical layer specification equations) and the actual transmitted signal (normalized by the intended signal magnitude). A geometrical representation of this is sketched in Figure 22.1. The EVM sets the maximum possible SNR of a radio link in the absence of any noise, interference, propagation loss and other distortions introduced by the radio channel; it therefore serves to determine the maximum useful modulation order and code rate.

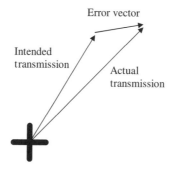

Figure 22.1 Geometrical representation of the EVM concept.

For the UE (uplink), the EVM is defined to measure the quality of the transmitted signal across all the allocated Resource Blocks (RBs), while for the eNodeB (downlink) the EVM measures the quality over one RB. The measurement duration is one slot for the UE (uplink) and one subframe for eNodeB (downlink) and takes into account all the symbols belonging to the modulation scheme under test.

Figures 22.2 and 22.3 show the EVM measurement points in the downlink and uplink cases respectively [5, 6]. The EVM measurement is taken after an equalizer in the test equipment, which carries out per-subcarrier channel correction [5,6]. The equalizer is used in

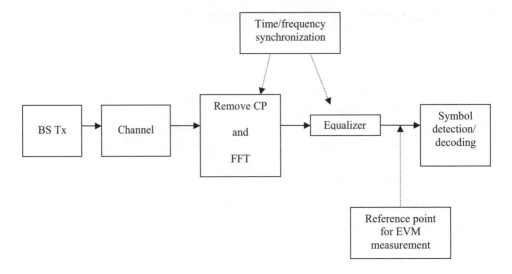

Figure 22.2 Measurement points for the EVM for the downlink signal.

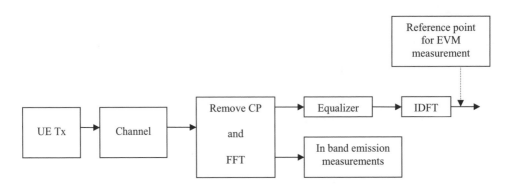

Figure 22.3 Measurement points for the EVM for the uplink signal.

order to arrive at a measurement which realistically shows what a receiver might experience. It is intended to reflect the fact that the equalizer in the receiver is capable of correcting some of the impairments of the transmitted signal to some extent. A zero-forcing equalizer is used for EVM measurement;[1] however, real receiver implementations may use different types of equalization technique (for example, to avoid the problem of noise enhancement associated with zero-forcing equalizers), and therefore the measured EVM may not exactly correspond to the actual signal quality experienced by all receivers.

It should be noted that before measuring the EVM, time and frequency synchronization must be carried out. The test equipment then computes the EVM for two extreme values of sample timing difference between the FFT processing window and the nominal timing of the

---

[1]Details of how to compute the coefficients of the equalizer can be found in [5, 6].

Table 22.4 EVM requirements.

	Modulation	EVM requirement (%)
Uplink	QPSK	17.5
	16QAM	12.5
Downlink	QPSK	17.5
	16QAM	12.5
	64QAM	8

ideal signal; these two extreme values correspond to the beginning and end of the window, the length of which is expressed as a percentage of the Cyclic Prefix (CP) length. Finally the measurement is averaged over 20 slots for UE (uplink) and 10 subframes for eNodeB (downlink) to reduce the impact of noise.

Hence, the maximum computed EVM can be defined as

$$EVM = \max\{\overline{EVM_l}, \overline{EVM_h}\}$$

where $\overline{EVM_l}$ and $\overline{EVM_h}$ are the averaged measurements at the low and high values of sample timing difference respectively. The specified values which the EVM must satisfy in LTE are given in Table 22.4 [5, 6].

These EVM values were selected using analysis by system simulation to verify that no more than a 5% performance loss would result in terms of average and cell-edge throughputs in typical deployment scenarios.[2] It has also been shown [10] that when EVM is less than 8%, it provides a good metric to predict the link level performance loss arising from transmitter impairments; for example, an EVM of 8% corresponds to an SNR loss of 0.5 dB with 16QAM rate 3/4.

A description of the main transmitter impairments which generally give rise to non-zero EVM is given in Section 22.5.1.

22.3.1.2 Transmit Output Power

The transmitted output power directly influences the inter-cell interference experienced to neighbouring cells using the same channel, as well as the magnitude of unwanted emissions outside the transmission band. This affects the ability of the LTE system to maximize spectral efficiency, and it is therefore important that the transmitters can set their output power accurately.

For the eNodeB, the maximum output power must remain within about 2 dB of the power declared by the manufacturer. In addition, the dynamic range in the frequency domain (computed as the difference between the power of a given Resource Element (RE) and the average RE power) must not exceed specified limits [6] depending on the modulation order, in order to avoid saturating the UE receivers.

For the UE, the maximum output power is 23 dBm, which must also be satisfied within a range of ±2 dB.

---

[2]The 'Case 1' and 'Case 3' deployment scenarios – see Section 8.1 and [9].

If multiple antennas are used for the transmitter as part of a MIMO scheme, these specifications apply to the total power output from all the antennas.

22.3.2   Requirements for Unwanted Emissions

Ideally, the radio should transmit nothing at all outside its designated transmission channel. However, in practice this is very far from being the case, as is evident from the example shown in Figure 22.4 of the measured Power Spectral Density (PSD) of a typical LTE uplink SC-FDMA waveform for a 5 MHz carrier bandwidth [11]. The figure shows the PSD when the UE is transmitting over the full bandwidth, as well as the cases for 2.16 and 1.08 MHz resource allocations.

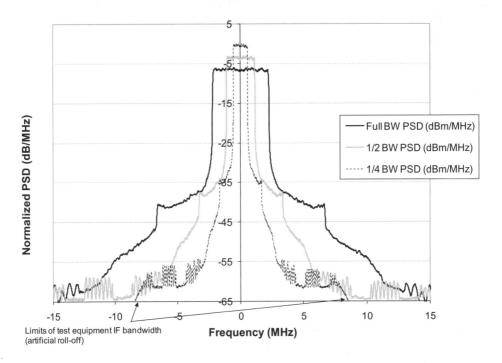

Figure 22.4 PSD for different occupied bandwidths. Reproduced by permission of © 2006 Motorola.

Outside the intended channel bandwidth, the LTE specifications define two separate kinds of unwanted emission: *Out-Of-Band (OOB) emissions* and *spurious emissions*. These are shown schematically in Figure 22.5.

Out-of-band emissions are those which fall in a band close to the intended transmission, while spurious emissions can be at any other frequency. The precise boundary between the OOB range and the spurious range is different for different aspects of the LTE specifications.

LTE defines requirements for both types of unwanted emission, with those for spurious emissions being the more stringent. The specified requirements must be fulfilled in response

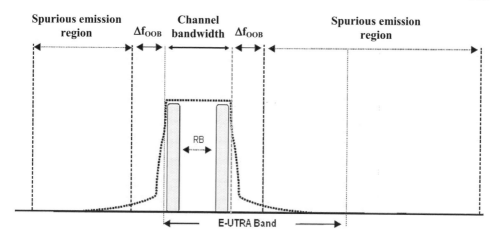

Figure 22.5 Transmitter spectrum. Reproduced by permission of © 3GPP.

to certain network signals or under particular conditions. This means that, unlike the relatively fixed RF requirements of earlier systems, the emission requirements may change in different scenarios. If the transmitter cannot satisfy a particular requirement, it can either switch the transmitter off, or adapt the transmitter characteristics (for example by reducing the transmitter power or using software-defined radio techniques) to alter the performance in the required manner. This is explained in more detail in Section 22.3.3.

22.3.2.1 Out-of-Band Emissions

Since OOB emissions occur close to the wanted transmission, increasing the power level of the wanted transmission will usually increase the level of the unwanted emissions. Conversely, reducing the transmitted power is usually an effective method for reducing the OOB emissions, thus providing one possible method for responding to network-signalled requirements as discussed above.

OOB emissions may be an almost inevitable by-product of the modulation process itself, and are also often caused by non-linearities in PAs. A summary of the main practical transmitter impairments which can cause OOB emissions is given in Section 22.5.1.

In previous fixed-bandwidth radio systems (including UMTS), the OOB emissions requirements have been defined with respect to the centre frequency of the transmission. As the bandwidth of the LTE system is variable, it is not particularly convenient to express the requirements in this way. The OOB specifications are therefore defined with respect to the edge of the 'occupied bandwidth' of the system, which is defined as the bandwidth containing 99% of the total integrated mean power of the transmitted spectrum.

In LTE, OOB emissions are defined by means of Spectrum Emission Masks (SEMs) and Adjacent Channel Leakage Ratio (ACLR) requirements. The SEM has a much narrower reference bandwidth than the ACLR, which is a stricter requirement.

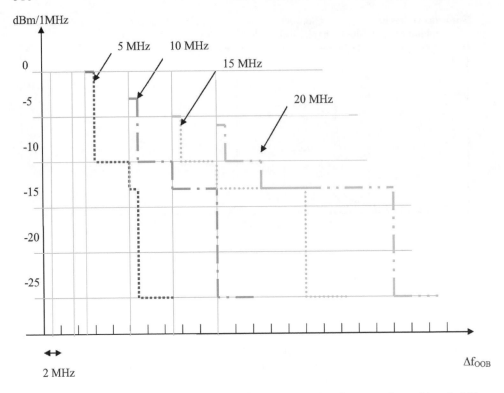

Figure 22.6 Spectrum Emission Mask (SEM) for a UE transmitter per channel bandwidth.

Spectrum Emission Mask

The Spectrum Emission Mask (SEM) is a mask defined for out-of-channel emissions relative to the in-channel power. The spectrum emission mask of the UE applies to frequencies within Δf_{OOB} (MHz) of the edge of the assigned LTE channel bandwidth, as shown diagrammatically in Figure 22.5. Figure 22.6 shows the basic SEM requirement for a UE transmitter.

LTE also provides a set of 'Additional SEMs' (A-SEMs) which the network may instruct the UE to use in particular deployment scenarios. Such an instruction may, for example, be signalled on handover to a new cell.

For the eNodeB, the SEM is defined in a similar manner, an example is given in Figure 22.7. It should again be noted that this represents only the basic requirement, and there are many conditions and additional cases to be taken into account in different circumstances [6].

Adjacent Channel Leakage Ratio

The second method used to measure OOB emissions in LTE is the ACLR. While the SEM measures the performance of the transmitter, the ACLR measures the power which

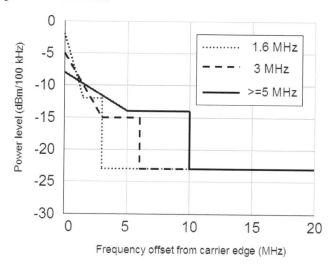

Figure 22.7 Spectrum emission mask for eNodeB transmitter. Reproduced by permission of © 3GPP.

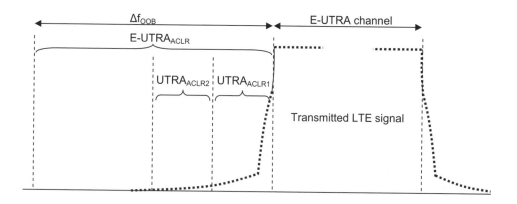

Figure 22.8 Adjacent carrier leakage ratio.

actually leaks into certain specific nearby radio channels, and thus estimates how much a neighbouring radio receiver will be affected by the OOB emissions from the transmitter.

The ACLR is defined as the ratio of filtered mean power in a bandwidth within the wanted channel divided by the filtered mean power in an adjacent channel.

The LTE specifications not only set ACLR requirements for adjacent 20 MHz LTE channels, but also for UMTS channels which may be located in two adjacent 5 MHz channels (i.e. a total of 10 MHz) as shown in Figure 22.8.

The ACLR of a UE for an adjacent LTE channel is required to be >30 dB for the whole 20 MHz bandwidth, >33 dB for the adjacent UMTS channel deployed in the first 5 MHz of the OOB portion of the spectrum (UTRA$_{ACLR1}$ in Figure 22.8) and >36 dB for the adjacent

Table 22.5 Spurious emissions limits.

Frequency range	Maximum level (dBm)	Measurement bandwidth
9 kHz ≤ 150 MHz	−36	1 kHz
150 kHz ≤ 30 MHz	−36	10 kHz
30 kHz ≤ 1000 MHz	−36	100 kHz
1 GHz ≤ 12.7 GHz	−30	1 MHz

UMTS channel deployed in the second 5 MHz portion of the OOB spectrum (UTRA$_{ACLR2}$ in Figure 22.8) [5].

22.3.2.2 Spurious Emissions

Spurious emissions occur well outside the bandwidth necessary for transmission and may arise from a large variety of non-ideal effects including harmonic emissions and intermodulation products.[3] The magnitude of the spurious emissions may or may not vary with transmitter power.

The basic spurious emissions requirements for an LTE UE are given in Table 22.5.

These basic requirements are more stringent than the OOB requirements, but are not usually difficult to meet. The main challenge for the transmitter designer lies in some 'additional spurious emissions' requirements which are specified for certain frequency bands. The additional spurious emissions requirements exist to protect the receiver on the UE, colocated receivers, or nearby receivers from the transmission. In the case of FDD radios, the receiver and transmitter operate at the same time, and the receiver relies on the duplex spacing to separate the weak received signal from its strong transmitted signal. If the transmitter emits anything significant on its own receiver frequency, then reception will be blocked or degraded. This results in the additional spurious emissions requirements being very stringent. In addition, some further additional spurious emissions requirements may be signalled by the network if needed for specific deployment scenarios.

The leakage of the transmitted signal into the receiver band is discussed further in Section 22.4.2.

22.3.3 Power Amplifier Considerations

One of the strongest constraints on the maximum transmission power level of a UE is the need to meet stringent OOB emission requirements set by the ITU for IMT2000-family equipment (see Chapter 1) and by the Federal Communications Commission (FCC) of the USA. These requirements are largely specified in terms of ACLR and spectrum masks.

The ACLR is typically directly related to the operating point of the power amplifier. In general, leakage into adjacent channels increases sharply as the PA is driven into its non-linear operating region at the highest output power levels, due in particular to the intermodulation products. Consequently it is important that the peak output power of the

---

[3]Intermodulation products are unwanted frequencies generated whenever two or more tones are present in a non-linear device, such as an amplifier or mixer. The possible combinations of generated frequencies can be described by a power series.

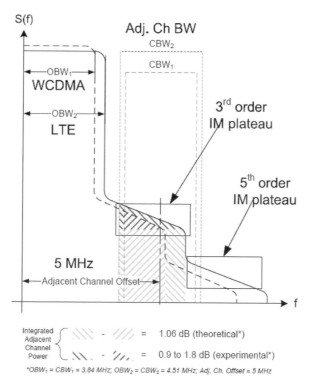

Figure 22.9 Adjacent channel power increase (2–3 dB) for 4.51 versus 3.84 MHz Occupied BandWidth (OBW) [12]. Reproduced by permission of © 2006 Motorola.

UE does not cause the PA to enter too far into this region. On the other hand, most PAs are designed to operate efficiently only in a small region at the top of the linear operating region. This corresponds to the *rated* output power of the PA, below which the efficiency falls sharply. Since high efficiency is crucial to ensuring a long battery life for the UE, it is therefore also desirable to keep the PA operating as close as possible to the top of the linear operating region.

If the ACLR and spectrum mask requirements cannot be met, typically the UE output power must be reduced to bring the leakage to acceptable levels. This can be achieved without undue loss of efficiency by reducing the peak output power of the PA – a process known as *de-rating*. The amount of de-rating required is highly dependent on the waveform of the transmitted signal. For example, for a given carrier bandwidth, transmissions with larger occupied bandwidth create more OOB emissions, resulting in larger adjacent channel leakage than transmissions with lower occupied bandwidth; this is shown in Figure 22.9, in which WCDMA and LTE occupied bandwidths and adjacent channel leakage are compared in a 5 MHz channel. The increase in OOB emissions from the larger occupied bandwidth of the LTE signal is mainly due to increased adjacent channel occupancy by 3rd and 5th order InterModulation products (called 3rd and 5th order IM plateau in Figure 22.9).

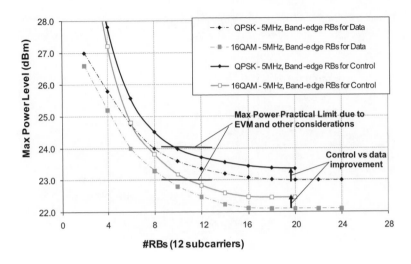

Figure 22.10 Maximum power level versus the number of RBs which can be allocated. Reproduced by permission of © 2006 Motorola.

For an LTE transmitted signal, certain combinations of RB allocations and modulation schemes create more OOB emissions than others. For QPSK and 16QAM the number of RBs which can be allocated for a given amount of power de-rating is shown in Figure 22.10. For applications such as VoIP, which typically use low-bandwidth transmissions with QPSK modulation, no power de-rating is required from the normal rated maximum UE output power (see Section 22.3.1.2). This helps to ensure wide-area coverage for such applications, since the full rated output power of the PA can be used to counteract path-loss at the cell edge. Figure 22.10 does not show 64QAM modulation, but it is expected that a power de-rating of approximately 6 dB is needed to meet OOB emission (and EVM) requirements.

A further example of the dependence of ACLR (and hence the amount of power de-rating required) on the transmitted waveform is seen in the structure of the LTE uplink signal, with control information being situated at the band edge, and high-bandwidth data transmissions toward the middle of the band, as discussed in Section 17.3. Reference [13] shows the reduction in ACLR which arises from this arrangement compared to allowing data transmissions at the band edge.

Taking the above considerations into account, the Total Power De-rating (TPD) required to meet a given ACLR requirement can be broken down into an element corresponding to the occupied bandwidth (as a proportion of the channel bandwidth), termed here the Occupied Bandwidth Power De-rating (OBPD), and an element corresponding to the waveform of the transmitted signal, termed here the Waveform Power De-rating (WPD):

$$\text{TPD} = f(\text{OBPD}, \text{WPD}) \tag{22.1}$$

Note that the function $f(\cdot)$ can be the simple summation of OBPD and WPD such that TPD = OBPD + WPD. OBPD can be approximately expressed as a ratio of the transmitted

Occupied Bandwidth (OBW) and the Reference Occupied Bandwidth (OBW$_{ref}$) [14]:

$$\text{OBPD} \propto 10 \log_{10}\left(\frac{\text{OBW}}{\text{OBW}_{ref}}\right) \quad (22.2)$$

For example, a transmission with 4.5 MHz occupied bandwidth centred on a 5 MHz LTE carrier with a fixed 5 MHz carrier separation will have a larger measured ACLR (e.g. −30.5 dBc instead of −33 dBc) relative to the adjacent 5 MHz carrier, than a transmission with only 3.84 MHz occupied bandwidth. To reduce the ACLR back to −33 dBc requires an OBPD of approximately 0.7 dB.

The other element of the required TPD, the WPD, accounts for the waveform attributes like modulation and the number of frequency or code channels. The relationship between a particular waveform and its resulting ACLR and required power de-rating arises in a large part from the dynamic range of the signal. This is usually quantified in terms of its *Peak to Average Power Ratio* (PAPR) and *Cubic Metric* (CM). These measures are typically used as indications of how much PA power headroom is required (i.e. how far from the rated power the PA generally has to operate) to avoid entering the non-linear region of operation. A high PAPR means that on average the operating point of the PA has to be lower in order to avoid the non-linear region and achieve the required ACLR. This results in a reduction in efficiency. A rule of thumb is that for each 1 dB increase in required power amplifier headroom, a corresponding 10 to 15% increase in PA current drain occurs, leading to a corresponding reduction in battery life. De-rating the PA is therefore usually a preferable strategy (as opposed to increasing the power headroom), since it achieves the required reduction in adjacent channel leakage by reducing the total maximum output power (including that of the wanted signal), while allowing the PA still to operate partly in the non-linear region. De-rating therefore enables a UE to meet the OOB emission requirements without a loss of efficiency, at the expense of a loss in coverage due to the reduced power in the wanted channel.

An LTE UE with a power class of 23 dBm can nominally support a rated maximum power level (P_{MAX}) of 23 dBm. In practice, the UE's instantaneous maximum power level is thus limited to the operational maximum power level given by $P_{MAX} - f(\text{OBPD}, \text{WPD})$.

In practice, it has been found that the CM of a waveform is a better indicator of the required power de-rating to meet a given ACLR than the PAPR. This is because the CM characterizes the effects of the 3rd order (cubic) non-linearity of a PA on the waveform of interest relative to a reference waveform, in terms of the power de-rating needed to achieve the same ACLR as that achieved by the reference waveform at the PA's rated maximum power level. The CM can be defined as follows:

$$\text{CM} = \frac{20 \log_{10}\{\text{r.m.s.}\lfloor v_{norm}^3(t)\rfloor\} - 20 \log_{10}\{\text{r.m.s.}\lfloor v_{ref,\ norm}^3(t)\rfloor\}}{K} \ \text{dB} \quad (22.3)$$

where $20 \log_{10}(\text{r.m.s.}\lfloor v_{norm}^3(t)\rfloor)$ is the *raw* CM (in dB) of the waveform in question, and $20 \log_{10}(\text{r.m.s.}\lfloor v_{ref,\ norm}^3(t)\rfloor) = 1.52$ dB is the raw CM of the reference signal, r.m.s.$(x) = \sqrt{(x^T x)/N}$, $v_{norm}(t) = |v(t)|(\text{r.m.s.}[v(t)])$ and x^T is the transpose of x. K is a slope factor which is determined empirically for different families of waveform. Different values are used for WCDMA and SC-FDMA. For example, the CM is 2.13 dB for a 16QAM-modulated SC-FDMA waveform, using a slope factor K of 1.56.

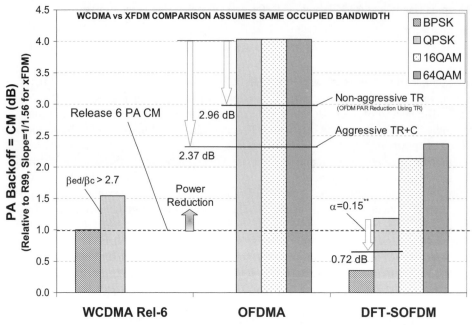

Figure 22.11 Cubic Metric (CM) for different waveforms. Reproduced by permission of © 2006 Motorola.

As discussed in Chapter 15, one criterion which was instrumental in selecting SC-FDMA instead of OFDMA for the LTE uplink was the low inherent CM of the SC-FDMA waveform compared to OFDMA, as shown in Figure 22.11, thus enabling more efficient operation for a given ACLR and maximum total output power.

Although WPD for an LTE QPSK waveform is better than for WCDMA QPSK (see Figure 22.11), the larger occupied bandwidth for LTE results in the OBPD for LTE being ~0.77 dB worse than for WCDMA in a 5 MHz bandwidth. Overall, the total CM (CM_{total} = TPD) for QPSK SC-FDMA is about 0.5 dB worse if the entire LTE bandwidth is occupied. However, a loaded cell rarely allocates the entire carrier bandwidth to a single UE, and with smaller resource allocations the total CM for LTE is the same as or better than WCDMA.

Figure 22.11 also shows the effect of some CM reduction techniques for OFDM (see Section 5.2.2.1), namely 'aggressive' Tone Reservation (TR) and Clipping (C) [15, 16]. However, the cost of such techniques is increased complexity and a degradation in EVM. Finally, the figure also shows that frequency-domain spectral shaping of the SC-FDMA waveform [17] using a Root-Raised-Cosine (RRC) filter with roll-off factor equal to 0.15 can reduce the CM, but at the expense of reduced spectral efficiency.

A further aspect of the LTE uplink signal design which directly originates from the desirability of a low CM is the fact that the d.c. subcarrier of the SC-FDMA signal is modulated in the same way as all the other subcarriers, as discussed in Section 15.3.3. It is well known that direct conversion (zero IF) transmitters and receivers can introduce significant distortion on the baseband signal components near zero Hz, such as the d.c.

subcarrier of an OFDM or SC-FDMA system. This is addressed in the LTE downlink by the inclusion of an unused d.c. subcarrier. However, for the uplink, it is not possible to avoid modulating the d.c. subcarrier without causing significant degradation to the low PA de-rating (CM) property of the transmitted signal. If a resource allocation were to span an unused d.c. subcarrier, the CM would be degraded by ~0.7 dB for QPSK or 0.5 dB for 16QAM.

22.3.4 Summary of Transmitter RF Requirements

The main requirements for the RF part of an LTE transmitter fall into two categories: those ensuring the quality of the transmitted signal in the wanted channel, and those protecting other frequencies from unwanted emissions. The former consist of the EVM and maximum transmitted power requirements, while the latter include a number of constraints depending on the frequency offset between the wanted channel and the channel of interest.

These requirements have particular impact on the dimensioning of the PA in the UE, which needs to operate linearly to avoid unwanted emissions, as well as operating efficiently. The signal structure of the LTE uplink has been specifically designed with these considerations in mind.

22.4 Receiver RF Requirements

Like the transmitter, the LTE receiver requirements are based largely on those of UMTS, with many of the RF requirements being identical or similar. The general intention is for the implementation of an LTE receiver not to be significantly more complex than UMTS, so as to reduce redesign effort. An underlying assumption is that there is no need for tightening the coexistence specifications, as the basic cellular scenarios remain similar.

The main differences between the LTE and UMTS RF receiver requirements arise from the variable channel bandwidth and the new multiple access schemes.

The emphasis in this section is on the UE receiver requirements for FDD operation.

22.4.1 Receiver General Requirements

The receiver requirements are based on a number of key assumptions for testing purposes:

- The receiver has integrated antennas with a 0 dBi gain.

- The receiver has two antenna ports. Requirements for four ports may be added in the future – see Chapter 24.

- Test signals of equal power level are applied to each port, with Maximum Ratio Combining (MRC) being used to combine the signals. It is assumed that the signals come from independent Additive White Gaussian Noise (AWGN) channels, so that signal addition gives a 3 dB diversity gain.

Although LTE supports a variety of Modulation and Coding Schemes (MCSs) (see Chapter 10), the RF receiver specifications are defined for just two MCSs (referred to as 'reference channels', near the extremes of the available range, in order to reduce the number of type approval tests which have to be performed. The 'low SNR' reference channel uses

QPSK with a code rate of 1/3, while the 'high SNR' reference channel uses 64QAM with a code rate of 3/4. For each of these reference channels the SINR requirements are specified at which 95% throughput should be achieved.

22.4.2   Transmit Signal Leakage

When an LTE receiver is tested in full-duplex FDD mode, the transmitter must also be operating so that signal leakage from the transmitter is taken into account. This does not apply to half-duplex FDD or to TDD operation, which are discussed in detail in Chapter 23.

An FDD receiver's exposure to the transmitted signal from the same equipment is largely due to insufficient isolation within the duplexer. The power of the leakage of the transmitted signal will be proportional to the transmitted power.

Not only can the power of the fundamental components of the transmitted signal interfere with the receiver, but also the OOB phase noise of the transmitted signal can fall into the receive band. In LTE the spurious emissions from the UE transmitter in its own receive band are required to be -50 dBm or lower, measured in a 1 MHz bandwidth. This corresponds to -110 dBm/Hz. The maximum transmit power for a mobile device is $+23$ dBm (power class 3), so the spurious emissions requirement is -133 dBc/Hz. UMTS has a tougher requirement of -60 dBm spurious emissions in 3.84 MHz bandwidth, which is -125.8 dBm/Hz or -149.8 dBc/Hz for Class 3 $+24$ dBm output power. However, for UMTS the toughest requirement is in one specific receive band ('Band 8'), with a requirement of -79 dBm measured in 100 kHz bandwidth, which is -129 dBm/Hz or -153 dBc/Hz (when considering the maximum output power). The transmitter may therefore be designed to achieve this more broadly, including its own receive band.

For both UMTS and LTE, receiver sensitivity is measured with the transmitter operating at the full power for its class. In order to allow the transmitter to degrade the receiver sensitivity by no more than 0.5 dB, the transmitter noise needs to be 9 dB below the Noise Floor (NF) at the antenna (NF -9 dB $= -183$ dBm/Hz). If we assume that the spurious emissions of the transmitter in its receive band are just at the limit of the specifications, it can be shown that the duplex isolation needs to be at least 71 dB for LTE and 52 dB for UMTS.[4]

However, making duplexers smaller and cheaper is often achieved at the expense of isolation, and typical duplexers might provide just 45 to 55 dB isolation. Potentially, therefore, the transmit signal could reduce receiver sensitivity substantially. This is not acceptable, so within the receive band the LTE transmitter will have to achieve spurious emissions about 20 dB lower than the transmitter specifications require.

The transmit power at the input to the Low Noise Amplifier (LNA) is equal to the transmit power at the antenna plus the duplexer loss from transmit port to antenna (about 2 dB) less the duplexer transmit-to-receiver isolation (assume 50 dB). Therefore, at the maximum LTE UE output power of 23 dBm, the mean transmitter power leaking back to the LNA, $P_{\text{Tx\_leak}}$ is

$$P_{\text{Tx\_leak}} \simeq 23 + 2 - 50 = -25 \text{ dBm} \tag{22.4}$$

UMTS will be 1 dB higher.

---

[4]The duplex isolation is obtained as NF $-$ 9 dB $-$ 2 dB $-$ SE, where NF is the noise floor (-174 dBm/Hz), 9 dB comes from the degradation of the receiver sensitivity, 2 dB comes from the loss from transmit port to antenna and SE is the Spurious Emission requirement.

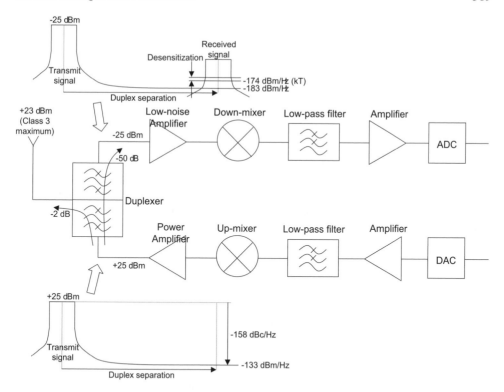

Figure 22.12 Illustration of transmit leakage problem in LTE FDD UE transceiver.

However, most of the receiver specifications are defined with the transmitter power at 4 dB below maximum, so then the transmit leakage could be −29 dBm. This value is for the average transmit power, but the transmitted signal contains amplitude modulation and therefore the peak signal will be higher. This scenario is illustrated in Figure 22.12 which shows a simplified block diagram of the transceiver with direct-conversion architectures for both receiver and transmitter.

The transmit leakage problem presents a particularly challenging requirement if applied to RF bands with a small duplex separation. In FDD, when performing a selectivity or blocking test with a single interfering signal in one of the adjacent channels, the receiver is also exposed to its own transmitted signal as a second interferer, which, in the worst scenario, can cause cross-modulation to occur, in which the interferers mix together to generate intermodulation distortion products in-band.

22.4.3 Maximum Input Level

The maximum input level is the maximum mean received signal strength, measured at each antenna port, at which there is sufficient SINR for a specified modulation scheme to meet a minimum throughput requirement. Some wireless standards also specify a higher level at which it is guaranteed that no damage will be done to the receiver, but this is not covered in the

3GPP specifications. Although in UMTS the maximum input level is defined for the high and low SINR reference channels, in LTE it is only specified for the high SINR reference channel, on the assumption that the high SINR reference channel is about 4 dB more demanding due to the higher PAPR of the signal. The downlink requirement is for a maximum average input level of −25 dBm at any channel bandwidth with full RB allocation. For FDD operation, the requirement must be met when the transmitter is set at 4 dB below the maximum output power. The corresponding UMTS requirement is the same, applied to the combination of all the channels rather than full RB allocation.

The peak maximum input level is higher than the maximum average input level by an amount corresponding to the PAPR of the signal. PAPR is a quantity which increases the longer the duration of the measurement. If PAPR is measured on a bit-by-bit basis and plotted as a complementary cumulative probability distribution, then for the LTE downlink it reaches a value of around 11 dB for QPSK and 11.5 dB for 64QAM, over a window of 10^6 bits. With OFDMA, the PAPR is dominated by the multicarrier nature of the signal, and therefore does not vary much with modulation. The UMTS downlink signal consists of a combination of CDMA codes which also leads to a relatively high PAPR of around 8 dB.

The downlink peak maximum input level is therefore about −13.5 dBm for LTE and −17 dBm for UMTS.

As explained above, the maximum input level requirement must be satisfied with the transmitter operating at 4 dB below the maximum output power. For a UMTS UE, this can result in a mean transmit signal leakage into the receiver of −28 dBm,[5] while the typical uplink PAPR of 3.4 dB results in a peak transmit signal leakage of of −24.6 dBm. For LTE, the mean leakage from the uplink transmitter can be −29 dBm, resulting in a peak transmitter leakage of up to about −21.5 dBm for QPSK (or −20 dBm for 64QAM[6]). The peak value of the total received power is the combination of the received input signal and the leakage from the transmitter and thus reaches a peak of about −12 dBm for LTE and −15.2 dBm for UMTS.

Thus the transmitter leakage raises the total received signal power by less than 2 dB, but if the duplexer isolation were less then the LTE total received power would be higher.

Some operating margin is needed between this total peak received signal power and the non-linear operating region of the amplifiers in the receiver; typically this margin is chosen relative to the 1 dB compression point of the amplifiers, at which the output power is 1 dB below the expected linear gain.

22.4.4 Small Signal Requirements

22.4.4.1 SINR Requirements for Adaptive Modulation and Coding

The LTE specifications will define requirements for the demodulation error rate of the different modulation and coding schemes. An extra Implementation Margin (IM) is included to account for the difference in SINR requirement between theory and practicable implementation. This can include degradation of the signal due to any (digital) processing of the signal before the demodulator (such as filtering and re-sampling) and the use of a non-ideal

---

[5]The mean transmit leakage is obtained from Equation (22.4) as the maximum output power (24 dBm for UMTS or 23 dB for LTE) minus 4 dB because the transmitter power is set at 4 dB below maximum, plus the duplexer loss from transmit port to antenna (about 2 dB) minus the isolation between transmitter and receiver (assume 50 dB).

[6]Note that only the highest category of LTE UE supports 64QAM transmission in the uplink.

Table 22.6 Downlink SINR requirements.

System	Modulation	Code rate	SINR (dB)	IM (dB)	SINR+IM (dB)
LTE UE	QPSK	1/8	−5.1	2.5	−2.6
		1/5	−2.9		−0.4
		1/4	−1.7		0.8
		1/3	−1		1.5
		1/2	2		4.5
		2/3	4.3		6.8
		3/4	5.5		8.0
		4/5	6.2		8.7
	16QAM	1/2	7.9	3	10.9
		2/3	11.3		14.3
		3/4	12.2		15.2
		4/5	12.8		15.8
	64QAM	2/3	15.3	4	19.3
		3/4	17.5		21.5
		4/5	18.6		22.6
UMTS UE	QPSK	1/3	1.2	2	3.2

demodulator, as well as the diversity gain being less than 3 dB. A 2.5 dB IM has been defined for QPSK with a 1/3-rate code in low SINR conditions; the IM will be higher for other modes. In general, 2.5 dB is a reasonable implementation margin for all QPSK modes, while 3 dB and 4 dB could be expected for 16QAM and 64QAM respectively.

Typical assumptions for the SINR values for different modulation and coding schemes are given in Table 22.6.

22.4.4.2 Thermal Noise and Receiver Noise Figure

In the LTE specifications the thermal noise density, kT, is defined to be −174 dBm/Hz where k is Boltzmann's constant (1.380662×10^{-23}) and T is the temperature of the receiver (assumed to be 15°C). No account is taken of the small variations in temperature over normal operating conditions (typically + 15° to +35°C) or extreme operating conditions (−10° to +55°C).

kTB represents the thermal noise level in a specified noise bandwidth B, where $B = N_{RB} \times 180$ kHz in LTE and N_{RB} is the number of RBs and 180 kHz is the bandwidth of one RB.

The receiver Noise Figure (NF) is a measure of the degradation of the SINR caused by components in the RF signal chain. This includes the antenna filter losses, the noise introduced by the analogue part of the receiver, the degradation of the signal due to imperfections of the analogue part of the receiver (such as IQ imbalance), the noise introduced by the Analogue to Digital Converter (ADC) and any other noise sources. The NF

Table 22.7 Reference sensitivity.

System	Modulation	Channel BW (MHz)	kTB (dBm)	NF (dB)	SINR (dB)	IM (dB)	REFSENS (dBm)
LTE UE	QPSK 1/3	5	−107.5	9	−1	2.5	−100
	QPSK 1/3	20	−101.4	9	−1	2.5	−94
	64QAM 3/4	5	−107.5	9	17.5	4	−80
	64QAM 3/4	20	−101.4	9	17.5	4	−74
LTE BS	QPSK 1/3	5	−107.5	5	1.5	2.5	−101.5
UMTS UE	QPSK 1/3	3.84	−108.2	9	1.2–21.1 (21.1 dB spreading gain)	2.5	−117

is the ratio of actual output noise to that which would remain if the receiver itself did not introduce noise.

LTE defines a NF requirement of 9 dB for the UE, the same as UMTS. This is somewhat higher than the NF of a state-of-the-art receiver, which would be in the region of 5–6 dB, with typically about 2.5 dB antenna filter insertion loss and a NF for the receiver integrated circuit of 3 dB or less. Thus, a practical 3–4 dB margin is allowed. The eNodeB requirement is for a NF of 5 dB.

22.4.4.3 Reference Sensitivity

The reference sensitivity level is the minimum mean received signal strength applied to both antenna ports at which there is sufficient SINR for the specified modulation scheme to meet a minimum throughput requirement of 95% of the maximum possible. It is measured with the transmitter operating at full power.

REFerence SENSitivity (REFSENS) is a range of values that can be calculated using the formula:

$$REFSENS = kTB + NF + SINR + IM - 3 \text{ (dBm)}$$

where kTB is the thermal noise level introduced above, in units of dBm, in the specified bandwidth (B), NF is the prescribed maximum noise figure for the receiver, SINR is the signal to interference plus noise requirement for the chosen modulation and coding scheme, IM is the implementation margin and the −3 dB represents the diversity gain.

For UMTS, the NF, SINR and IM are not specified separately, but the total is 12 dB. Typical REFSENS values can be calculated using the assumptions outlined above for a few example cases, as in Table 22.7.

REFSENS is plotted for an LTE UE over the complete range of bandwidths and likely modulation and coding schemes in Figure 22.13.

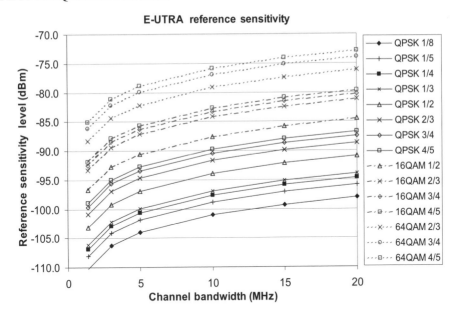

Figure 22.13 LTE UE reference sensitivity.

Maximum Sensitivity Reduction (MSR) for each RF band

It is envisaged that in FDD mode the receiver sensitivity will be more substantially degraded by transmitter noise falling into the receive band when operating in some RF bands than others. To allow for this, the REFSENS requirement is relaxed by an amount which can be referred to as a Maximum Sensitivity Degradation (MSD). MSD is not yet specified but can be inferred from the varying REFSENS requirement for each band. The requirement for MSD arises from insufficient isolation in the duplexer; it is band-specific because it is roughly proportional to the duplex separation.

Effect of receiver sensitivity on coverage

Using the path-loss equations described in Section 21.2, and knowing the downlink transmit power and receiver sensitivity, the practical operating coverage can be estimated. For the downlink, assuming an eNodeB power of +46 dBm for a 5 MHz channel and ~900 MHz RF frequency (Band 5), the coverage can be estimated as shown in Figure 22.14 using the parameters shown in Table 22.8.

The modulation and coding scheme can be varied to maximize the throughput for any given radio channel, so throughput is also a function of range. This relationship is illustrated in Figure 22.15.

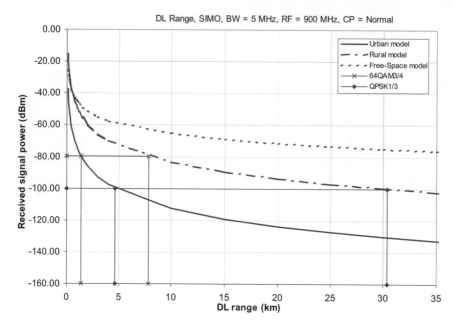

Figure 22.14 Downlink coverage taking receiver sensitivity into account.

Table 22.8 Downlink range.

Modulation	QPSK 1/3	64QAM 3/4
Reference sensitivity (dB)	−100.0	−80.0
Maximum path loss (dB)	146.0	126.0
Maximum range urban (km)	4.7	1.4
Maximum range rural (km)	30.3	7.8

22.4.5 Selectivity and Blocking Specifications

Selectivity and blocking tests measure a receiver's ability to receive the wanted signal at its assigned channel frequency in the presence of interfering signals in adjacent channels and beyond. As usual, this requirement must be met when the transmitter is set to 4 dB below the supported maximum output power. A low SINR is assumed. A summary of the selectivity and blocking tests for the LTE UE in a 5 MHz channel is illustrated in Figure 22.16. The requirements generally scale with bandwidth. As with UMTS, the selectivity and blocking requirements include the case of a close Continuous-Wave[7] (CW) blocking signal, some cases with modulated interferers in the first three adjacent channels, and some OOB blocker requirements with a spurious response allowance [5].

---

[7] A 'continuous wave' signal is an unmodulated tone.

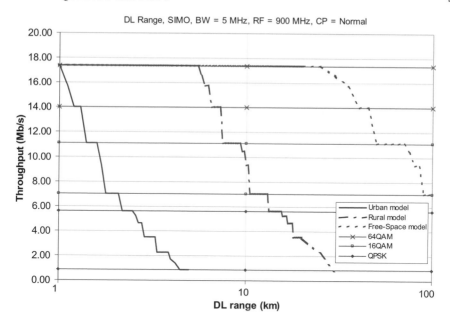

Figure 22.15 Throughput versus coverage.

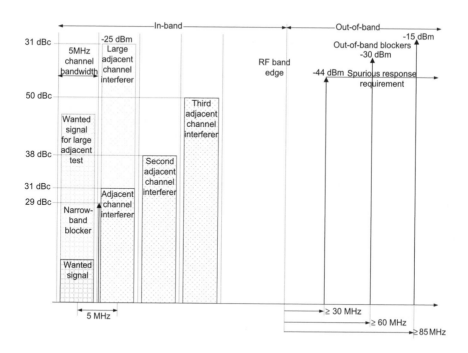

Figure 22.16 Summary of selectivity and blocking requirements for a 5 MHz UE.

22.4.5.1 Adjacent Channel Selectivity

Adjacent Channel Selectivity (ACS) is a measure of a receiver's ability to receive a wanted signal at its assigned channel frequency in the presence of an adjacent channel interfering signal at a given frequency offset from the centre frequency of the assigned channel, without the interfering signal causing a degradation of the receiver performance beyond a specified limit. ACS is predominantly defined by the ratio of the receive filter attenuation on the assigned channel frequency to the receive filter attenuation on the adjacent channel.

For LTE, ACS is defined following the same principles as UMTS, and requiring similar performance capability up to a 10 MHz bandwidth, but is more relaxed for 15 and 20 MHz bandwidths. The LTE ACS is defined for each bandwidth using a modulated LTE signal as the interferer, and is only defined for low SINR conditions.

In order to check the ability of the receiver to handle the full required dynamic range, the ACS requirement is specified for two cases – a small adjacent channel interferer power and a large adjacent channel interferer power. In both cases the Carrier to Interference ratio (C/I) is set at −31.5 dB. These cases are explained in more detail below.

ACS with a small adjacent interferer

In this ACS case, the wanted signal is, like UMTS, 14 dB above REFSENS (given in Table 22.7) and therefore takes a different absolute level for each bandwidth. For bandwidths of 10 MHz or less the C/I is −31.5 dB and the ACS is quoted as being 33 dB including a 2.5 dB implementation margin. At higher bandwidths, the C/I is reduced. Thus, for a 5 MHz bandwidth, with REFSENS = −100 dBm, the wanted signal is −86 dBm and the adjacent channel interferer −54.5 dBm.

Up to a bandwidth of 5 MHz, the bandwidth of the interferer is the same as the bandwidth of the wanted signal. Above 5 MHz, the bandwidth of the interferer stays at 5 MHz, which means that the RF test equipment does not need to be able to generate a wide bandwidth interferer. For bandwidths above 5 MHz the interferer is positioned at the near edge of the channel. The consequence of this is that the interferer power is concentrated at the edge of the adjacent channel at which the filters used in the receiver have least selectivity. The digital filters will have a sharp cut-off because they are designed for OFDM, but the analogue filters in the RF front-end of the receiver could have much less attenuation at the near edge of the channel, which will push up the dynamic range at the ADC. To compensate, the ACS is relaxed by 3 dB and 6 dB for the 15 MHz and 20 MHz modes respectively.

Level diagrams for ACS requirements for two bandwidths are shown in Figure 22.17.

The gap between the edge of the wanted signal and the edge of the interferer is a function of the used bandwidth of both signals. Given that both use a full allocation of RBs, this gap can be calculated as 1.25 MHz for 20 MHz bandwidth reducing to a minimum of 300 kHz for 3 MHz bandwidth. Below this bandwidth the channel usage reduces, so the gap increases. Note that in percentage terms, a 1.25 MHz gap adjacent to a 20 MHz channel is actually smaller than a 300 kHz gap adjacent to a 3 MHz channel (6.25% compared to 10%). The channel filters should be designed to scale with bandwidth, so it is this percentage ratio which determines the filter roll-off requirements. Comparing the 20 and 5 MHz modes we see that the percentage has roughly doubled but the specification is relaxed by 6 dB, so the filter roll-off requirements should be similar.

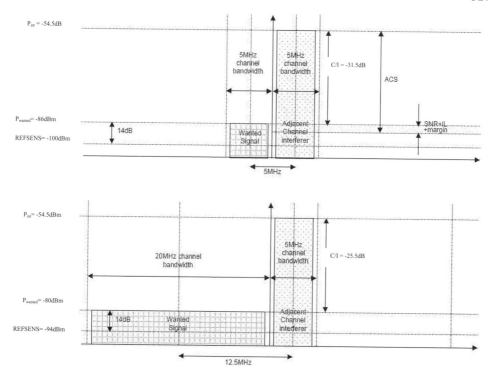

Figure 22.17 Adjacent channel selectivity illustrated.

ACS with a large adjacent interferer

The large adjacent interferer requirement uses an adjacent channel interferer power of -25 dBm. The wanted signal power is fixed at -31.5 dB below -25 dBm for bandwidths of 10 MHz and below (i.e. -56.5 dBm). The margin between the wanted signal power and REFSENS varies for each bandwidth in the range 41.0 to 50.2 dB, which is well above the noise floor and therefore not of much significance.

In practice, the toughest test for the receiver is likely to be somewhere between the small interferer and large interferer ACS tests. Assuming that the dynamic range of the receiver is limited, there is a point at which the interferer power first becomes high enough to require that the front-end gain needs to be reduced. This gain reduction will degrade the noise figure of the receiver at a point at which the wanted signal power is not very high. The gain control algorithm used by the receiver must be well planned to avoid such problems.

Adjacent Channel Interference Ratio (ACIR)

The ACS and ACLR (see Section 22.3.2.1) together give the Adjacent Channel Interference Ratio (ACIR). The ACIR is the ratio of the total power transmitted from a source to the total interference power affecting a victim receiver, resulting from both transmitter and receiver imperfections.

It follows that

$$ACIR \cong \frac{1}{\dfrac{1}{ACLR} + \dfrac{1}{ACS}}$$

ACLR and ACS have been extensively used for coexistence studies.

22.4.5.2 Narrowband Blocking (in Adjacent Channel)

The blocking characteristic is a measure of the receiver's ability to receive a wanted signal at its assigned channel frequency in the presence of an unmodulated unwanted interferer on frequencies other than those of the spurious response or the adjacent channels, without this unwanted input signal causing a degradation of the performance of the receiver beyond a specified limit. The blocking performance applies at all frequencies except those at which a spurious response occurs.

The LTE 'narrowband' blocking performance is a measure of blocking performance with an interferer very close to the wanted signal at an offset less than the nominal channel spacing. The interferer is a CW signal.

The CW blocker is positioned at 200 kHz from the near edge of the adjacent channel. For example, for a 5 MHz bandwidth the offset is 2.7 MHz, which is 450 kHz from the edge of the wanted signal. The offset of the blocker from the edge of the wanted signal reduces with bandwidth, reaching just 350 kHz for the 3 MHz bandwidth mode. Below 3 MHz bandwidth the channel occupancy reduces, which compensates for the reduced gap.

For this test, for bandwidths of 10 MHz and below, the power of the wanted signal is about -84 dBm, the blocker power is around -55 dBm and the C/I is about -29 dB. Compared to the ACS test, the C/I is 3 dB less, and the powers of the wanted and interfering signals are similar. However, the gap between the wanted and interfering signals is between 30 and 50 kHz less, which makes the narrowband blocking test a little more demanding than the ACS test.

A summary of the narrowband blocking requirements for LTE and UMTS UEs is shown in Table 22.9. For UMTS, the wanted signal power includes 21 dB spreading gain.

Comparing the UMTS narrowband blocking specification to the 5 MHz LTE case, the wanted signal is 6 dB lower at 10 dB above REFSENS and the blocker is positioned at a comparable offset of 2.7 MHz and similar power of -56 dBm (or 2.8 MHz offset and -57 dBm for some RF bands). More importantly, the blocker for UMTS is actually a GMSK[8] modulated signal, not CW. A GMSK signal is a little narrower than QPSK (although clearly not as narrow as a CW signal), and the modulation has a constant envelope so there is no PAPR variation which could increase non-linear distortion. On balance, it can be concluded that the UMTS and LTE narrowband blocking specifications are similarly demanding.

The narrowband blocking specification is a severe test of the receiver's ability to reject 3rd order intermodulation products resulting from cross-modulation of the transmitter leakage which appears around the narrowband blocker. The frequency of the unwanted cross-modulation product depends only on the narrowband blocker frequency and not on the frequency of the transmitter, or any other modulated blocker. For the small offsets specified by UMTS and LTE, nearly half of the transmitter leakage will appear in-band.

---

[8]Gaussian Minimum-Shift Keying, as used in GSM.

Table 22.9 Narrowband blocking.

System	LTE						UMTS
Bandwidth of wanted signal (MHz)	1.4	3	5	10	15	20	3.84
Own signal power above Ref. Sens (dB)	22	18	16	13	14	16	13
Power of interferer (dBm)	−55	−55	−55	−55	−55	−55	−56
Carrier to Interference ratio (dB)	−29.2	−29.2	−29	−29	−26.2	−22.9	−29
Frequency Offset of interferer (MHz)	±0.9	±1.7	±2.7	±5.2	±7.7	±10.2	±2.7
Offset from edge of wanted (MHz)	0.36	0.35	0.45	0.7	0.95	1.2	—

22.4.5.3   Non-Adjacent Channel Selectivity

Non-adjacent Channel Selectivity (NACS) is a measure of the receiver's ability to receive a wanted signal at its assigned channel frequency in the presence of unwanted interfering signals falling into the receive band beyond the adjacent channel or into the first 15 MHz below or above the receive band. The interfering signals are modulated and occupy the same bandwidths as specified for ACS. The LTE specifications refer to this test as 'in-band blocking', although, unlike the other blocking tests, NACS does not use CW signals.

There are two requirements to be met, the first with an interferer of −56 dBm in the second adjacent channel or further, and the second with an interferer of −44 dBm in the third adjacent channel or any larger frequency offset up to 15 MHz out of band. These interferer powers are the same as in UMTS.

Unlike the ACS tests, NACS does not need to be repeated at higher signal levels, so it does not test dynamic range to the same extent. However, the wanted signal level is much lower, at just 6 dB above REFSENS for 10 MHz bandwidth and below. Consequently the C/I ratios are much lower, falling to −56.2 dB for the third adjacent channel. The total filtering requirement will therefore be of the order of −60 dB at three times the bandwidth.

A summary of the NACS requirements for LTE and UMTS UEs are shown in Tables 22.10 and 22.11. For UMTS, the wanted signal power includes 21 dB spreading gain.

22.4.5.4   Out-of-Band Blocking

The LTE OOB blocking tests measure the receiver's ability to receive a wanted signal at its assigned channel frequency in the presence of unwanted interfering signals falling outside the receive band, at 15 MHz or more offset from the edge of the band.

The wanted signal power is 7 dB above REFSENS for bandwidths of 10 MHz or below, and relaxed to 7 or 9 dB above REFSENS for bandwidths of 15 MHz or 20 MHz respectively.

Table 22.10 Non-adjacent (N ± 2) channel selectivity.

System	LTE						UMTS
Bandwidth of wanted signal (MHz)	1.4	3	5	10	15	20	3.84
Own signal power above Ref. Sens (dB)	6	6	6	6	7	9	3
Power of interferer (dBm)				−56			
Carrier to Interference ratio (dB)	−44.2	−40.2	−38.0	−35.0	−32.2	−28.9	−37
Frequency offset of interferer (MHz)	±2.8	±6	±10	±12.5	±15	±17.5	±10
Bandwidth of interferer (MHz)	1.4	3	5	5	5	5	3.84

Table 22.11 Non-adjacent (N ± 3) channel selectivity.

System	LTE						UMTS
Bandwidth of wanted signal (MHz)	1.4	3	5	10	15	20	3.84
Own signal power above Ref. Sens (dB)	6	6	6	6	7	9	3
Power of interferer (dBm)				−44			
Carrier to Interference ratio (dB)	−56.2	−55.5	−52.2	−50.0	−44.2	−40.9	−49
Frequency offset of interferer (MHz)	±4.2	±9	±15	±17.5	±20	±22.5	±15
Bandwidth of interferer (MHz)	1.4	3	5	5	5	5	3.84

The blocker is a CW signal. The blocker power is −44 dBm at 15 to 60 MHz offset, −30 dBm at 60 to 85 MHz offset and −15 dBm at 85 to 1275 MHz offset. These values are all the same as UMTS. The actual offset of the blocker is the specified offset from the band edge plus half the wanted signal bandwidth plus the RF guard band, which is 2.4 MHz. So, for example, for a 5 MHz bandwidth, the −44 dBm blocker is at a minimum offset of $15 + 2.5 + 2.4 = 19.9$ MHz, which is close to four times the bandwidth.

Like UMTS, there is an additional requirement for RF bands 2 (DCS1800), 5 (GSM850), 12 and 17, comprising a −15 dBm blocker coming from the transmit band, which is at an offset of just 20 MHz for bands 2 and 5, 12 MHz for band 12 and 18 MHz for band 17,

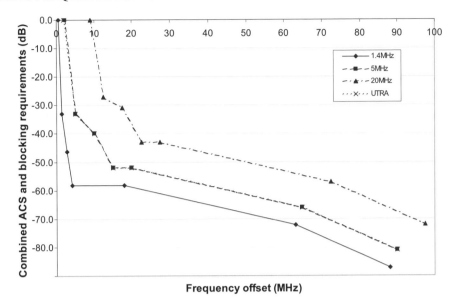

Figure 22.18 Selectivity and blocking requirements.

see Table 22.1. This latter requirement is clearly far tougher than all the other blocking specifications.

A summary of the selectivity and blocking requirements is shown in Figure 22.18. This includes the specified 2 or 3 dB margin added to the C/I requirements. It can be seen that LTE 5 MHz requirements are very similar to UMTS. A more illuminating summary of the channel filter requirements can be found by normalizing all frequency offsets to the wanted signal bandwidth. This demonstrates that the 20 MHz LTE bandwidth requires the most severe filter frequency response relative to its bandwidth (and also to the digital sampling frequency). Additionally, the LTE 5 MHz selectivity requirement is relatively tougher than UMTS.

22.4.5.5 Spurious Response Specifications

Frequencies for which the throughput does not meet the requirements of the OOB blocking test are called spurious response frequencies. Spurious responses occur at specific frequencies at which an interfering signal mixes with the fundamental or harmonic of the receiver local oscillator and produces an unwanted baseband frequency component. Spurious responses are measured by recording when the out-of-band blocking test is not passed.

At the spurious response frequencies the LTE receiver must still achieve the required throughput with a −44 dBm blocker and a wanted signal level as in the out-of-band blocking test; for example, if a spurious response is measured with a −30 dBm blocker, then the required throughput must be met by the time the blocker level is reduced to −44 dBm.

22.4.6 Spurious Emissions

The spurious emissions power is the power of emissions generated or amplified in a receiver which appear at the antenna connector. For LTE, limits are set covering the range 30 MHz to 12.75 GHz. The LTE specification is the same as UMTS, namely not more than −57 dBm between 30 MHz and 1 GHz (measured in 100 kHz bandwidth), and −47 dBm from 1 GHz to 12.75 GHz (measured in 1 MHz bandwidth); these requirements correspond to −107 dBm/Hz across the full range.

For UMTS, some additional spurious emissions requirements are specified for the cellular bands. Generally the requirement is −60 dBm measured in 100 kHz bandwidth (−110 dBm/Hz), but there are some more demanding requirements, of which the toughest is for the 935–960 MHz band, where the requirement is −79 dBm measured in 100 kHz (−129 dBm/Hz). This requirement is intended to protect the GSM downlink receive band.

Spurious emissions are caused by power from the local oscillator or the amplified received signal leaking back to the antenna. The required isolation from the amplified received signal can be estimated as follows. The maximum received power is −25 dBm, and the minimum received bandwidth 1.08 MHz; thus at the maximum input signal density the power in the 100 kHz measurement bandwidth is −35.3 dBm. The isolation required to prevent the amplified received signal from breaking the spurious emission limit will depend on the amplifier gain, which will vary with the signal and is likely to be at its minimum at the maximum input signal level. Assuming a 20 dB amplifier gain, then the isolation would need to be about 64 dB to reach the −79 dBm spurious emission limit. This requirement could be tougher than for UMTS since signal levels for UMTS are lower because of spreading. However, the toughest isolation requirement is likely to be for the local oscillator signal which drives the mixer, since it typically operates at a higher power, for example around −4 dBm. The worst case isolation requirement for the local oscillator is thus about 75 dB, higher than the received signal isolation requirement.

22.4.7 Intermodulation Requirements

22.4.7.1 Intermodulation Distortion

As mentioned in Section 22.3.2.2, when two or more tones are present in a non-linear device, such as an amplifier in a receiver, intermodulation products are created. A power series describes all of the possible combinations of generated frequencies. Intermodulation products generated from mixing with a tone outside the wanted band may themselves fall into the wanted band. The problem is more severe for in-band interfering tones, since in that case the RF filter provides no attenuation.

As a device is driven further into its non-linear region, the amplitudes of the intermodulation products increase, while the power of the original tones at the output decreases. If the device was not limited in output power, then the powers of the intermodulation products would increase to the level of the original tones.

Intermodulation response rejection is a measure of the capability of the receiver to receive a wanted signal on its assigned channel frequency in the presence of two or more interfering signals, where the frequencies of the interfering signals relative to the wanted signal are such that the InterModulation Distortion (IMD) products fall into the wanted signal band.

The strongest interference is caused by the third-order intermodulation products. The standard procedure for measuring the third-order intermodulation performance of a receiver is with two CW interfering tones offset from the wanted carrier frequency, with the offset of one tone being twice that of the other. Positioning one tone with twice the offset of the other results in the IMD product which occurs at the difference frequency falling into the wanted band. Such a test scenario is defined for UMTS, where it is referred to as a *narrowband intermodulation requirement*. UMTS also includes a *wideband intermodulation* test, with interferers at larger offsets, where one of the interferers is a modulated UMTS signal. This is designed to assess the effect of IMD products arising from UMTS transmissions in other channels.

For LTE, a narrowband intermodulation requirement could be defined following the same principles as in UMTS, using two CW signals as interferers. However, as a narrow bandwidth IMD product falling on top of the wanted signal would only degrade the performance of few subcarriers, the impact on throughput would be minimal. Hence, this is not defined in the current specification.

For the LTE wideband intermodulation performance requirement, the modulated interferer is defined to have the same bandwidth as the ACS test, denoted here 'BW$_{int2}$'.[9] The CW interferer is at offset BW$_{channel}$ + 1.5 × BW$_{int2}$, where BW$_{channel}$ is the channel bandwidth; so for the 5 MHz bandwidth case the CW interferer is at a 10 MHz offset and the modulated interferer at 20 MHz offset, twice the offset of the CW interferer, as for UMTS. The interferer powers, are both −46 dBm, the same as for UMTS. The wanted signal is at a variable level between 6 and 12 dB above REFSENS. This variation of the wanted signal level is designed to keep the wanted signal at an absolute value of close to −94 dBm and thus ensure a consistent IMD requirement for each bandwidth. For the 5 MHz bandwidth case, the wanted signal is 6 dB above REFSENS, 3 dB higher than UMTS (because of the 3 dB diversity gain).

For bandwidths of 10 MHz and above the modulated interferer is always 5 MHz. The consequence of this is that the power of the interferer is spread over a narrower bandwidth than the wanted signal, so the interfering power on each subcarrier is higher than the wanted power of each subcarrier by the ratio of the bandwidths. This increases the IMD requirement: for example, for the 20 MHz bandwidth, a value of $10 \times \log_{10}(\mathrm{BW}_{channel}/\mathrm{BW}_{int2}) = 6$ dB should be taken into account when calculating IMD requirements.

The usual way to quantify the IMD performance of a receiver is to determine the *third-order intercept point*, or IP3. IP3 is a theoretical point where the amplitudes of the intermodulation tones are equal to the amplitude of the fundamental tones. The third-order IMD products, IM3, increase as the cube of the power of one of the input tones (i.e. at 60 dB/decade, three times the rate of the fundamental). Therefore, at some power level the distortion products will overtake the fundamental signal; the point at which the curves of the fundamental signal and the IM3 intersect on a log-log scale is the IP3. At this point, by definition, IM3 = 0 dBc. The corresponding input power level is known as Input IP3 (IIP3), and the output power when this occurs is the Output IP3 (OIP3) point, which is higher than IIP3 by an amount equal to the gain of the receiver. The IIP3 can be computed as follows:

$$\mathrm{IIP3} = (3P_{in} - P_{\mathrm{IMD3,in}})/2 = \mathrm{OIP3} - G$$

---

[9]BW$_{int2}$ is equal to 1.4, 3 and 5 MHz for channel bandwidth equal to 1.4, 3 and 5 MHz, and it is limited to 5 MHz for higher channel bandwidths.

where P_{in} is the power of a single tone at the input of the system, $P_{IMD3,in}$ the input-referred third-order intermodulation distortion product and G the passband gain (all in dB).

It is not possible to measure IP3 directly, because by the time the non-linear amplifier reached this point it would be heavily overloaded. Instead the amplifier is measured at a lower input tone power and IP3 is found by extrapolation.

The wide-band intermodulation test can be used to derive an equivalent IIP3 requirement. The IM3 products fall in-band and add to the existing receiver noise. The maximum power of the IM3 products that can be tolerated, referred back to the input (i.e. so that the gain of the receiver is factored out), $P_{IMD3,in}$, is at a level such that when combined with the existing noise, it reaches a level that is equal to the maximum tolerable noise floor. This noise floor is below the wanted signal power by the SINR requirement plus implementation margin. Thus, if the wanted signal is 3 dB above REFSENS the IM3 products and existing noise can be the same, but as the wanted signal level is further raised above the REFSENS level the existing noise floor will be less significant and the IM3 products a little higher.

As an example, we can calculate the IIP3 requirement for a 15 MHz bandwidth. The SINR requirement plus implementation margin is assumed to be 1.5 dB. The wanted signal is 7 dB above REFSENS at -88.2 dBm, so it can be calculated that the IM3 products must be 0.97 dB[10] below the maximum tolerable noise floor. The factor of three difference in bandwidth must be accounted for by a factor of 4.77 dB. Thus, $P_{IMD3,in} = -88.2 - 1.5 - 0.97 - 4.77 = -95.4$ dBm and IIP3 $= [(3 \times (-46)) + 95.4]/2 = -21.3$ dBm. IIP3 can be calculated for the other bandwidths and reaches a maximum of -20.6 dBm for 5 or 10 MHz bandwidth. Further calculations are shown in Table 22.12.

The first version of the LTE specifications uses a simpler calculation of $P_{IMD3,in}$, assuming that the pre-existing noise floor (i.e. without IM3) and the power of the IM3 products are the same. This results in a more consistent but pessimistic calculation of IIP3 of close to -19.8 dBm for most bandwidths.

The IIP3 requirement for UMTS can be calculated in a similar manner.

22.4.7.2 Out-of-Band Intermodulation Distortion

As explained above, the LTE transmit signal may leak into the receiver. If a blocker is also present, IM3 products may result. If the blocker is half-way between the transmit and receive bands (i.e. between 15 and 200 MHz from the wanted band, depending on the duplex spacing) then the IM3 products will fall into the wanted band. For Band 5, the duplex spacing is 45 MHz, so a blocker at a 22.5 MHz offset, which can have a power of -44 dBm, would be the worst case. For the out-of-band blocking tests, the wanted signal is 3 dB above REFSENS. Therefore $P_{IMD3,in}$ can be calculated as described above, for example -101.5 dBm for the 5 MHz bandwidth. If we assume the transmitter leakage is -29 dBm and the blocker is at -44 dBm, then the average input power is -31.9 dBm. Thus the out-of-band IP3 requirement is IIP3 $= [(3 \times (-31.9)) + 101.5]/2 = +2.9$ dBm and the maximum IIP3 is $+6$ dBm for the 1.4 MHz bandwidth.

For Band 1 the duplex spacing is 190 MHz, so a blocker at 95 MHz, which can be at -15 dBm, would cause IMD. If the RF filter gives less than 29 dB attenuation of the blocker, then this band would present an even tougher out-of-band IIP3 requirement.

---

[10]The IM3 noise plus the thermal noise combines to use up the 7 dB margin, $0.97 = -10 \cdot \log_{10}(1 - 10^{-7/10})$, therefore $10 \cdot \log_{10}(10^{-7/10} + 10^{-0.97/10}) = 0$.

Table 22.12 Third-order intermodulation distortion.

System	LTE						UMTS
Bandwidth of wanted signal (MHz)	1.4	3	5	10	15	20	3.84
Own signal power above REFSENS (dB)	12	8	6	6	7	9	3
Power of each interferer (dBm)				−46			
C/I (dB)	−48.2	−48.2	−48.0	−45.0	−42.2	−38.9	−47
Frequency offsets of interferers (±MHz)	2.8, 5.6	6, 12	10, 20	12.5, 25	15, 30	17.5, 35	10, 20
Bandwidth of mod. interferer (MHz)	1.4	3	5	5	5	5	3.84
$BW_{channel}/BW_{int2}$ (dB)	0	0	0	3	4.8	6	0
IMD Noise margin (dB)	0.28	0.75	1.26	1.26	0.97	0.58	3.02
Input referred IMD power req't P_{IM3} (dBm)	−95.9	−96.4	−96.7	−96.7	−95.4	−93.1	−98.0
IIP3 (dBm)	−21.0	−20.8	−20.6	−20.6	−21.3	−22.5	−20.0

A similar calculation for UMTS for Band 5, assuming +20 dBm transmit signal leaking into the receiver at −28 dBm and a −44 dBm blocker, would yield a $P_{IMD3,in}$ requirement of −98 dBm and an out-of-band IIP3 of +2.7 dBm.

Since the receiver's IP3 is dominated by that of the mixer, the problem of cross-modulation can be reduced for the most demanding bands by using a band-pass filter between the low-noise amplifier and the mixer, so that most of the cross-modulation occurs in the low-noise amplifier.

22.4.8 Dynamic Range

The input dynamic range is a key factor affecting the cost of the receiver. The larger the dynamic range, the larger must be the linear operating region of the receiver components.

There are many ways that the dynamic range and resulting signal handling requirements can be analysed for a receiver, with widely differing results.

The simplest definition of the dynamic range of a receiver is the ratio of the maximum and minimum signal levels required to maintain a specified throughput or error rate. For UMTS the maximum and minimum wanted signal powers are −44 and −117 dBm respectively, giving a dynamic range of 73 dB. For an LTE UE, the maximum input is −25 dBm (assumed to be applicable to any modulation), while the minimum signal level for a 5 MHz bandwidth is −100 dBm , giving a maximum dynamic range of 75 dB.

Dynamic range can also be specified as the ratio between the maximum signal and the noise floor. Even more usefully, the maximum signal level could include a margin to allow for the variation of the peak signal power above its average (PAPR). Such a measure of dynamic range gives an indication of the total signal handling requirements of the receiver in the absence of gain control. In Section 22.4.3, the total peak received signal power was derived, which includes the PAPR and the contribution from the transmitter leakage; the values were -12.0 dBm for LTE and -15.2 dBm for UMTS. For the LTE 5 MHz bandwidth the noise floor at the antenna is -107.5 dBm. Thus this dynamic range estimate is $-12.0 - (-107.5) = 94.9$ dB. For UMTS the dynamic range is $-15.2 - (-108.2) = 92.9$ dB. For UMTS, the REFSENS is below the noise floor.

In practice, gain control for large signals is normally used in a receiver to reduce the dynamic range. The minimum dynamic range that the receiver should handle linearly, measured at the antenna, is the maximum SINR requirement, plus implementation loss, plus NF, plus a margin for PAPR (assumed to be 11.5 dB). For LTE this gives $17.5 + 4 + 9 + 11.5 = 42$ dB, while for UMTS using QPSK it is only $12.2 + 8 = 20.2$ dB.

However, this does not take account of any interferers which could be present, nor leakage from the transmitter. At REFSENS, for an FDD LTE UE, there could be transmitter leakage at -25 dBm (peak about -16 dBm); the difference between the peak transmitter leakage and the noise floor is 91.5 dB for LTE and 83.6 dB for UMTS, although the frequency offset is large so the analogue filters will reduce the interference. Another useful measure of receiver dynamic range is the Spurious-Free Dynamic Range (SFDR) which is the input power range in which the received signal can be detected in the presence of noise, and amplified, without being exposed to intermodulation distortion from the non-linear amplification of interfering signals. The lower bound of the SFDR is set by the Minimum Discernible Signal (MDS), which is defined to be 3 dB above the equivalent noise power of the receiver.[11] The upper bound is defined as the interferer power level at which the IM3 (typically the dominant source of non-linear interference) equals the noise power. The IM3 is incorporated into the SFDR expression by using the 2-tone IP3 value. More specifically, SFDR is two-thirds the difference between the IIP3 and the MDS. For LTE UEs, MDS is -104.5 dBm for a 5 MHz bandwidth, and the in-band IIP3 requirement is -20.6 dBm. Hence the SFDR is 55.9 dB. For UMTS, the MDS is -105.2 dBm and the in-band IIP3 is -20 dBm, leading to a similar SFDR at 56.8 dB.

In conclusion, is clear that the higher modulation, higher SINR and higher PAPR requirements of LTE present a higher signal-handling dynamic range requirement than for UMTS, but the linearity requirements due to interferers are similar.

22.4.9   Summary of Receiver Requirements

The LTE RF receiver must combat a wide range of interfering signals in order to enable reliable demodulation of the wanted signals and to avoid undue susceptibility to extraneous transmissions.

The sources of interference are many and varied, including leakage from the equipment's own transmitter, legitimate transmissions in adjacent or non-adjacent channels, narrowband blocking signals, and the products of non-linear distortions arising within the receiver itself.

---

[11] It is the noise floor in the receiving bandwidth.

Although in general the requirements for rejection of such interference in LTE are not significantly more demanding than in UMTS, LTE brings additional challenges, especially in relation to the wide and variable bandwidths and duplex spacings.

22.5 RF Impairments

The RF parts of the transmitter and receiver are comprised of non-ideal components which can have a strong impact on the ultimate demodulation performance. Different access technologies have different sensitivities to these RF non-idealities. For example, an OFDM-based systems are particularly sensitive to any distortion which may remove the orthogonality between the subcarriers, resulting in Inter-Carrier Interference (ICI), as discussed in Section 5.2.3. The RF impairments can have a non-negligible impact on BER, as shown for example in [10, 18] and references therein, where some compensation algorithms are also discussed.

The goal of this section is to show how the most common RF impairments introduce errors in an OFDM signal. First a simplified model of the impairments is discussed for the purpose of establishing SINR limitations for LTE performance assessment. This is followed in Section 22.5.2 by a mathematical framework for analysing the sensitivity of the LTE to particular impairments.

22.5.1 Transmitter RF Impairments

22.5.1.1 RF model

A generic model of typical transmitter impairments is given in Figure 22.19. It comprises the following components:[12]

- **Digital to Analogue Converter (DAC).** Quantization noise source, assumed to arise from a uniform linear quantizer;

- **Up-sampling function.** Up-samples the OFDMA/SC-FDMA sample rate process by a factor (not shown in diagram);

- **Baseband equivalent filter.** Equivalent to the concatenation of linear filtering components in the transmit path, including digital and analogue elements;

- **Quadrature error component.** Corresponding to the loss of I/Q orthogonality in the frequency conversion process;

- **d.c. offset.** Arising from, for example, direct conversion Local Oscillator (LO) leakage effects;

- **Phase noise.** Corresponding to the LO phase noise process;

- **Frequency error.** Due to LO long-term frequency error;

- **InterModulation (IM).** Attributable to, for example, PA non-linearity.

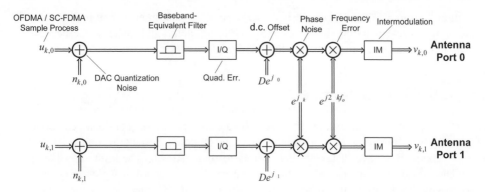

Figure 22.19 Multi-antenna transmitter impairment model. Reproduced by permission of © 2006 Motorola.

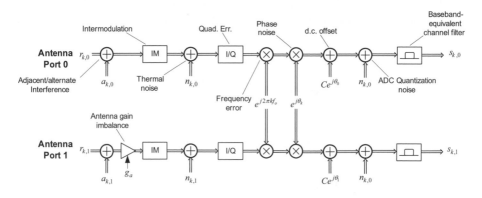

Figure 22.20 Multi-antenna receiver impairments model. Reproduced by permission of © 2006 Motorola.

A representative model of the RF impairments in the receiver, shown in Figure 22.20 for two receiver antennas, comprises the following conceptual components:

- **Adjacent Channel Interference (ACI).** Important when establishing deployment and coexistence guidelines and for estimating total system spectral efficiency. The effect of ACI may be neglected, however, in an initial LTE system performance analysis by assuming that Co-Channel Interference (CCI) is the dominant interference-limiting effect.

- **Antenna Gain Imbalance (AGI).** In practical multiantenna UE designs, the antenna gains can be affected by multiple factors including user grip pattern, orientation and proximity to the user's body. Nevertheless, for this analysis, we neglect these factors,

---

[12]It is not implied that a real implementation would necessarily use exactly the components specified; the aim here is to give a representative view of the typical impairments.

and the antenna gain specifications of Tables A.2.1.6-1 and A.2.1.8-1 of reference [9] are uniformly applied to each receiving antenna.

- **InterModulation (IM).** Receiver non-linearity in general, is a critical consideration when operating at the upper limits of the receiver dynamic range, and in the presence of strong adjacent channel interference. However, for the purposes of the present analysis, this effect is neglected.

- **Thermal Noise.** Applicable noise figures for LTE UE and eNodeB devices are specified in reference [9]. One further practical consideration is the potential for non-uniform noise figures applicable to each LTE antenna port. Again, it is proposed that this aspect be neglected in an initial analysis.

- **Quadrature error component.** As with the transmitter, this element models the loss of quadrature in the frequency conversion process. As an initial assumption, quadrature error may be neglected in eNodeB receivers, but is an essential element in direct conversion UE receiver modelling.

- **Frequency error.** The eNodeB receiver frequency error attributed to eNodeB LO error may be neglected since the UE uses the downlink waveform as a frequency reference. Clearly, in some circumstances there can be a significant frequency shift between the downlink signal received by the UE and the resulting uplink signal observed by the eNodeB. For simplicity, however, uniform azimuthal scattering at the UE is assumed and this effect is neglected. The effect of UE frequency estimation errors on receiver performance should, however, be considered.

- **Phase noise.** This corresponds to the eNodeB and UE LO phase noise process.

- **d.c. offset.** As discussed earlier for the transmitter model, this can arise due to LO leakage effects. However, this impairment can be neglected in an initial analysis.

- **Analogue to Digital Converter (ADC).** This can be modelled as a quantization noise source assumed to arise from a uniform linear quantizer.

- **Baseband equivalent filter.** Equivalent to the concatenation of linear filtering components in the receive path, including digital and analogue elements.

- **Down-sampling function (not shown).** Decimates the OFDMA/SC-FDMA sample rate process by a factor.

Often, a simple additive white Gaussian model can be considered to simplify the analysis of the impact of the RF impairments on link performance (see, for example, [10]). In fact the distortion generated by any non-linear device present for example in the downlink transmitter[13] can be modelled using Bussgang's theorem [19] as follows:

$$\widetilde{\mathbf{x}} = \alpha \mathbf{x} + \mathbf{d} \tag{22.5}$$

---

[13]The analysis here considers the downlink OFDM transmitter, but it can equally well be applied to the uplink transmitter.

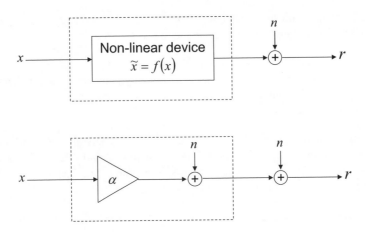

Figure 22.21 Non-linear device equivalent model.

where $\widetilde{\mathbf{x}}$ is the OFDM signal vector distorted by non-linearities, $\mathbf{x}$ is the column vector of the ideal OFDM symbol, $\mathbf{d}$ is the vector of the equivalent interference term due to distortions which are uncorrelated with the signal $\mathbf{x}$ and α is a complex gain factor accounting for the attenuation and phase rotation. This is illustrated in Figure 22.21.

The parameter α can be derived as

$$\alpha = \frac{\mathbb{E}[\widetilde{\mathbf{x}}^H \mathbf{x}]}{\mathbb{E}[\mathbf{x}^H \mathbf{x}]} \tag{22.6}$$

where $\mathbb{E}[\cdot]$ is the expectation operator. The covariance matrix of the output signal can therefore be written as

$$\mathbf{C}_{\widetilde{\mathbf{x}}\widetilde{\mathbf{x}}} = \mathbb{E}\{\widetilde{\mathbf{x}}\widetilde{\mathbf{x}}^H\} = |\alpha|^2 \mathbf{C}_{\mathbf{xx}} + \mathbf{C}_{\mathbf{dd}} \tag{22.7}$$

with $\mathbf{C}_{\mathbf{xx}} = \mathbb{E}\{\mathbf{xx}^H\}$ and $\mathbf{C}_{\mathbf{dd}} = \mathbb{E}\{\mathbf{dd}^H\}$.

Assuming, for the sake of simplicity, an AWGN[14] propagation channel, the distorted signal $\widetilde{\mathbf{x}}$ generated by the RF transmitter is further corrupted by complex white circular Gaussian noise $\mathbf{n} \sim \mathcal{N}(0, \sigma_n^2 \mathbf{I})$ such that the signal received by the OFDM receiver can be expressed as

$$\mathbf{r} = \widetilde{\mathbf{x}} + \mathbf{n} = \alpha\mathbf{x} + \mathbf{d} + \mathbf{n} \tag{22.8}$$

Equation (22.8) can be straightforwardly extended to the case of a frequency-selective convolutive channel and an OFDM system with CP.

As a result, the signal at the output of the FFT in the OFDM receiver can be written as

$$\mathbf{R} = \alpha\mathbf{Fx} + \mathbf{Fd} + \mathbf{Fn} \tag{22.9}$$

Consequently, the covariance matrix of the frequency-domain received signal is

$$\mathbf{C}_{\mathbf{RR}} = |\alpha|^2 \mathbf{F}\mathbf{C}_{\mathbf{xx}}\mathbf{F}^H + \mathbf{F}\mathbf{C}_{\mathbf{dd}}\mathbf{F}^H + \sigma_n^2 \mathbf{I} \tag{22.10}$$

---

[14] Additive White Gaussian Noise.

Equation (22.9) shows how the OFDM received signal, distorted by non-linearities, can be expressed as a function of the transmitted signal $\mathbf{x}$, scaled by a complex gain factor α, corrupted by AWGN and an additional interference term. This results in an error floor which is irreducible as the variance of the AWGN σ_n^2 is decreased.

Assuming coherent detection in the OFDM receiver,[15] the phase rotation induced by the scaling factor α has no effect as it will be ideally compensated in the receiver's channel estimation process. However, in both coherent and non-coherent OFDM systems, the impact of the magnitude of α translates into a power penalty for the useful signal. The reader is referred to [19] for further details.

Assuming FFT processing of large blocks and applying the central limit theorem, the frequency-domain distortion term $\mathbf{D} = \mathbf{Fd}$ can be closely approximated by complex white circular Gaussian noise $\mathbf{D} \sim \mathcal{N}(0, \sigma_d^2 \mathbf{I})$, and the SINR of the distorted received OFDM signal can be simply expressed as

$$\text{SINR}_d = \frac{|\alpha|^2 \text{tr}\{\mathbf{C_{XX}}\}}{\text{tr}\{\mathbf{C_{DD}}\} + \sigma_n^2} = \frac{|\alpha|^2}{\sigma_d^2 + \sigma_n^2} \tag{22.11}$$

where $\mathbf{C_{XX}} = \mathbf{F C_{xx} F^H}$ and $\mathbf{C_{DD}} = \mathbf{F C_{dd} F^H}$, assuming that the signal power $\mathbf{X}$ is normalized.

This model can be used in determining the impact of RF impairments on the LTE performance, with the additive distortion term being modelled as a transmitter EVM source.

22.5.2 Model of the Main RF Impairments

In order to gain insight into the impact of typical impairments, not only in terms of the impact on a single-user's communication link but also in terms of their overall effect on system-level spectral efficiency, it is instructive to develop a more precise analytical model. In particular, three sources of errors are included here:

- Phase noise;

- Modulator image interference (amplitude and phase imbalance);

- Non-linear distortion and intermodulation products which fall into adjacent resource blocks.

The model here is developed considering an OFDM signal (downlink) but it can be easily generalized for an SC-FDMA signal (uplink). The OFDM signal can be written as the sum of closely-spaced tones which do not interfere with each other due to their orthogonality. In particular the discrete-time signal for symbol ℓ is written as

$$x_\ell[n] = \sum_{k=-N/2}^{N/2-1} S_{k,\ell} \exp\left[j 2\pi k \Delta f \frac{n}{N}\right] \tag{22.12}$$

where N is the number of subcarriers, $S_{k,\ell}$ is the constellation symbol sent on the kth subcarrier for the ℓth OFDM symbol and n represents the component within the ℓth OFDM symbol, $n \in [0, N-1]$. In the following we can drop the dependency from the variable ℓ without loss of generality.

---

[15]Coherent detection means that the phase of the received signal is known to the receiver a priori, for example from reference signals as discussed in Chapter 8.

22.5.2.1 Amplitude and Phase Imbalance

Let us consider the continuous time version of the signal $x(t)$ modulated to frequency f_c, i.e.

$$r(t) = \text{Re}\{x(t)e^{j2\pi f_c t}\} = x(t)e^{j2\pi f_c t} + x^*(t)e^{-j2\pi f_c t} \tag{22.13}$$

When all the (de-)modulator components are ideal the signal $x(t)$ is recovered as

$$y(t) = \text{LP}\{r(t)e^{-j2\pi f_c t}\} = x(t) \tag{22.14}$$

where $\text{LP}\{\cdot\}$ means Low-Pass filtering. With a non-ideal IQ (de-)modulator, the recovered signal can be written as follows:

$$
\begin{aligned}
y_{IQ}(t) &= \text{LP}\{r(t)(\cos(2\pi f_c t) - j\beta \sin(2\pi f_c t + \phi))\} \\
&\overset{(1)}{=} \text{LP}\{(x(t)e^{j2\pi f_c t} + x^*(t)e^{-j2\pi f_c t})(\gamma_1 e^{-j2\pi f_c t} + \gamma_2 e^{j2\pi f_c t})\} \\
&= \text{LP}\{(x(t)\gamma_1 + x^*(t)\gamma_2 + \gamma_2 x(t)e^{j4\pi f_c t} + \gamma_1 x^*(t)e^{-j4\pi f_c t})\} \\
&\overset{(2)}{=} x(t)\gamma_1 + x^*(t)\gamma_2
\end{aligned}
\tag{22.15}
$$

where (1) is because of Equation (22.13) with $\gamma_1 = (1 + \beta e^{-j\phi})/2$ and $\gamma_2 = (1 - \beta e^{-j\phi})/2$ and (2) is obtained by low-pass filtering the signal, i.e. deleting all the components at frequencies $\pm 2f_c$. By substituting Equation (22.12) into (22.15), after a variable change it follows that

$$
\begin{aligned}
y_{IQ}(t) &= \sum_{k=-N/2}^{N/2-1} (\gamma_1 S_{k,\ell} + \gamma_2 S^\star_{-k,\ell}) \exp\left[j2\pi k \Delta f \frac{t}{N}\right] \\
&= \sum_{k=-N/2}^{N/2-1} \left(\frac{1 + \beta e^{-j\phi}}{2} S_{k,\ell} + \frac{1 - \beta e^{-j\phi}}{2} S^\star_{-k,\ell}\right) \exp\left[j2\pi k \Delta f \frac{t}{N}\right]
\end{aligned}
\tag{22.16}
$$

By letting $\beta = 1 + q$ and $Q(\beta, \phi) = (q - j\phi - qj\phi)/2$, using the approximation that $\cos(\phi) = 1$ and $\sin(\phi) = \phi$ for small angle ϕ, it follows that

$$
\begin{aligned}
y_{IQ}(t) &= \sum_{k=-N/2}^{N/2-1} S_{k,\ell} \exp\left[j2\pi k \Delta f \frac{t}{N}\right] \\
&\quad + Q(\beta, \phi) \sum_{k=-N/2}^{N/2-1} (S_{k,\ell} - S^\star_{-k,\ell}) \exp\left[j2\pi k \Delta f \frac{t}{N}\right]
\end{aligned}
\tag{22.17}
$$

Equation (22.17) shows clearly that the I/Q imbalance creates two different error terms: the first is the self-interference created by the same signal at the same subcarrier frequency, while the second is the signal at the frequency mirror-image subcarrier. The ideal case corresponds to $q = 0$ and $\phi = 0$, such that

$$y_{IQ}(t) = y(t) = \sum_{k=-N/2}^{N/2-1} S_{k,\ell} \exp\left[j2\pi k \Delta f \frac{t}{N}\right]$$

Note also that the amplitude imbalance and the phase imbalance create the same effect, i.e. only the factor $Q(\beta, \phi)$ changes, in case of only amplitude imbalance being present ($\phi = 0$) $Q(\beta, \phi) = q/2$, and when only phase imbalance is present the multiplicative coefficient becomes $Q(\beta, \phi) = j\phi/2$.

For high-order modulation (increasing number of states in the constellation) the error term acts as noise and it spreads the constellation points as shown in [20].

In general the amplitude and phase imbalance can vary depending on the frequency, on a subcarrier basis. This happens when there is a timing offset between the in-phase and quadrature signal paths. In this case Equation (22.17) becomes

$$
\begin{aligned}
y_{IQ}(t) = & \sum_{k=-N/2}^{N/2-1} S_{k,\ell} \exp\left[j2\pi k \Delta f \frac{t}{N} \right] \\
& + \sum_{k=-N/2}^{N/2-1} Q_k(\beta_k, \phi_k)(S_{k,\ell} - S^{\star}_{-k,\ell}) \exp\left[j2\pi k \Delta f \frac{t}{N} \right]
\end{aligned}
\tag{22.18}
$$

where the amplitude mismatch and the phase mismatch depends on the subcarrier index k.

22.5.2.2 Phase Noise

When the LO frequency at the transmitter is not matched to the LO at the receiver, the frequency difference implies a shift of the received signal spectrum at the baseband. In OFDM, this creates a misalignment between the bins of the FFT and the peaks of the sinc pulses of the received signal. This results in a loss of orthogonality between the subcarriers and a leakage between them. Each subcarrier interferes with every other (although the effect is strongest on adjacent subcarriers), and as there are many subcarriers this is a random process equivalent to Gaussian noise. Thus a LO frequency offset lowers the SINR of the receiver. An LTE receiver will need to track and compensate for this LO offset and quickly reduce it to substantially less than the 15 kHz subcarrier spacing. This is a tougher requirement than for UMTS. The LO phase offset presents a similar issue, resulting in a constant phase rotation of all the subcarriers.[16] The phase noise impairment has been widely studied both for a single antenna [20–22] and for the MIMO case [23], especially in the context of WiMAX.

The ideal baseband transmitted signal, neglecting the additive white Gaussian noise, is given in Equation (22.12). The received baseband signal, in the presence of only phase noise can be written as

$$
y_\theta(t) = \sum_{k=-N/2}^{N/2-1} S_{k,\ell} \exp\left[j2\pi k \Delta f \frac{t}{N} \right] e^{j\theta(t)}
\tag{22.19}
$$

where the transmitted signal is multiplied by a noisy carrier $e^{j\theta(t)}$. In [10] it is shown that the single-sideband phase noise power follows a Lorentzian spectrum [10, 24]:

$$
L(f) = \frac{2}{\pi \Delta f_{3dB}} \frac{1}{1 + [2f/(\Delta f_{3dB})]^2}
\tag{22.20}
$$

---

[16]This does not result in a loss of orthogonality, since RSs experience the same degree of rotation and can be used to compensate for the offset.

where $\Delta f_{3\text{dB}}$ is the two-sided 3 dB bandwidth of phase noise. The power spectrum given in Equation (22.20) can be considered as an approximation to practical oscillator spectra.

The received signal in Equation (22.19) is passed through the FFT in order to obtain the symbol transmitted on the mth subcarrier in the ℓth OFDM symbol, i.e.

$$z_{m,\ell} = \frac{1}{N} \sum_{n=-N/2}^{N/2-1} y_\theta(n) \exp\left[-j2\pi m \Delta f \frac{n}{N}\right]$$

$$= \frac{1}{N} \sum_{n=-N/2}^{N/2-1} \sum_{k=-N/2}^{N/2-1} S_{k,\ell} \exp\left[j2\pi \Delta f \frac{n}{N}(k-m)\right] e^{j\theta(n)}$$

$$= S_{m,\ell} \frac{1}{N} \sum_{n=-N/2}^{N/2-1} e^{j\theta(n)} + \frac{1}{N} \sum_{k=-N/2, k\neq m}^{N/2-1} S_{k,\ell} \sum_{n=-N/2}^{N/2-1} \exp\left[j2\pi \Delta f \frac{n}{N}(k-m)\right] e^{j\theta(n)}$$

$$(22.21)$$

By defining $C(k) = (1/N) \sum_{n=-N/2}^{N/2-1} e^{j2\pi \Delta fnk/N + j\theta(n)}$, as in [21], Equation (22.21) can be rewritten as

$$z_{m,\ell} = S_{m,\ell} C(0) + \sum_{k=-N/2, k\neq m}^{N/2-1} S_{k,\ell} C(k-m) \qquad (22.22)$$

Equation (22.22) shows that the effect of phase noise is twofold: the useful symbol transmitted on m^{th} subcarrier is scaled by a coefficient $C(0)$ which depends on the phase noise realization on the ℓth OFDM symbol, but it is independent of the subcarrier index – i.e. all the subcarriers are rotated by the same quantity $C(0)$. This is referred to as the Common Phase Rotation (CPR). This term can be estimated from the Reference Signals (RSs) and removed. In LTE, four symbols per subframe contain RS (see Section 8.2), so that in theory low frequency phase noise up to about 2 kHz can be compensated. However common loop bandwidths of PLLs[17] are in general in the order of 10 to 100 kHz, so that a major part of the phase noise energy is outside the region which can be compensated. The second term causes ICI,[18] the amount of which is given by the coefficient $C(k-m)$ for subcarrier k interfering on subcarrier m. The ICI due to phase noise creates a fuzzy constellation as shown in [20]. In [23] the above equations are generalized for the MIMO case, showing that phase noise has similar effects in that case.

Time-domain offsets

Any timing synchronization error results in a misalignment between the samples contained in the OFDM symbol and the samples actually processed by the FFT. If the timing error is large some samples could be from the wrong symbol, which would cause a serious error. It is more likely that the timing error will be just a few samples and the presence of the cyclic prefix should give enough margin to prevent samples from the wrong symbol being used. Assuming this is the case, the shift in time is equivalent to a linearly-increasing phase rotation of the constellation points.

---

[17]Phase-Locked Loop.
[18]Inter-Carrier Interference.

The use of higher modulation schemes and wider channel bandwidths mean that timing synchronization needs to be performed more accurately for LTE than UMTS.

Another source of timing error is a sampling frequency error. This can occur when the baseband sampling rate of the receiver is offset from that of the transmitter, or if there is jitter of the receiver's clock. When the sampling rate is too high this contracts the spectrum of the signal; similarly, slow sampling widens the spectrum. Either of these effects misaligns the FFT bin locations, resulting in a loss of orthogonality and hence reduced SINR. A sampling rate converter, perhaps driven by an error estimate, could correct for a systematic sampling rate error or sampling rate drift. Sampling jitter should be kept low by the choice of a clean crystal reference oscillator.

Group delay distortion

Filter group delay and amplitude ripple variation create deviation from the wanted impulse response of the receiver and cause inter-symbol interference. Unfortunately analogue filters cannot have both flat group delay and flat amplitude response, so a compromise is needed. To achieve this the analogue channel filter is usually slightly larger than the channel to allow reduced in-band distortion at the expense of out-of-band attenuation and in-band noise suppression.

One of the inherent advantages of multicarrier OFDM is that it is quite resilient to group delay variation and amplitude ripple, more so than the single-carrier QPSK modulation used by UMTS.

Additionally, the OFDM symbol rate is relatively low and so there are greater margins for delays introduced by large filters. Overall there is more freedom with LTE to design the receiver analogue and digital filters to achieve high selectivity.

Reciprocal mixing

An unwanted phase offset of the received signal may include a large fixed (or slowly varying), phase offset, which can be corrected by the equalizer. However, the LO also introduces small random variations and jitter of the frequency and phase, which manifests itself as shaped phase noise, tending to reduce with offset from the carrier frequency. This cannot be corrected by the equalizer. Crystal oscillators have low levels of phase noise, but synthesizers, especially PLL synthesizers, are not so clean.

Each of the elements of a frequency synthesizer produce noise which contributes to the overall noise appearing at the output. The noise within the PLL loop bandwidth arises from the phase detector and the reference, whereas outside the loop bandwidth the VCO[19] is the main noise source. The phase noise tends to reduce from the edge of the PLL loop bandwidth until eventually it reaches a flat noise floor.

A serious problem for receivers caused by phase noise is reciprocal mixing. Reciprocal mixing occurs when the phase noise on the receiver LO mixes with a strong interfering signal to produce a signal which falls inside the pass-band of the receiver. Intermediate-Frequency (IF) filters, if used, do not give any rejection of such signals; instead the problem must be solved by keeping the LO phase noise at a sufficiently low level. The various interference requirements (see Sections 22.4.6 to 22.4.7.2) together create an overall requirement for the

---

[19]Voltage-Controlled Oscillator.

receiver LO phase noise. The interferer which gets mixed onto the received signal needs to be weaker by a margin of the SINR requirement plus the implementation margin, plus a further margin of about 10 dB if it is to have no impact on the received signal. Therefore, the LO noise, L, needs to satisfy the following requirement:

$$L \leq \text{SINR} + \text{IM} - C/I + 10 + 10 \times \log_{10}(B) \ \text{(dBc/Hz)} \tag{22.23}$$

The requirement at the smallest offset comes from the LTE narrowband blocking test, which for a 5 MHz bandwidth is given by

$$L \leq 0 + 2 - (-28) + 10 + 10 \times \log_{10}(4.5 \cdot 10^6) \leq -106.5 \ \text{dBc/Hz at a 450 kHz offset.}$$

The out-of-band 15 dBm blocker test defines the phase noise at the farthest offset; for a 5 MHz bandwidth it is given by

$$L \leq 0 + 2 - (-78.3) + 10 + 10 \times \log_{10}(4.5 \cdot 10^6) \leq -156.8 \ \text{dBc/Hz at an 85 MHz offset}$$

(or 20 MHz for RF bands 2 or 5).

It should be noted, however, that these figures have not taken into account the amplitude response of the LNA and mixer. A reduced gain at an offset will reduce the impact of reciprocal mixing. Additionally, the RF filter should help to reduce distant blockers. In practice, the requirement for the transmitter LO signal, which was -153 dBc/Hz at a greater offset, could be tougher.

22.5.2.3   Non-Linear Distortion and Intermodulation Products

Intermodulation distortion results from non-linearities. This creates OOB emissions which cause interference with co-channel users, and in-band distortion which causes self-interference. The high PAPR (see Section 5.2.2) associated with multicarrier signals is one of the principal challenges in the implementation of OFDM systems. It requires linear operation of the PA over a large dynamic range, and this imposes a considerable implementation cost and reduced efficiency. The linearity and efficiency of a power amplifier are mutually exclusive specifications.

In the transmitter, the linearity requirement must be met at reasonably high power levels. With the associated requirement for high efficiency, this places extreme design challenges. In the receiver, the power levels are much lower but the linearity requirement is even more challenging because an adjacent channel blocker is often received with a significantly stronger signal power level than the desired signal. The linearity requirement for the receiver components must include this anticipated level of blocker power above the desired signal. Two direct consequences of this are difficulty in achieving acceptable noise figure performance for the overall receiver, and difficulty in achieving sufficient dynamic range for the ADCs.

As already discussed, practical power amplifiers have a non-linear response. Numerous models exist, a selection of which can be found in [25]. Here we describe a simple polynomial memoryless model, by which the output signal can be written as

$$y_{PA}(t) = a_1 y(t) + a_2 y^2(t) + a_3 y^3(t) \tag{22.24}$$

where a_1, a_2, a_3 are independent coefficients which can be found by measurement. From Equations (22.14) and (22.12), (22.24) can be written as

$$y_{\text{PA}}(t) = a_1 \sum_{k=-N/2}^{N/2-1} S_{k,\ell} \exp\left[j2\pi k\Delta f \frac{t}{N} \right]$$

$$+ a_2 \sum_{k=-N/2}^{N/2-1} \sum_{p=-N/2}^{N/2-1} S_{k,\ell} S_{p,\ell} \exp\left[j2\pi (k+p)\Delta f \frac{t}{N} \right]$$

$$+ a_3 \sum_{k=-N/2}^{N/2-1} \sum_{p=-N/2}^{N/2-1} \sum_{v=-N/2}^{N/2-1} S_{k,\ell} S_{p,\ell} S_{v,\ell} \exp\left[j2\pi (k+p+v)\Delta f \frac{t}{N} \right] \quad (22.25)$$

The non-linear response of the PA creates ICI. In particular the interference on the kth and $(k+1)^{\text{th}}$ subcarriers (prior to FFT) is given by

$$I(k) = a_2 S_{k,\ell} S_{0,\ell} + a_3 \sum_{v=-N/2}^{N/2-1} S_{k,\ell} S_{-v,\ell} S_{v,\ell}$$

$$I(k+1) = a_2 S_{k,\ell} S_{1,\ell} + a_3 \sum_{v=-N/2+2}^{N/2-1} S_{k,\ell} S_{1-v,\ell} S_{v,\ell}$$

$$\ldots \quad (22.26)$$

These intermodulation products contribute to a noise-like cloud surrounding each constellation point.[20] For higher-order modulation in particular (such as 64QAM), these constellation clouds contribute to an increase in error rate for each subcarrier. Thus, in an OFDM modem design, linearity must be carefully controlled.

22.6 Conclusion

In this chapter, we have reviewed the RF aspects of the practical implementation of LTE equipment, especially the UE. We have seen how the effects of typical RF impairments of the transmitter and receiver can be analysed, including the use of a mathematical model for quantitative analysis.

The LTE RF requirements can be compared to those of UMTS. In many aspects, the RF requirements of LTE are similarly demanding as for UMTS, including for example narrowband blocking of the receiver and self-interference in case of full-duplex FDD operation. However, the LTE specifications are more challenging for other aspects, such as

- The channel bandwidth is variable over a wide range;

- At least two receive antennas are expected for receive diversity, and only the lowest UE category does not have to support reception of MIMO spatial multiplexing;

---

[20]Note that the interference is correlated with the symbol transmitted on subcarrier k, and hence strictly it cannot be considered as white noise.

- There is a large variety of modulation and coding schemes, with some of them requiring a high SNR;

- OFDM is more sensitive to phase noise.

On the other hand, LTE is more robust than UMTS against amplitude and phase distortions from receiver and transmitter filters.

References[21]

[1] 3GPP Technical Specification 25.101, 'User Equipment (UE) Radio Transmission and Reception (FDD) (Release 8)', www.3gpp.org.

[2] 3GPP Technical Specification 25.104, 'Base Station (BS) Radio Transmission and Reception (FDD) (Release 8)', www.3gpp.org.

[3] 3GPP Technical Specification 25.102, 'User Equipment (UE) Radio Transmission and Reception (TDD) (Release 8)', www.3gpp.org.

[4] 3GPP Technical Specification 25.102, 'Base Station (BS) Radio Transmission and Reception (TDD) (Release 8)', www.3gpp.org.

[5] 3GPP Technical Specification 36.101, 'User Equipment (UE) Radio Transmission and Reception (Release 8)', www.3gpp.org.

[6] 3GPP Technical Specification 36.104, 'Base Station (BS) Radio Transmission and Reception (Release 8)', www.3gpp.org.

[7] 3GPP Technical Report 36.803, 'User Equipment (UE) Radio Transmission and Reception', www.3gpp.org.

[8] 3GPP Technical Report 36.804, 'Base Station (BS) Radio Transmission and Reception', www.3gpp.org.

[9] 3GPP Technical Report 25.814, 'Physical Layer Aspects for Evolved UTRA, (Release 7)', www.3gpp.org.

[10] B. E. Priyanto, T. B. Sorensen, O. K. Jensen, T. Larsem, T. Kolding and P. Mogensen, 'Assessing and Modeling the Effect of RF Impairments on UTRA LTE Uplink Performance', in *Proc. IEEE Vehicular Technology Conference*, Baltimore, MD, USA, September 2007.

[11] Motorola, 'R1-061922: CQI Coding Schemes', www.3gpp.org, 3GPP TSG RAN WG1, LTE Adhoc, Cannes, France, June 2006.

[12] Motorola, 'R1-060144: UE Power Management for E-UTRA', www.3gpp.org, 3GPP TSG RAN WG1, LTE Adhoc, Helsinki, Finland, January 2006.

[13] Motorola, 'R1-070084: Coexistance Simulation Results for 5MHz E-UTRA → UTRA FDD Uplink with Revised Simulation Assumptions', www.3gpp.org, 3GPP TSG RAN WG4, meeting 42, St. Louis MO, USA, February 2007.

[14] Motorola, 'R1-060023: Cubic Metric in 3GPP-LTE', www.3gpp.org, 3GPP TSG RAN WG1, LTE Adhoc, Helsinki, Finland, January 2006.

[15] Nortel, 'R1-050891: OFDMA UL PAPR Reduction', www.3gpp.org, 3GPP TSG RAN WG1, meeting 42, London, UK, August 2005.

[16] Motorola, 'R1-081032: Tone Reservation for OFDMA PAPR Reduction', www.3gpp.org, 3GPP TSG RAN WG1, meeting 42bis, San Diego, USA, October 2008.

[17] Ericsson, 'R1-050765: Some Aspects of Single-carrier Transmission for E-UTRA', www.3gpp.org, 3GPP TSG RAN WG1, meeting 42, London, UK, August 2005.

---

[21] All web sites confirmed 18th December 2008.

[18] M. Moonen and D. Tandur, 'Compensation of RF Impairments in MIMO OFDM Systems' in *Proc. IEEE International Conference on Acoustics, Speech and Signal Processing*, Las Vegas, NV, USA, April 2008.

[19] J. Bussgang, 'Crosscorrelation Function of Amplitude Distorted Gaussian Signals', Technical Report 216, Research laboratory of electronics, Massachusetts Institute of Technology, Cambridge, Massachusetts, USA, March 1952.

[20] Agilent Technologies, 'R1-061276: Effects of Physical Impairments on OFDM Signals', www.3gpp.org, 3GPP TSG RAN WG4, meeting 33, Riga, Latvia, November 2006.

[21] F. Munier, T. Eriksson and A. Svensson, 'Receiver Algorithms for OFDM Systems in Phase Noise and AWGN' in *Proc. IEEE International Symposium on Personal, Indoor and Mobile Radio Communications*, Barcelona, Spain, September 2004.

[22] S. Wu and Y. Bar-Ness, 'Performance Analysis on the Effect of Phase Noise in OFDM Systems', in *Proc. IEEE International Symposium on Spread Spectrum Techniques and Applications*, Prague, Czech Republic, 2002.

[23] T. C. W. Schenk, T. Xiao-Jiao, P. F. M. Smulders and E. R. Fledderus, 'On the Influence of Phase Noise Induced ICI in MIMO OFDM Systems'. *IEEE Communications Letters*, August 2005.

[24] E. Costa and S. Pupolin, 'M-QAM-OFDM System Performance in the Presence of a Nonlinear Amplifier and Phase Noise'. *IEEE Trans. on Communications*, pp. 462–472, March 2002.

[25] S. C. Cripps, *RF Power Amplifiers for Wireless Communications*. Artech House, Norwood, MA, USA, 1999.

23

Paired and Unpaired Spectrum

Nicholas Anderson

23.1 Introduction

Expansion of consumer demand for cellular communications, as well as for the multitude of other wireless applications, places a corresponding strain on the basic physical resource needed to support them: spectrum. The suitability of spectral resources to a particular application is governed by a range of inter-related factors of a technological, commercial and regulatory nature. Technological considerations influencing the choice of frequency band include propagation characteristics, antenna size and separation, the viability of Radio Frequency (RF) circuitry, and design implications resulting from the need to coexist with systems operating in neighbouring spectrum without causing (or suffering from) undue interference. These considerations determine in part the commercial viability of a system. Signal range and spectral efficiency govern coverage and capacity, and hence determine the required number of base station sites and the capital outlay, whilst terminal costs and form factor affect the acceptability of products in the marketplace.

From a regulatory perspective, in spite of significant coordination at an international level through the International Telecommunications Union (ITU), regional variations in governmental policy regarding spectrum and technology are inescapable due to the differing needs of each region and the different historical factors which have shaped their present-day spectrum allocations. Thus, the availability of globally-harmonized spectrum for a particular technology cannot be guaranteed. Furthermore, in many regions, a tendency towards technology neutrality within viable spectrum assignments (see, for example, [1]) is beginning to liberalize the traditional one-to-one mappings between technologies and their addressable spectrum and is hence opening up new markets.

Against this background, a key design goal of LTE has been to enable deployment in a diverse range of spectrum environments in terms of bandwidth, uplink-downlink duplex

LTE – The UMTS Long Term Evolution: from Theory to Practice   Stefania Sesia, Issam Toufik and Matthew Baker
© 2009 John Wiley & Sons, Ltd

spacing[1] and uplink-downlink asymmetries. By supporting paired spectrum allocations (separate uplink and downlink carriers) in addition to stand-alone unpaired allocations (uplink and downlink operating on the same carrier frequency), wastage of valuable spectrum can be avoided.

It is also the case that traffic is increasingly data-centric and often asymmetric. In unpaired spectrum, asymmetry may be provided through the use of unequal duty cycles in the time domain for uplink and downlink, which may further be adapted according to demand. In paired spectrum asymmetry is also possible via the deployment of unequal bandwidths for uplink and downlink although this is less likely to be dynamically changeable.

Considering the nature of the regulatory environment, an increase in demand and competition for spectrum and the trends of mobile traffic and service usage, it is advantageous for the LTE system to be designed such that it may be flexibly adapted to diverse spectral assignments including both paired and unpaired bands. The market addressable by the technology is thereby increased.

23.2 Duplex Modes

The term 'duplex' refers to bidirectional communication between two devices, as distinct from unidirectional communication which is referred to as 'simplex'. In the bidirectional case, transmissions over the link in each direction may take place at the same time ('full duplex') or at mutually exclusive times ('half-duplex').

It should be noted that a communication link being half-duplex need not imply that only half-duplex user services are supported (such as 'push-to-talk' voice applications). Full-duplex services such as a normal telephone conversation may of course be carried over half-duplex communication systems. The key factor differentiating the duplex nature of the service from that of the underlying communication link is the timescale over which the communication direction is cycled in relation to the service duplex timescale: if the link direction is cycled at a sufficiently high rate then the duplex nature of the link can be concealed when viewed from the perspective of the user application.

In the case of a full-duplex transceiver, the frequency domain is used to separate the inbound and outbound communications; that is, different carrier frequencies are employed for each link direction. This is referred to as Frequency Division Duplex (FDD). As discussed in Section 22.4.2, the ability of a full-duplex transceiver to transmit and receive at the same time instant is enabled via the use of a duplexer – a tuned filter network able to provide a high degree of isolation between the inbound and outbound signals on the different carrier frequencies (often sharing the same antenna). However, these filter networks incur some signal attenuation. For the receiver, this attenuation occurs before the low-noise amplifier in the signal path and hence contributes directly to the receiver noise figure and degrades its sensitivity. For the transmitter, the duplexer follows the high power amplification stage, requiring either a higher-powered amplification device to overcome the loss, or tolerance of the corresponding reduction in communication range.

Conversely, in the case of a half-duplex transceiver, the time domain provides the necessary separation between the inbound and outbound communications. When the same

---

[1] 'Duplex spacing' is the term used to describe the size of the frequency separation between uplink and downlink carriers.

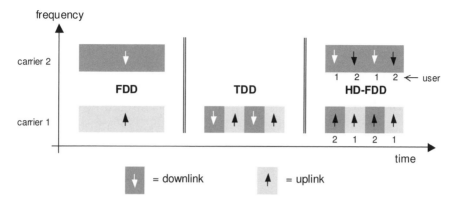

Figure 23.1 FDD, TDD and HD-FDD duplex modes.

carrier frequency is used for each link direction the system is said to be purely Time Division Duplex (TDD) and incorporates half-duplex transceivers at each end of the radio link. Alternatively different carrier frequencies may be used, in which case the system is often known as 'Half-Duplex FDD' (HD-FDD). The popular GSM system utilizes HD-FDD, with uplink and downlink communications taking place for a particular user not only on distinct carrier frequencies but also at different times. HD-FDD can therefore be seen as a hybrid combination of FDD and TDD. By scheduling users (or sets of users) at mutually exclusive times, full occupancy of the transmission resources may be achieved without requiring simultaneous transmission and reception at each mobile terminal. Thus for HD-FDD the base station is full-duplex whilst the terminals are half-duplex.

The three duplex modes outlined above are represented diagrammatically in Figure 23.1.

Both HD-FDD and TDD carry the advantage that terminals may be developed without the need for duplexers, thus simplifying the design. HD-FDD operation is also potentially useful for paired bands in which the duplex spacing is relatively narrow (e.g. less than only a few times the system bandwidth) since it avoids the need for a technically challenging duplexer design in the terminal and can thereby help in reducing its cost. At the base station there is more scope to implement such technically challenging designs due to less onerous size and cost constraints and because base station products are usually targeted towards a specific frequency band.

The initial specifications for LTE identify a total of 23 applicable spectrum bands (15 paired and eight unpaired), as listed in Section 22.2. The paired bands are all currently symmetrically dimensioned between uplink and downlink, although this may change in the future.

23.3 Interference Issues in Unpaired Spectrum

As discussed in the previous chapter, the spectral emissions from an eNodeB or from a UE are unfortunately not strictly band-limited within the desired carrier bandwidth. This is due to practical limitations on filter technology and on the linearity of amplification.

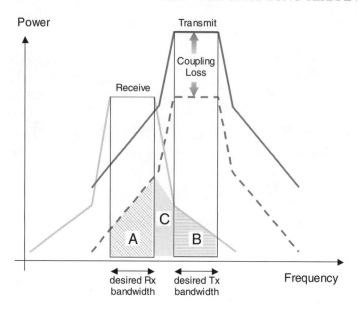

Figure 23.2 Frequency-domain interference between transmitter and receiver.

In addition, receivers suffer from non-ideal suppression of signals falling outside the desired reception band. There is thus the potential for interference between transmitter and receiver as shown in Figure 23.2. In Figure 23.2, all sources of attenuation of the signal between transmitter and receiver due to phenomena such as propagation, antenna radiation patterns and cable losses are denoted by a single 'coupling loss'. In the region marked 'A', energy from the transmitter falls directly within the desired passband of the receiver, while in region 'B' the non-ideal characteristic of the receiver collects energy from inside the passband of the transmitter. In region 'C' transmitted energy falling outside both the transmit and receive passbands is collected by the non-ideal receiver characteristic. The aggregated effects of regions A, B and C represent the total unwanted energy from the interfering transmitter that is captured by the receiver.

Interference arising from regions A and B is controlled by performance requirements imposed on the transmitting and receiving devices respectively, while in region C both transmitter performance and receiver performance have an impact. The interference effects are generally small for large frequency separations (such as in the case of FDD), and when there is a large coupling loss between the interfering transmitter and a 'victim' receiver, as is the case when there is a large physical separation between the two).

Naturally, the interference effects are most pronounced when the transmitter and receiver operate on adjacent carrier frequencies (excepting the co-channel case) and with low coupling loss (e.g. due to small spatial separation). These worst-case adjacent-channel scenarios form the basis upon which key requirements for transmitter and receiver are generally set.

The overall 'leakage' (whether at the transmit side or the receive side) from a transmission on one carrier into a receiver operating on an adjacent carrier is described by the *Adjacent Channel Interference Ratio* (ACIR), which is derived from the transmitter's *Adjacent*

Channel Leakage Ratio (ACLR) and the receiver's *Adjacent Channel Selectivity* (ACS) as defined in Section 22.3.2.1. It is worth noting that it is often the case that the ACIR is dominated by either the ACS or the ACLR. For example, it is technologically difficult to design a mobile transmitter with an ACLR that approaches or exceeds the ACS of a base station receiver.

As is discussed in the following section, the presence of imperfect ACIR has implications for the deployment of systems at a boundary between unpaired and paired spectrum allocations, and also for unsynchronized systems operating in closely-spaced unpaired allocations.

23.3.1 Adjacent Carrier Interference Scenarios

For an FDD cellular system, adjacent channel frequency separation of an interfering transmitter and a victim receiver naturally implies that the interferer and victim are of differing equipment types (i.e. one is a mobile terminal whilst the other is a base station). Transmitter-receiver interference between one User Equipment (UE) and another, or between one eNodeB and another is avoided by virtue of the duplex spacing.

The same is also generally true in TDD systems if they are time-synchronized so that overlap between uplink and downlink transmission periods is avoided. However, when synchronization is not or cannot be provided, or when TDD systems operate on carriers adjacent to an FDD system, the possibility arises for interferer and victim to be of the same device type. Figure 23.3 depicts a relatively common scenario in which an unpaired spectral allocation is located in a region between an FDD downlink band and an FDD uplink band (as is the case for the 2.5–2.6 GHz UMTS extension band, for example).

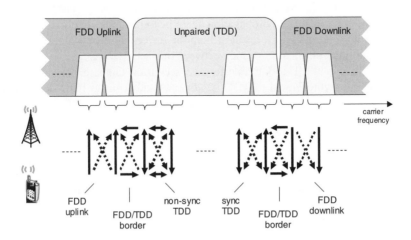

Figure 23.3 Possibilities for Adjacent Carrier Interference (ACI) between adjacent FDD and TDD systems.

The vertical arrows in Figure 23.3 represent the desired communication between the base station and the mobile terminal, unidirectional on a per-carrier basis for FDD and bi-directional for TDD. The diagonal dotted lines represent base-station-to-mobile and mobile-to-base-station adjacent channel interference that results from imperfect ACIR.

Two TDD scenarios are shown, synchronized and non-synchronized – the latter encompassing any general possibility for partial or full uplink/downlink overlap in time between two systems. For the synchronized case the interference scenarios between the base station and mobile (and vice versa) are the same as for the corresponding FDD-to-FDD adjacent channel cases. However, for the non-synchronized case additional interference paths exist between TDD mobiles and between TDD base stations, represented by the horizontal bidirectional arrows. At the FDD/TDD border regions, these same 'horizontal' interference paths exist but are unidirectional in nature.

There are many facets of a deployment which affect the severity of these various interference paths. For example, the locations of the interfering transmitter and victim receiver, as well as the characteristics of the propagation between them, clearly influence the overall coupling that exists. Macrocellular deployments typically use base station transmit antennas mounted on masts located above roof-top level, thereby resulting in an increased likelihood of Line-Of-Sight (LOS) propagation between base stations, with a correspondingly low path-loss exponent. The common use of macrocell base station antennas with vertical directivity and hence high gain can further worsen this situation. These aspects are less problematic for microcellular and dense-urban deployments in which LOS propagation between base stations is less likely due to their antennas being located below rooftop level.

Coupling between mobile terminals is often mitigated by the surrounding local clutter, and due to the lower antenna gain in the terminals. One typical scenario is depicted in Figure 23.4 in which non-co-located macro base stations have some potential to exhibit stronger mutual coupling than between the mobiles which they respectively serve. However, the figure also shows that it is not always possible to rely on local clutter to provide the necessary isolation between terminals, due to the fact that when terminals are closely-spaced (for example, in the same office or café), LOS propagation again becomes more likely and the potential for interference is increased – as a result of both the lower path loss exponent between the terminals and the small physical separation between them.

The base-to-base and UE-to-UE interference scenarios that are particular to unsynchronized TDD deployments and to TDD deployments adjacent to FDD deployments are reviewed in more detail in the following two subsections.

23.3.1.1   Base-Station to Base-Station Interference

Base stations of relevance to a particular base-to-base interference scenario may be either co-located (i.e. antennas mounted at the same cell-site), or non-co-located. Nevertheless, base-to-base interference is generally deterministic. This is because the locations of the base stations are fixed, and furthermore the link adaptation strategy typically employed for the LTE downlink usually results in all available transmit power being used to maximize the throughput of the link. It is therefore reasonable to analyse the interference assuming full transmit power from each base station.

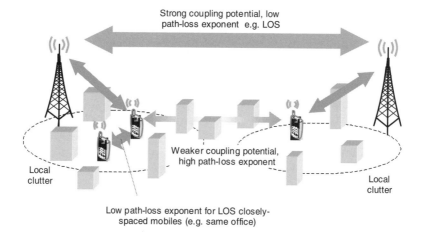

Figure 23.4 Typical RF interference scenario for a TDD system.

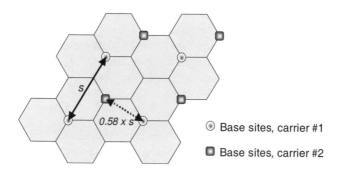

Figure 23.5 An idealized deployment of two cellular networks with non-co-located base stations.

A favourable scenario is one in which the interfering base stations are optimally non-co-located. Figure 23.5 shows an idealistic example of such a deployment, in which base stations operating on adjacent carriers utilize the same inter-site spacing of s metres on an equilateral triangular grid, and the base stations using one carrier are located at the edges of the cells of those using the other.

A crude assumption is that the base-to-base interference level observed at each victim receiver is dominated by only one of the first tier of three interfering base stations (the others in the first tier arriving outside the receiver's antenna beam width, and those outside the first tier being of secondary significance due to the increased path distance). The dominant interfering site in this case is therefore at a (best-case) distance of $0.58s$.

However, in general, one cannot expect systems on adjacent carriers to be similar in cell size and even if they are, maximal spacing between base-sites is unlikely. Thus in the generalized case one can consider the worst-case base-to-base distance only, and can assume that at some point in the network the transmit and receive antenna patterns are aligned to provide maximum gain at this worst-case distance. The co-channel Power Spectral Density (PSD) of the received interference at the victim base station antenna connector then becomes

$$\mathrm{PSD_{Rx}} = \mathrm{PSD_{Tx}} + 2G_{\mathrm{BS}} - \rho_{\mathrm{BS\text{-}BS}}(x_0) \qquad (23.1)$$

where $\mathrm{PSD_{Tx}}$ is the transmitted PSD at each interfering base station antenna connector, G_{BS} is the antenna gain at each base site, $\rho_{\mathrm{BS\text{-}BS}}(x)$ is the path-loss between base sites as a function of distance x in metres and x_0 is the worst-case (smallest) distance between base sites of different carriers.

In order for the inter-system interference to have only a minor effect, one can assume that the PSD of the interference after benefiting from any available ACIR should be of the order of $\mathrm{PSD_N} + \mathrm{NF_{BS}} - 6$ dB or less if it is to produce no more than a 1 dB desensitization of the base station receiver (where $\mathrm{NF_{BS}}$ denotes the base station noise figure and $\mathrm{PSD_N}$ is the power spectral density of thermal noise, e.g. -174 dBm/Hz at typical temperatures):

$$(\mathrm{PSD_{Rx}} - \mathrm{ACIR}) \leq (\mathrm{PSD_N} + \mathrm{NF_{BS}} - 6)\ \mathrm{dB} \qquad (23.2)$$

Note that in the case of an SC-FDMA uplink victim receiver, consideration needs to be paid not only to the ACIR averaged over the system bandwidth, but, more challengingly, to the localized frequency resource blocks located closest to the interfering carrier, especially if the important uplink control signalling on the Physical Uplink Control Channel (PUCCH) is to be protected (see Section 17.3.1). In general however, the ACIR requirement varies directly with $\mathrm{PSD_{Tx}}$:

$$\mathrm{ACIR} \geq \mathrm{PSD_{Tx}} + 2G_{\mathrm{BS}} - \rho_{\mathrm{BS\text{-}BS}}(x_0) - \mathrm{PSD_N} - \mathrm{NF_{BS}} + 6\ \mathrm{dB} \qquad (23.3)$$

In order to arrive at an ACIR requirement we must therefore know the transmitted PSD and the intervening path loss. To do so, it is reasonable to assume that the base station transmit power capabilities are dimensioned in order that each of the two systems are interference-limited on the downlink (at least this can apply for small- and medium-sized cells without exceeding the eNodeB output power capabilities). We therefore assume here that the spectral density of the downlink signal for 95% of the total area is γ_{DL} dB larger than the spectral density of the thermal noise in the UE receivers. Thus

$$\mathrm{PSD_{Tx}} = \mathrm{PSD_N} + \mathrm{NF_{UE}} + \gamma_{\mathrm{DL}} + \rho'_{\mathrm{BS\text{-}UE}} - G_{\mathrm{BS}} \qquad (23.4)$$

where $\mathrm{NF_{UE}}$ is the noise figure of the User Equipment (UE) receiver, γ_{DL} is the received downlink signal to noise ratio in dB which is exceeded at 95% of the UE receivers, and $\rho'_{\mathrm{BS\text{-}UE}}$ is the 95-percentile path-loss between UEs and their serving eNodeBs in the interfering network.

For ease of representation, an empirical approximation to $\rho'_{\mathrm{BS\text{-}UE}}$ is applied here specific to this particular example deployment: let σ be the standard deviation of the log-normal shadow fading between eNodeB and UE and L_{b} represent the additional building penetration loss (assuming indoor coverage); then with $\rho_{\mathrm{BS\text{-}UE}}(x)$ denoting the path-loss between base stations and UEs separated by distance x metres, we can write

$$\rho'_{\mathrm{BS\text{-}UE}} \approx \rho_{\mathrm{BS\text{-}UE}}(0.58s) + 0.7\sigma + L_{\mathrm{b}} \qquad (23.5)$$

Substituting Equations (23.5) and (23.4) into (23.3) we obtain an approximate expression for the necessary ACIR to maintain an acceptable adjacent channel interference level at a victim base station:

$$\text{ACIR} \geq \text{NF}_{\text{UE}} - \text{NF}_{\text{BS}} + \gamma_{\text{DL}} + \rho_{\text{BS-UE}}(0.58s) + 0.7\sigma + L_b + G_{\text{BS}} - \rho_{\text{BS-BS}}(x_0) + 6 \text{ dB}$$
$$(23.6)$$

This function is plotted in Figure 23.6 as a function of the smallest eNodeB-eNodeB separation x_0 for several selected values of inter-site spacing s under the following assumptions:

- Free-space propagation between eNodeBs $\rho_{\text{BS-BS}}(x)$;

- $\rho_{\text{BS-UE}}(x) = 128.1 + 37.6$ dB, from [2];

- $\gamma_{\text{DL}} = 6$ dB, $\text{NF}_{\text{UE}} = 9$ dB, $\text{NF}_{\text{BS}} = 5$ dB, $G_{\text{BS}} = 14$ dBi, $L_b = 20$ dB.

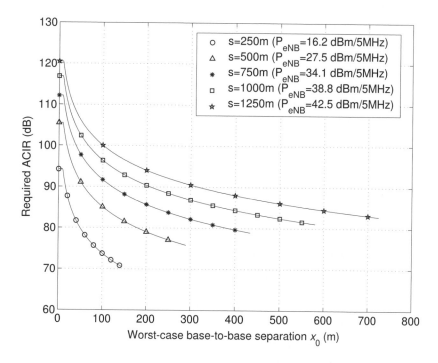

Figure 23.6 Required ACIR for 1 dB desensitization in an eNodeB-to-eNodeB interference scenario.

The transmit PSD of the eNodeB (P_{eNB}) for 95% coverage is also listed in the legend for each Inter-Site Distance (ISD).

For a given ISD, the required ACIR naturally decreases as the worst-case separation between interfering eNodeBs is increased. Notice, however, that the base-to-base problem worsens significantly as the ISD in the interfering network increases, due to the fact that the path-loss exponent from eNodeB to UE is higher than the path-loss exponent between eNodeBs. The transmit power needed by the interfering eNodeB to reach UEs at its cell edge increases at a faster rate than can be compensated by the path-loss to a victim eNodeB receiver at the same cell-edge location. It should be remembered, however, that this is representative of a macrocellular scenario with eNodeB antennas mounted above rooftop level, and that for the smaller cell sizes (characteristic of microcells) the eNodeB-to-eNodeB situation will be significantly improved by the higher propagation exponent between eNodeBs whose antennas are mounted below rooftop level.

For co-sited eNodeBs (i.e. very small x_0), a Minimum Coupling Loss (MCL) value of 30 dB has been used to replace $2G_{BS} - \rho_{BS\text{-}BS}(x_0)$ in Equation (23.1), resulting in

$$\text{ACIR}_{\text{co-siting}} \geq \text{PSD}_{Tx} - \text{MCL} - \text{PSD}_N - \text{NF}_{BS} + 6 \text{ dB} \qquad (23.7)$$

For wide area base stations, the ACIR required for co-siting can rise towards a challenging 120 dB or so. This problem, however, is not new or specific to LTE and has been encountered previously for FDD/TDD coexistence WCDMA. Some practical solutions to this problem have been documented in [3], in which RF bandpass cavity resonator filters were used to improve greatly the ACLR and ACS of base station transmitters and receivers respectively either side of a TDD/FDD boundary. These significantly exceed the standardized minimum requirements which were not intended to cope unaided with the case of co-sited base stations.

Similar techniques also apply to LTE yet remain significantly challenging. With careful design, however, adjacent channel deployment of FDD and TDD LTE base stations, or of two non-synchronized LTE TDD base stations, should be feasible, even for co-sited arrangements, provided that appropriate measures are adopted in both the interfering and victim base stations.

23.3.1.2 Mobile-to-Mobile Interference

The UE-to-UE interference scenario requires a more probabilistic approach than for base-to-base interference for the following reasons:

- the locations of the interferers and victims are variable and dynamic;

- the physical resources assigned by the scheduler to the UEs are variable;

- the transmit powers of the interfering UEs are a function of their channel conditions and of the power control policy implemented;

- the received levels of the wanted base station signals at the victim UEs are also variable as a function of the UEs' channel conditions.

It is difficult, therefore, to formulate a definitive analysis of UE to UE interference for LTE. Nonetheless, a basic analysis is presented here together with some discussion of the attributes of the LTE system which have some bearing on the magnitude of the interference effects.

We consider two similar overlaid tri-sectored LTE deployments on adjacent 5 MHz carriers. One deployment contains the interfering UEs while the other contains the victim UEs. Both carriers have regions in which uplink transmissions overlap in time with downlink transmissions. The base stations of the deployments are either co-located or maximally spaced non-co-located. Cells in the interfering network each schedule groups of four contiguous uplink Resource Blocks (RBs) to a number of randomly-selected UEs, resulting in six simultaneously-scheduled interferers per subframe.

The impacts of the scheduled interferers' transmissions on a UE receiver in the victim network are calculated for the case in which a randomly selected victim UE is scheduled a downlink transmission resource in the 1 RB next to the band edge separating the two carriers. The impact caused by the interfering adjacent-carrier UEs is analyzed in terms of the mean percentage reduction in victim UE downlink throughput, R_{loss}, caused by the presence of the interfering UEs.

As assumed in [2], the ACIR increases by an additional 13 dB for localized SC-FDMA interferer transmissions located anywhere other than the 4 RBs next to the band edge of the interferer network. Additionally, for the purposes of this analysis an uplink power control strategy is employed whereby the transmit PSD of each UE is set such that it is not received at any co-channel non-serving eNodeB receiver any higher than 6 dB above the eNodeB receiver's thermal noise floor.

The transmit PSD of the eNodeB in the victim network is set via Equations (23.4) and (23.5) in the same manner as for the eNodeB-to-eNodeB analysis, such that the downlink is in an interference-limited region of operation (but is not excessively 'over-powered'). The same path-loss model between eNodeBs and UEs is assumed. The path-loss between UEs is assumed to be given by Equation (23.8) (with a carrier frequency f_c of 2000 MHz), based upon a simple two-slope microcellular model from [4] with break point at $x_b = 45$ m to reflect the likelihood of free-space-like propagation (with exponent 2) for low separation distances (i.e. $x \leq x_b$), and increased attenuation exponent $z = 6.7$ due to local clutter at higher distances ($x > x_b$).

$$
\rho_{\text{UE-UE}}(x) = \begin{cases} -27.56 + 10\log_{10}(f_c^2 x^2) \text{ dB} & \text{for } x \leq x_b \\ -27.56 + 10\log_{10}\left(\dfrac{f_c^2 x^z}{x_b^{z-2}}\right) \text{ dB} & \text{for } x > x_b \end{cases} \qquad (23.8)
$$

A microcellular model is considered applicable to UE-to-UE interference as it reflects the case where both the transmitting and receiving antennas are below rooftop level. Other system parameters assumed for this analysis are generally in line with those of [4] for macrocell simulation. With these assumptions, and for $L_b = 20$ dB, the results of Figure 23.7 are obtained, displaying the relationship between the band-edge ACIR and the throughput loss R_{loss} for various values of inter-site distance s.

The results in Figure 23.7 are given for the case of co-located base stations. Those for the non-co-located case are very similar showing only a small further degradation for low ACIR values; the absence of a significant difference is a result of the uplink power control strategy employed as described above (whereby the mobile transmit power is correlated with the strongest non-serving cell path-loss rather than with the path-loss of the serving cell).

The LTE specifications are based upon a 30 dB ACLR which would provide an ACIR of 28 dB assuming an ACS of 33 dB. In this case, it can be seen from Figure 23.7 that the worst-case throughput loss for the band-edge downlink resource block would be between 7% and 16% depending on cell size.

This analysis is, however, rather sensitive to certain assumptions, especially the value of the in-building penetration loss L_b. When the building loss is increased, the serving eNodeB may instruct the UE to increase its transmit power by the same amount in order to maintain

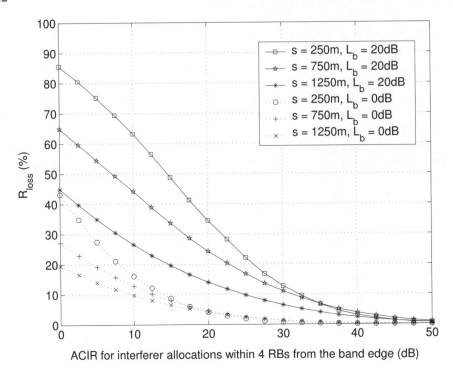

Figure 23.7 Throughput reduction due to UE-to-UE interference.

its received SINR at the serving eNodeB (subject to maximum UE output power constraints) without causing additional interference to non-serving eNodeBs which are also protected by the same building penetration loss. However, the path-loss to a worst-case nearby victim UE (e.g. with LOS propagation between the UEs) is not affected by the increased building penetration loss. Thus, increased building penetration loss can have the effect of increasing worst-case UE-to-UE interference levels. This effect is clearly evident from the fact that the curves of Figure 23.7 for the case of $L_b = 20$ dB show a greater R_{loss} than the dotted curves for $L_b = 0$ db.

This suggests that the susceptibility of the system to UE-to-UE interference is lowered considerably in an outdoor scenario. In these cases the system's need to control the uplink inter-cell interference between UEs and neighbouring eNodeBs in a frequency-reuse-1 network constrains the quantity of interference power that is injected by those UEs into adjacent carriers. System throughput loss is then minimal ($\sim$2%) for commonly-expected ACIR levels. The fact that reduction in-building loss mitigates UE-to-UE interference also points towards possible cell-planning solutions to alleviate the problem, for example using picocells or home base stations rather than macrocells to provide in-building coverage.

The statistical nature of UE-to-UE interference is also of relevance when assessing the impact of UE-to-UE interference. LTE allows for randomization of the allocated radio resources for both the interferer and the victim in both the time and frequency domains. Uplink frequency hopping is able to provide the necessary randomization in frequency, and in the time domain a degree of randomization can be provided by different resource scheduling strategies, as well as the possibility for differing retransmission delays due to the fact that the downlink retransmissions to the victim UE are dynamically scheduled rather than synchronous. The use of Hybrid Automatic Repeat reQuest (HARQ) also provides robustness against those instantaneous events in which high interference levels are experienced. Thus, in the case of frequency-adjacent LTE systems, UE-to-UE interference may be heavily randomized. Its effects can therefore be 'smoothed' and shared amongst all users of the system on a probabilistic basis, avoiding persistent effects on specific pairs of users with close RF coupling. This helps to maintain a more consistent user experience.

There are also other means by which UE-to-UE interference may be alleviated. The assumption that all LTE UEs will possess more than one receive antenna (which is generally required to meet the receiver demodulation performance specifications) enables exploitation of the spatial characteristics of the wanted and interfering signals. The Interference Rejection Combining (IRC) receiver [5] calculates and applies a set of antenna weights in the receiver to maximise the SINR of the signal post-combining, taking into account the instantaneous direction of arrival of the wanted and interfering signals. This is in contrast to a Maximum Ratio Combining (MRC) receiver which does not consider the spatial characteristics of the interference when calculating the antenna weights. Forms of the IRC receiver are possible that do not require explicit channel estimation of the interfering sources and instead make use of averaged correlation (e.g. in time or in frequency) of the received signals across antennas (e.g. [6]). These forms are more applicable to the adjacent-channel UE-to-UE interference scenario in which explicit channel estimation of interferers is likely to be impractical.

The ability of an IRC receiver to suppress interference is a function of many factors including the number and strength of the interfering signals and the number of receive antennas. For a single dominant source of interference the mean SIR gain (effective interference suppression) that IRC is able to offer relative to an MRC receiver improves with the Interference to thermal-Noise (I/N) ratio, and not with the wanted Signal-to-thermal-Noise ratio (S/N). In typical cellular applications, significant interference often arrives from more than one source (i.e. more than one cell), from more than one direction, and often with similar powers, especially in networks with frequency reuse factor of 1. In such cases, an IRC receiver with only a small number of antennas (typical of a mobile terminal) is not able to provide as much spatial separation between the wanted and interfering signals as one would like and performance then tends more towards that of an MRC receiver. However, for the case in which there is one dominant interference source at a mobile receiver (for example as could originate from a nearby closely-coupled UE on an adjacent channel), the IRC receiver should be able to provide a gain when the instantaneous I/N ratio is relatively high. Thus, one could anticipate that use of the IRC receiver may provide some additional robustness against UE-to-UE interference.

Figure 23.8 follows the same format as Figure 23.7 and reveals that the anticipated result is indeed the case (here shown for a 500 m ISD). Although the magnitude of the mean gains is relatively modest (5 dB for lower ACIRs, reducing to around 2 to 3 dB for higher ACIRs), these could at least in theory allow for non-trivial system benefits.

An IRC receiver in a cellular system is also of course able to deliver benefits for mean user throughput in the absence of UE-to-UE interference (due to its ability also to remove interference from eNodeBs), as can be seen from the plots of absolute mean user rate in Figure 23.9. The curves of Figure 23.8 are, however, already normalized to take this

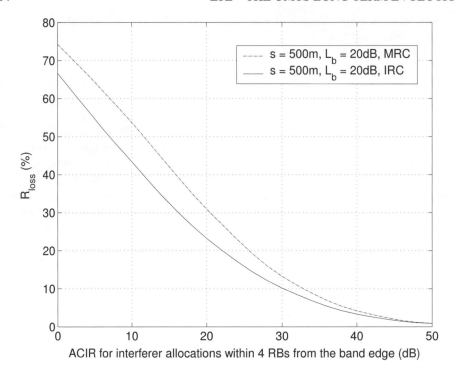

Figure 23.8 Comparison of IRC and MRC receivers for mitigating UE-to-UE interference.

into account; hence the conclusion is that the ideal IRC receiver delivers not only a rate improvement for the serving carrier but also some additional robustness against UE-to-UE interference from an adjacent carrier.

More generally, Figure 23.9 reveals that for the scenario considered, a system using ideal IRC receivers in the presence of UE-to-UE interference with 29 dB ACIR could perform as well as a system using MRC receivers with no UE-to-UE interference at all. Thus, the detrimental effects of UE-to-UE interference may be mitigated via a combination of moderate ACIR together with the deployment of IRC receivers.

23.3.2   Summary of Interference Scenarios

The preceding sections have provided some discussion of the issues facing the deployment of LTE TDD on carriers adjacent either to other non-synchronized TDD systems, or to FDD downlink or uplink carriers. In these deployment scenarios, one must consider not only eNodeB-to-UE and UE-to-eNodeB interference but also the nature and severity of eNodeB-to-eNodeB and UE-to-UE interference.

In the case of eNodeB-to-eNodeB interference, the discussions of Section 23.3.1.1 have shown that very stringent ACLR and ACS are needed, especially in the case of co-siting. Nonetheless, co-siting would appear to be technically feasible, for example by means of cavity-based low-loss RF filtering solutions at the base stations.

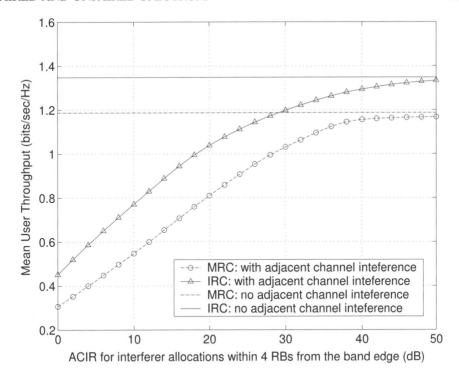

Figure 23.9 Comparison of IRC and MRC receivers with and without adjacent channel UE-UE interference.

In the case of UE-to-UE interference, system-level analysis suggests that by using appropriate radio resource management on the uplink, together with interference randomization and possibly also interference suppression at the receiver, the overall system throughput loss and user experience in the presence of the adjacent carrier interference are likely to remain acceptable for many deployment scenarios.

Although it is generally desirable to enable alignment of uplink/downlink switching points between the LTE TDD system and LTE or legacy TDD systems on an adjacent-carrier, it should be remembered that this cannot always be relied upon.

23.4 Half-Duplex System Design Aspects

For half-duplex (including TDD and HD-FDD) operation, the restriction that a mobile terminal may not transmit and receive at the same instant in time has consequences for the physical layer (and to a lesser extent, higher-layer) design and specification of the LTE system.

In order to enable the exploitation of economies of scale, LTE has generally followed a design principle in which differences between the duplex modes are introduced only where necessary for correct system operation, or where they offer a significant performance advantage when used for a particular duplex mode. With careful design, the extent of the required changes can be curtailed and only a relatively small set of attributes need to be modified. Nonetheless, these have the potential to alter certain behaviours and structures of the physical layer primarily in terms of frame structure, HARQ operation and control signalling, which are discussed in the following sections.

23.4.1 Accommodation of Transmit/Receive Switching

For TDD systems, switching between transmit and receive functions occurs on the transition from uplink to downlink (for the UE) and on the transition from downlink to uplink (for the eNodeB). For half-duplex FDD systems, switching only occurs at the UE, as the eNodeB is assumed to be full-duplex.

In order to preserve the frequency-domain orthogonality of the LTE uplink multiple access scheme, propagation delays between an eNodeB and the UEs under its control are compensated by means of timing advance as explained in Section 20.2.

At a half-duplex terminal, the timing-advanced uplink transmission cannot be allowed to overlap with reception of any preceding downlink. For TDD, to prevent the overlap, a transmission gap or 'Guard Period' between transmission and reception at the eNodeB is created (T_{G1}) to accommodate the greatest possible timing advance and any required switching delay (including power amplifier ramp-up or ramp-down to avoid excessive wideband emissions). A further guard period (T_{G2}) is also required at the TDD eNodeB transition between uplink and downlink to cater for switching and power ramping delays only (this being independent of the propagation delay or timing advance). These are illustrated in Figure 23.10.

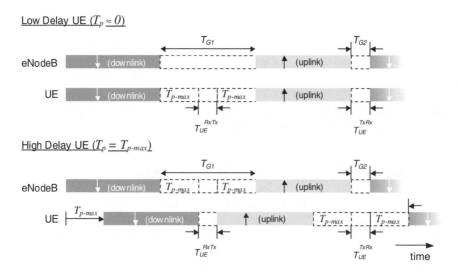

Figure 23.10 TDD signal timings in the presence of uplink timing advance.

Four switching times are therefore of relevance in the case of TDD operation. These correspond to the transmit-to-receive and receive-to-transmit delays at the UE (denoted as time intervals $T_{\mathrm{UE}}^{\mathrm{TxRx}}$ and $T_{\mathrm{UE}}^{\mathrm{RxTx}}$ respectively) and likewise at the base station (denoted $T_{\mathrm{eNB}}^{\mathrm{TxRx}}$ and $T_{\mathrm{eNB}}^{\mathrm{RxTx}}$). Figure 23.10 depicts two cases corresponding to the two extremes of propagation delay T_{P} within a TDD cell ($T_{\mathrm{P}} = 0$ for a UE physically close to the eNodeB and $T_{\mathrm{P}} = T_{\mathrm{P\_max}}$ for a UE at the border of a cell, where $T_{\mathrm{P\_max}} = d_{\mathrm{max}}/c$ corresponds to the maximum one-way propagation delay supported by the cell, occurring at distance d_{max}). Note that the switching delays are exaggerated for diagrammatical clarity and that UE switching delays are assumed to be longer than those at the eNodeB.

It is apparent from Figure 23.10 that the time available at the UE for downlink to uplink transition is a function of the propagation delay T_{P} (most stringent for the case of high delay) whereas the time available at the eNodeB for the same transition is constant and equal to T_{G1}:

$$T_{G1} = 2T_{\mathrm{P\_max}} + T_{\mathrm{UE}}^{\mathrm{RxTx}} \tag{23.9}$$

The time interval T_{G2} at the eNodeB is independent of the propagation delay. To support the case for which $T_{\mathrm{P}} \to 0$ (i.e. a UE close to the eNodeB), T_{G2} needs to be dimensioned such that

$$T_{G2} = \max(T_{\mathrm{UE}}^{\mathrm{TxRx}}, T_{\mathrm{eNB}}^{\mathrm{RxTx}}) \tag{23.10}$$

In the case of HD-FDD, $T_{\mathrm{eNB}}^{\mathrm{RxTx}} = 0$, so T_{G2} is determined only by the time $T_{\mathrm{UE}}^{\mathrm{TxRx}}$. In order to support the case of low T_{P}, the (full duplex) eNodeB must still allow sufficient time for this UE switching delay if the uplink and downlink subframes surrounding the switching point are both active for a particular user. Hence in practice the uplink frame timing at the eNodeB should be advanced for the whole cell by an amount $T_{\mathrm{UE}}^{\mathrm{TxRx}}$ relative to the downlink frame timing at the eNodeB even for a full-duplex eNodeB if it supports HD-FDD UEs in the cell. Full duplex FDD UEs communicating with the same eNodeB will likewise need to have their timing advanced to maintain uplink orthogonality with the HD-FDD UEs.

It is important to note that d_{max} may be significantly larger than the notional cell radius r_0 (i.e. half the ISD) due to propagation effects such as shadow fading. This effect is shown for one example of a tri-sectored deployment with frequency reuse factor 1 in Figure 23.11 (the shadow fading is assumed to be log-normal with standard deviation σ). It can be observed that in order to accommodate, for example, at least 98% of UE locations, the guard period should be dimensioned in accordance with $d_{\mathrm{max}} \geq \gamma r_0$ where γ is a factor between approximately 1.5 and 3 depending on the degree of shadow fading.

Overall, the total guard time T_G at a TDD eNodeB per uplink-downlink cycle is equal to the sum of T_{G1} and T_{G2}, as given by Equation (23.11). In the LTE specifications this is represented by a single guard period amalgamating both parts (with the uplink subframe timing advanced by an amount T_{G2} at the eNodeB with respect to the downlink timing). This is only a matter of representation, however, and the end result is essentially identical to the presence of the two separate guard periods:

$$T_G = T_{G1} + T_{G2} = 2T_{\mathrm{P\_max}} + T_{\mathrm{UE}}^{\mathrm{RxTx}} + \max(T_{\mathrm{UE}}^{\mathrm{TxRx}}, T_{\mathrm{eNB}}^{\mathrm{RxTx}}) \tag{23.11}$$

By assuming reasonable values for the UE and eNodeB switching times (typically of the order of 10 to 20 μs) the length of this amalgamated guard period can be dimensioned (in multiples of the OFDM[2] symbol duration) for a particular deployment. Thus although

---

[2]Orthogonal Frequency Division Multiplexing.

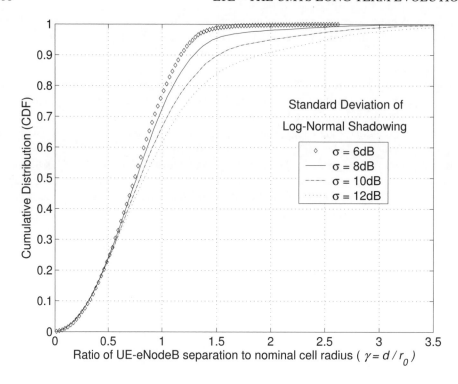

Figure 23.11 Cumulative distribution of serving cell propagation delays relative to nominal cell radius.

guard periods represent an undesirable overhead for TDD, by allowing for a flexible and configurable guard period duration, systems can be tailored to the topology of the deployment whilst minimizing the spectral efficiency loss.

The LTE specifications [7] support a set of guard period durations ranging (non-contiguously) from 1 to 10 OFDM symbols for the normal CP (or from 1 to 8 OFDM symbols for the extended CP). A duration of 1 OFDM symbol should be sufficient for many of the anticipated cellular deployments of LTE (up to around 2 km nominal cell radius for $\gamma = 2$), whereas at the other end of the scale, guard period durations of the order of 700 µs support one-way propagation-path delays of the order of 100 km.

The guard period in LTE TDD is located within a mixed uplink/downlink subframe as shown in Figure 6.2 of Section 6.2 and further discussed in Section 23.4.2.

23.4.2   Coexistence between Dissimilar Systems

There are scenarios in which an LTE TDD system could need to coexist with other (non-LTE) radio access technologies within the same frequency bands. As mentioned in Section 23.3.2, the ability to time-align the uplink and downlink transmissions between neighbouring systems can be used to avoid base-to-base and UE-to-UE interference paths and hence to alleviate the dependency on other aforementioned coexistence measures. This also applies to the case of dissimilar neighbouring systems.

As shown in Figure 6.2 in Section 6.2, the LTE TDD system has a 10 ms radio frame supporting either one pair of switching points per 5 ms, or one pair of switching points per 10 ms, which we denote here as '5 ms switching' and '10 ms switching' respectively.

The provision for both 5 ms and 10 ms switching options in LTE TDD was introduced to enable switching point alignment between LTE and the UTRA TDD modes. Ideally, for maximum flexibility of alignment, the location of the downlink to uplink transition within the 5 ms or 10 ms cycle would be fully adjustable with symbol-level granularity, although certain practical considerations must also be taken into account. The overall intention is first to align the location of the uplink to downlink transitions between the dissimilar systems by means of a frame-timing offset, and then to adjust the position of the LTE downlink to uplink transition such that it is approximately aligned with that of the other system.

The TDD variant of Mobile WiMAX (based upon the IEEE 802.16e amendment) also utilizes 5 ms switching, and can itself accommodate a variable position of the switching point with symbol-level granularity (i.e. with adjustments of approximately ± 102.8 μs). Hence switching point alignment between TDD WiMAX and TDD LTE is possible with sufficient resolution.

The presence of a switching point that is adjustable at the OFDM symbol level results in a mixed subframe, potentially containing both downlink and uplink regions (referred to as 'DwPTS' and 'UpPTS' respectively[3]) as shown in Figure 23.12 (see also Figure 6.2).

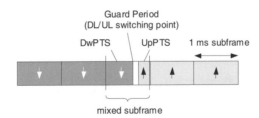

Figure 23.12 Mixed subframe for downlink-uplink switching in TDD operation.

The mixed subframe is the most significant difference between the FDD and TDD physical layers in LTE, giving rise to a number of details which must be considered in the design. Uplink and downlink transmission durations in this irregular subframe are effectively reduced compared to a normal subframe, implying that less forward error correction redundancy can be employed for a given transport block size. Alternatively, the transport block size itself can be reduced, although this smaller transport block size must then also be used for HARQ retransmissions which may occur in a regular 1 ms subframe. In general, the use of HARQ with incremental redundancy, as described in Section 10.3.2.5, limits the impact of the mixed subframe having less downlink transmission resource than a normal subframe.

---

[3]DwPTS and UpPTS is terminology inherited from (TD-SCDMA) where the terms denoted Downlink Pilot Time Slot and Uplink Pilot Time Slot respectively. This, however, does not well reflect the usage of the fields bearing the same names in LTE.

For the downlink control signalling on the Physical Downlink Control CHannel (PDCCH), the mixed subframe does not have a significant effect, as the PDCCH is anyway contained within the first few OFDM symbols of a subframe. Hence PDCCH transmission in the 'DwPTS' region is possible, with the exception that it is constrained in this case to a duration of a maximum of two OFDM symbols (instead of up to three in normal subframes, or even four in cases of narrow system bandwidths, as explained in Section 9.3).

The length of the uplink 'UpPTS' field in the mixed subframe is constrained in the LTE specifications to support lengths of only 1 and 2 SC-FDMA symbols. This field does not therefore support uplink data transmission and is instead used only for a shortened random access preamble (suitable only for small cells – see Section 19.6) and for transmission of uplink Sounding Reference Signals (SRSs) – see Section 16.6. One downside of the absence of support for 'UpPTS' lengths other than one or two symbols is that it restricts the set of possible configurations for switching point alignment with other TDD systems. Nonetheless, Figure 23.13 shows some examples of uplink/downlink alignment that are possible between LTE TDD and other TDD systems.

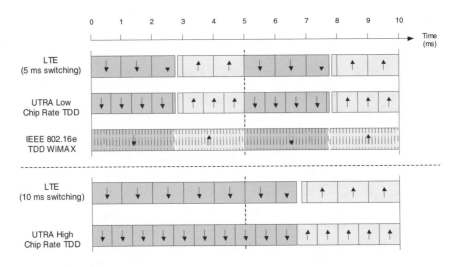

Figure 23.13 Examples of switching point alignment between LTE and non-LTE TDD systems.

23.4.3 HARQ and Control Signalling Aspects

As explained in Section 10.3.2.5, transmission of downlink or uplink data with HARQ requires that an acknowledgement (ACK or NACK) be sent in the opposite direction to inform the transmitting side of the success or failure of the packet reception.

In normal operation, a one-to-one mapping exists between transport block transmissions and acknowledgements. However, in the case of TDD, subframes are designated on a cell-specific basis as uplink or downlink (with the exception of the mixed subframe), which

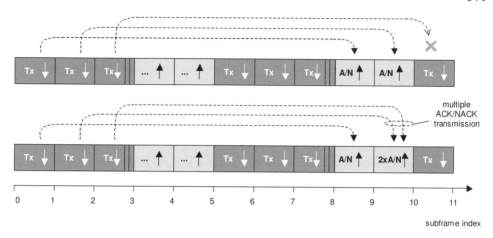

Figure 23.14 ACK/NACK transmission for multiple subframes in TDD operation.

constrains the times at which resource grants, data transmissions, acknowledgements and retransmissions can be sent. In asymmetric uplink/downlink cases the situation can arise in which acknowledgements for more than one received TTI need to be sent in the same subframe. Consider, for example, the situation in the upper part of Figure 23.14. In each 5 ms period, three downlink TTIs require three acknowledgement instances, yet only two uplink TTIs are available for sending them. The lower part of Figure 23.14 shows how this situation could be overcome, by sending multiple acknowledgements in one of the uplink subframes.

It should also be appreciated that the overall HARQ round trip times will be larger for TDD than for FDD as a result of the additional 'waiting time' for the appropriate link direction to become available for transmission of an acknowledgement, grant or retransmission. This results in some marginal degradation to the overall latency performance of the LTE physical layer.

The fact that some TDD configurations contain more downlink subframes than uplink subframes leads to a requirement to transmit (in one uplink subframe) acknowledgements of multiple downlink transport blocks. Explicit transmission of multiple acknowledgements in the uplink for TDD requires support for a PUCCH design capable of carrying a larger volume of uplink control data than the PUCCH in FDD. This is known as ACK/NACK multiplexing. An alternative mechanism that allows for reuse of the FDD PUCCH design is also provided in LTE TDD, whereby acknowledgements corresponding to a group of downlink TTIs are combined prior to transmission on the uplink. In this latter case the individual downlink transport block Cyclic Redundancy Check (CRC) results are passed through a logical 'AND' operation to form a single acknowledgement indicative of whether zero or more than zero blocks in the group were received in error. The obvious disadvantage for this scheme is that the base station does not know which block(s) failed and all must be resent, reducing the link performance of the retransmission scheme. A more subtle impact is that the average HARQ round trip time (and hence latency) is increased due to the fact that some blocks cannot be acknowledged until the remainder of the group have been received. Furthermore, the PDCCH control signalling is not 100% reliable and there is some possibility that the UE will miss some downlink grants. This introduces the possibility for HARQ protocol errors, including the erroneous transmission of ACK for an incompletely-received group of downlink assignments. In order to assist the UE in the detection of missed downlink grants, an indicator is included in the PDCCH to communicate to the UE the number of downlink transmissions associated with the transmission group, although this mechanism alone cannot safeguard against all possible error conditions.

In terms of other aspects of the downlink control signalling there are only relatively minor aspects that require specific attention for TDD. Contrary to the situation for downlink HARQ, the sending (in one downlink subframe) of multiple acknowledgements on the Physical Hybrid ARQ Indicator CHannel (PHICH) for uplink HARQ is not problematic since when viewed from the eNodeB this is not significantly different to the case in which single acknowledgements are sent to multiple simultaneous uplink users. One further consideration concerns signalling of the granted uplink resources. For FDD, the location in time at which an uplink grant is sent implicitly also signals the specific uplink subframe that has been assigned (located four subframes later), whereas for TDD this relationship cannot always hold due to the various uplink/downlink configurations. An alternative linkage is therefore formulated for each specific uplink/downlink configuration to associate each uplink subframe with one preceding downlink subframe that controls it (maintaining the same four-subframe spacing as FDD wherever possible). For configuration number 6 of Figure 6.2, there are more uplink subframes per 10 ms than the number of subframes available for PDCCH, and here it is necessary to employ additional procedures to handle the fact that certain subframes containing PDCCH must be associated with more than one uplink subframe [8].

23.4.4   Half-Duplex FDD (HD-FDD) Physical Layer Operation

In principle, one of two paths could be taken for the physical layer design of a HD-FDD system, namely either:

- a derivative of the FDD system, in which the scheduling is arranged such that the terminals are not required to simultaneously transmit and receive; or

- a derivative of the TDD system, in which the uplink and downlink happen to reside on different carrier frequencies.

The scheme selected for HD-FDD operation in LTE is in accordance with the FDD derivative. This has helped to maintain the desirable commonality with FDD and has reduced the overall complexity of the solution. Unless otherwise informed, a UE in the FDD-derived HD-FDD scheme has no a priori knowledge of the uplink/downlink transmission pattern. Instead the UE checks any subframe which has not otherwise been pre-assigned to uplink transmission for the presence of PDCCH control signalling addressed towards it. The eNodeB is of course aware of any uplink transmission grants it has sent using the PDCCH, and does not therefore expect the UE to be able to receive downlink transmissions in the corresponding uplink subframe(s).

In this scheme, the fixed one-to-one association between downlink and uplink subframes (arising from the timing relationships between resource grants, data transmission and HARQ acknowledgements) is retained from that in normal FDD operation. It is worth noting, however, that this approach results in the situation that for a given user, no more than 50% of subframes may be used for any one link direction for a given UE.

Nonetheless, the perceived impacts to a user in anything other than an unloaded system are likely to remain small. The instantaneous (i.e. in one subframe) peak data rate is not affected, and in any normally loaded or even partially loaded system the base station is in any case unable to dedicate its full time resources to a user for any lasting period due to its need to service other users. Furthermore, at least for the downlink, scheduling strategies may be adapted to increase the amount of frequency resource allocated to a HD-FDD UE at each scheduling instant in order to alleviate reliance on frequent transmissions in the time domain. For the uplink this is also possible except for situations in which the UE is transmission-power-limited. In such cases the application of wider bandwidths may not allow for increased instantaneous data rates and therefore is also unlikely to allow for a reduction in the fraction of time-domain resources needed to achieve a targeted aggregate uplink rate.

In order to allow sufficient time for downlink-to-uplink switching in HD-FDD operation the UE is not expected to be able to receive the last symbol(s) of a downlink subframe that precedes an uplink subframe in which the UE is active. The length of the switching period is very similar to that previously given in Equation (23.11), with the exception that $T_{\text{eNB}}^{\text{RxTx}} = 0$ and the cell-specific maximum propagation delay value $T_{\text{P\_max}}$ may be replaced with a value T_P that is applicable to the particular UE in question. Thus for half-duplex FDD:

$$T_G = T_{G1} + T_{G2} = 2T_\text{P} + T_{\text{UE}}^{\text{RxTx}} + T_{\text{UE}}^{\text{TxRx}} \tag{23.12}$$

During this time, the eNodeB may or may not continue to transmit the remainder of the data towards the UE. In order to simplify the system and to minimize the differences from full-duplex FDD the transport channel coding structure continues to output the same number of channel-coded bits as for a normal-length subframe. Correct reception of the data in spite of the UE's lack of reception during the switching period is therefore reliant on the presence of sufficient forward error correction. For larger switching times the effects on performance obviously become more pronounced due to the increasing code rate, and in these instances the scheduler could attempt to reduce the transport block size mapped to the downlink subframe containing the switching period.

Due to the fact that the eNodeB is full-duplex, common reference signals, common signalling and any other data directed towards other UEs may also continue to be transmitted by the eNodeB throughout a particular UE's switching period and uplink subframes; that is to say, the switching periods and designated uplink subframes are UE-specific in the case of HD-FDD, rather than cell-specific as is usually adopted for TDD.

23.5 Reciprocity

Reciprocity is a general phenomenon encountered in wave propagation over linear media. The details of the underlying theory are not reiterated here, and the interested reader is referred to the comprehensive treatment which can be found elsewhere – see, for example, [9–12]. We cover here only the practical implications for cellular radio communication systems, and more specifically those for LTE TDD.

The efficiency of an antenna in converting between electrical and electromagnetic energy is equal in both directions at a given frequency. The result is that the directional sensitivity of an antenna (when acting as a receiver) is also equal to its far-field radiated pattern (when acting as a transmitter).

For a simple source and a simple receiver in a linear medium separated by sufficient distance such that one does not affect the load of the other, the reciprocity theorem states that the locations of the source and receiver may be interchanged yet the transfer function between them remains unchanged. Therefore when two transceivers A and B each utilize the same antenna for transmission and reception, the frequency-domain transfer function $H(f)$ (or equivalently the time domain impulse response $h(t)$) observed at B when A is transmitting will be the same as the transfer function observed at A when B is transmitting. This is illustrated in Figure 23.15.

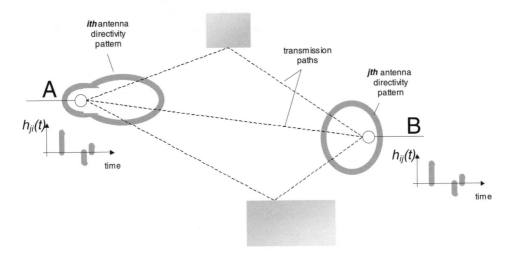

Figure 23.15 Reciprocity between a pair of transceivers.

This holds for any pair of antennas in each other's far field, and hence is easily generalized to the MIMO case in which each transceiver has multiple antennas ($i = 1, \ldots, I$ for transceiver A and $j = 1, \ldots, J$ for transceiver B):

$$H_{ij}(f) = H_{ji}(f) \tag{23.13}$$

Considering for simplicity a particular frequency k, and denoting the MIMO channel matrix between A and B at frequency k as $H_{k,\mathrm{AB}}$ where $H_{k,\mathrm{AB}}(i, j) = H_{ij}(k)$, it is evident that

$$H_{k,\mathrm{AB}} = H_{k,\mathrm{BA}}^{\mathrm{T}} \tag{23.14}$$

In order to exploit reciprocity in a radio communications link, the set of frequencies used for transmission and reception at each transceiver must overlap, as is the case for TDD systems. Reciprocity cannot be exploited for FDD systems unless the nature of each constituent channel $H_{ij}(f)$ is frequency-invariant over a range of frequencies spanning both distinct transmit and receive bands, which is rarely the case.

23.5.1 Conditions for Reciprocity

Reciprocity can be used to control the characteristics of an outbound transmission in anticipation of the channel response that the transmission will experience. In these instances, knowledge of the outbound channel is inferred at the transmitter from measurement of the inbound channel responses. For such techniques to be successful however, one must first consider the following:

- the topology/configuration of the MIMO antenna system;

- the rate of change in the mobile radio channel;

- calibration of the transmitters and receivers involved.

23.5.1.1 Antenna Configuration

Equation (23.14) holds for any number of antennas at each end of the radio link, yet it may be the case that not every antenna element is used for both the transmit and receive functions at a given transceiver. For example, implementation of a transmit chain (and associated power amplifier) per receive antenna element may be feasible at an eNodeB whereas it may not be so in a UE for reasons of form-factor, cost and power consumption.

In these cases knowledge of the Channel State Information (CSI) of the outbound channel is incomplete if obtained only using inbound reference signals. Additional measures would then need to be adopted to determine the missing CSI components. These could include the use of supplementary and explicit feedback of the CSI from the remote receiver for any of its receive antennas that do not have transmit capability. Alternatively, the inbound MIMO channel may be sounded using reference signals sent for example from only one transmit antenna at a time. This can then allow for a UE with only one transmit chain and power amplifier to sound the channel from all of its receive antennas by means of a time-controlled switch.

23.5.1.2 Time Variations in the Radio Channel

Mobile radio channels are of course time-varying in nature, as discussed in Chapter 8. For TDD systems, uplink and downlink transmissions occur at different times, thereby violating the reciprocity principle in a time-varying channel. Fortunately, the channels can still be considered as approximately reciprocal when viewed over a time span short enough to preclude any appreciable change. The duration between the instant at which the CSI is measured and the instant at which it is applied thus becomes of paramount importance. Note that this applies generally to feed-back and feed-forward radio link control systems, and not only to those attempting to exploit uplink and downlink channel reciprocity. It is simply the case that for control schemes reliant upon reciprocity the CSI is measured at the output of the inbound link rather than at the output of the outbound link.

The time over which a channel's impulse response $h(t)$ remains relatively constant is termed its *coherence time* and is defined as the delay δt at which the squared magnitude of its autocorrelation first drops below a certain fraction of its value at zero delay (a fraction of 50% is often used to define the coherence time).

Clearly in order to benefit from channel reciprocity, the delay between the time at which the observation of the inbound channel is made and the time at which this knowledge is

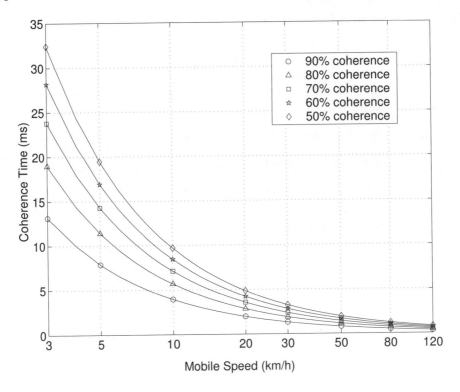

Figure 23.16 Channel coherence times for flat Rayleigh fading at 2 GHz.

applied to the outbound signal, must be shorter than the coherence time of the channel. In order to provide some feel for the magnitude of the values involved in typical cellular applications, the relationships between mobile speed and coherence time (in this instance for a carrier frequency of 2 GHz) are shown in Figure 23.16.

From a practical perspective, the processes to be executed within the channel coherence time include receiving the necessary reference signals, estimating the CSI, computing the required action and encoding and processing the outbound signals. With present-day technology this turnaround time is unlikely to be much lower than 1 ms. Even if it were, the LTE TDD frame structure would need to support uplink-downlink transitions at this rate to ensure continual updating of the CSI on the inbound link (interspersed by transmission on the outbound link); however, due to the minimum switching period of 5 ms, this is not possible. The CSI cannot be applied earlier than the next occurrence of a subframe in the return direction. It may also be the case that a particular UE will not be scheduled in the first available subframe of the return direction, potentially further increasing the delay between measurement of the CSI and its application.

The reference signals commonly used for exploiting reciprocity in LTE are the common reference signals in the downlink and the SRSs in the uplink. The common downlink reference signals are present in each downlink subframe as shown in Section 8.2, while the

SRSs are located either in the short UpPTS field of the mixed subframe (see Figure 23.12) or on one symbol of a normal uplink subframe where configured.

Under the assumption that one would prefer to use UpPTS for sounding, then with a 5 ms switching cycle and a balanced DL/UL configuration (TDD configuration 1 in Figure 6.2), the shortest possible turnaround times to exploit reciprocity for the two downlink subframes and the DwPTS would be 2 ms, 3 ms and 4 ms respectively, as shown in part (a) of Figure 23.17. These correspond to mobile speeds of up to around 24 km/h at a 50% coherence level at 2 GHz.

For the uplink, the ability of the system to take advantage of reciprocity is not limited by the occurrence interval of the wideband common reference signals (as they are present in every downlink subframe), but primarily by the length of the contiguous uplink period within the frame and by the turnaround processing time at the UE. Part (b) of Figure 23.17 gives one example in which the processing delay at the UE is assumed to be of the order of 2 ms, resulting in similar turnaround latencies to the downlink case. With faster processing at the UE, these could be reduced a little further.

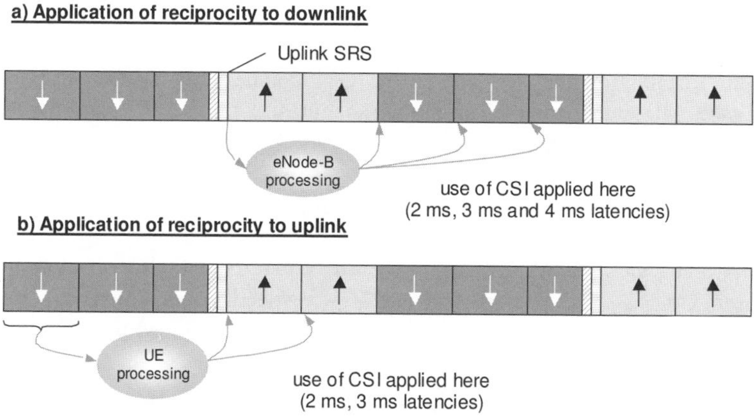

Figure 23.17 Example turnaround times for exploitation of reciprocity in LTE TDD: (a) application of reciprocity to downlink; (b) application of reciprocity to uplink.

23.5.1.3 Calibration

Under conditions of channel reciprocity, the forward and reverse responses between the antennas at two transceivers A and B are the same, although this does not mean that the overall responses between the baseband processors at those transceivers are also equivalent in both directions. The responses of the individual transmitters and receivers themselves must also be taken into account, as shown for a 2×2 system transmitting from baseband A to B in Figure 23.18.

At a given frequency the responses of the multiple transceiver elements may be represented together in matrix form (assuming here that variations within the desired system bandwidth are small). Without coupling effects between elements, these are simply diagonal

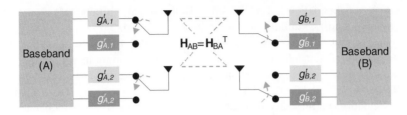

Figure 23.18 Transceiver elements in a reciprocal channel.

matrices containing the per-element complex gains, such that for I antennas at A and J antennas at B:

$$G_{\text{Tx,A}} = \begin{pmatrix} g^t_{\text{A},1} & \cdots & 0 \\ \vdots & \ddots & \vdots \\ 0 & \cdots & g^t_{\text{A},I} \end{pmatrix}, \; G_{\text{Rx,B}} = \begin{pmatrix} g^r_{\text{B},1} & \cdots & 0 \\ \vdots & \ddots & \vdots \\ 0 & \cdots & g^r_{\text{B},J} \end{pmatrix} \quad (23.15)$$

A similar formulation may be applied to $G_{\text{Tx,B}}$ and $G_{\text{Rx,A}}$ in the reverse direction assuming each antenna is capable of both transmit and receive. The composite transfer functions between A and B (and vice versa) at a given frequency then become

$$Z_{\text{AB}} = G_{\text{Tx,A}} H_{\text{AB}} G_{\text{Rx,B}} \quad \text{and} \quad Z_{\text{BA}} = G_{\text{Tx,B}} H_{\text{BA}} G_{\text{Rx,A}} \quad (23.16)$$

It is clear that Z_{AB} and Z_{BA} are not necessarily equal. Nonetheless, if the time between the measurements $\hat{Z}_{\text{AB}}$ and $\hat{Z}_{\text{BA}}$ of Z_{AB} and Z_{BA} respectively is sufficiently smaller than the coherence time, then using Equation (23.14) it can be deduced that (for invertible G matrices):

$$\hat{Z}_{\text{AB}} = G_{\text{Tx,A}} (G^{\text{T}}_{\text{Rx,A}})^{-1} \hat{Z}^{\text{T}}_{\text{BA}} (G^{\text{T}}_{\text{Tx,B}})^{-1} G_{\text{Rx,B}} \quad (23.17)$$

The general goal therefore of a system wishing to exploit reciprocity is to infer $\hat{Z}_{\text{AB}}$ from the observation $\hat{Z}_{\text{BA}}$ using knowledge of the relevant transceiver gains in G.

There are numerous methods by which this can in theory be achieved. One such method is via an explicit signalling exchange between transceivers of the measured composite channel responses ($\hat{Z}_{\text{AB}}$ and $\hat{Z}_{\text{BA}}$), as observed within the coherence time of the channel. By analysing the relation between these, an active transmitter may then derive the necessary transfer function needed to correctly infer the composite outbound channel from future measurements of the composite inbound channels [13]. Note that the update rate of such feedback to track variations in the transceiver responses should be relatively slow – otherwise a pure feedback-based approach that is not reliant on reciprocity becomes a more natural choice. Fortunately, the variations in transceiver responses are predominantly driven by temperature changes, and therefore a low update rate is usually possible (although other factors can also arise, such as interaction between the antenna and close-by objects such as human hands at the UE side). In theory, self-calibration methods are also possible whereby the transceivers make use of accurate reference transmitters or receivers, or of auxiliary calibration signals.

Calibration is not generally required for link control systems using measured CSI from the output of the outbound link (i.e. those not reliant upon reciprocity, including both feedback-based and certain feed-forward approaches[4]). In these cases, the aggregated transmitter and receiver responses for the link under control already form an integral part of the measured CSI and hence any variations are accommodated as a natural part of the ongoing link control process without the need for specific calibration procedures.

One noteworthy exception to the above, however, is the case of beamforming in which the ability of an array to steer energy towards (or collect energy from) a given direction is impaired if the phase between the elements is not adequately controlled. In this instance calibration would generally only be required at the transceiver attached to the array (such as at the eNodeB); an opposing transceiver without an array (such as a mobile terminal) would not require calibration.

In general therefore, it can be said that systems attempting to utilize channel reciprocity require calibration of their constituent transceivers in order to restore the overall reciprocity between them at baseband. Calibration is also required for antenna arrays reliant upon directionality, such as transmit and receive beamforming, although this applies irrespective of the presence of channel reciprocity.

23.5.2 Applications of Reciprocity

For the purposes of this discussion, we define here the 'outbound link' as that which is being actively controlled in response to the channel.

Active control techniques may be generally classified as feed-forward (open-loop) or feed-back (closed-loop). Note that both may involve feedback of information from the receiving side to the transmitting side, the distinction lying in the particular nature of this information. Receivers in closed-loop feedback systems return information regarding how well the transmitting side is performing in relation to a target. Conversely, feed-forward transmitters do not know how well their transmissions are performing, and simply use whatever information is available (via fed-back information or otherwise) to try to pre-empt one or more aspects of the radio channel.

There also exists a class of passive open-loop systems in which the transmitting side takes no pre-emptive action, and exploitation of the channel is confined solely to the receiving side. Figure 23.19 shows this classification of link control methods.

For active feed-forward control systems, the information required by the transmitter often takes the form of CSI which supplies information regarding the response between transmitter and receiver. It is really only this class of system for which reciprocity may be used – the outbound link CSI being inferred from the inbound link CSI given that the conditions discussed in Section 23.5.1 are met and that inbound reference signals covering the desired range of frequencies are available for this purpose.

However, for wireless communication, it is often not only the channel between the controlling transmitter and the receiver that is of interest; information relating to interference caused by other transmitters is also relevant and cannot be obtained via reciprocity.

Some link control methods make use of only amplitude information from the channel (e.g. open-loop uplink power control and frequency-domain scheduling), whereas others, such as MIMO precoding, also need phase information. Even in a TDD deployment, it cannot always

---

[4]Feed-back and feed-forward control mechanisms are discussed in the context of reciprocity in Section 23.5.2.

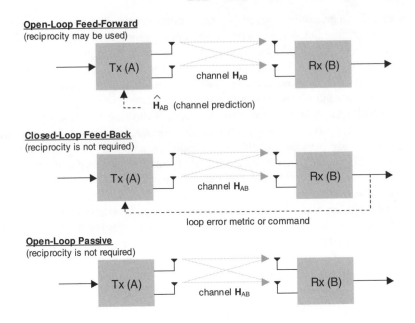

Figure 23.19 Feed-forward, feed-back and open-loop exploitation of the radio channel.

be assumed that reciprocity can be freely exploited to achieve this. In this regard, we discuss below some link control examples of relevance in a TDD LTE system.

23.5.2.1 Downlink Frequency-Domain Scheduling

The eNodeB scheduler can exploit the frequency-selective nature of the channel by attempting to schedule downlink data to a user using frequency resources that are currently experiencing higher than average SINR (see Chapter 12). Although reciprocity allows for the frequency response of the downlink radio channel to be inferred via the observed response of the uplink channel, it does not provide information concerning the downlink interference and noise levels observed at the UE. Hence knowledge of the channel provides only partial knowledge of the SINR.

23.5.2.2 Uplink Frequency-Domain Scheduling

As for the downlink case, this is also an interference responsive mechanism. Contrary to the downlink, however, the eNodeB in this instance does have some knowledge of the interference and noise levels affecting the uplink transmission at its own receiver. Together with the required CSI of the uplink channel gleaned from the SRS transmissions, this provides the eNodeB with all the information required. However, at no point is reciprocity required in order for this process to function, due to the fact that all involved signals exist only in one link direction.

23.5.2.3 Downlink Power Control and Link Adaptation

This function requires interference feedback from the UE if it is to operate effectively. Reciprocity can be used by the eNodeB to infer the downlink CSI, and thereby to provide partial information, but pure feedback-based techniques tend to be better suited to this application.

23.5.2.4 Uplink Power Control and Link Adaptation

In theory, reciprocity can be put to good use in the feed-forward sense to adjust the UE's transmit power and coding/modulation. In this instance, the fact that the interference levels may be controlled by the eNodeB means that the received SINR can be better-correlated with the channel response, thereby improving the usefulness of reciprocity. This may be applied to the LTE TDD system to reduce the variance in received power of an uplink transmission and to compensate for fast fading at slow to moderate UE speeds.[5] For uplink link adaptation, however, the LTE system deliberately does not allow autonomous selection of the modulation and coding rate by the UE (in order to avoid the need to signal the selected format to the eNodeB, and to enable the eNodeB to exercise full control of the uplink and its optimization). Reciprocity therefore does not have a specific application to uplink link adaptation in LTE.

23.5.2.5 Phase-Passive MIMO

These MIMO schemes are those in which the transmitter does not attempt to adapt to the phase of the channel and responds only to amplitude information. Two techniques in LTE that fall into this category are open-loop downlink MIMO (i.e. transmission mode 3 not using Precoding Matrix Indicator (PMI) feedback – see Section 9.2.2.1) with per-stream Modulation and Coding Scheme (MCS) selection, and switched transmit antenna diversity for the uplink. The former essentially comprises parallel instances of downlink rate control and hence the discussion of Section 23.5.2.3 is applicable.

Switched antenna diversity for the uplink in LTE is directed by the eNodeB as described in Section 17.5.1, and does not require reciprocity to function. However, in TDD operation antenna switching could alternatively be based on reciprocity (without requiring calibration).

23.5.2.6 Short-Term MIMO Precoding

As is discussed in Chapter 11, precoding refers to the application of a set of antenna weights per transmitted spatial layer (the precoding matrix) which, when applied to a forthcoming transmission, interact beneficially with the radio channel in terms of improved reception or separation of the layers at the intended receiver(s). Both feed-back and feed-forward techniques are possible in terms of precoding matrix selection strategy. Feed-back approaches (using PMI in LTE) do not rely on reciprocity, only on the feedback loop delay being shorter than the channel coherence time. For feed-forward approaches, the necessary CSI can in theory be obtained in one of two ways:

---

[5]For higher speeds, the UE should ideally disable attempts to track the instantaneous channel as this can worsen the variance of the uplink received power when applied outside of the channel coherence time.

(1) Using so-called Direct Channel Feed-Back (DCFB), in which the CSI observed by the receiver is explicitly signalled back to the transmitter;

(2) Using channel sounding reference signals (e.g. SRS on uplink) to sound the link inbound to the precoding transmitter.

DCFB obviates the need for reciprocity, whereas the reference-signal assisted method is reliant upon it. LTE does not support DCFB, and hence reciprocity and TDD are prerequisites for feedforward-based short-term precoding whereas the feedback-based (PMI) scheme is applicable to all duplex modes.

As described in Chapter 11, explicit control signalling is required for the PMI-based scheme, whereas for a reciprocity-based scheme the role of control signalling is undertaken by UE-specific reference signals in the downlink (see Section 8.2) and SRS in the uplink.

Although a reciprocity-based scheme carries the advantage that the applied precoding matrix is not constrained to a finite set of possibilities[6] (a codebook), the support for UE-specific reference signals is limited in the first release of LTE to allow only a single spatial layer.[7] This correspondingly limits the applicability of reciprocity in terms of short-term precoding in the first release of LTE.

23.5.2.7   Long-Term MIMO Precoding (Beamforming)

At a beamforming transmitter, energy is focused in the directions which are more commonly observed in the inbound channel than others, assuming that a similar angular distribution also applies in the mean sense to the outbound channel. Beamforming is reliant, therefore, only on longer-term statistics of the radio channel and does not require short-term correlation between the inbound and outbound channels in order to function. As such, this scheme is applicable to both FDD and TDD; channel reciprocity in the sense described above is not required.

23.5.3   Summary of Reciprocity Considerations

Exploitation of channel reciprocity for LTE TDD is possible in certain areas. For the initial LTE release, those most useful would appear to be uplink power control (using amplitude information) and some limited applicability to single-layer short-term non-codebook-based precoding (using also phase information from the channel). The benefits arising from the use of reciprocity are generally dependent upon the intended deployment scenario, and in some cases pure feedback-based approaches are sufficient.

In order to realize a true state of reciprocity between a base station and a UE, the feed-forward loop delay must be within the coherence time of the channel. Calibration of the transceivers is also required and each receive antenna element must generally be capable of transmission.

---

[6]This can potentially improve performance by removing mismatch between the best-fit precoding matrix in the codebook and the precoding matrix optimal for the channel.

[7]The initially-intended application of UE-specific reference signals is to support beamforming.

References[8]

[1] Ofcom, 'Spectrum Framework Review: Implementation Plan', Ofcom consultation document, www.ofcom.org.uk, January 2005.

[2] 3GPP Technical Report 36.942, 'Evolved Universal Terrestrial Radio Access (E-UTRA); Radio Frequency (RF) System Scenarios (Release 8)', www.3gpp.org.

[3] T. Wilkinson and P. Howard, 'The Practical Realities of UTRA TDD and FDD Co-Existence and their Impact on the Future Spectrum', in *Proc. 15th IEEE International Symposium Personal and Indoor Mobile Radio Communications (PIMRC 2004)* (Barcelona, Spain), September 2004.

[4] 3GPP Technical Report 25.814, 'Physical Layer Aspects for Evolved Universal Terrestrial Radio Access (UTRA)', www.3gpp.org.

[5] J. H. Winters, 'Optimum Combining in Digital Mobile Radio with Cochannel Interference'. *IEEE Journal on Selected Areas in Communications*, Vol. 2, pp. 528–539, July 1984.

[6] E. G. Larsson, 'Model-Averaged Interference Rejection Combining'. *IEEE Trans. on Communications*, Vol. 55, pp. 271–274, February 2007.

[7] 3GPP Technical Specification 36.211, 'Evolved Universal Terrestrial Radio Access (E-UTRA); Physical Channels and Modulation (Release 8)', www.3gpp.org.

[8] 3GPP Technical Specification 36.213, 'Evolved Universal Terrestrial Radio Access (E-UTRA); Physical Layer Procedures (Release 8)', www.3gpp.org.

[9] J. W. S. Rayleigh, *The Theory of Sound* [Macmillan & Co., 1877 and 1878]. Reprinted by Dover, New York, 1945.

[10] H. A. Lorentz, 'Het theorema van Poynting over de energie in het electronmagnetisch veld en een paar algemeene stellingen over de voortplanting van licht'. Verhandelingen en bijdragen uitgegeven door de Afdeeling Natuurkunde, Koninklijke Nederlandse Akademie van Wetenschappen te Amsterdam, Vol. 4, pp. 176–187, 1895–1896.

[11] J. R. Carson, 'Reciprocal Theorems in Radio Communication' in *Proc. Institute of Radio Engineers*, June 1929.

[12] S. Ballantine, 'Reciprocity in Electromagnetic, Mechanical, Acoustical, and Interconnected Systems', in *Proc. Institute of Radio Engineers*, June 1929.

[13] M. Guillaud, D. Slock and R. Knopp, 'A Practical Method for Wireless Channel Reciprocity Exploitation through Relative Calibration', in *Proc. Eighth International Symposium on Signal Processing and Its Applications*, August 2005.

---

[8]All web sites confirmed 18th December 2008.

Part V

Conclusions

24

Beyond LTE

François Courau, Matthew Baker, Stefania Sesia and Issam Toufik

Through the chapters of this book, it will have become clear that LTE is more of a revolution than an evolution. By embracing the technological fruits of decades of academic research and practical experience, LTE represents a momentous step forward in the possibilities for communication on the move. Wider bandwidths, multicarrier modulation, multiple antennas and packet scheduling, to name but a few of the technologies finding exploitation in LTE, together provide the basis and opportunity for exciting new services and applications.

Yet, however great the advance made by LTE, further steps of enhancement must surely follow. The path from theoretical breakthrough to commercial deployment is a long one, and while the first version of the LTE specifications is solidifying and equipment is being developed, it is important that consideration is already being given to LTE's continuing evolution.

Such ongoing enhancement is fuelled on the one hand by the need of the mobile telecommunications industry for the ability to offer new services, and on the other hand by the unabating zeal of technologists and researchers to improve on what has gone before.

The provision of new services in the sphere of mobile communications is often driven by the possibilities in the fixed networks. An ever-present desire exists to achieve wirelessly the same as can be done via copper cable or optical fibre. This can in principle be done in two ways – either by using more radio spectrum, or by using the available spectrum more efficiently. The latter is heavily constrained by the laws of physics, and many aspects of LTE come close to achieving the fundamental limits of what is theoretically possible within a given bandwidth. Against this background, the International Telecommunication Union (ITU) has taken steps to ensure that more radio spectrum will be available, globally if possible, for systems beyond LTE. The World Radiocommunication Conference (WRC) 2007 resulted in some new spectrum bands being earmarked for mobile services. In order to ensure that effective use is made of such new spectrum allocations, the ITU invited proposals for candidate radio technologies to be submitted which would satisfy the perceived needs. Successful proposals would come under the name 'IMT-Advanced', following on from the IMT-2000 family of systems which became known as the 3rd Generation of mobile wireless systems, including UMTS. The key areas of enhancement to be targeted by IMT-Advanced systems are further enhanced data rates, interworking/compatibility with other technologies, and worldwide roaming – the latter two being aimed at an increasingly seamless user experience. The targets for enhanced data rates are ambitious – up to 1 Gbps in the downlink[1] and up to 500 Mbps in the uplink.

From a technological perspective, this provides both challenges and opportunities, which 3GPP moved quickly to address. Mirroring the LTE inauguration workshop held in Toronto in 2004, 3GPP organized a workshop in Shenzhen in April 2008 (Figure 24.1) to consider the requirements and technologies for the systems beyond LTE under the name of 'LTE-Advanced'. LTE-Advanced, whose specifications should be fully complete by 2011, aims not just to meet, but in some respects to exceed, the demanding requirements set out by the ITU for systems to be designated as 'IMT-Advanced'.

Figure 24.1 The 3GPP LTE-Advanced inauguration workshop in Shenzhen, China, in April 2008. Reproduced by permission of © 3GPP.

One clear challenge is in adapting to the characteristics of the available frequency bands. While data rates of 1 Gbps might ideally be achieved using contiguous bandwidths of 100 MHz or more, it is clear that competition for spectrum, and the fragmentation of the available spectrum, makes it unrealistic to expect such large contiguous bandwidths in most cases. The systems of the future will therefore have to rely on non-contiguous spectrum allocation and bandwidth aggregation methods (as illustrated in Figure 24.2) if

---

[1] At least for low mobility; 100 Mbps may be expected for high mobility.

Aggregation of available spectrum

frequency

Figure 24.2 Aggregation of noncontiguous carriers.

they are to achieve the target peak data rates. This is potentially even more challenging for implementation and system design than the variable bandwidth characteristics of LTE. The design of suitable control signalling, already difficult in LTE, is likely to become still more critical, as scheduling decisions have to be made across multiple carriers and increasing amounts of channel state information are required in order to inform these decisions.

LTE-Advanced will not, however, be able to start from a clean sheet of paper. Investments already made in LTE have to be taken into account, and the next enhancements will be required to be backward-compatible with the first version of LTE. This means that the first LTE terminals should be able to operate in LTE-Advanced cells just as if they were LTE cells. This is analogous to the enhancement of the first version of UMTS by means of High Speed Packet Access (HSPA), whereby network operators are able to continue serving existing customers while their network equipment is progressively upgraded to enable users with newer equipment to benefit from the very latest technology. Many aspects of LTE already facilitate such backward-compatible enhancement. For example, the basic multiple access schemes and many aspects of the physical layer and protocol design (such as modulation schemes and channel structures) are easily extensible to higher data rates and wider bandwidths. Most of the characteristics of the LTE radio interface will therefore be retained, so that from operational and performance perspectives the effort to upgrade to the advanced version is kept reasonable.

It is important to note that although the use of larger bandwidths (whether contiguous or aggregated from separate carriers) may allow for higher peak data rates, it does not result in improved spectral efficiency (or reduced cost). It is in this area that further advances in theoretical understanding will continue to provide the key to enhancing the value of a given section of spectrum. Some advances can be made by extension of schemes already initiated in LTE, especially in the area of Multiple-Input Multiple-Output (MIMO) transmission. The first version of LTE assumes the presence of up to four antennas at the eNodeB and either two or four at the User Equipment (UE) (depending on the UE category). Higher order spatial multiplexing, with larger numbers of parallel data flows may be considered. One strategy consists of increasing the number of antenna elements (for example, up to eight transmit and four receive antennas) in order to increase the multiplexing gain by a factor of at least two. Such methods will be dependent on achieving sufficient decorrelation between the antennas – a potentially significant challenge on a small mobile terminal, but more feasible on a laptop computer. Another strategy would be to support more advanced MIMO techniques, perhaps making better use of knowledge of the radio channel state, in order to approach more closely the theoretical performance limits than is possible in the first version of LTE.

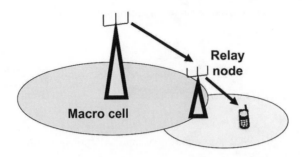

Figure 24.3 The use of a relay node to extend cell coverage.

Other steps forward in spectral efficiency may be derived from as-yet untapped technologies. Two prime examples of these are coordinated MIMO schemes and the use of relay nodes.

Coordinated MIMO schemes involve multiple cells cooperating to improve the overall spectral efficiency. Multiple eNodeBs may schedule transmissions simultaneously to one or more UEs with various aims depending on the chosen strategy. For example, coordinated scheduling may be used to reduce interference or to achieve spatial multiplexing gain by benefiting from macro-diversity resulting from the low correlation between geographically diverse base station sites. With an even higher degree of coordination, multisite beamforming approaches may be considered. Typically these techniques require a high degree of synchronization and communication between eNodeBs.

The use of relay nodes, as shown in Figure 24.3, is a promising idea to increase the data rates available to edge-of-cell users, or to increase coverage at a given data rate. Relaying technology has been much studied in academia, resulting in improved understanding of its potential impact on overall system spectral efficiency.

Other perhaps more classical approaches may also be considered, such as the support of higher order modulation schemes. This could extend the dynamic range of the link adaptation techniques already available in the first version of LTE. Implementation challenges must also be addressed, though, as ever higher order modulation schemes impose increasingly stringent constraints on the RF components of the transceivers.

The challenge for LTE-Advanced will be to deliver these improvements without an unacceptable increase in equipment cost.

It is clear that a range of techniques exist which, individually or in various combinations, can be expected to bring significant further improvements beyond the dramatic advances already made by the first version of LTE. These enhancements are likely to be introduced in steps following an evolutionary approach. This is fitting for a system developed by an organization like 3GPP, whose main goal is to define functional systems which are able to evolve gracefully and provide useful service to consumers.

The targets for LTE-Advanced are challenging, both in terms of the timescale and expected data rates and performance. To a large extent, theoretical solutions exist. As with LTE, the transition of these solutions from academic theory to real-world networks will be exciting.

Index
